1st International Ruminant Seminar (IRS 2017)

Eco-Friendly Livestock Production for Sustainable Agriculture

IOP Conference Series: Earth and Environmental Science
Volume 119

Semarang, Indonesia
24 October 2017

ISBN: 978-1-5108-5919-7
ISSN: 1755-1307

Printed from e-media with permission by:

Curran Associates, Inc.
57 Morehouse Lane
Red Hook, NY 12571

Printed by Curran Associates, Inc. (2018)

For permission requests, please contact the Institute of Physics
at the address below.

Institute of Physics
Dirac House, Temple Back
Bristol BS1 6BE UK

Phone: 44 1 17 929 7481
Fax: 44 1 17 920 0979

techtracking@iop.org

Additional copies of this publication are available from:

Curran Associates, Inc.
57 Morehouse Lane
Red Hook, NY 12571 USA
Phone: 845-758-0400
Fax: 845-758-2633
Email: curran@proceedings.com
Web: www.proceedings.com

Table of contents

Volume 119

International Ruminant Seminar: "Eco-friendly livestock production for sustainable agriculture"

24 October 2017, Diponegoro University, Indonesia

Accepted papers received: 7 February 2018
Published online: 1 March 2018

Preface

Papers

Plenary

Oral

IOP Conf. Series: Earth and Environmental Science **119** (2018) 011001 doi:10.1088/1755-1315/119/1/011001

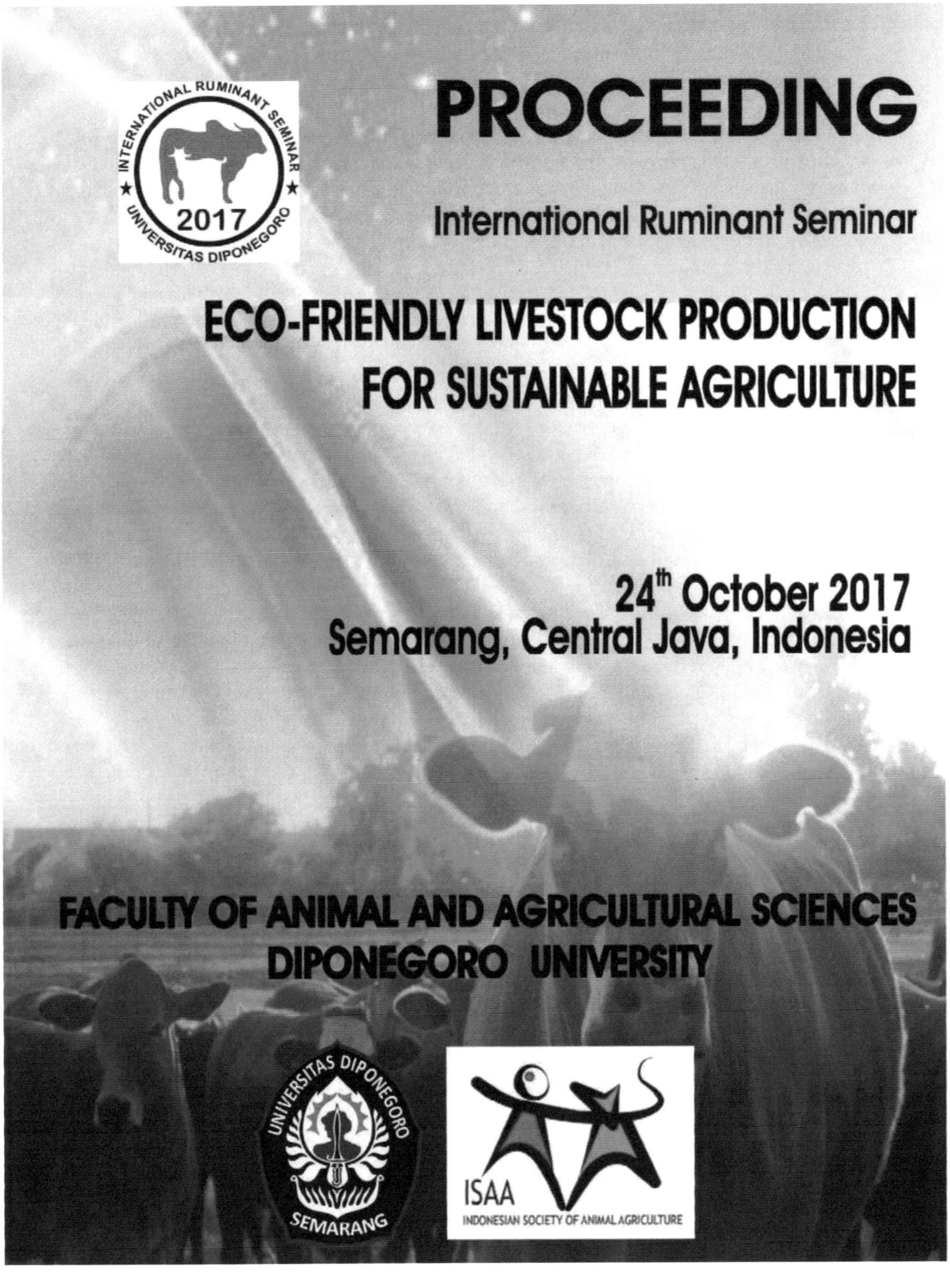

IOP Conf. Series: Earth and Environmental Science **119** (2018) 011001 doi:10.1088/1755-1315/119/1/011001

Preface

Faculty of Animal and Agricultural Sciences, Diponegoro University has actively participated in disseminating of research outcome from academicians, researcher as well as livestock producers. In ruminant science scope, following the success of the first, second and the third national ruminant seminar, it has been held successfully the First International Ruminant Seminar (IRS) in Semarang Central Java, Indonesia at 24 October 2017.

The primary objective of IRS 2017 were to provide a perspective and insight into environmentally friendly livestock production for sustainable agriculture, to discuss the solutions on mitigating the livestock's contributions to climate change and to provide a venue in order to exchange information and ideas on eco-friendly livestock production system for academicians, researchers, administrators, livestock producers as well as related stakeholders and to share their experiences and develop collaborations and to enhance development of livestock production and environmental friendly concerns in their respective countries. We also welcomed our colleagues from the related fields to contribute to the above objectives. More than 70 papers have been presented and discussed during the seminar by either keynote speakers or participants from different countries. The papers cover ruminant feed and technology, ruminant genetic and breeding, ruminant physiology, health and production, environmental issues, post-harvest technology, social and economic issues and other issues related to the theme of conference.

The selected papers of this seminar are published in this proceeding. It is expected that this proceeding would provide valuable information and contribution for readers in improving the productivity and sustainability of ruminant production.

To follow up the seminar and for regular and continuous discussion on the related aspects of ruminant sustainable production, it is the committee's great honours and pleasures to invite again the participants (academics, scientist, practitioners, as well as industries and related stakeholder) to attend and actively support for the next success of the next IRS seminar.

Semarang, December 2017

Editors

IOP Conf. Series: Earth and Environmental Science **119** (2018) 011001 doi:10.1088/1755-1315/119/1/011001

Speech from Chairman of International Ruminant Seminar 2017

Assalamualaikum Wr. Wb

Our sincerely Prof. Yos Johan Utama, Rector of Diponegoro University; Prof. Mukh Arifin, Dean of the Faculty Animal and Agricultural Sciences, Diponegoro University Distinguished quests, speakers and all participants of International Ruminant Seminar 2017.

On behalf of the organizing committee, I would like to express my deeply thanks and welcoming all of you at this International Ruminant Seminar 2017. The ruminant seminar is a seminar about the ruminant livestock production and its related topic. The seminar is organized every three year and has been held for three times at the Faculty of Animal and Agricultural Sciences, Diponegoro University. Previously, the Ruminant Seminar was a national seminar, in that the speakers and participants came only from Indonesia. Start from this year, the ruminant seminar is organized as an international seminar, so that the seminar is attended by international speakers and participants.

The theme of this present seminar is **"Eco-friendly Livestock Production for Sustainable Agriculture"** and has been honoured by the attendance of four keynote speakers: Prof. Takuro Oikawa (Faculty of Agriculture, University of the Ryukyus, Japan), Prof. Sang Jip Ohh (Department of Animal Science, Kangwon National University, South Korea), Dr. Renato S.A. Vega (College of the Agriculture, University of The Philippines Los Banos) and Prof. Agung Purnomoadi (Faculty of Animal and Agricultural Sciences, Diponegoro University-Indonesia).

By end of the registration period, the seminar committee received 77 abstracts, in which 66 papers will be presented in oral presentation and 11 papers will be presented in the poster format. Selected papers of this seminar will be published in an open access reputable international proceeding IOP Conference Series: Earth and Environmental Sciences (EES).

I would like to express my sincerest gratitude to keynote speakers, oral and poster presenters, distinguished guests, participants and our partner institutions who have contributed to this seminar. I thank all of the colleagues, organizing committee, student technical committee and all parties who have worked hard to make this international seminar possible. Finally, good luck and I wish you enjoy the seminar and your time in Semarang.

Thank you very much
Wassalamualaikum Wr. Wb.

Chairman,
Sutaryo, Ph.D.

IOP Conf. Series: Earth and Environmental Science **119** (2018) 011001 doi:10.1088/1755-1315/119/1/011001

Rector Speech

Assalamualaikum Wr. Wb

Keynotes speakers, Distinguished participants, Ladies and Gentlemen

It gives me great privilege and pleasure to extend to you all my warmest welcome on behalf of Diponegoro University and to say how honored we are to have your presence here at ICT Centre International Convention Hall of Diponegoro University for **International Ruminant Seminar 2017.**

Diponegoro University is one of the best universities in Indonesia. We are on the stage of transformation from full public funded university to legal entity university which encourage us to be more autonomus. Diponegoro University or usually called as Undip has eleven faculties and two schools that consist of 49 undergraduate programs, three of them are International Undergraduate Program /IUP, 35 master programs and thirteen doctoral programs. Undip pays a big concern to elevate the number of scientific research in particular area to achieve our goal to be a World Class Excellent Research University.

The theme of this seminar is **Eco-friendly Livestock Production for Sustainable Agriculture.** It is well known that livestock production particularly ruminant production is one of the largest contributors to greenhouse gases from agriculture sector. In addition to greenhouse gases production, livestock sector increase deforestation due to expansion of pasture land and arable land used to grow feed crops. However, since animal protein consumption particularly in developing country is much lower than that in developed country and that consumption has very important role to human being growth, development and health, therefore we should respond that issue wisely since this industries also have some functions such as it can utilize agriculture by product to produce nutritious food (meat, egg and milk), it can prolonged carbon cycle as organic material rather than back to nature as carbon dioxide and it can provide land nutrients in the form of animal manure.

This is truly an impending challenge which needs all key players' contribution such as government, researchers, academics and also industries. Mitigating the livestock sector's contributions to climate change is crucial for sustainable agriculture. Technical solution such as reformulation of ruminant diets to reduce enteric fermentation and some methane emissions can be undertaken. Other solutions may be proposed by researchers to support the sustainable livestock farming, from the aspect of feed, genetics, physiology, socio-economic, environment, etc. Hence, as this seminar theme goes, Eco-friendly Livestock Production for Sustainable Agriculture and this seminar is designed to disseminate of such study result.

Indonesia is a developing country that struggling toward developed nation with more than 250 million population must take part on this endeavour, and we extend our hand to the international community to work together toward the same goal. Therefore, I on behalf of Diponegoro University, am deeply honoured for the privilege to host this International Ruminant Seminar. I believe that this seminar will be a good place to share your recent research discoveries, to build partnership in national and international level, so that we can move closer to answer the world especially in climate change challenge together.

IOP Conf. Series: Earth and Environmental Science **119** (2018) 011001 doi:10.1088/1755-1315/119/1/011001

To conclude, I would like to congratulate the committee for well-done preparing this seminar. My special gratitude goes to keynote speakers, all participants, partner institution and all other parties supporting this event.

By the name of Almighty Allah Swt, I declare that the International Ruminant Seminar 2017 is officially opened. Finally, welcome and enjoy your time in Semarang, Indonesia. I wish you a very fruitful seminar.

Thank you very much
Wassalamualaikum warohmatullahi wabarokatuh

Semarang, October 24, 2017

Prof. Dr. Yos Johan Utama, SH, M.Hum
Rector of Diponegoro University

Committee
International Ruminant Seminar 2017

Steering Committee:

- Prof. Dr. Ir. Mukh Arifin, M.Sc
- Prof. Ir. Agung Purnomoadi, M.Sc., Ph.D
- Agus Setiadi, S.Pt., M.Si., Ph.D
- Dr. Ir. Bambang Waluyo H.E.P., M.S., M.Agr
- Dr. Limbang Kustiawan N., S.Pt., M.P

Organizing Committee:

- Sutaryo, Ph.D (Chairman)
- Sugiharto, Ph.D (Secretary)
- Dr. Baginda Iskandar, M.T, M.S
- Ir. Surono, M.P
- Dr. Ir. Sutopo, M.Sc
- Daud Samsudewa, Ph.D
- Dr. Ir. Endang Purbowati, M.P
- Dr. Ir. Titik Ekowati, M.Sc
- Dr. Yoyok Budi Pramono, S.Pt., M.P
- Dr. Ir. Sri Mukodiningsih, M.S
- Dr. Ir. Anis Muktiani, M.Si
- Dr. drh. Enny Tantini Setiatin, M.Sc
- Wahyudi, S.Kom
- Dr. Sri Sumarsih, S.Pt., M.P
- Karno, Ph.D
- Siwi Gayatri, Ph.D
- Dian Harjanti, Ph.D
- Drh. Fajar Wahyono

Scientific Committee:

- Dr. Ni Wayan Kurniani Karja (Faculty of Veterinary Medicine, Bogor Agricultural University, Bogor, Indonesia)
- Prof. Dr. Cece Sumantri (Faculty of Animal Science, Bogor Agricultural University, Indonesia)
- Prof. Dr. Chalong Wachirapakorn (Department of Animal Science, Khon Kaen University, Thailand)
- Dr. Arda Yildirim (Faculty of Agriculture, Gaziosmanpasa University, Turkey)
- Prof. Dr. Wan Zahari Mohamed (Faculty of Veterinary Medicine, Universiti Malaysia Kelantan, Malaysia)

- Dr. Atien Priyanti (Indonesian Centre for Animal Research and Development, Bogor, Indonesia)
- Prof. Dr. Edy Kurnianto (Faculty of Animal and Agricultural Sciences, Diponegoro University, Semarang, Indonesia)
- Prof. Dr. Joelal Achmadi (Faculty of Animal and Agricultural Sciences, Diponegoro University, Semarang, Indonesia)

Acknowledgement

International Ruminant Seminar "Eco-friendly livestock production for sustainable agriculture" was supported by the Institute for Research and Community Services (LPPM) Diponegoro University through "World Class University" program. The Faculty of Animal and Animal Agricultural Sciences, Diponegoro University was also acknowledged.

IOP Conf. Series: Earth and Environmental Science **119** (2018) 011005 doi:10.1088/1755-1315/119/1/011005

Peer review statement

All papers published in this volume of *IOP Conference Series: Earth and Environmental Science* have been peer reviewed through processes administered by the proceedings Editors. Reviews were conducted by expert referees to the professional and scientific standards expected of a proceedings journal published by IOP Publishing.

IOP Conf. Series: Earth and Environmental Science **119** (2018) 012001 doi:10.1088/1755-1315/119/1/012001

Improvement of indigenous cattle to modern Japanese Black (Wagyu) cattle

T. Oikawa

Faculty of Agriculture, University of the Ryukyus, Nishihara, Okinawa, Japan.

Correspeonding E-mail: tkroikawa@gmail.com

Abstract

Wagyu cattle have been improved from indigenous cattle raised in Japan since the country was opened 100 years ago. Characteristics of the breed were formed during that period. Here, the process of the breeding is described, and recent topics about breeding studies are discussed.

1. Introduction

The Japanese Black cattle (Black Wagyu) provides a distinct quality of meat characterized by well-marbled beef. The origin of this characteristic is unknown, however. For example, how and when the characteristic became manifest in Wagyu cattle is an unresolved mystery in terms of meat traits of this breed. Another mystery is the prolificacy of Japanese quail!

Modern breeding of Wagyu cattle began 150 years ago after the opening of the country, when numerous exotic breeds were introduced from Europe and North America and crossbred with indigenous cattle, which over time developed into the Wagyu breed. Wagyu cattle is one of the examples of newly developed Asian breeds. The objective of this study is to introduce the process of breeding and to discuss recent topics on genetic improvement of the breed.

2. Dawn of breed development

The roots of indigenous cattle in Japan can be traced back to the Heian period, around one thousand years ago. Before that, records of cattle feeding were based only on archeological evidence. On the other hand, documented records of breeding were found later in 1700s, the mid-Edo period (1603-1868), when the lineage of animals was recorded in several areas of the Chugoku region (western part) of Japan. Each valley or river course had its unique lineage of animals. The practice of breeding seems to have been a repeating cycle of inbreeding and outbreeding based on lines of male animals.

After the opening of the country, a large number of foreign customs were introduced into Japan, exotic breeds being one of them. Because such breeds had distinct characteristics lacking in domestic cattle such as large bodies, high milk yield and muscular power, crossbreeding of domestic and foreign animals became very popular during that period.

3. Establishing Wagyu breeds

The drawbacks of crossbreeding were recognized by cattle breeders because the diverse characteristics of crossbred animals, such as their large bodies, handling difficulty, lack of legerity and meat quality, were proving unfavorable in that crossbreeding was practiced without a well-organized policy. Crossbred animals were then gradually integrated into four Wagyu breeds, and a registry system was introduced to record and establish breed characteristics. At the same time, important social changes were taking place in Japan: from animal power to the combustion engine as an energy resource. Accordingly, the object of breeding changed from stamina to meat value. To achieve this goal, performance, and later, progeny tests were initiated to genetically improve meat quantity and quality. Information technology (IT), such as best linear unbiased prediction (BLUP) once called state of the art, was then developed by Henderson (1975) and applied to dairy cattle first, then to beef cattle including Wagyu cattle, thus accelerating the rate of genetic improvement. During that period, the carcass grading system was formalized because various grading systems were used throughout Japan. Improvement of Wagyu cattle was propelled by a station-testing system; however, because of IT, field testing based in production farms has been the mainstream of the testing method since 1980's.

4. Recent topics in Wagyu cattle breeding

Since the introduction of genomic selection methodology proposed by Meuwissen et al. (2001), the focus has been on the application of genomic information to animal breeding. Genomic selection quickly replaced genome-wide association study (GWAS) in terms of methodology by incorporating genomic information into animal breeding. Trials in dairy cattle breeding have been carried out for over ten years with genomic selection, resulting in 10-20% higher genetic response by genomic selection than by evaluation without genomic information. Generation interval can be shortened for traits the measurements of which are time consuming or are possible only in the latter half of the animal's productive life. Nonetheless, recent research has revealed that progress in terms of inbreeding has doubled compared with non-genomic breeding. Single-nucleotide polymorphism (SNP) genotyping is still costly. In Wagyu cattle breeding, the use of genomic selection is still under investigation.

5. Change of traits targeted in Wagyu breeding

Traits targeted in Wagyu breeding have been changing for over 100 years. The first major change from size/conformation traits to carcass traits occurred with a change in objective use of cattle: from draft animals to beef producing ones. Next, for Wagyu cattle priority was given to genetic improvement of meat quality in an environment of market opening and global competition. Visual evaluation of marbling promoted improvement of meat quality around 2005. Nonetheless, well-marbled beef did not necessarily satisfy consumers' taste. For example, the score for savor (umami) evaluated by panel tests reached a peak on beef with a modest content of fat (Iida et al., 2015). Although new traits of meat quality have been studied, no definite candidates have as yet been found. Among promising factors are fatty and amino acids, Suzuki et al. (2017) have demonstrated that inosinic and glutamic acids as promising candidates.

Secondary traits, other than meat quantity and quality, include important ones for the production of Wagyu cattle: feed efficiency and reproduction, both of which have a conductive effect on the production of Wagyu cattle, resulting in more profitability for cattle farmers. Such traits have been investigated for genetic improvement.

Feed efficiency has for long been measured as the rate of feed conversion; however, it involves deficiency of ratio traits. Therefore, residual feed intake was investigated as an alternative trait because of its linear index in nature. Heritability and genetic correlation of residual feed intake were studied to determine genetic improvement by feed efficiency of Wagyu cattle. Our studies demonstrated that the trait of residual feed intake was replaceable to feed conversion ratio because it showed higher heritability and favorable genetic correlation with weight gain and feed intake. Furthermore, residual feed intake shows favorable correlation with the marbling score and the depth of subcutaneous fat.

Recent performance of reproductive traits has shown a downward trend in cattle. Reproduction by females is the basis of stable beef production. To investigate possible genetic improvement of reproductive performance, reproductive traits were collected from registry records. Heritability of age at first calving was higher (0.17) than that of calving interval between parities

(0.01-0.05). Nonetheless, heritability of the calving interval based on a repeatability model with pooled calving intervals was higher (0.18) than individual heritability. Thus, pooled records show potential as a tool for genetic evaluation of reproductive traits.

Reproductive performance of Wagyu cows shows seasonal fluctuations, and heat stress is one of the factors lowering reproductive performance of Wagyu cows in Okinawa, the southernmost region of Japan, in an oceanic sub-tropical climate. The phenotypes of age at first calving and the calving interval were analyzed by a linear model. Responses of age at first calving and calving intervals against heat stress were significant on different days. A negative response was observed against an upward trend in heat stress and a reverse one against a downward trend. For example, in age at first calving, day -2 showed an upward trend, whereas both day -3 and day +33 showed an upward trend after the first calving interval. In later parities, the responses were more complex, probably due to the effects of previous parities. The reasons may be availability of feed resource, physiological adaptability of the animal and seasonal variations of heat detected by the farmer.

Throughout studies on Wagyu cattle, breeding objectives have changed with the demands of the end user, ex animal, with more power and legerity to animals with high meat values. In terms of method, evaluation of animals has changed from visual appearance to core values. The future breeding objective of Wagyu cattle will be determined depending on the definition of the real value of an animal.

References

[1] Henderson CR 1975, Best linear unbiased estimation and prediction under a selection model, Biometrics, **31. 423**

[2] Iida F, Saitou K, Kawamura T, Yamaguchi S and Ni T 2015 Effect of fat content on sensory characteristics of marbled beef from Japanese Black steers J. Anim. Sci. **86** 707

[3] Meuwissen THE, Hayes BJ, and Goddard ME 2001 Prediction of total genetic value using genome-wide dense marker maps Genetics **157. 1819**

[4] Suzuki K, Shioura H, Yokota S, Katoh K, Roh S, Iida F, Komatsu T, Syoji N, Sakuma H and Yamada S 2017 J. Anim Sci **88** 421

IOP Conf. Series: Earth and Environmental Science **119** (2018) 012002 doi:10.1088/1755-1315/119/1/012002

Early Fattening Lamb Could Mitigate Methane Production-an Example of Climate Smart Livestock Farming System in Indonesia

A. Purnomoadi[1*]

[1]Faculty of Animal and Agricultural Sciences, Diponegoro University, Semarang, Indonesia

*Corresponding E-mail: agung194@yahoo.com

Abstract

Ruminants, with a presence of rumen in their digestive tract, will produce methane during his life. Therefore, shortening the rearing period to produce a meat or milk is an alternative to reduce methane emissions. In Indonesia, in the last decades the tendency to slaughter the sheep at young age (around 5 months old) is increasing. This tendency is due to the young sheep (lamb) provide a tender meat and low in fat, as well as a faster in economical return. This study was aimed to evaluate whether shortening (and early age) fattening can reduce methane emissions. Sixteen data from two experimental sheep, each 8 heads of young sheep (aged 3 months old; initial weight 14.32+/- 1.25 kg), and 8 mature sheep (aged 12 m.o.; initial weight 20.65 +/- 1.89 kg) were used in this study. They were fed the pelleted diet formulated to give at least CP 12% and TDN 60% ad libitum. The results showed that the dry matter intake of young sheep was higher than mature sheep (1.33 vs. 1.08 kg), as well as daily gain (156 vs. 83 g/d) and the methane production (41.4vs 36.0 L/d), or methane production per kg daily gain (269 vs 434 L/kg body weight gain), although the production of methane per DMI was similar (32.0 vs 33.5 L/kg DMI). With assumption that sheep commonly weaned at 2 months old at body weight of 10 kg, and was slaughter at 22 kg of body weight, it could be calculated that shortening (and early age) fattening could reduce totally 1,984 L per head and 10 months shortened.

1. Introduction

The demand on lamb meat in Indonesia was increasing as shown by the growing of lamb used culinary or restaurant. Traditionally, it called *Balibu*, abbreviation of *Domba Lima Bulan* – an Indonesian words that literally means five months old sheep. This growing is caused by the quality of the lamb meat that is more tender and –the the important thing is- low in fat and low cholesterol due to the consideration of consumers on healthy meat. The meat characteristics can be obtained from the flesh of young animals. Ponnampalam et al. (2007) [1] reported that young sheep carcass have a higher lean meat and lower fat content than the carcass from old sheep. Moreover, [2] reported that the meat of sheep slaughtered at the age of 8 months was tenderer than of 10 months.

This condition is economically interesting to fasten the economical return and in improving the benefit of farmer. Traditionally, Indonesian sheep is slaughtered at mature weight (26.7 - 31.3 kg) at 15 months old with growth rate of 60.8 - 75.6 g/d [3]. Specifically, pre-weaning sheep grew faster than mature ones [4] as shown by the ADG of the age categories 0–3, 3–6, 6–12 and older than 12 months were 96.5, 77.9, 52.7 and 34.7 g/d, respectively. Another study showed that sheep gained around 120 g/d by allowing high levels of energy intake [5]. But, the most interesting is on methane mitigation that possibly occurred due to the shortened lifetime to reach the slaughter weight. Therefore, this study was aimed to evaluate the possibly of methane mitigation by this sheep production system.

2. Materials and Methods

This study used sixteen data from two experimental sheep, each 8 heads of young sheep (aged 3 months old; initial weight 14.32+/- 1.25 kg), and 8 mature sheep (aged 12 m.o.; initial weight 20.65 +/- 1.89 kg). Those young and mature sheep were fed pelleted diet formulated to give (at least) CP 12% and TDN 60% ad libitum. All sheep were adapted to their diets for a month prior to data collection periods.

Parameters measured were dry matter intake (DMI) and methane production as well as body weight gain. The DMI was measured by weighing the feed given and the residual, and was calculated in 100% dry matter base. Dry matter was measured by oven-dried feed (and residuals) samples at 135 C for two hours. Methane emission was measured using facemask method [6] for two days (2x24 h) using a mask which is connected to a methane analyzer (Horiba Ltd., Japan) and airflow meter. Methane emissions were recorded automatically by IBM PC for 10 minutes at 3 hours intervals. The results obtained are converted into weight units (1 liter CH_4 equal to 0.716 g CH_4; [7]. Bodyweight gain was calculated by subtracted initial body weight to the final body weight after 10 weeks rearing period. The data were analyzed using t-test.

3. Results and Discussion

The dry matter intake, body weight gain and methane production (daily, per kg DMI, per kg daily body weight gain; BWG) are presented in Table 1. Dry matter intake in young sheep was higher ($P<0.05$) than in mature either in kg (1.33 vs 1.076 kg) or in percentage of body weight (6.23 vs 4.41%BW). These DMI lead to body weight gain of young was higher than of mature (156 vs 83 g/d; $P<0.05$). The level of DMI observed in this study was higher than normally found in rural raising system (around 3-4%BW), and it might be pointed to a pelleted diet used in this study. Pelleted diets may have a positive effect on increasing feed intake [8]. But, another two reasons could be considered, first is due to the fast and higher body weight gain in lamb than of sheep, and the second it may indicated that lamb has a better adaptation to pelleted feeding since both lamb and sheep were just purchased from the farmers prior to the study.

The methane production in this study was linearly correlated to DMI. Daily methane production in young sheep was higher than of mature (41.45 vs 36.05 L/d; $P<0.05$), even though the methane production per DMI was similar (32.04 vs 33.51 L/kg DMI). Moreover, those phenomena lead to the methane production per kg body weight gain in young was fewer than the mature (269 vs 434 L/kg BWG; $P<0.05$). The methane production per kg body weight gain in young sheep was lower than the result from Garut sheep [9] given napier grass ad libitum and concentrate feeding at 2%BW resulted BWG 0.11 kg per day, being 440 L/kgBWG. This result also better than of buffalo fed with napier grass supplemented with soy-sauce cake (1%BW) which is given BWG 0.45 kg per day [10] being 413 L/kg BWG, but still higher than 177 L/kg BWG reported from Brahman cattle fed high grain [11] in tropical regions.

Table 1. The dry matter intake, body weight gain and methane production (daily, per kg DMI, per kg daily gain) in young and mature sheep

	Young (Lamb)	Mature (Sheep)	Sign.
Initial age, month	3	12	
Number of animal, head	8	8	
Initial body weight, kg	14.32 $\pm$ 1.25	20.65 $\pm$ 1.89	*
Dry matter intake, kg/d	1.33	1.08	*
Daily BW gain, g/d	156	83	*
Methane production			
L CH_4/d	41.45	36.05	*
L CH_4/kgDMI	32.04	33.51	ns
L CH_4/kgBWG	269.08	434.37	*

ns) not significant; *) significant at 5%

Based on practical, early age fattening sheep is slaughtered less than 5 months old, while the conventional raising systems is slaughtered at 15 months, so it can be calculated that early fattening system may reduce 10 months raising day. With assumption that sheep commonly weaned at 2 months old with body weight of 10 kg, and slaughter weight is 22 kg, by using the data of methane production per kg body weight gain, it could be estimated to reduce totally 1,984 L per head in sheep lifetime, or equal to 38% reduction. This value is very promising since the Indonesian government targeted to cut emission for 20%.

4. Conclusion

This study concluded that early fattening lamb could be considered as an alternative method to reduce methane emission, and it could be calculated to give 38% reduction.

References

[1] Ponnampalam, E.N., D. L. Hopkins, K.L. Butler, F. R. Dunshea, M. G. Kerr, and R. D.Warner. 2007. Genotype and age effects on sheep meat production carcass composition. Aus. J. of Exp. Agric. **47**: 1172.

[2] Veiseth E., S. D. Shackelford, T. L. Wheeler, and M. Koohmaraie. 2004. Factors regulating lamb longissimus tenderness are affected by age at slaughter. Meat Science, **68** (4): 635.

[3] Budisatria, I.G.S. 2006.Dynamics of small ruminant development in Central Java – Indonesia. PhD Thesis, Animal Production Systems Group, Wageningen University

[4] Budisatria, I.G.S., H.M.J. Udo, C.H.A.M. Eilers, E. Baliarti and A.J. van der Zijpp. 2010. Preferences for sheep or goats in Indonesia. Small Ruminant Res. **88**: 16.

[5] Batubara, L.P., J. Sianipar, P.M. Horne and K.R. Pond. 1994. Growth responses of ram lambs from four sheep breed types to rations varying in energy content. Proc. 7th AAAP Congress, Bali, 11-16 July 1994.

[6] Kawashima, T., Sumawal, W., Terada, F., Shibata, M., 2001. Respiration trial system using ventilated flow-through method using face mask. Jircas J. **9**, 53 .

[7] Brouwer, E. 1965. Report of sub-committee on constants and factors. In Proceedings of the 3rd Symposium on Energy Metabolism, pp. pp. 441–443 [KL Blaxter, editor]. London: Academic Press.

[8] Peterson, S.J., B.L. Nuttelman, D.B. Burken, M.K. Luebbe, G.E. Erickson, and J.C. MacDonald. 2015. Use of a pelleted corn-residue complete feed in calf receiving diets. The Professional Animal Scientist, **31 (3)**: 201.

[9] Purnomoadi, A., M. Yusman, E. Rianto and M. Kurihara. 2005. Methane production from Garut rams fed tofu cake as concentrate substitution. Proceedings of AHAT/BSAS International Conference, Vol. 2. (P. Rowlinson, C. Wachirapakorn, P. Pakdee and M. Wanapat, Eds). November 2005, Khon Khaen. P6.

[10] Purnomoadi, A., E. Rianto, F. Terada and M. Kurihara. 2003. Reduction of methane production from buffalo heifers fed soy-sauce by-product in South East Asia. Proceedings of the 3rd International Methane and Nitrous Oxide Mitigation Conference (II), Beijing. 135-142.

[11] Kurihara M., T. Magner, R.A. Hunter, and G.J. McCrabb. 1999. Methane production and energy partition of cattle in the tropic. Br. J. Nutr. **81**: 227.

IOP Conf. Series: Earth and Environmental Science **119** (2018) 012003 doi:10.1088/1755-1315/119/1/012003

An Agenda for Growth and Metabolism Research in Farm Animals: Healthy Food for a Healthy Nation

R. S. A. Vega[1*], D. D. Manalo[2] and J. N. M. Garcia[2]

[1*]Institute of Animal Science, College of Agriculture & Food Science, U.P. Los Baños, 4031 College, Laguna, Philippines
[2]Researcher & Associate Professor, Agricultural Systems Institute, College of Agriculture & Food Science, U. P. Los Baños, 4031 College, Laguna, Philippines

Corresponding E-mail: rsvega@up.edu.ph

Abstract

The aim of this paper is to present a paradigm shift from meat demand to health driven understanding of the effects of livestock production and meat consumption in human. Included are comments on animal nutrition research with the purpose of getting more relevant health information. The growth and metabolism research covers the principles from feed intake regulation, nutrient digestion, absorption and growth. Meat preservation provides enough food for the urban population, and its major ingredient includes the combination of salt, sugar and fats, suspected to cause food addiction in man. The reported effect of processed red meat causing systemic inflammation is a major consideration in food preparation, diet selection and reasonable control for meat consumption. In livestock "Pharming" the danger of heavy metal contamination of commercial feed poses threat to human health aside from drug residues. It is proposed that the 50% of global greenhouse gas is due to the lifestyle of the rich and epidemic of obesity-related diseases are the effect of livestock "Pharming" and addiction to fast food and or processed meat.

1. Introduction

There has always been an increasing demand in meat products justified by increasing population, increasing income and urbanization ([1], [2]). From 1980 onwards, the growth in the per capita consumption of livestock in developing countries outgrow the other major food commodities, such as cereals, root crops/tubers and milk [3]. According to FAO the per capita consumption of meat in developing country in 1980 and 2005 is 14.1 and 30.9 kg, respectively while in developed countries it is 76.3 and 82.1 kg. Although the increase in developing country is 16.8 kg and only 5.8 kg in the developed country for 25 years, the provoking question is, "Does developing country need to consume meat the same amount the developed nation consume?". They should not, because the body size of the people in the developing country is just half or more than half the size of the developed nation.

Obesity is defined as individuals with body mass index above 30. Obesity – related diseases, such as diabetes, cardiovascular, cancer and arthritis are mostly degenerative diseases. According to WHO (2017b) [4], in 2016 the world obesity tripled since 1975. WHO (2017c) [5] reported that the global obesity in 2016 has been mapped, listed from lowest to highest are India (19.7%), Philippines (27.6%), Indonesia (28.2%), Canada (64.1%), Australia (64.5%) and USA (67.9%). India per capita meat consumption is 3.7 and 5.1 in 1980 and 2005. The country's average per capita meat consumption and average incidence of obesity apparently shows direct association.

2. Discussion

This paper goes beyond the capitalist idea of raising livestock to meet the meat demand of the increasing population in the urban and economically able consumer of the developing countries. This paper attempts to present relevant principles on the (1) feed intake regulation; (2) compensatory growth mechanisms; (3) animal nutrition research; (4) heavy metals in commercial feeds and (5) the framework for human health agenda.

Feed Intake Regulation

The Voluntary Feed Intake (VFI) covered in this paper is classified into two, (a) the physical regulation and (b) the metabolic regulation. The physical regulation is composed of the oropharyngeal metering and the stomach distention or the gutfill. The metabolic regulation is composed of the following: (a) Volatile Fatty Acid (VFA) regulation in ruminants; (b) hormonal regulaton and (c) the Hepatic Oxidation Theory. Although voluntary feed intake regulation principles maybe associated to man, there is no 100% assurance that the same follows in human. A more troubling concern in humans is the unravelling of the incidence of eating or food addiction intentionally included because of the research agenda of providing healthy food towards a healthy the nation.

The oropharyngeal metering theory of food regulation is the concept that there is something in the mouth and pharynx that counts or measures the food that passes until it gradually slow down and stop. This idea was derived from surgically implanted balloon beyond the pharynx that collects the feed that enters the mouth. The possible reason for this oropharyngeal feed intake regulation could be the emptying of the saliva from the salivary gland that initially digest the feed and or the muscles for mastication and chewing get tired, such as the digastricus, masseter and pterygoid muscles.

The stomach distention or gutfill is simply attributed to the capacity of the stomach to accommodate feeds [6]; hence surgically reducing the size of the stomach may surely result to reduction in the voluntary feed intake and body weight consequently. In 2005, Kogima et al. published in nature magazine the linked between the stomach/gastrointestinal tract and the brain by the secretion of Ghrelin, a 28 amino acid peptide considered as orexigenic (hunger hormone). Ghrelin secretion is high during hunger or underfed state, transported by the blood, passes the blood brain barrier, and acts on the receptor (GHR) in the hypothalamus to stimulate appetite and also in the anterior pituitary gland to stimulate secretion of growth hormone.

The Volatile Fatty Acid and gutfill regulation in ruminants is considered anorexigenic ([7] [8]). Both VFA and gutfill regulations are connected because the infusion of VFA in the rumen of cannulated ruminants that results to linear depression in voluntary feed intake is in the form of acid. Increasing amount of salt resulted when increasing VFA solution was prepared to attain similar pH with the rumen environment. Therefore, the linear depression in feed intake attributed to increasing VFA concentration is in combination with salts. Hence this feed intake reduction due to VFA cannot just be attributed to metabolic regulation but to physical regulation as well because of the increase in osmotic concentration on the lining of the rumen due to salt combined with VFA. Separating the associated feed intake depression due to specific components of VFA such as acetate, propionate and butyrate reveals differing amount of linear increase in feed intake reduction [7]. Therefore the VFA infused depression in feed intake is a combination of both salt and VFA.

The Hepatic Oxidation Theory (HOT) in ruminants was proposed and presented by [9], which states that feed intake is controlled by oxidation of fuels in the liver. According to the authors feed and energy intake of ruminant animal can change dramatically due to change in diet composition and metabolic state. The hepatic oxidation reduction in feed intake is made possible by the vagal reflex due to increase in fuel oxidation or generation of ATP. This has been found consistent with propionate higher feed intake reduction relative to acetate, butyrate and lactate which has been found negligible to very low [9]. The control of feed intake by hepatic oxidation becomes higher when nutrient demand decreases, hence gutfill increases. Hence in times of increasing milk production in dairy animals VFI regulation is influenced more by the gutfill and less of the metabolic or hepatic oxidation. Therefore the application of the combined hepatic oxidation and gutfill regulation can be used in the dairy production by increasing the proportion of concentrate over the roughage in the diet as the milk production rises to its peak and vice versa as the milk production declines up to the dry period.

The hormonal regulation of feed intake has been understood to be controlled by orexigenic Ghrelin and anorexigenic insulin, cholecystokinin (CCK) and leptin. Orexigenic stimulates appetite and anorexigenic hormones reduce the voluntary feed intake in farm animals as well as in man. Leptin influence of feed intake reduction in cattle is only 25% at a condition when plasma leptin is elevated through lipogenic action at fattening stage in cattle (Vega et al, 2013). The eating and food addiction in man may have two possibilities that need further research, i.e. the stimulation of the Cocaine Amphetamine Receptor Transcript (CART), Proopiomelanocortin (POMC) and the Dopamine D2 Receptor (D2R). The D2R is the likely link in the addiction – like reward dysfunction reported in rats by [10].

Compensatory Growth Mechanisms

Growth Rate & Feed Conversion Ration with age decrease and follow a curvilinear pattern in swine ([11]; [12]) in dairy beef [13]. The principle that ADG and Feed Conversion Efficiency decrease with age can be explained by the fact that maintenance energy increase with increase in metabolic body weight. Therefore it seems economical if livestock are produced and slaughtered at earlier age. Researches are needed to have this validated across farm species.

Compensatory Growth, Realimentation, and Catch-Up Growth are interchangeable used and it is defined by [14] as physiological process whereby an organism accelerate its growth after the period of restricted development caused by energy and protein restriction. In Holstein Steers, feed restriction from 38% of the mature weight by withdrawing concentrate feed allowed the animals to undergo compensatory growth [15] [16] This was similar to the findings that animals exposed to feed restriction 25 to 30% of the mature weight can rebound and gain weight similar to the control or ad libitum fed. Compensatory growth research should further be explored as a research topic because its application may be naturally occurring, e.g. during summer and winter for grazing cattle. In compensatory growth principle implies that growing animals beyond 35% of mature weight should be allowed to undergo nutritional stress.

Animal Nutrition Research

Animal nutrition research is conducted with the aim of increasing the body weight of farm animals. The principle involve in digestibility study is nutrients consumed minus nutrients excreted is equals to digestible nutrients. In the conduct of feeding trial, most parameters taken are feed consumption, body weight gain and average daily gain. Other parameters are computed as feed efficiency or feed conversion efficiency. This is driven by the fact that the product sold is live weight in kg., the measure of economics is basically the cost of feed offered and income from live weight sold. If the ration/diet is economically feasible they are applied in the large scale commercial production. Food safety issues in animal nutrition research are not considered, only when the animals died. How about the issue of systemic inflammation? Systemic inflammation is the increase in body weight because toxins and residues in feed causes inflammation of the villi or linings of the small and large intestines that allows the unwanted substances to pass through, this is the case of leaky gut syndrome. Residues that enter the animal's body can cause inflammation where they are deposited. After the successful rat or mouse experiment, pigs for biomedical research are now becoming necessary before human clinical trials is conducted for a successful drug application. Therefore a "One Health" program is becoming necessary which can be done not just by the medical, veterinary, researcher/ scientist of the drug companies but also by the animal scientist. Understanding the total effect of the diet in swine research extending it to assay pesticide and heavy metals in feed, gut microflora, small intestines villi, visceral condition and other blood parameters for possible endocrine disruption may provide useful information to extend the benefit of research not only for capitalist interest but for human health.

Heavy Metals in Commercial Feed and Edible Tissues

In ducks heavy metal Cd beyond 0.025 ppm [17] is 80% of all commercial feeds shown in Table 1 (Octura et al, 2017). Based on this exploratory survey and sampling of feed and ducks it was found out that visceral organs, i.e. the liver and ovary poses Cd threat. Liver and ovary have very high amount of Cd ranging from 0.40 to 1.17ppm, i.e. beyond weekly tolerable intake of 0.025ppm by EFSA (2009).

In pigs, Table 2 shows Cd content in ration and very high in pre starter (1.33), starter (1.12), grower (1.18), breeder (1.35) and finisher (1.43 ppm). The 0.46ppm Cd are released in manure. Hence it is not advisable for use as organic fertilizer. The liver, blood and meat have 0.849, 0.104 and 0.069 ppm Cd, respectively. Analyzing the each feed ingredients used, on the highest concentration of Cd in ppm is biofos (12.03), salt (7.1), mineral premix (5.32), antioxidant (4.48), zeolite (3.74), mono Calcium Phosphate (2.71), vitamin premix (1.62), US Soya (0.68), Rice Bran (0.64), Cassava (0.39) and Ground Corn (0.19). Aside from drug residues in meat of livestock "Pharming", heavy metal Cd is another threat to health of consumers. Cadmium is not biodegradable and it accumulates in the environment, in the air, soil and water through time. Therefore control in the consumption of processed meat and meat per se must be dealt with if quality and healthy life is our goal in.

Framework for Human Health Agenda

It is the moral responsibility of teachers to be healthy, because students do not just listen to what they hear but in what they see. The message is to disseminate a healthy lifestyle so that we will live life in happiness and in health. The proposed framework for human health agenda is directed towards reducing the emission of greenhouse gases and control of food addiction specially processed meat for health reasons. To make this proposed framework a reality, the foundation must be grounded on established facts and carefully derived principles.

The prevalence of addiction to fast food products and processed meat is growing worldwide coupled with the prevalence of obesity and its related diseases. According toRabobank International (unpublished), the projected livestock demand is 40% in 2030 and 70% in 2050. Looking at their graph clearly it is not the lamb (4%) and beef (20%) that increase in demand, but the pork (37%) and poultry (39%). Therefore, it is the urbanization, fast food chain, processed meat in the supermarket that drives the demand for meat.

Table 1. Mean Cd content in mg/kg (ppm) of commercial feed samples in selected farms of Laguna and Pampanga. (EMDL = 0.04 mg/kg)

Commercial Feeds	LOCATION				MEAN
	Laguna		Pampanga		
	mg/kg	% Users	mg/kg	% Users	
A	<EMDL	9.0	0.16	26.7	0.08
B	0.44	27.3	0.29	6.7	0.37
C	0.46	9.1	0.16	20	0.31
D	-	0	0.12	6.7	0.12
E	-	0	0.84	26.7	0.84
F	-	0	0.83	13.3	0.83
G	0.14	9.1	-	0	0.14
H	<EMDL	27.3	-	0	<EMDL
I	0.12	9.1	-	0	0.12
J	<EMDL	9.1	-	0	<EMDL
Average Cd mg/kg	0.186		0.437		0.311

The Guardian (2017) [18] an independent investigative journal publisher report that 10% of the richest people produce half of the global CO_2 emission based on lifestyle consumption emission. The lifestyle of the rich would mean just a taste-then-waste, extravagance and overly designed appliances/ cars.

Based on the rule of thumb on dietary requirements, our plate should have 25% rice, 25% fish or meat, 25% vegetable and 25% fruits. This balanced diet is not served in almost all fast food. Most children addicted to fast food and processed meat turned obese early in life and failed to live life in happiness and health. The many unwanted residues in commercial feeds and the added preservative in meat to keep its keeping quality for a long period of time are the possible reasons why meat consumption is linked to unhealthy life.

Looking at Figure 1, the framework clearly states that the rich are the contributor in global greenhouse gas emission and in both rich and poor the higher per capita meat consumption the greater the possibility that an individual is developing an unhealthy lifestyle. The framework tells us that the extravagant lifestyle of the rich contributes a lot in global greenhouse gas. Moreover, regardless of our economic status, addiction to fast food and or processed meat brings us to unhappy and unhealthy life. Therefore our action must be based on reason and not on emotion, for the rich to live simple life, share their wealth and for all to control addiction to fast food and or processed meat.

Table 2. Concentration of Cd ± SD[1] (ppm or mg/kg) of formulated rations[2], manure and edible parts/ tissues taken from farmers field (February to September 2014)

CLASSIFICATION	SAMPLE	N	MEAN ± SEM[3]
FORMULATED RATION	Pre-Starter	2	1.33 ±0.27
	Starter	2	1.12 ±0.03
	Grower	2	1.18 ±0.06
	Breeder	4	1.35 ±0.12
	Finisher	1	1.4275 ±0.0
ORGANIC FERTILIZER	Manure	8	0.46 ±0.125
EDIBLE PARTS	Liver	6	0.849 ±0.22
	Blood	4	0.104 ±0.07
	Meat	6	0.069 ±0.03
FEED INGREDIENTS	Rice Bran	2	0.635 ±0.13
	US Soya	2	0.68 ±0.0
	Ground Corn	2	0.19 ±0.21
	Cassava Meal	1	0.39 ±0.0
	Mono Ca-P	1	2.71 ±0.0
	Biofos	1	12.03 ±0.0
	Zeolite	1	3.74 ±0.0
	Antioxidant	2	4.475 ±2.55
	Mineral Premix	2	5.315 ±1.72
	Vitamin Premix	1	1.62 ±0.0
	Salt	2	7.1 ±0.07
	DL – Methionine	1	ND
	Choline Chloride	1	0.04 0.0
	L- Threonine	1	ND
	L - Lysine	1	ND
	Limestone	2	8.22 ±0.07

[1] Heavy metal Cd was analysed using AAS-Flame method at the Analytical Service Laboratory of the National Institute of Molecular Biology and Biotechnology, UP Los Baños;

[2] The feed companies available in Sta. Cruz watershed were LIMCOMA, UNAHCO, Primera, San Miguel, Masagana and Robina;

[3] ND= Non-detectable; Provisional Tolerable Monthly Intake (TDI) is less than 0.025ppm (25ug/kg b.w.) as imposed by EFSA 2009

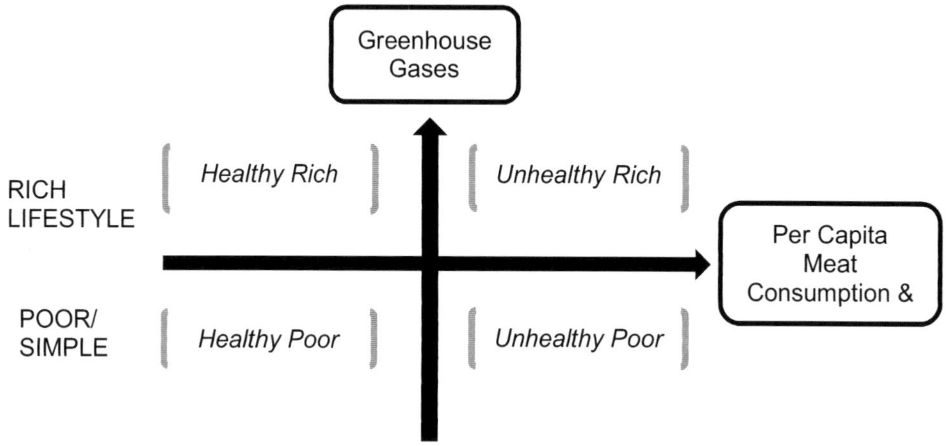

Figure 1. Framework for Understanding the Effects of Livestock "Pharming" and Lifestyle of the Rich to Greenhouse Gas Emission and The Food Addiction to Fast Food and or Processed Meat to Human Health

Acknowledgement

Gratitude is given to Chancellor Fernando C. Sanchez, Jr of the University of the Philippines Los Baños for giving the opportunity to serve as keynote speaker in the, "International Seminar on Ruminant Production" held at Central Java, Indonesia. Also to Dr. Daud Samsudewa, the Dean and Administrator of the Faculty of Animal and Agricultural Sciences, Diponegoro University for the invitation and providing financial support.

References

[1] World Health Organization (WHO). 2017a. Global and Regional Consumption Pattern. http://www.who.int/nutrition/topics/3_foodconsumption/en/index4.html. pp 1 – 9. Accessed 1 December 2017.

[2] Delgado, Christopher L. 2003. Rising Consumption in Meat and Milk in Developing Countries has Created a New Food Revolution. J. Nutr. 133. 3907S–3910S. http://jn.nutrition.org/content/133/11/3907S.full.pdf+html. Accessed 2 December 2017.

[3] Food and Agricultural Organization (FAO). Livestock in Balance. In Chp 2. Change in Livestock Sector. http://www.fao.org/docrep/012/i0680e/i0680e02.pdf. pp 1–23. Accessed 1 December 2017.

[4] World Health Organization (WHO). 2017b. Obesity and Overweight. Fact Sheet, Updated 2017. http://www.who.int/mediacentre/factsheets/fs311/en/.

[5] World Health Organization (WHO). 2017c. Global Health Observatory Data. Overweight and Obesity. http://www.who.int/gho/ncd/risk_factors/overweight/en/. Accessed 2 December 2017.

[6] Campling RC and Bulch CC. 1961. Factors Affecting the Voluntary Feed Intake by Cows. 1. Preliminary Observations on the Effect on the Voluntary Intake of Hay, of Changes in the Amount of Reticulo – Ruminal Contents. Brit. J. of Nutr. 15:523 – 530.

[7] Baile CA, Mayer AJ, McLaughlin CI. 1969. Feed Intake of Goats during Volatile Fatty Acid Injections into Four Gastric Areas. J. Dairy Sci., 55:1058 – 1063.

[8] Choi Byung–Ryul and Michael S. Allen. 2009. Intake Regulation by Volatile Fatty Acid and Physical Fill. African Journ. of Anim. Sci. 29 (ISRP): pp. 40- 41.

[9] Allen MS, BJ Bradford and M. Oba. 2009. Board Invited Review: The Hepatic Oxidation Theory and its Application in to Ruminants. Journal of Animal Science 87: pp. 3317–3334. http://www.uky.edu/Ag/AnimalSciences/instruction/asc684/PDF/JAS87_3317.pdf. Accessed in December 2011.

10] Johnson PM and Kenny PJ. 2010. Addiction – like Reward Dysfunction and Compulsive Eating in Obese Rats. Role for Dopamine D2 Receptors. Nat. Neurosci. 13(5): 635 – 641.

[11] Octura JER, Manalo DD and Vega RSA. Cadmium in Feed and Tissues of Female Mallard Ducks (*Anas platyrhynchos* L) in Selected Farms of Victoria and Candaba. The NRCP Research Journal. 16(2): 46 – 61.

[12] Peñalba FF, Capitan SS, Vega RSA, Estrella CAS and Garcia BR. 2005. Physiological Growth Stages of Female Progenies of Backcrosses and Two – Breed Cross for Smallhold Pig Raising. Proceedings of the 42nd Phil. Society of Animal Science and Annual Scientific Convention. p 41.

[13] Vega R.SA., Hidari H., Suzuki M., Kuwayama H. and Manalo D. 2004. The relationships of plasma leptin, backfat thickness and TDN intake across finishing stages of Holstein Steers. Asian-Australasian Journal of Animal Science 17 (3):330 – 336.

[14] Hornick JL, Van Eanaeme C, Gerard O, Dufrasne I, Istasse L. 2000. Mechanism of Reduced and Compensatory Growth. Domestic Animal Endocrinology, 19: 121 – 132.

[15] Vega, R.SA., H.G. Lee and H. Kuwayama. 2009. Changes in plasma leptin and IGF-1 (Insulin-like Growth Factor – 1) during compensatory growth in Holstein steers. Phil. J. of Vet. Med. and Anim. Sci. 35(1) 1 – 14.

[16] Vega RSA, Lee H-G, Kuwayama H and Hidari H. 2012. Body Fat & Plasma Leptin Involvement in the Voluntary Feed Intake in Cattle: Study on Plasma Leptin Profiles in Cattle. Lambert Academic Publishing. pp. 44 – 49.

[17] European Food Safety Authority. 2009. Cd in Food Scientific Opinion of the Panel on Contaminants of Food Chain. The EFSA Journal (2009) 980, 1-139. Retrieved:http://www.efsa.europa.eu/en/scdocs/doc/980.pdf . Accessed: October 15, 2014.

[18] The Guardian. 2017. World Richest 10% Produce Half of the Global CO_2 Emission says Oxfam. https://www.theguardian.com/environment/2015/dec/02/worlds-richest-10-produce-half-of-global-carbon-emissions-says-oxfam. Accessed October 2017.

Biotechnological Approaches on Ruminant Feed Nutrition - With emphasis on Indonesian Situation -

S. J. OHH[1*]

[1]Kangwon National University, Chuncheon, Korea

*Corresponding E-mail : sjohh@kangwon.ac.kr

1. Introduction to feed biotechnology

Feed biotechnology can be defined as 'biotechnology onto the area of feed nutritional manipulation'. Recently the feed biotechnology has emerged as *de novo* tools for improving the nutritive value of feed resources, with more emphasis on poor quality ingredients, and modulation of gut microbiota for better feed nutritional metabolism. Although many biotechnological tools are available now, most of feed biotechnological tools in feed biotechnology are more or less, associated with microorganism and their fermentation products. Matter of fact, many live microorganisms and their fermentation products like enzyme, amino acids, peptides, organic acids, vitamins, bioplexes, and even antibiotics have been developed and employed for the feed nutritional improvement. In addition, use of various genetically modified organisms(GMO), mostly grains and oilseeds became part of feed biotechnological application. However, development of new GMO itself is technically driven by biotechnological scientist, which means there is not much room available for feed nutritional application of those biotechnological GMO products. Therefore, this article mainly discussed about the modulation of rumen microbiota and the improvement of feed nutritional values by biotechnological products.

2. Biotechnological application for the better rumen development

2.1. Factors affecting rumen development

Well-developed rumen is pivotal for the better feed nutritional metabolism throughout the life of ruminant animal [1]. One of the primarily important factor for the good rumen development is known as the appropriate intake of good quality calf starter [2]. Ruminant animal with well-developed rumen could utilize even poorer quality local feed resources and relatively perform better than less-developed animal. However, it is physiologically difficult for calf to voluntarily increase the intake since consumption of the calf starter is the first and huge challenge for calf itself. Many approaches have developed and applied though there still no simple and easy-to-apply techniques are available especially for the small holder farms since many factors affect the feed intake. Factors affecting the intake of calf starter feed are calf starter formulation and physical form, type and addition level of hay, grain and protein sources, physical processing and fermentation method, and feed supplements [3]. Thereby, the best possible approaches would be the combination of both physical processing of feed and use of suitable biotechnological products.

2.2. Modulation of physical forms

First of all, one of the simple and straight forward tool is the modulation of physical form. The physical form of calf starter feed was known to affect palatability and intake [4]. Calves do not prefer either finely-ground feed and too much hard pellet. Instead, calves usually prefer 'coarse and texturized' feed along with consistency in the physical form. However, what would be the suitable degree of coarseness and texturization is not known yet since the degree is varied by the type of the ingredient and local condition. It should be noted that most of the globally published recommendation about the physical form came from studies done with corn and soybean meal otherwise soybean. This means the optimal physical form could be different once employed different type or quality ingredients. Therefore, it is fundamentally needed to develop optimal physical form by applying individual local condition. Most practical comparison study showed that other texture form, compared to pellet form, were preferred. However, complete pellet calf starter pellet could exert its own advantage [5] [6] since it is easy for the animal care-taker to feed them and also convenient for animals to consume especially when and where the appropriate feed consumption is challenged due to local environmental situation.

2.3. Complete calf starter pellet processing method

Making pellet would be simple, however making good quality calf starter pellet is not simple both technically and nutritionally. Therefore, many trials need to be preceded to establish the custom formula and processing method to prepare the best possible quality complete pellet starter. One of the contrasting challenge in processing pellet would be the particle size of the ingredients. It is difficult to agglomerate pellet when the mash are too coarse. Therefore, many processors prepared the finer particle size mash regardless of ingredients quality. Finer particles of grains and concentrate feed may speed up the rate of rumen hydrolysis, which is not recommended for the balanced rumen metabolism. In addition, finer particles of soft fiber sources may also not appropriate for imparting physical stimulation onto rumen wall. Therefore, the particle size of each ingredients in the mash should be carefully controlled and compromised in view of both optimum speed of hydrolysis and optimum physical stimulation.

2.4. Ingredients selection and quality

It is also needed to pay keen attention to the quality of local forages and roughages. Physical stimulus potential, rumen hydrolysis and retention time of the similar forage sources could be varied even by harvesting season, location, weather and post-harvest handling. Therefore, both type and addition level of a hay or forage should be carefully selected and pretested before application. Harsh and lignocellulosic roughage are not always recommended no matter how any better sources are not available. Good quality grass could be generally acceptable and beet pulp would be a good choice if available.

Among grain sources, corn(maize) is preferred since the hydrolyzing speed of corn starch is appropriate, meaning not too fast like wheat starch. Soybean(meal) is preferred as a protein source. There is a big difference in speed of rumen hydrolysis between oilseed protein and grain protein. Soybean protein would be preferred because protein in the soybean is mostly salt-water soluble protein whereas grain protein are mostly soluble in rather strong solution like alkali solution. Therefore, for the young calf, it is recommended to supply proteins which can be readily hydrolyzed even in underdeveloped rumen fluid of young calf.

2.5. Biotechnological approaches for increasing feed intake

Biotechnological approaches should be incorporated as well to increase the intake. Fermentation of calf starter has tested although the result was not promising. Once poor quality and poor taste ingredients were used, however, the fermentation may exert benefit. If unavoidable, mild fermentation would be recommended since strong flavor of fermentation reduce the appetite of calf.

To improve the taste of calf starter and to supply readily available carbon sources to rumen microbe, molasses has been widely used for the calf starter. However, the addition levels of molasses need to be carefully controlled. It is always needed to pay attention to the degree of sweetness. Calf does not prefer strong sweetness and certainly not the bitterness. Molasses is able to generate big buffering capacity compare to other ingredients. Therefore, appropriate addition level of molasses need to be selected by compromising the effect between buffering potential and taste enhancement. If other biotechnological taste enhancer is available, then try to more focus on buffering potential. Certainly, it is needed not to impart enormous potential since certain degree of pH fluctuation is needed for the rumen development and balanced rumen microbiota.

Biotechnological products like amino acids, probiotic microbes, prebiotics, enzymes, ionophores, and essential oils have been incorporated individually and with combination to the calf starter. Many researches are undergone to find and to test those biotechnological products. However, the benefit is not always economically feasible and the responses are varied by unknown reason. It would be evident at this moment to recommend the supplementation of any biotechnological product at only pre-tested then justified condition. Properties of microorganisms and their products have been known very environment sensitive. Therefore, it is needed to develop custom type of the products instead of using commercial products developed elsewhere. For example, it is needed to isolate appropriate rumen bacteria in good calf in each region and each breed. This is also true for isolating the same lactate utilizing bacteria like *Megasphaera elsdenii* [7] from the rumen of local breed at local condition.

Prebiotic also can be applied to the calf starter. However, it is needed not to accept the reported benefit of a prebiotic in a type of animal for applying the same prebiotic into calf starter. Oligosaccharides such as plant mannan oligosaccharide (MOS) and fructo oligosaccharide(FOS) may not exert benefit as reported through non ruminant animal application [8] [9]. Probiotic originated microbial cell wall such as β-glucan, microbial MOS are more likely impart benefit on ruminants.

2.6. Other factors affecting intake of calf starter

Water is also essential for giving rumen bacteria the appropriate aqueous environment. Therefore, intake and quality of water need to be always monitored to guarantee the sufficient intake of the clean water. Good quality water promotes the intake of dry feed and decrease selective consumption by minimizing segregative sorting of the feed although this sorting is not likely in good quality pellet. In addition, the water could impart both buffer potential and homogeneity of rumen status. Certainly, feed intake tendency and potential is also affected by genetics, gender, birth weight of calf. Housing and feeder condition and environmental condition also affect the feed intake.

3. Biotechnological modulation of rumen microbiota

Feed utilization efficiency in ruminant animal is largely affected by rumen microbiota and vice versa [10] [11]. Therefore, many biotechnological approaches have been implemented. Those approaches could be categorized by the following five routes. Those are ① use of fiber hydrolyzing enzyme ② lactate utilizing microbial supplementation ③ supplementation of commensal bacteria utilizing low quality roughage ④ competitive inhibition by supplementing substances suppressing methanogenic fermentation ⑤ supplementation of ionophores to suppress certain microbiota.

3.1. Positive modulation by fiber hydrolyzing enzymes

Although rumen microbiota have enormous potential to hydrolyze fiber sources, its rate and potential of the hydrolysis is varied by not only microbiota but also quality of roughage. Therefore, type and quality of fiber sources during developmental stage is able to stimulate the expression of microbial types. Once the amount of poor quality fiber is relatively huge, then the expression of rumen microbiota would be changed to the way to adapt that load. Supplementation of fiber hydrolyzing enzyme into the feed is able to reduce the load of rumen microbe to hydrolyze the fiber sources, more likely poor quality fiber. Therefore, rumen microbiota could be balanced to be more efficient for feed utilization.

3.2. Positive modulation by supplementing microbes

Higher proportion of lactate accumulation generally leads to poor feed utilization and metabolic disorder. To prevent or relieve this problem, extrinsic supplementation of lactate utilizing microbe has tried although the result are varied by the trial. Before implementing this approach, it is needed to know what would be the typical pattern of fermentation and typical rumen microbiota of the target animal in each region. This would give the idea to select the better type of lactate utilizing microbes.

This strategy would be also needed for selecting other beneficial probiotic supplement [12]. Since rumen microbiota is so diverse, there is always the better approach to select a customized type of probiotic for the target ruminant fed its locally abundant feedstuffs [13]. All fiber hydrolyzing microbes are not equal for hydrolyzing a local poor quality fiber sources. Therefore, to improve poor quality feed utilization, it would be useful to isolate a microbe that are commensal and beneficial for the local breeds in local condition.

3.3. Negative modulation through competitive inhibition

Although major rumen microbes are bacteria, protozoa, methanogens, and fungi, proportion of those could be changed by suppressing the growth of certain microbes. For this purpose, ionophores have been used and their effect was generally successful. Ionophores mitigate ions, thereby disrupt bacteria cell membrane and suppress the gram-positive bacteria, which changed the proportion of VFA. This supplement is also appropriate where the feed utilization efficiency is relatively poor. However, use of the ionophores is forbidden in some region.

Therefore, many biotechnological products and functional component have introduced to change the rumen microbiota by inhibiting certain microbes. Some flavonoids, tannins, and essential oils extracted from plant and microbes are able to inhibit the growth of microbes more likely methanogenics. Reduction of the methanogenic fermentation is one of the way to improve feed utilization efficiency. However, methanogenic fermentation and methanogenic microbes have stabilized through the adaptation process to the local condition. It is also needed to understand the characteristic of rumen fermentation and microbiota before selecting components to suppress target microbes.

4. Other biotechnological approaches

Biotechnological approaches should not be limited only to modulate rumen microbiota and to stimulate rumen development. There are many areas still needing improvement and change toward the better feed nutrition in ruminant animal. Biotechnological tools could be employed for those approaches.

One of the fundamental thing for improving ruminant feed nutrition is the correct information on local feed resources. It is needed as early as possible to have the data on the rumen hydrolyzing rate and buffering capacity of local feed ingredients. It is also needed to establish the each regional feed status by season(dry vs rainy). That could lead to the right development of both concentrate mix and feed supplements customized for each season and to the development of season specific TMR for each region.

Once the correct nutritional information and availability of local feed resources are available, then need to establish proper on-farm particle sizing method/tool, including cutting, chopping, grinding, etc. Manufacturing homogenous quality rough diet in terms of particle size and mixture uniformity is not a simple and easy task. Developing local measure to avoid post-harvest mold on feed resources is also crucial for improving feed utilization since mycotoxin contamination disturbs the proper rumen fermentation and rumen microbiota. Throughout the both rainy and dry season, it is also better to establish the proper hay making process and storage method. For the grazing in certain location and in small holder farm, it is also needed to develop locally suitable block feed for grazing animal.

For improving feed nutritional efficiency in long term basis, biotechnological tools should be implemented as well. If poor quality forages are naturally dominating in a certain region, then apply selection and mutation techniques for developing better quality forage crop in each region. Enzymes and beneficial microbes can be developed by applying biotechnological tools. Isolation and culture of locally available microbes which is high in NSP and fiber hydrolyzing enzymes could be another way to improve feed utilization for long term basis. For example, anaerobic bacteria and white rot fungi (*Pleurotus ostreatus*) that can degrade poor quality forage would be beneficial where poor quality forages is unavoidable.

Using locally abundant resources for developing a biotechnological products can emanate big potential. MOS and functional oligosaccharides from palm kernel meal and copra meal would be one example. Develop how to apply tapioca starch for better feed nutritional function, such as nutritional binder material for agglomeration, and modulator to control the rumen release of nutrients including ammonia and energy for bacterial protein synthesis. In addition, try to find how to compromise the use of sugarcane molasses for improving early rumen development, increasing buffer capacity and enhancing taste. Where volcano eruption occurred, it is needed to develop how to apply or use volcanic scoria as a supplement to improve rumen buffering potential.

5. Conclusion

Today, biotechnological approaches worth *de novo* commitment and become powerful and economical even in feed nutrition area. In fact, microbial biotech tools and products are practically meaningful feed biotechnology in the market as well as in the farm. However, the biotechnological advances and application developed for other region cannot be the customized appropriate biotechnology for animals in any other local regions. Do not dream the fancy biotechnology for tomorrow but apply the fundamental biotechnology today.

References

[1] Malmuthuge N, Griebel PJ, Guan LL. 2015. The gut microbiome and its potential role in the development and function of newborn calf gastrointestinal tract. Frontiers Vet. Sci. **2**: 36.

[2] Govil K, Yadav DS, Patil AK, Nayak S, Baghel RS, Yadav PK, Malapure CD, Thakur D. 2017. Feeding management for early rumen development in calves. Journal of Entomology and Zoology Studies **5**: 1132

[3] Chester-Jones H, Broadwater N. 2009. Calf starters. Proceeding of Minnesota Dairy Days. University of Minnesota Extension Service.

[4] Mirzaei M, Khorvash M, Ghorbani GR, Kazemi-Bonchenari M, Riasi A, Soltani A, Moshiri B, Ghaffari MH. 2016. Interactions between the physical form of starter (mashed versus textured) and corn silage provision on performance, rumen fermentation, and structural growth of Holstein calves. J Anim Sci. **94**: 678

[5] Mukodiningsih S, Budhi SPS, Agus A, Haryadi, Ohh SJ. 2010. Effect of molasses addition level to the mixture of calf starter and corn fodder on pellet quality, rumen development and performance of Holstein-Friesian calves in Indonesia. J. Anim. Sci. Technol. **52**:167

[6] Mukodiningsih S, Achmadi J, Wahyono F, Sung KI, Ohh SJ. 2016. Effect of feeding pellet type complete calf starter combined with maize fodder and molasses on the rumen development of dairy calves. Livestock Res. Rural Develoment. 28

[7] Weimer PJ and Kohn RA. 2016. Impacts of ruminal microorganisms on the production of fuels: how can we intercede from the outside? Appl. Microbiol. Biotechnol. **100**:3389

[8] Kara C, Cihan H, Temizel M, Catik S, Meral Y, Orman A, Gencoglu H. 2015. Effects of supplemental mannanoligosaccharides on growth performance, fecal characteristics and health in dairy calves. Asian-Australasian J. Anim. Sci. 28: 1599

[9] Hill TM, Bateman II HG, Aldrich PAS JM, Schlotterbeck RL. 2008. Oligosaccharides for dairy calves. The Professional Anim. Scientist. **24**:460

[10] Mizrahi I. 2012. The Role of the Rumen Microbiota in Determining the Feed Efficiency of Dairy Cows. In: Rosenberg E., Gophna U. (eds) Beneficial Microorganisms in Multicellular Life Forms. Springer, Berlin, Heidelberg

[11] Ellison MJ, Conant GC, Lamberson WR, Cockrum RR, Austin KJ, Rule DC, Cammack KM. 2017. Diet and feed efficiency status affect rumen microbial profiles of sheep. Small Ruminant Research. **156**:12

[12] Kobayashi Y, Oh S, Myint H, Koike S. 2016. Use of Asian selected agricultural byproducts to modulate rumen microbes and fermentation. J. Anim. Sci. Biotechnol. **7**:70

IOP Conf. Series: Earth and Environmental Science **119** (2018) 012005 doi:10.1088/1755-1315/119/1/012005

Nutrient Intake and Digestibility in Merino Sheep Fed Peanut Straw

C. Hanim[1*] and Muhlisin[1]

[1]Faculty of Animal Science, Gadjah Mada University, Yogyakarta, Indonesia.

*Corresponding E-mail: c.hanim@ugm.ac.id

Abstract

This study aimed to compare nutrient intake and digestibility between male and female Merino fed peanut straw as sole feed. Four male and five female Merino sheep ages 10 to 12 month with average weight of 40 kg were used in this study. All animals were confined in individual metabolism cages and nylon nets were fitted bellow the cages for faecal collection. This experiment was done with 14 d adaptation period and 7 d collection period. Representative sample of feed, refusal feed, and faeces were analysed proximate including dry matter (DM), organic matter (OM), crude protein (CP), crude fibre (CF), ether extract (EE), and extract-free nitrogen (EFN). Data collected were used to calculate nutrient intake and digestibility. All nutrient intake in male Merino was higher ($P<0.05$) than female one, whereas digestibility of DM, CP, and CF in male Merino sheep were higher ($P<0.05$) than female sheep. It could be concluded male Merino sheep consumed feed more than female, but they had similar value of digestibility.

1. Introduction

Merino is Indonesian local sheep breed which has high potencies to be developed, they commonly reared for fattening because their good performances. The main problem arise in animal fattening is feed supply which is less than the requirement. Therefore, this study on feed stuffs that can meet the nutrient requirement, nutrient contents and its quality is needed. One of the potential feedstuff for sheep was agriculture by products such as peanut straw which is easy to find during harvesting time. Usually farmers use agriculture by products such as rice straw, maize straw, and peanut straw as sole diet for ruminant. Ruminant require protein in the diet to nitrogen supply for rumen microbial activity and animal tissue metabolism. Microbial requirement are met at 6-8% crude protein in the diet while animal requirement range from 7-20% in the diet depending upon species, sex, and physiological status [1].

The performance of ruminants fed forage diets depends largely on feed consumption and its digestibility. It is well known that feed intake and its digestibility is main factor affecting nutrient utilization [2]. Substitution of legumes for low quality forage has been shown to increase forage consumption compared with diets consisting only of low quality forage [3]. A major problem in diet formulation is predicting the effect of changes in composition of diet on intake. For diets containing long forage, intake is positively correlated with digestibility [4]. Digestion of plant material is a time-dependent process. An increase of feed intake accelerate the passage of digesta through the tract, resulting reduce digestive efficiency [5,6].

Based on the review, it is necessary to evaluate a single feed from peanuts straw to supply protein requirement of Merino, either male or female, using nutrient intake and digestibility experiment.

2. Materials and Methods

2.1. Animals and Feeding Treatment

In this experiment, peanut straw as sole diet was obtained from the Yogyakarta province. Peanut straw was served without the empty root neither the hard woody straw, only the edible portion which was used as the feedstuff. The chopped straw was offered in fresh condition, they collected after harvesting and stored in protected shed. Four male and five female Merino sheep, age 10-12 months and average body weight 40-50 kg were used in this experiment. The animals were confined in the individual metabolism cages, and nylon nets were fitted bellow the cages for faecal collection to get good separation of the urine and faeces.

2.2. Animal Digestion Trial

The animals were kept in individual metabolism cages and had free access to water. The experiment consisted of a 14 days adaptation period, and a 7 days measurement of voluntary intake and total faecal collection. The amount of feed offered was adjusted so that feed refusals were in excess of 10% of the amount offered during the preliminary and voluntary intake period. Feed refusals were removed and weighed once daily. During each collection period, samples of feed were collected and subsampled. Faeces were collected daily and stored at 5°C until the end of the collection period. After each collection period, total faeces were weighed, mixed by hand and subsampled. Subsamples of feed and faeces were dried at 65°C for 2 days for chemical analysis.

2.3. Chemical Analysis

During the collection period, 200 g of daily feed offered samples were collected, refusal feed were also collected daily, early in the morning before feeding time and then were weighed. Individually, 10% of the refusal feed samples were taken. Representative samples of feed, refusal feed and faeces were subjected to proximate analysis. After drying at 55°C, dried subsamples of feed and refusal feed were ground through a 1-mm screen and conducted for chemical analysis, including dry matter (DM), organic matter (OM), crude protein (CP), crude fibre (CF), and ether extract (EE), according to AOAC procedures [7], and then nitrogen-free extract (NFE) was calculated. Meanwhile, faeces were analysed in fresh condition. Nutrient intake and digestibility value were calculated based on the result of chemical analysis of feed and faeces.

2.4. Experimental Design and Statistical Analysis

Treatments were arranged in t-test design, withthe main factor being the sex of sheep, male and female. The data in the main study was analysed as t-test arrangement, the differences of mean value were analysed by Duncan's new multiple range test [8].

3. Results and Discussion

3.1. Chemical composition of peanut straw and nutrient intake of sheep

Nutrient contents of peanut straw was dry matter (DM) 20.14%, organic matter (OM) 84.25%, crude protein (CP) 14.55%, ether extract (EE) 1.55%,crude fibre (CF) 17.67%, and nitrogen free extract (NFE) 50.16%. Based on those data, the amount of feedstuff offered and the amount of refusal feed, as well as nutrient intake could be calculated, and the result of the study was presented in Table 1. It showed that there were significant differences in the nutrient intake of male compared with female Merino. All nutrient intake in male Merino sheep was higher (P<0.05) than female sheep.

Dry matter intake of male and female Merino sheep was 2.84% and 2.09% of their live weight, respectively. The DM intake of sheep was enough for maintenance need as mentioned it should be 2.0% of animal body weight [9]. Protein intake in Merino sheep also met their requirement, as known that protein intake for sheep was 63 g/day in weight 40 kg [9]. Dry matter and protein intake for maintenance in sheep was 2% of body weight and 95% in weight 50 kg, respectively [10]. The sex of animals influence the nutrient intake, the male sheep has higher nutrient intake than those female. Male sheep body weight was heavier than those female, thus, they need more nutrient for met their requirement.

Table 1. Nutrient Intake in Merino Fed by Peanut Straw as Single Diet

Nutrient intake	The sex of Merino	
	Male	Female
In g/day		
Dry matter	1,089.27 ± 65.88^{a}	713.86 ±110.07^{b}
Organic matter	912.34 ± 54.39^{a}	597.28 ± 93.50^{b}
Crude protein	132.85 ± 7.53^{a}	87.74 ± 12.92^{b}
Ether extract	17.82 ± 1.01^{a}	11.69 ± 1.68^{b}
Crude fibre	170.87 ± 11.87^{a}	109.21 ± 15.93^{b}
EFN	442.35 ± 24.96^{a}	290.93 ± 48.97^{b}
In g/(d·kg $BW^{0.75}$)		
Dry matter	70.64 ± 1.75^{a}	50.46 ± 6.43^{b}
Organic matter	12.93 ± 1.06^{a}	11.82 ± 0.96^{b}
Crude protein	10.29 ± 0.26^{a}	7.41 ± 0.87^{b}
Ether extract	1.73 ± 0.14^{a}	1.58 ± 0.13^{b}
Crude fibre	98.58 ± 2.21^{a}	69.32 ± 8.73^{b}
EFN	4.49 ± 0.33^{a}	4.19 ± 0.40^{b}

^{a,b} significant (P<0.05)

3.2. Nutrient Digestibility

All nutrient intake in male Merino was higher (P<0.05) than female one, as presented in Table 1. In contrast, several nutrient digestibility in female sheep was higher than male. The digestibility of DM, CP, and CF in male Merino sheep was lower (P<0.05) than those female (Table 2). However, no differences in digestibility of OM, EE, and EFN were observed for female compared with their digestibility in male sheep, as well as nutrient digestibility based metabolic body weight.

Even though male sheep consumed feed more than female, but they also excreted nutrient more than female. Nutrient digestibility was influenced by rumen microbe activities. It could be different among the sex of animals, because digestion process in rumen was affected by microbial enzyme activities, which is absolutely species specific. Nutrient intake in animal could also affected coefficient of digestibility of nutrient. It seems probable that the primary cause of lower digestibility at higher intakes was a faster rate of passage of feed through the digestive tract. Animals having an increase feed intake spent more time eating and ruminating and consequently less time resting. The

increase of feed intake reduce retention time of digesta, it means increase rate of passage of digesta through the gastrointestinal tract in cattle and sheep [2]. Digestion of plant material is a time-dependent process. An increase of feed intake accelerate the passage of digesta through the tract, resulting reduce
digestive efficiency [5].

Table 2. Nutrient digestibility in Merino fed by peanut straw as single diet

Nutrient intake	The sex of Merino	
	Male	Female
In %		
Dry matter	60,57 ± 3,06^{a}	69,98 ± 4,56^{b}
Organic matter ^{ns}	66,05 ± 2,54^{a}	74,47 ± 4,56^{b}
Crude protein	68,05 ± 1,79^{a}	73,27 ± 3,87^{b}
Ether extract ^{ns}	73,53 ± 5,17^{ns}	75,62 ±14,63^{ns}
Crude fibre	40,58 ± 7,09^{ns}	55,82 ±14,62^{ns}
EFN ^{ns}	83,78 ± 2,39^{a}	88,28 ± 1,72^{b}
In g/(d·kg BW$^{0.75}$)		
Dry matter	39.93 ± 5.42^{ns}	35.27 ± 4.65^{ns}
Organic matter	8.00 ± 0.47^{ns}	8.78 ± 0.59^{ns}
Crude protein	6.67 ± 0.73^{a}	5.43 ± 0.73^{b}
Ether extract	1.30 ± 0.14^{ns}	1.19 ± 0.22^{ns}
Crude fibre	34.20 ± 6.75^{ns}	38.33 ± 3.69^{ns}
EFN	3.59 ± 0.20^{ns}	3.20 ± 0.33^{ns}

^{a,b} significant (P<0.05)
^{ns} not significant

4. Conclusion

It could be concluded male Merino sheep consumed feed more than female, but they had similar value of digestibility.

Acknowledgments

This study was supported by Ministry of Research, Technology and Higher Education in Fundamental Project Research.

References

[1] Huston J E and Pinchak W E Range animal nutrition Grazing Management. An ecological perspective, ed R K Heitschmidt and J W Stuth chapter 2 http://cnrit.tamu.edu/rlem/ textbook/htm

[2] Colucci P E, Chase L E and Van Soest P J 1982 Feed Intake, Apparent Diet Digestibility, and Rate of Particulate Passage in Dairy Cattle J. Dairy Sci. **65.** 1445

[3] Hunt C W, Paterson J A, and Williams J E 1985 Intake and Digestibility of Alfalfa-Tall Fescue Combination Diets Fed to Lambs J. Anim. Sci. **60.** 1 301

[4] Thorlacius S O 1977Prediction of Digestible Energy Intake by Sheep of Diets Containing Ground Roughage and Grain Can. J. Anim. Sci. **57.** 279

[5] Clauss M, Streich W J, Schwarm A, Ortmann S and Hummel J 2007 The relationship of food intake and ingesta passage predicts feeding ecology in two different megaherbivore groups Oikos **116.** 209

[6] Dias R S, Patino H O, López S, Prates E, Swanson K C and France J 2011 Relationships Between Chewing Behavior, Digestibility, and Digesta Passage Kinetics in Steers Fed Oat Hay at Restricted and Ad Libitum Intakes J. Anim. Sci. **89.** 1873

[7] AOAC 2005 Official Method of Analysis of the Association of Official Analytical Chemists 18th ed, Maryland: AOAC International, ed William Harwitz (United States of America)

[8] Steel R D G and Torrie J H 1980 Principles and Procedures of Statistics 2nd ed (New York: McGraw-Hill Book Company)

[9] NRC 1981 Nutrient requirements of goats: angora, dairy, and meat goats in temperate and tropical countries Nutrient Requirements of Domestic Animal (Washington D C: National Academy Press)

[10] Cheeke P R 2005 Applied Animal Nutrition Feeds and Feeding 3rd ed (USA: Pearson Prentice Hall)

IOP Conf. Series: Earth and Environmental Science **119** (2018) 012006 doi:10.1088/1755-1315/119/1/012006

Microbiological Quality of *Panicum maximum* Grass Silage with Addition of *Lactobacillus* sp. as Starter

S. Sumarsih[1*], B. Sulistiyanto[1], C. S. Utama[1]

Faculty of Animal and Agricultural Sciences, Diponegoro University, Semarang, Central Java, Indonesia.

*Corresponding E-mail: ssumarsih71@gmail.com

Abstract

The aim of the research was to evaluate microbiological quality of *Panicum maximum* grass silage with addition *Lactobacillus sp* as starter. The completely randomized design was been used on this research with 4 treaments and 3 replications. The treatments were P0 (*Panicum maximum* grass silage without addition *Lactobacillus sp*), P1 (*Panicum maximum* grass silage with 2% addition *Lactobacillus sp*), P2 (*Panicum maximum* grass silage with 4% addition *Lactobacillus sp*) and P3 (*Panicum maximum* grass silage with 6% addition *Lactobacillus sp*).The parameters were microbial populations of *Panicum maximum* grass silage (total lactic acid bacteria, total bacteria, total fungi, and *Coliform* bacteria. The data obtained were analyzed variance (ANOVA) and further tests performed Duncan's Multiple Areas. The population of lactic acid bacteria was higher ($P<0.05$) and the total bacteria, fungi and *Coliform* were lower ($P<0.05$) with addition *Lactobacillus sp.* Microbiological quality of *Panicum maximum* grass silage with addition *Lactobacillus sp* was better than no addition *Lactobacillus sp.*

1. Introduction

Panicum maximum grass is one of the popular grasses in Indonesia. *Panicum maximum* grass have been used for silage production because mainly because of their elevated dry mass productivity between 15 and 30 ton / ha year [1]. The main aim of silage making is to conserve the plants with a minimum loss of nutritive value by fermentation of soluble carbohydrates in ananaerobic environment into organic acids, preferably lactic acid, which reduce pH. Lactic acid bacteria (LAB) produce lactic acid as a result of water soluble carbohydrate fermentation and play an important role in feed technology by production and preservation of silage [2]. By using selected wild-type strains, the large-scale production of fermented silage can bedeveloped without losing their unique flavor and particular characteristics [3]. The aim of the research was to evaluate microbiological quality of *Panicum maximum* grass silage with addition *Lactobacillus sp* as starter.

2. Materials and Methods

2.1. Materials

Research was done in Laboratory Feed Technology, Faculty of Animal and Agricultural Sciences, Diponegoro University. Materials research were *Lactobacillus sp* that isolated from ceca of Pengging duck [4], *Panicum maximum* grass and molasses.

2.2. *The Methods of silage productions and microbials analysis*

The *Panicum maximum* grass silage were done by the methods of Shah *et al* [2].The microbial populations in the *Panicum maximum* grass silage (total lactic acid bacteria, total bacteria and total fungi) were analyzed by the method of Li *et al* [5]. The Most Probable Number (MPN) was used for determination of total *Coliform*.. Brilliant Green Lactose Bile Broth (BGLBB) as a cultured media used MPN method was incubated at at 37 °C for 24 – 48 hours. Afterwards, a confirmatory test was done using Lauryl Sulphate Tryptose Broth (LSTB) as a culture media which had incubated at 37 °C for 48 hours [6]. There were two steps enrichment for qualitative *Salmonella* bacteria determination. Pre enrichment using Lactose Broth (LB) as a culture media had been incubated at 35 °C for 24 hours, followed by enrichment step using Selenite Cystine Broth (SCB) as a culture media that had been incubated at 43 °C for 24 hours. Isolation and identification method used Triple sugar Iron Agar (TSIA) and Lysine Iron Agar (LIA) that had been incubated at 35 °C for 24 hours.

2.3. *Preparation of test sample*

Panicum maximum grass silage (1 g) was diluted in sterile diluents peptone water solution (9 ml) to make primary dilutions (10^{-1}). Then a series up to 10^{-5} dilution was prepared by transferring primary dilution (1 ml) into test tube containing test diluents (9 ml) to obtain 10^{-2} dilution and repeating the operations with sterile diluents (9 ml) using the 10^{-2} and further dilutions to obtain 10^{-3}, 10^{-4}, 10^{-5} $10^{-6}$and/or 10^{-2}. The samples from each group were used to enumerate the total bacteria, total fungi, total lactic acid bacteria and *Coliform* by selective culture medium. One milliliter of appropriate serial dilutions of each sample was pour plate onto the sterile media. After 24 h incubation at 37 °C, the colonies that appeared on the plates were counted and the cfu ml^{-1} was calculated.

2.4. *Counting of population bacteria*

The dishes containing more than 30 and/or fewer than 300 colonies were selected and counted using colony counter. The result was calculated using following formula **:**

$$N = \frac{\Sigma c}{(n_1 + 0.1 \times n_2)\, d}$$

Σc : sum of colonies counted on all the dishes retained
n_1 : Number of dishes retained in the first dilution
n_2 : Number of dishes retained in the second dilution
d : dilution factor corresponding to the first dilution

2.5. *Statistical Analysis*

The parameters were populations of bacteria (total bacteria, total fungi, total lactic acid bacteria and total *Coliform* bacteria) in the *Panicum maximum* grass silage. Experiment was conducted according to Completely Randomized Design (CRD) consisted of 4 treaments and 3 replications with 5 experimental units. The treatments applied were P0 (*Panicum maximum* grass silage without addition *Lactobacillus sp*), P1 (*Panicum maximum* grass silage with 2% addition *Lactobacillus sp*), P2 (*Panicum maximum* grass silage with 4% addition *Lactobacillus sp*) and P3 (*Panicum maximum* grass silage with 6% addition *Lactobacillus sp*). The data obtained were analyzed variance (ANOVA) to determine the influence of treatment and further tests performed Duncan's Multiple Areas to know the difference between treatments [7].

3. Results and Discussion

Significant changes in microbial populations of *Panicum maximum* grass silage were found. The effect of *Lactobacillus sp* on population bacteria in *Panicum maximum* grass silage can be seen in Table 1.

Table 1. The Effect of *Lactobacillus* sp B12 on microbial populations of *Panicum maximum* grass silage

Parameter	Treatments			
	P0	P1	P2	P3
	------cfu / gram------			
Total Bacteria	$5.4 \times 10^{7a} \pm 0.1$	$4.9 \times 10^{6b} \pm 0.1$	$5.2 \times 10^{6b} \pm 0.2$	$6.2 \times 10^{6b} \pm 0.2$
Total LAB	$4.5 \times 10^{1b} \pm 0.1$	$1.6 \times 10^{6a} \pm 0.2$	$2.4 \times 10^{6a} \pm 0.1$	$3.6 \times 10^{6a} \pm 0.1$
Total Fungi	$6.4 \times 10^{6a} \pm 0.1$	$3.8 \times 10^{2b} \pm 0.2$	$1.5 \times 10^{2b} \pm 0.1$	$1.4 \times 10^{2b} \pm 0.1$
Colliform	$2.1 \times 10^{3a} \pm 0.1$	$4.6 \times 10^{1b} \pm 0.2$	$1.6 \times 10^{1b} \pm 0.1$	$1.4 \times 10^{1b} \pm 0.1$

Values with different superscripts show significant ($p<0.05$) differences among treatmens

The counts of total bacteria, total fungi and *Coliform* were decreased ($p<0.05$) with *Lactobacillus sp* treatments. Sumarsih *et al.* [8] reported that addition of *Lactobacillus sp* can reduce ($p<0.05$) total fungi of fish silage. The same effect showed with addition of antibiotic treatments decreased total microbe include fungi [9]. Similar findings have been reported by Choi *et al.* [10] and Murry *et al.* [11]. Ivanof [12] has also found that after feeding lactic acid bacteria can decreased *Coli* counts and increase in gram-positive bacteria in the large intestine of broilers. The reduction of pathogenic bacteria in the intestinal tract can be expected to improve nutrient digesbility and growth performance, alleviating weaning stress and lowering the imflammatory responses to sub-clinical infection [13].

Murry *et al.* [14] observed *in vitro* that media with culture of *Lactobacillus salivarius* and *Lactobacillus plantarum* produce more acetic and lactic acid and the pH was lower. *Lactobacillus salivarius* and *Lactobacillus plantarum* can ferment carbohydrates to produce pH levels and concentrations of lactic and acetic acid that inhibit the growth of pathogenic bacteria. The greater population on lactic acid bacteria with application *Lactobacilli* might inhibit harmful bacteria by blocking possible intestinal receptors of these pathogens or by secreting toxic metabolites againts gram negative bacteria [10].

The statistical analysis showed that application of *Lactobacillus sp* as starter affects the lactic acid bacteria in *Panicum maximum* grass silage significantly ($p<0.05$) increased. The presented results are agreement with the findings of Li *et al.* [5]. These changes in total lactic acid bacteria populations showed favourable effect of added on *Lactobacillus* concentrations.

4. Conclusion

The conclusion was addition of *Lactobacillus sp* as starters affects the lactic acid bacteria in *Panicum maximum* grass silage increased but decreased the counts of total bacteria, total fungi and *Coliform*.

References

[1] Viana, P. T., F. A. Teixeira, A. J. V. Pires, G. Giordano, P. De Carvalho, M. P. de Figueiredo and H. A. de Santana Júnior 2013 Losses and nutritional value of elephant grass silage with inclusion levels of cottonseed meal. Acta Scientiarum. Anim. Sci. **35** (2) : 139.

[2] Shah, A. A,,Y. Xianjun, D. Zhihao, W. Siran and S. Tao 2017 Effect of lactic acid bacteria on ensiling characteristics, chemichal composition and aerobic stability of King grass. The Journal of Animal & Plant Sciences, **27**(3): 747.

[3] Ammor, S., G. Tauveron, E. Dufour and I. Chevallier 2006 Antibacterial activity of lactic acid bacteria against spoilage and pathogenic bacteria isolated from the same meat small-scale facility: 1—screening and characterization of theantibacterial compounds. Food control, **17**(6): 454.

[4] Sumarsih, S., B. Sulistiyanto and E. S. Rahayu 2014 Characteristic of *Lactobacillu*s isolated from Pengging duck's intestines as probiotics. Inter. J. Poultry Sci. **13** (1) : 47.

[5] Li W. F., I. R. Rajput, X. Xu, Y. L. Li, J. Lei. Q. Huang and M. Q. Wang 2011 Effect of probiotic (*Bacilllus subtilis*) on laying performance, blood biochemical properties and Intestinal microflora of shaoxing duck. Int. J. Poult. Sci. **10** (8) : 583.

[6] Standard Nasional Indonesia (SNI) 2897 2008 Metode cemaran mikrobia dalam daging, Telur dan Susu, serta hasil olahannya. Badan Standarisasi Nasional (BSN), Jakarta

[7] Steel, R. G. D. dan J. H. Torrie 1981 Principles and Procedures of Statistics. McGraw-Hill Company Inc., New York

[8] Sumarsih, S., B. Sulistiyanto, H. S. Adi and C. S. Utama 2010 Pengaruh aras starter *Lactobacillus sp* terhadap performa mikrobiologi silase ikan dilihat dari total bakteri, bakteri asam laktat dan Fungi. J. Kesehatan **3** (1) : 43.

[9] Gunal, M., G. Yayli, O. Kaya, N. Karahan and O. Sulak 2006 The effects of antibiotic growth promoter, Probiotic or organic acid supplementation on performance, intestinal microflora and tissue of broilers. Int. J. Poult. Sci. **5** (2) : 149.

[10] Choi, J. Y., P. L. Shinde, I. K. Kwon, Y. H. Song and B. J. Chae 2009 Effect of woovinegar on the performance, nutrient digestibility and intestinal microflora in weanling Pigs. Asian-Aust. J. Anim. Sci. **22** (2) : 287.

[11] Murry, Jr., A. C., A. Hinton Jr. And R.J. Buhr 2006 Effect of botanical probiotic containing *Lactobacilli* on growth Performance and populations of bacteria in the ceca, cloaca and carcass rinse of broiler chichens. Int. J. of Poult. Sci. **5** (4) : 344.

[12] Ivanof, I. 2004. Testing a probiotic mixture for broiler chickens. Poultry Int. 43 : 44 – 47Li W. F., I. R. Rajput, X. Xu, Y. L. Li, J. Lei. Q. Huang and M. Q. Wang 2011 Effect of probiotic (Bacilllus subtilis) on laying performance, blood biochemical properties and Intestinal microflora of shaoxing duck. Int. J. Poult. Sci. **10** (8) : 583.

[13] Dibner, J.J., C. Knight. G. F. Yi and J. D. Richerds 2007 Gut development and health in the absence of antibiotic growth promoters. Asian-Aust. J. Anim. Sci. **20** : 1007.

[14] Murry, Jr., A. C., A. Hinton Jr. And H. Morrison 2004 Inhibition of growth of Escherichia coli, Salmonella typhimurium and Clostridia perfringens on chicken feed media by *Lactobacillus salivarius* and *Lactobacillus plantarum.* Int. J. Poult. Sci. **3** : 603.

Protein Binding Capacity of Different Forages Tannin

L. M. Yusiati*[1], A. Kurniawati[1], C. Hanim[1] and M. A. Anas[1]

[1]Laboratory of Nutritional Biochemistry, Department of Animal Nutrition and Feed Science, Faculty of Animal Science, Gadjah Mada University, Yogyakarta, Indonesia 55281

*Corresponding E-mail: liesmira@ugm.ac.id

Abstract

Eight forages of tannin sources*(Leucaena leucocephala, Arachis hypogaea, Mimosa pudica, Morus alba* L, *Swietenia mahagoni, Manihot esculenta, Gliricidia sepium,* and *Bauhinia purpurea)*were evaluated their tannin content and protein binding capacity. The protein binding capacity of tannin were determined using precipitation of bovine serum albumin (BSA). *Swietenia mahagoni*has higest total tannin level and condensed tannin (CT) compared with other forages ($P<0.01$). The *Leucaena leucocephala* has highest hydrolysable tannin (HT) level ($P<0.01$). The total and condensed tannin content of *Swietenia mahagoni* were 11.928±0.04 mg/100 mg and 9.241±0.02mg/100mg dry matter (DM) of leaves. The hydrolysable tannin content of *Leucaena leucocephala* was 5.338±0.03 mg/100 mg DM of leaves. Binding capacity was highest in *Swietenia mahagoni* and *Leucaena leucocephala* compared to the other forages ($P<0.01$). The optimum binding of BSA to tannin in *Leucaena leucocephala* and *Swietenia mahagoni*were1.181±0.44 and 1.217±0.60mg/mg dry matter of leaves. The present study reports that *Swietenia mahagoni* has highest of tannin content and *Leucaena leucocephala* and *Swietenia mahagoni* capacity of protein binding.

1. Introduction

Forages are important for ruminant livestock as nutrient source and keep continuous rumen physiology. Forages contain high polyphenolic compounds including tannin. This compound is plant secondary metabolites which are generally categorized into two subgroups based on their chemical structures, being hydrolysable (HT) and condensed (CT) tannins [1]. Tannins give beneficial effect for ruminant for methane reducing agent, rumen protein protection (bypass), and improvement of productivity. [2;3]. However, tannins in ruminants may act as anti-nutritional compounds which can reduce feed intake and nutrient digestibility [4]. Supplementation forages as tannins sources should be appropriate to prevent negative effect on ruminant.

The complex bonds between proteins and tannins can be formed by four types of binding, hydrogen bonding, hydrophobic interactions, covalent bonds and ionic bonds. Tannin is a polyphenol component that can bind protein feed to inhibit the transport of nutrients into rumen microorganisms. The addition of tannin extract from *Acacia mearnsii* 2.5% (dry matter, DM) decreased methane production and increased total volatile fatty acids (VFA) of ruminant [3;5].

Tannins of different forages may differently affect nutrient availability and utility, even when consumed at the same concentration, because of the very heterogeneous chemical structure of these phenolic compounds. The tannins of different plant species have different physical and chemical properties and therefore they have very diverse biological properties [6]. The high affinity of tannins for proteins is due to the presence of great number of phenolic groups. Forages as tannins sources that were often given by farmers include *Artocarpus heterophyllus* L, *Calliandra calothyrsus* and *Paraserianthes falcataria* L. The forages can reduce methane production and protein digestibility in the rumen [4]. It is considered that tannin content and its characteristic of protein binding from different forages as ruminant feed need to be evaluated.

2. Materials and Methods

2.1. Samples Collection and Preparation

Forages were collected from Banyusoco Village, Gunung Kidul District, Yogyakarta Province. Nine forages (1 kg each samples in as fed) were used for analysis in laboratory. Samples were dried at 55°C for 48 h and ground to pass a 0.5-mm screen for chemical composition and tannins essay. Samples were proximate analyzed using AOAC [7].

2.2. Tannins Assays: Extraction, Total Phenolic, Condensed Tannins, and Hydrolysable Tannin

Dried samples (0.2 g) were extracted in 10 ml aqueous acetone (70%; acetone: water, 7:3) for 20 min in an ultrasonic at room temperature. The samples extract were centrifuged (3000 g, 10 min, 4°C). Supernatant were used for tannins assay and protein-binding capacity.

Folin-Ciocalteu method was used for determination of total phenols and total tannins. Polyvinylpolypyrrolidone (PVPP) was used to binds tannin and separate tannin phenols from non-tannin phenols. Condensed tannins were determined by the butanol-HCl method [8;9]

2.3. Protein-Binding Capacity by Tannins

Increasing levels of the tannin-containing extract and methanol (50%) were added to 2 ml of BSA solution (containing 1 mg BSA/ml acetate buffer) to made 3 ml of total volume. For example, by using 0.95, 0.90, 0.85, 0.80, 0.75, 0.70 ml methanol with 0.05, 0.10, 0.15, 0.20, 0.25, 0.30 ml of the extract. The mixture was incubated in refrigerator (4°C) overnight and was centrifuged at about 3000 g for 10 min [8].

Total 1.5 ml of 1% SDS solution was added to the precipitate. An aliquot (1 ml) was added to SDS-triethanolamine solution and 1-ml of the ferric chloride reagent. The condensed tannins were measured by absorbance at 510 nm after 15–30 min. The absorbance were converted to tannic acid equivalent, using a standard curve.

3. Results and Discussion

The content of tannins of different forages used in this study was summarized in Table 1. Total phenol, total tannin, condensed tannin and hydrolysable tannin varied widely. Total phenol (TP) content in *Leucaena leucocephala,Gliricidia sepium* and *Swietenia mahagoni* was > 8 mg/100mg DM and total tannin (TT) 4 mg/100mg DM. Condensed tannin (CT) in *Swietenia mahagoni* highest than in other forages. Compared with others forages, hydrolysable tannins (HT) in *Leucaena leucocephala* was highest. Condensed and hydrolysable tannin content in *Morus alba* L were lower than other forages.

Table 1. Tannins Content (mg/100mg) of Different Forages

Sample	Total Phenol	Total Phenol Non Tannin	Total Tannin	Condensed Tannin	Hydrolysable Tannin
Leucaena Leucocephala	8.299±0.01^{d}	1.726±0.01^{e}	6.572±0.02^{g}	1.235±0.05^{c}	5.338±0.03^{g}
Arachis hypogaea	2.284±0.25^{b}	0.294±0.00^{a}	1.992±0.25^{c}	0.043±0.00^{a}	1.946±0.02^{c}
Mimosa pudica	4.114±0.01^{c}	0.653±0.00^{c}	3.462±0.00^{e}	1.234±0.04^{c}	2.229±0.05^{d}
Morus alba L	0.746±0.00^{a}	0.288±0.00^{a}	0.458±0.00^{a}	0.080±0.01^{a}	0.379±0.00^{a}
Swietenia mahagoni	13.930±0.11^{e}	1.982±0.06^{f}	11.928±0.04^{h}	9.241±0.02^{e}	2.707±0.06^{e}
Manihot esculenta	4.163±0.02^{c}	1.017±0.04^{d}	3.146±0.03^{d}	1.918±0.03^{d}	1.227±0.00^{b}
Gliricidia sepium	8.518±0.03^{d}	4.221±0.00^{g}	4.296±0.04^{f}	0.476±0.00^{b}	3.821±0.04^{f}
Bauhinia purpurea	2.083±0.02^{b}	0.512±0.00^{b}	1.571±0.03^{b}	0.489±0.00^{b}	1.082±0.03^{b}

^{a,b,c} Means in the same rows with different letters differ at P<0.05

The results of the protein precipitation assay showed in Table 2. *Swietenia mahagoni* and *Leucaena leucocephala has* highest protein binding (P<0.05). Protein binding capacity was similar in *Mimosa pudica, Morus alba* L, *Manihot esculenta,*and *Gliricidia sepium,* however the forages had have heterogeneous of tannins content.

Table 2. Protein Binding Capacity of Different Forages Tannin

Sample	Bovine Serum Albumin (mg BSA/mg DM of leaves)
Leucaena leucocephala	1.181±0.44^{c}
Arachis hypogaea	0.414±0.01^{b}
Mimosa pudica	0.232±0.02^{a}
Morus alba L	0.238±0.03^{a}
Swietenia mahagoni	1.217±0.60^{c}
Manihot esculenta	0.322±0.04^{ab}
Gliricidia sepium	0.336±0.03^{ab}
Bauhinia purpurea	0.377±0.03^{b}

^{abc} Means in the same rows with different letters differ at P<0.05

Tannins content in forages have intensive research because of positive function that reducing methane production and protecting protein source [10;11]. Utilization of tannins in ruminant feed can increase productivity. However, high dietary concentration of tannins can reduce feed efficiency [3].

The differences of tannins content in various plant species were, tannins content of seventeen forages varied widely, from 4.0-209.3 g/kg DM [12]. The production of plant secondary metabolites (tannins) were effected by composition of culture media, genetic (cell genotypes), and environmental stress factors, age, season and soil type [13; 14; 15].

There was a correlation between tannin content and protein binding capacity, such as *Swietenia mahagoni* and *Leucaena leucocephala,* both of forages have high tannins content which increasing protein binding capacity. However, higher tannin content of *Leucaena leucocephala* in hydrolysable tannin.

The high affinity of tannins for proteins is due to the presence of great number of phenolic groups. These provide many points at which bonding may occur with the carbonyl groups of peptides. The formation of such complexes is specific, both in terms of the tannin and protein involved, the degree of affinity between the participating molecules residing in the chemical characteristics of each [6]. Hydrolysable tannins in the extracts contributed considerably toward the protein binding capacity of the extracts [16].

Binding reaction between protein and tannins were influenced by pH [17]. The partial dissociation of protein from the complex in low pH condition (pH<7), abomasum compartment [18]. The binding ability of tannins will be beneficial as protein protectant and increasing availability of protein feed can for directly used by the host animals. The ability of tannins in binding proteins in ruminants can reduce protein digestibility. Soybean meal was treated with 15 g and 30 g/kg of a tannins extract from *C. ladanifer* L. reducing the degradability of soybean meal protein in the rumen [11].

4. Conclusion

Forages have different tannins content and protein binding capacity. Forages as tannins resources have potential to protecting protein in ruminant, but its need evaluation.

Acknowledgments

This study was supported by grants from Ministry of Research, Technology and Higher Education of the Republic of Indonesia.

References

[1] Frutos, P., G. Hervás, F. J. Giráldez and A. R. Mantecón 2004 Tannins and ruminant nutrition Spanish J. Agric. Res. **2** 191

[2] Yusiati, A. Kurniawati, C. Hanim 2015 Int. Sem. on Trop. Anim. Prod.

[3] Carulla, J. E., M. Kreuzer, A. Machmuller, dan H. D. Hess 2005 Supplementation of Acacia mearnsii tannins decreases methanogenesis and urinary nitrogen in forage-fed sheep Aust. J. of Agri. Res.**56**:961

[4] Yusiati, L.M., C. Hanim and C S Setyawati 2016 Nutritiveevaluation of pineapple peel fermented by cellulolytic microbe and lactic acid bacteria by invitro diges production technique Int. Conf. on Biodiversity, Food Security and Health. OFT-30

[5] Kumar, R. dan M. Singh 1984 Tannins: their adverse role in ruminant nutrition J. of Agri. and Food Chem. **32**: 447.

[6] Bunglavan, S. J. and N. Dutta 2013 Use of tannins as organic protectants of proteins in digestion of ruminants J. of Livestock Sci. **4** 67

[7] Association of Official Analytical Chemists (AOAC), 2005 Association of Official Analytical Chemists, Arlington, VA, USA

[8] Makkar, H.P.S., 2003 Joint FAO/IAEA, Division of Nuclear Techniques in Food and Agriculture. Kluwer Academic Publishers, Dordrecht, The Netherlands

[9] Porter, L.J., Hrstich, L.N. Chan, B.G 1986 The conversion of procyanidins and prodelphinidins to cyanidin and delphinidin Phytochemistry **25.** 223

[10] Hess, H.D., T.T Tiemann, F. Noto, J.E. Carulla and M. Kreuzer 2006 Strategic use of tannins as means to limit methane emission from ruminant livestock Int. Congress Series. **1293.** 164

[11] Dentinho, M.T.P., A. T. Belo and R. J. B. Bessa 2014 Digestion, ruminal fermentation and microbial nitrogen supply in sheep fed soybean meal treated with Cistus ladanifer L. tannins J. of Small Rumminant Res.**119.** 57

[12] Jayanegara, A., N. Togtokhbayar, H. P. S. Makkar, and K. Becker 2009 Tannins determined by various methods as predictors of methane production reduction potential of plants by an in vitro rumen fermentation system Anim. Feed Sci. Technol. **150.** 230

[13] Martin, A., A. Cabrera, and J. Lopez Medina. 1991 Antinutritional Factors in Faba Bean. Tannin content in Vicia faba: possibilities for plant breeding Options Mediterraneennes **10**, 105

[14] Kähkönen, M.P., A.I. Hopia, dan M. Heinonen. 2001 Berry phenolics and their antioxidant activity J. of Agri. and Food Chem. **49.** 4076

[15] Berard, N. C., Y. Wang, K. M. Wittenberg, D. O. Krause, B. E. Coulman, T. A. McAllister and H. Ominski 2011 Condensed tannin concentrations found in vegetative and mature forage legumes grown in western Canada Can. J. Plant. Sci .**91.** 669-675

[16] Dawra, R. K., H. P. S. Makkar and B.Singh 1988 Protein-binding capacity of microquantities of tannins Analytical biochemistry **170.** 50

[17] Nack, M., D. Oickle, D. Pink and F. Shahidi 1996 Protein precipitating capacity of crude canola tannins: effect of pH, tannin, and protein concentrations J. Agric. Food. Chem. **44.** 2144

[18] Diaz-Hernandez, A., M. J. Nixon, F. M. Ball, R. A. Leng and J. B. Rowe 1997 Protein Tannin complexes are susceptible to proteolytic degradation Proc. of The New Zea Society of Anim Production **57.** 116

Effectiveness of Accelerator and Inoculum in Fermentation of Goat's Rumen Contents as Animal Feed Ingredients

B. Bakrie[1*], U. Sente[1], K. Mayasari[1] and R.F. Syah[2]

[1] Assessment Institute for Agricultural Technology, Jakarta. Indonesia.
[2] University of Respati Indonesia, Jakarta. Indonesia.

[*]Corresponding E-mail: bachtarbakrie@yahoo.com

Abstract

The goat's rumen contents is slaughterhouse waste that has potential to be used as animal feed, but it has to be first processed into silage. This study aims to determine the type of accelerator and to investigate whether the addition of inoculum was required during the fermentation process. The research was conducted using a Completely Randomized Factorial Design, consisting of 2 treatment factors and 6 replications. The treatment factors were: a) Accelerator (rice bran or cassava pomade/onggok); b) Inoculum *Lactobacillus plantarum* (with or without using inoculant). Results showed that there was an increase in crude protein (CP) content with the use of rice bran at after fermentation compared to before fermentation. The CP contents with the use of onggok almost the same at after and before fermentation. Increase in the content of crude fiber (CF) after fermentation was both for using rice bran or onggok. However, the content of CF using onggok was much higher than with rice bran. There was no significant effect for both types of accelerators used in CP and CF contents at after fermentation with or without the addition of *Lactobacillus plantarum* as the inoculant. It can be concluded that for the fermentation of goat's rumen contents it is better to use rice bran rather than onggok as the accelerator and inoculant is not required during the fermentationprocess.

1. Introduction

Number of recorded goats slaughtered during 2016 in the Province of DKI Jakarta was 91,184 heads, while the number of sheep slaughtered was only 13.405 heads [1]. The goats are slaughtered either at the government-owned slaughterhouses or at unofficial places owned by individuals. The government-owned slaughterhouses are situated in 2 locations, i.e. at Cakung slaughterhouse, East Jakarta, and Tanah Abang slaughterhouse, Central Jakarta. Meanwhile, the individual slaughtering places are found in almost all areas with locations within or around residential areas. With regard to the slaughterhouses that are mostly located in the residential areas, the wastes generated from the slaughterhouses are serious problems. The wastes have not been well utilized and are generally just thrown away into waterways or rivers. Therefore, it can cause environmental pollution, as it has a stinging smell and easy to decay. Various technologies have been available for processing wastes from slaughtering places made into fertilizer, planting media or animal feed ingredients.

One type of waste that can be utilized for animal feed is in the form of goat's rumen contents. This type of waste is a collection of feed ingredients consumed by goat and present inside the stomach, but they are not fully fermented or ingested by the goat. The amount of fresh rumen contents of a goat is about 7.67-12% of the body weight [2, 3], so the amount of goat's rumen contents as wastes in the area of DKI Jakarta Province can reach up to 1.0 ton/day. As more goat is slaughtered during the Eid Adha, so that more goat's rumen contents are also available at that time. Therefore, if all of this type of waste can be processed into animal feed ingredients, the shortage of animal feed in this area can be partly solved.

IOP Conf. Series: Earth and Environmental Science **119** (2018) 012008 doi:10.1088/1755-1315/119/1/012008

Due to the presence of various types of microbial which aids the digestion, as well as the growth stimulants and some enzymes, the rumen contents contain nutrients of high quality as results of microbial synthesis in the rumen. Therefore, the nutritional value of the rumen content is much better and it also has a lower CF content compared to the field grass. The nutritional composition of the rumen contents varies greatly depending on the type and age as well as the type of food consumed by the animal [4]. Some studies reported that the CP content of the goat's rumen contents ranges between 13.8-27.3% and the CF content is between 28.3-34.91% [5]. Therefore, the goat's rumen contents are potential to be used as animal feed ingredients.

The rumen contents cannot be directly used as animal feed ingredients when given in the fresh form because it has a very strong odor that is not favored by the animal. Apart from that the rumen contents also has fairly high water content, ranges from 80 to 90%, so it will quickly become rotten and the nutrients it contains will also quickly be damaged [5, 6, 7]. The decay can be prevented by means of preservation, such as drying or processing into silage. Drying is one method of dry preservation, while the making of silage is a fresh preservation. However, drying will cause the occurrence of air pollution, due to the distinctive smell derived from volatile fatty acid (VFA).

One advantage of silage making is that the silage can be stored for a longer period of time as long as it is placed in a dry place. Fermentation can also increase protein digestibility, decrease levels of fiber, improve flavor and increase the aroma of feed ingredients [8, 9].

The fermentation process requires the addition of soluble carbohydrate fraction as accelerator and inoculant to optimize lactate acid production and minimize nutrient loss during ensilage [10, 11]. Commonly used accelerators are rice bran, milled corn, cassava pomade (onggok), molasses, glucose and fibrolytic enzymes [12, 13]. While the type of inoculant commonly used includes *Lactobacillus plantarum, Enterococcus faecium, Pediococcus spp.* and *Saccharomyces cerevisiae* [14, 15].

This study was conducted in order to determine the type of accelerator and to learn whether the addition of *Lactobacillus plantarum* as an inoculum is required during the fermentation process in the making of goat's rumen contents silage.

2. Materials And Methods

2.1. Experimental Design

This research was conducted using a Completely Randomized Factorial Design, consisting of 2 treatment factors and 6 replications. The treatment factors were: a) Accelerator (rice bran or cassava pomade/onggok); b) Inoculum *Lactobacillus plantarum* (with or without using inoculant). Therefore, there were 4 treatments altogether, namely: a) rumen contents + rice bran, b) rumen contents + rice bran + *Lactobacillus plantarum*, c) rumen contents + onggok, and d) rumen contents + onggok + *Lactobacillus plantarum*.

2.2. Inoculum Preparation

The propagation of isolate *Lactobacillus plantarum* was carried out in two stages according to the method developed by [16, 17] as described by [18].The first stage was to re-grow the isolate in an agar medium slant, while the second stage was to propagate it in a liquid medium. Then, it was diluted to reach the density of 10^6 cells per ml. Number of cells that are formed was counted using a haemocytometer.

IOP Conf. Series: Earth and Environmental Science **119** (2018) 012008 doi:10.1088/1755-1315/119/1/012008

2.3. Fermentation Process

The goat's rumen contents used in this study were obtained from an individual goat slaughterhouse in Cilincing, North Jakarta. The fermentation process was conducted at the Laboratory of Jakarta Assessment Institute for Agricultural Technology (Jakarta AIAT), South Jakarta. The material used was the entire contents of the rumen of the goat slaughtered. To fulfil the treatment, it was required to collect the rumen contents from as many as 9 goats. The weights of goats before slaughter, carcass and rumen contents were also measured as the additional data. Chemical analyses were also carried out for the rumen contents, rice bran, onggok and their mixtures, prior to the fermentation process.

All of the collected rumen contents were then thoroughly mixed and divided into two parts; one part was mixed with rice bran and another part with onggok. The amount of accelerators mixed with the rumen contents was calculated on the basis of the dry matter (DM) contents. The final DM content of the mixture should be around 35% [7]. It was found that the respective DM contents were 12.3% for the rumen contents, 86.3% for rice bran and 89.6% for onggok (Table 1). Therefore, to make a mixture with 35% DM content required 69.32% fresh rumen contents mixed with 30.68% rice bran and 70.63% fresh rumen contents mixed with 29.37% onggok. Thereafter, 4% (w/w) of molasses were added into each mixture and then stirred until they were of the accelerator and the rumen contents evenly mixed.

Table 1. Chemical Analyses of Goat's Rumen Contents, Rice Bran, Onggok and The Mixtures at Before Fermentation.

Materials	Dry matter (%)	Crude protein (%)	Crude fiber (%)	Gross energy (%)
Rumen contents	12.3	18.1	25.2	4,692
Rice bran	86.3	11.3	9.75	4,163
Onggok	89.6	2.63	13.6	3,007
Rumen contents + Rice bran + Molasses	34.9	12.93	15.3	4,307
Rumen contents + Onggok + Molasses	35.0	7.05	17.1	3,667

Each mixture was then divided into 12 parts, 6 parts were then mixed evenly with inoculum *Lactobacillus plantarum* (10^6 cfu/gram), and another 6 parts were not mixed with the inoculum. Finally each part of the mixture was transferred into a 10-liter plastic box, which was sealed with sticky tape to ensure an anaerobic condition. The boxes were then placed in a room temperature for a 3 weeks period of fermentation.

2.4. Chemical and Statistical Analyses

Observation on the silage quality was done both physically and chemically. Physical observation included: pH, aroma, color, texture and the presence or absence of fungi. While chemical analysis included: contents of DM, CP, CF and gross energy (GE). Chemical analysis was partly done at the Laboratory of Jakarta AIAT and partly in the Chemical Laboratory of Livestock Research Institute in Ciawi, Bogor. The chemical analyses were conducted according to [19] as described by [18]. The DM content was determined by drying the samples at 105 ^{0}C for 24 hour. The CP content was measured through destruction with sulfuric acid and the subsequent measurement of Kjeldahl Nitrogen using auto analyzer. The CF was determined by extraction in acid and alkali solution, whereas the GE was measured using a Bomb Calorimeter.

All data collected was processed using analysis of variance (ANOVA) with a computer package program of SPSS (Statistical Package for Social Sciences) version 21.0. If there was a difference between treatments they were then tested further by Duncan's multiple range tests [20].

3. Results and Discussion

3.1. *Weights and Chemical Compositions of Fresh Goat's Rumen Contents*

The goats that were slaughtered were purchased from Subang district, West Java province. All of them were females with ages ranging from 1.5 to 2.0 years. They were small size type goats which is a cross between the Kacang goat and Local goat.

The live weights of slaughtered goats were between 14.1 to 21.8 kg, with the average value of 17.2 kg (Table 2). The average carcass weight was 8.54 kg or equivalent to 49.8% of the live weight. While the weight of rumen contents ranged from 1.35 to 3.70 kg with an average value of 2.20 kg or equivalent to 12.6% of the live weight. The percentage of rumen contents obtained is in accordance with published data by [3] which states that the rumen contents of goat is 10-12% of the live weight. However, the result obtained in this study was much higher than [2] which stated that goats fed with various types of protein sources included soybean meal, toasted soybean, corn gluten meal or cottonseed meal having rumen content ranging from 7.57- 9.04% of the live weight.

Table 2. Weight of Live Animals, Carcass and Rumen Contents of Goat Slaughtered In North Jakarta.

Live animals (kg)	Carcass		Rumen contents	
	kg	%	kg	%
19.2	9.90	51.6	2.15	11.2
15.5	7.45	48.1	2.10	13.6
17.3	8.00	46.2	2.55	14.7
17.0	8.20	48.2	2.10	12.4
18.2	9.10	50.0	2.25	12.4
16.8	9.40	56.0	1.80	10.7
21.8	9.75	44.7	3.70	17.0
14.1	7.70	54.6	1.35	9.57
15.2	7.35	48.4	1.80	11.8
17.23 ± 2.32	8.54 ± 1.04	49.76 ± 3.74	2.20 ± 0.73	12.60 ± 2.22

The CP content of the fresh goat's rumen contents was 18.1% and the CF content was 25.2% (Table 1). The CP content was consistent with results of studies reported by [5, 21] stated that the CP content was in the range of 13.8-27.3%. However, the CF content was lower than obtained by those studies which were in the range of 28.3 to 34.91%. The CP content of the goat's rumen contents was much higher than the cow's rumen contents which generally range from 8.10 to 16.20%. However, the CF content of the goat's rumen contents found in this study was similar to the cow's rumen contents reported by [6].

Differences in weight or percentage as well as the nutritional contents of the rumen contents may be due to differences in the type and age of the animals and differences in the type of feed given before slaughtered. As it was informed that the goats in this study were fed with fermented rice straw and concentrate before they were slaughtered. As the consequences, the amount of the rumen contents will be much less than those given only grasses or tree leaves and without addition of concentrate.

For the purposes of fermentation then the contents of rumen goat must first be mixed with the accelerator material, which is in accordance with the type of the treatment, using rice bran or onggok and the addition of molasses. The mixtures obtained had a lower CP and CF contents compared to the fresh rumen contents (Table 1). This is may be due to the low CP and CF contents of both types of accelerators. The respective CP content of rice bran and onggok were 11.3% and 2.63%. The corresponding CF contents were 9.75% and 13.6%. The CP content of the fresh goat's rumen contents mixed with rice bran (12.91%) was much higher than when it was mixed with onggok (7.05%). In the contrary, the CF content of the fresh goat's rumen contents mixed with rice bran (15.3%) was much lower than when it was mixed with onggok (17.90%).

3.2. Physical Characteristics of Goat's Rumen Contents

The pH of the fresh goat's rumen contents at the time before fermentation was 5.0 and there was no effect of the addition of rice bran or onggok to the pH value (Table 3). After being fermented for 3 weeks it was seen that the pH of all fermented mixtures decreased to 4.0. This indicates that the fermentation activity was successful. It is associated with the development and growth of lactic acid bacteria during the ensilage process resulting in increased production of lactic acid and causing acidic conditions and characterized by decreased pH [9, 15]. This condition will also inhibit the growth of fungi and limit the development of bacterial decomposition.

The fresh goat's rumen contents mixed with rice bran and molasses did not cause changes in the aroma, but the color changed from green to brownish green. However, if mixed with onggok and molasses it caused change in the aroma which was dominated by the aroma of molasses and the color also changed to become brownish green. Differences in the aroma and colors of the mixed materials may be due to differences in the water absorption capacity of rice bran and onggok. Onggok is higher hygroscopic or has more water absorption capacity compared to rice bran, so with the addition of onggok, more water in the rumen contents will absorbed.

After fermented for 3 weeks, it was found that all mixtures had a similar acidic aroma which was the distinctive aroma of fermented materials (Table 3). This suggests that there has indeed been a perfect fermentation process for all such materials, shown by the decrease in pH, as described above. But there were differences in the colors of the fermented materials, where the goat's rumen contents mixed with onggok had a bright green color, while with the addition of rice bran the color became yellowish green. The texture of the fermented goat's rumen contents mixed with onggok was wetter than when it mixed with rice bran. Furthermore, the addition of inoculum *Lactobacillus plantarum* had no significant effect on the aroma and colors of the goat's rumen contents mixtures after fermentation, but the textures. The inoculum caused the textures of the mixtures to become drier. In addition, there was no visible fungus that grows on all fermented materials. This indicates that there was an anaerobic atmosphere and sufficient accelerator and inoculum during the fermentation process, creating an acidic atmosphere that does not allow for the growth of fungi.

3.3. Nutritional Composition of Fermented Rumen Contents

Measurement of DM content at before and after fermentation is necessary to know that during fermentation there is enough water soluble carbohydrate (WSC) available for the lactic acid bacteria, which is shown by no reduction in the DM content during the fermentation process [7]. The results obtained in this study indicated that there was no sufficient WSC material available during fermentation, resulting in a decrease in DM content, from 35.0% to an average of 29.64% (Table 4). The decrease in the DM content was greater if the inoculum *Lactobacillus plantarum* was added to the goat's rumen contents. This may be due to that the addition of the inoculum resulted in a more intensive fermentation process compared to without inoculum addition.

Table 3. Physical characteristics of goat's rumen contents with various mixtures at before and after 3 weeks of fermentation period.

Materials	pH	Aroma	Color	Texture	Mold
Before Fermentation:					
Rumen contents	5	Specific rumen contents	Green	Soft and watery	-
Rumen contents + Rice bran + Molasses	5	Specific rumen contents	Brownies green	Rather wet	-
Rumen contents + Onggok + Molasses	5	Specific molasses	White green	Rather dry	-
After Fermentation:					
Rumen contents + Rice bran + Molasses	4	Sour	Yellow green	Wet and sticky	No mold
Rumen contents + Rice bran + Molasses + LP	4	Sour	Yellow green	Dry and sticky	No mold
Rumen contents + Onggok + Molasses	4	Sour	Light green	Wet and sticky	No mold
Rumen contents + Onggok + Molasses + LP	4	Sour	Light green	Dry and sticky	No mold

After fermentation, it was found that the CP content of the goat's rumen contents mixed with the rice bran (14.72%) was much higher ($P<0.01$) than when it was mixed with onggok (7.63%). There was a tendency for an increase in CP content for all fermented materials; however, a higher increase was found when the goat's rumen content was mixed with the rice bran. The CP content of the goat's rumen contents fermented using rice bran, inoculum and molasses in this study (14.88%) was much higher than cow's rumen contents fermented with a similar mixture (12.73%) as reported by [22]. However, the CF content found in this study (18.74%) was lower than that study (27.16%). At the time after fermentation there was an increase in CF content in all fermented materials when compared to before fermentation, with a similar rate of increase. The CF content of the goat's rumen contents mixed with onggok (19.45%) was higher than when it mixed with the rice bran (17.90%).

There was no significant effect of inoculum addition to the contents of CP and CF of the mixtures after fermentation. This showed that the microorganisms contained in the goat's rumen contents have a capacity to assist the fermentation process. As it is known that there are various types of microorganisms present in the rumen contents, including bacteria, protozoa and fungi. The respective concentrations of microbiota formed by ruminal bacteria, protozoa and fungi are 10^{10}, 10^6 and 10^4 cells/ml [23].

The results of this study indicate that the goat's rumen contents fermented using rice bran with or without the use of inoculum *Lactobacillus plantarum* had a fairly good nutritional quality. The CP content of fermented goat's rumen contents is much better than the rice straw, field grass, elephant grass or sugar cane top [24, 25, 26]. Therefore, the fermented goat's rumen contents may be used, as good as the cow's rumen contents, as substitutes for various type of forages or as an ingredient in the concentrates for various types of animals. The cow's rumen contents have been successfully used as feeds for cattle, goats, chickens, rabbits and fish [27, 28, 29, 30].

Table 4. Chemical Analyses of Rumen Contents with Different Treatments after 3 Weeks Fermentation Period.

Accelerator	Inoculum	Dry matter (%)	Crude protein (%)	Crude fiber (%)	Gross energy (kkal/kg)
Rice bran		29.53[a]	14.72[a]	17.90[a]	4,395[a]
Onggok		29.75[a]	7.63[b]	19.45[b]	3,684[a]
	Without LP	30.91[a]	11.21[c]	18.15[ab]	4,041[a]
	With LP	28.35[a]	11.12[c]	19.20[b]	4,038[a]
Rice bran	Without LP	30.32[a]	14.52[a]	17.08[a]	4,337[a]
	With LP	28.71[a]	14.88[a]	18.74[ab]	4,453[a]
Onggok	Without LP	31.53[a]	7.91[b]	19.23[b]	3,745[a]
	With LP	28.04[a]	7.34[b]	19.67[b]	3,622[a]
Main effect:					
Accelerator		Ns	<0.01	<0.01	ns
Inoculum		Ns	Ns	ns	ns
Accelerator x Inoculum		Ns	Ns	ns	ns

[a-b] Means within a column with no common superscript differ significantly (P<0.05)

4. Conclusion

For fermentation of goat's rumen contents it is better to use rice bran rather than onggok as the accelerator, as it will produce better quality silage with a higher CP content. The addition of inoculum *Lactobacillus plantarum* is not required in the fermentation process of the goat's rumen contents either with rice bran or onggok as the accelerator, because the quality of the resulting silage is the same as without the addition of the inoculum.

Acknowledgements

We would to thank Head of the Indonesian Agency for Agricultural Research and Development and Head of the Jakarta AIAT who had provided funds for the implementation of this research. High appreciation is also conveyed to the Rector of the University of Respati Indonesia (Urindo), Jakarta and the Dean of the Faculty of Agrotechnology of Urindo for their cooperation and support in conducting this research. Furthermore, we would like to express our gratitude to Mr. H. Ismail, the owner of the goat slaughterhouse in North Jakarta for his kindly provision of goat's rumen contents for research materials. Our appreciation also goes to Dr. Yudi Sastro, S.P., M.P., Ir. Syamsu Bahar, M.Sc and M.A. Wahyuningrum, S.Pt., M.Si for their assistance in conducting this research activity.

References

[1] Ditjen Nak and Keswan 2016 Statistics of Livestock and Animal Health (Jakarta: Indonesian Ministry of Agriculture) p 215

[2] Felisberto N R O, Rodriguez M T, Bomfim M A D, Matos R S, Cordeiro A G P C and Silva M C 2011 Effects of different sources of protein on digestive characteristics, microbial efficiency,

and nutrient flow in dairy goats R. Bras. Zootec. **40.** (10) 2228

[3] Yenni, Dewilda Y and Sari S M 2012 Study of biogas production from vegetable and fruit waste with bovine rumen waste co-substrate J. Environ. Tec. Andalas University **9.** (1) 26

[4] Dlamini A M and Dube M A 2014 Contribution of animal agriculture to greenhouse gases production in Swazilan Am. J. Climate Change **3.** 253

[5] Agbabiaka L A, Madubuike F N and Amadi S A 2012 Studies on nutrients and anti-nutrients of rumen digesta from three most domesticated ruminants in NigeriaPakistan J. of Nutr. **11.** (7) 580

[6] Hanum Z and Usman Y 2011 Proximate analyses of ammoniated rice straw with the addition of rumen contents Agripet **11.** (1) 39

[7] Utomo R, Budhi S P and Astuti I F 2013 Effect of onggok level as an additive on the silage quality of cow's rumen contents Buletin Peternakan **37.** (3) 173

[8] Muhakka, Wijaya A and Ammar M 2015 Nutritional dried matter, crude protein and crude fiber on lowland tidal grass fermented by probiotic microorganisms as Bali cattle feed Animal Production **17.** (1) 24

[9] Naifa R, Oktovianus R, Nahak T B and Dethan A 2015 Nutritive quality of Elephant grass silage with addition of a different level of rice bran and cornmeal J. Anim. Sci. **1** (1) 6

[10] Amer S, Hassanat F, Berthiaume R, Seguin P and Mustafa A F 2012 Effects of water soluble carbohydrate content on ensiling characteristics, chemical composition and in vitro gas production of forage millet and forage sorghum silages Anim. Feed Sci.Technol.**177.** 23

[11] Santoso B, Hariadi B T and Jeni 2015 Fermentation quality of king grass silage treated with liquid or dried inoculant of lactic acid bacteria J. Indonesian Trop. Anim. Agric.**40.** (4) 208

[12] Li M, Zi X, Zhou H, Hou G and Cai Y2014 Effects of sucrose, glucose, molasses and cellulase on fermentation quality and in vitro gas production of king grass silage. Anim. Feed Sci. Technol.**197.** 206

[13] Nurjana D J, Suharti S and Suryahadi 2016 Improvement of Napier grass silage nutritive value by using inoculant and crude enzymes from Trichoderma reesei and its effect on in vitro rumen fermentation Media Peternakan **39.**(1) 46

[14] Hapsari S, Suryahadi and Sukria H A 2016 Improvement on the nutritive quality of Napier grass silage through inoculation of Lactobacillus plantarum and formic acid Media Peternakan **39** (2) 125

[15] Sofyan A, Widyastuti Y, Utomo R and Yusiati L M 2017 Improving physicochemical characteristics and palatability of King grass (Pennisetum hybrid) silage by inoculation of Lactobacillus plantarum - Saccharomyces cerevisiae consortia and addition of rice brand Buletin Peternakan **41** (1)61

[16] Steubing P M 1993 Isolation of an unknown bacterium from soil in: Tested studies for laboratory teaching Proc. of the 14th Workshop/Conference of The Association for Biology Laboratory Education (ABLE)**14** ed. C A Goldman (Toronto: Canada)

[17] Bujalance C, Jiminez-Velera M, Moreno E andRuiz-Bravo A 2006 A selective differentialmedium for Lactobacillus plantarumJ.Microbiol. Methods **66.** 572

[18] Bakrie B, Sastro Y and Sudolar N R 2017 Effect of different accelerators and inoculums used in fermentation on quality of dead chicken silage flour as feed ingredient for catfish J.

Indonesian Trop. Anim. Agric. **42(2)** 99

[19] AOAC 2000 Official Methods of Analyses17thEdition Association of Official Analytical Chemist (Washington DC: USA)

[20] Kusriningrum R S 2010 Experimental Design2nd Edition (Surabaya: Airlangga University Press)

[21] Purwati E, Rusfidra, Juliyarsi I and Depson R 2010 Biotechnology application for rumen contents of cattle, buffalo and goat as energy source for the production of environmentally friendly biogas Proc. Nat. Seminar of Faculty of Tec. Riau Univ. (Pekan Baru: Riau Province)

[22] Yakin E A, Ngadiyono N and Utomo R 2012 Effect of rumen content silage substitution to King grass basal feed on crossbreed cattle performance Buletin Peternakan **36** (3) 174

[23] Castillo-Gonzalez A R, Burrola-Barraza M E, Dominguez-Viveros J and Chavez-Martinez A 2014 Rumen microorganisms and fermentation. Arch. Med. Vet.**46** (3) 349

[24] Khuluq A D 2012 The potential of probiotic fermentation of sugarcane top as feed for animals Buletin Tanaman Tembakau, Serat dan Minyak Industri **4** (1) 37

[25] Rochana A, Indriani N P, Ayuningsih B, Hernaman I, Dhalika T, Rahmat D and Suryanah S 2016 Feed forage and nutrition value at altitudes during the dry season in West Java Animal Production **18** (2) 85

[26] Yanuartono, Purnamaningsih H, Indarjulianto S and Nururrozi A 2017 The potential of rice straw as feed for ruminants J. Ilmu-Ilmu Peternakan **27** (1) 40

[27] Rios-Rincon F G, Bermudez-Hurtado R M, Estrada-Angulo A, Juarez-Reyes A S dan Pujol-Manriquez C 2010 Dried ruminal content as a substitute for Alfalfa hay in growing-finishing diets for feedlot cattle J. Anim. Vet. Adv.**9** (10) 1526

[28] Agbabiaka L A, Anukam K U and Nwachukwu U N 2011 Nutritive value of dried rumen digesta as replacement for soybean in diets of Nile Tilapia (Oreochromis niloticus) fingerlings Pak. J. Nutr **10** 568

[29] Faradonbeh O P, Bagheri H and Soleimani R 2011 Replacing fermented rumen content flour as rice substitution: effects on broilers performance Res. Opin. Anim. Vet. Sci. **1** (12) 760

[30] Mohammed G, Igwebuike J U and Alade N K 2011 Performance of growing rabbits fed graded levels of Bovine blood-rumen content mixture Agric. Biol. J. N. Am.**2** (4) 720

Invitro Study on the Fluid From Banana Stem Bioprocess as Direct Fed Microbial

B.K. Mutaqin[1*], U.H. Tanuwiria[1] and E. Hernawan[1]

Animal Nutrition and Feeding Departement. Faculty of Animal Husbandry. Universitas Padjadjaran

*Corresponding Email: bambang14002@mail.unpad.ac.id

Abstract

The purpose of this research was to study the liquid produced by the bioprocess of banana stem as a Direct Fed Microbial (DFM) in order to enhance local sheep productivity invitro. Studying was the use of DFM in two invitro feeds. The object observed in this research was fermentability and digestibility value. The method was experimental with the experimental design, i.e. factorial experimental design with two factors. The first factor was DFM, the levels of which were 0, 0,2, 0,4 and 0,6%, while the second factor was two feed types (complete feed and *Pennisetum purpureum* only) with the treatment of threefold repetition. This research showed that fermentability and digestibility value were influenced by the DFM in the invitro complete feed. The research result analyzed using MANOVA with further testing using Duncan Test. The conclusion of the research result were shows the interaction DFM in the complete feed improve fermentability and digestibility values and DFM 0,6% shows the highest value.

1. Introduction

Livestock commodities need to be developed to support the achievement of food self-sufficiency in Indonesia. At present, the condition of national meat needs fulfillment is still dependent on imported. Meanwhile, local livestock resources have not been able to meet the fulfillment of national needs.

Sheep farming has enormous potential as an alternative to meet the national meat needs. However, most of the sheep are still maintained traditionally. Farmers do not really take into account the factors of production. Meanwhile, the feed is a major factor affecting the success of a sheep farming business. Feed quality greatly affects the basic needs and productivity of livestock. Efforts to overcome the lack of nutrients in meeting the basic needs and productivity of livestock require complete feed.

Proper provision of the feed will increase the sheep's weight gain. The increase in body weight will increase in line with the increasing digestibility. Increased digestibility can be achieved by manipulating the fermentation of fiber components in the rumen, which is directed to improve the efficiency of feed utilization. This manipulation can be done with the help of probiotics.

Probiotics are living microbes that are given directly to livestock with the aim of improving the balance of microbes in the digestion and reducing undesirable microbes. Probiotics given directly to ruminants are called Direct Fed Microbial (DFM).

Direct Fed Microbial (DFM) can be obtained from various media formed from a mixture of natural ingredients from various natural resources available. The microorganisms formed are called Indigenous Microorganism (IMO). Plant waste around the environment such as the remnants of the plants like banana cobs, rice straw, and others can be utilized as the growing media of the microorganisms. Banana stem is a source of natural material that has large size and weight compared to rice straw or other materials for indigenous microorganism growth media.

The work mechanism of DFM is to balance the rumen ecosystems, suppress the production of lactic acid, and help break down cellulose. The balance of rumen ecosystems occurring will stimulate the increase in growth and productivity of livestock whose growth indicator is the increase in body weight. This is known from the production of volatile fatty acids, ammonia, and fiber digestibility.

2. Materials and Methods

Feed A consisted of *Pennisetum purpureum*, while Feed B consisted of *Pennisetum purpureum* and concentrate mixture. The *Pennisetum purpureum* used was that of Taiwan cultivar. Feed A and Feed B were given the addition of DFM. The addition consisted of 4 levels, i.e. 0%, 0.2%, 0.4%, and 0.6%. The compositions of food substance and feed material used for making Feed B [1], Table 1 and Table 2. The result of proximate analysis of feed formulation [2], used can be seen in Table 3.

Table 1. Composition of Sheep Food Substance.

No.	Feed Material	Food Substance			
		DM	CF	CP	TDN
1	*Pennisetum purpureum*	22.20	38.20	9.20	52.40
2	Molasses	82.40	0.40	3.94	70.70
3	Cassava dregs (*onggok*)	79.80	0.32	1.87	78.30
4	Soy sauce dregs	26.83	22.77	25.74	66.00
5	Tofu dregs	75.00	26.17	23.39	60.00

Table 2. Sheep Material Composition

No.	Feed Material	Food Substance				
		%	DM	CF	CP	TDN
1	*Pennisetum purpureum*	42	9.32	16.04	3.86	22.00
2	Molasses	3	2.47	0.01	0.12	2.12
3	Cassava dregs (*onggok*)	25	19.95	0.08	0.47	19.58
4	Soy sauce dregs	16	4.29	3.64	4.12	10.56
5	Tofu dregs	14	10.50	3.66	3.27	8.40
	Complete feed	100	46.53	23.43	11.84	62.66

Table 3. Proximate Analysis Result of Research Feed

No.	Proximate Analysis		
1	Water (%)	10.07	31.48
2	Ash (%)	9.46	10.67
3	Crude Protein (%)	5.99	13.12
4	Crude fiber (%)	32.86	23.02
5	Crude fat (%)	0.62	3.71
6	Extract materials without nitrogen (%)	51.07	49.48
7	Total Digestible Nutrients (TDN) (%)	52.87	65,48
8	Metabolic energy (kcal)	2342	3199

2.1. Ammonia Concentration

The measurement of rumen fluid concentration and that of NH_3 concentration were done by using Conway Micro-diffusion method [3]. Before Conway dish was used, the lip of the dish had been smeared with Vaseline. 1 mL of the liquid produced from the fermentation process with 4 hours incubation was taken, and then placed at one groove end of Conway dish, while at the other end of which 1 mL of saturated Na_2CO_3 was added.

Boric acid as much 1 mL with indicator was placed in a small space located in the center of the Conway dish, and then the Conway dish was sealed tightly until it was airtight. After that, the Conway dish was shaken until the liquid and $NaCO_3$ were well blended, and left at room temperature for 24 hours. After 24 hours, the boric acid indicator was titrated with 0.005 N of H_2SO_4 until the color changed from blue to red. The NH_3 concentration was calculated by the formula:

$$NH_3\ (mM) = \frac{\text{Volume } H_2SO_4 \text{ x N. } H_2SO_4}{\text{Sample Weight x DM}_{\text{Sample}}} \text{ x } 1000$$

2.2. Volatile Fatty Acids (VFA)

The VFA was measured using steam distillation technique [4]. The 5 mL liquid (derived from the same tube as the liquid for the NH_3 analysis) was put into the distillation tube, and then the tube was added one mL of 15% H_2SO_4. The tube wall was rinsed with Aquadest and immediately covered with a rubber plug that had been connected to a distillation pipe of ± 0.5 cm diameter. Then, the other end of the pipe was connected to the Laibig cooler. The distillation tube was put into a boiling flask that had been filled with boiling water without touching the water surface. The distillation product was accommodated with a 500 mL Erlenmeyer flask filled with 5 mL of NaOH 0.5 N.

The distillation process was completed when the number of distillates accommodated reached 300 mL. The accommodated distillates was added with 2-3 drops of phenolphthalein (PP) indicator, then titrated with HCl 0.5 N until a change from pink to colorless (transparent). The concentration of VFA_{total} was measured by the formula:

$$VFA_{\text{total}} = \text{(a-b) x N. HCl x 1000 x 5 mL}^{-1}$$

Description:
a : blank titrant volume b : sample titrant volume

2.3. Dry Matter Intake and Organic Matter Intake

Measurement of digestibility was done by using in vitro method from [5], with the following stages:
a. Preparing rumen liquid.
b. Creating an artificial saliva solution referring to McDougall's guidance where the artificial saliva-making procedures followed McDougall's instructions [6].
c. Making fermentation media by mixing artificial saliva and rumen content fluid with ratio of 4:1.
d. The digestibility values of dry materials and organic materials were measured using [5]. The calculation formula was as follows.
1) Dry Matter Intake (DMI)
The value of dry matter intake coefficient can be calculated by the formula:

$$\text{DMI}\ (\%) = \frac{\text{DM}_{\text{Sample}} - (\text{DM}_{\text{Residue}} - \text{DM}_{\text{Blank}})}{\text{DM}_{\text{Sample}}\ (\text{gram})} \times 100\%$$

2) Organic Matter Intake (OMI)
The value of organic matter intake coefficient can be calculated by the formula:

$$\text{OMI}\ (\%) = \frac{\text{OM}_{\text{Sample}} - (\text{OM}_{\text{Residue}} - \text{OM}_{\text{Blank}})}{\text{OM}_{\text{Sample}}\ (\text{gram})} \times 100\%$$

Description:
$\text{OM}_{\text{Sample}}$: the feed dry /organic matter weight (gram) before incubation.
$\text{OM}_{\text{Residue}}$: the feed dry / organic matter weight (gram) after incubation.
OM_{Blank} : rumen liquid + artificial saliva.

3. Results and Discussion

The NH_3 content based on Table 4 shows the interaction of DFM with the feed types. The interaction of Feed A, the feed with low protein content in 0.2% DFM had the decrease of NH_3. The decrease of NH_3 concentration was inseparable from the substrate of Feed A, which was low in nutrient contents interacting with DFM. The NH_3 concentration in high protein feed with 0.6% DFM resulted in the decrease in NH_3. This happened because of the interaction between DFM factor and feed factor. It made the NH_3 not accumulated into the rumen environment due to the lack of the carbon source supply for the synthesis of the microbes in Feed A. High NH_3 concentration can show the process of feed protein degradation faster than the process of microbial protein formation, so that the resulting ammonia accumulated in the rumen [7].

Table 4. NH_3 content with DFM treatment and the Feed Types

Feed	DFM (%)				Average
	0	0.2	0.4	0.6	
---------------- mM ----------------					
Feed A	6.64^{D}	5.30^{G}	5.62^{F}	6.92^{B}	6.12^{a}
Feed B	6.34^{E}	6.72^{C}	7.42^{A}	4.44^{H}	6.23^{a}
Average	6.49^{a}	6.01^{b}	6.52^{a}	5.68^{b}	6.18

Description: Different superscripts lowercase letters within the average columns represent significant difference. Superscripts with significantly different capital letters significantly within the rows and columns of Feed A and Feed B represent the interaction of the two factors.

Table 5. VFA $_{Total}$ Contents with DFM and the Feed Types

Feed	0	0.2	0.4	0.6	Average
---------------- mM ----------------					
Feed A	139.00^{A}	146.67^{A}	108.33^{B}	107.13^{B}	125.28^{b}
Feed B	112.67^{B}	127.00^{B}	133.67^{A}	154.67^{A}	132.00^{a}
Average	125.83^{bc}	136.83^{a}	121.00^{c}	130.90^{ab}	128.64

Description: Different superscripts lowercase letters within the average columns represent significantdifference. Superscripts with significantly different capital letters within the rows and columns of Feed A and Feed B represent the interaction of the two factors.

Based on Table 5, the VFA $_{Total}$ content refers to the VFA $_{Total}$ value of the result of interaction between the 2 different feed types, i.e. Feed A and Feed B with the addition of DFM, affecting the VFA $_{Total}$ value. The interaction of Feed B, which was a high protein feed, with DFM had increased VFA $_{Total}$ concentration to 0.6% DFM. VFA contents in rumen fluid can be used as one of the fermentability measures of feed, and they were closely related to rumen microbial activities [8]. The addition of probiotics (DFM) through feeding can increase the microbial activities directly proportional to the increase of VFA content of the rumen [9], since the amount of VFA in the rumen characterizes microbial activities with the amount of VFA absorbed or out of the rumen [10].

Table 6. Dry Matter Intake (DMI) Values with DFM Treatment and the Feed Types

Feed	DFM (%)				Average
	0	0.2	0.4	0.6	
---------------- % ----------------					
Feed A	23.08^{H}	27.12^{G}	29.97^{F}	38.22^{E}	29.60^{b}
Feed B	55.67^{D}	59.21^{C}	62.36^{B}	68.48^{A}	61.43^{a}
Average	39.38^{d}	43.17^{c}	46.17^{b}	53.35^{a}	45.52

Table 7. Organic Matter Intake (OMI) Value with DFM Treatment and the Feed Types

Feed	DFM (%)				Average
	0	0.2	0.4	0.6	
---------------- % ----------------					
Feed A	20.61^{E}	25.54^{D}	$26,34^{D}$	34.79^{C}	26.82^{b}
Feed B	57.39^{B}	60.45^{B}	64.06^{A}	66.11^{A}	62.00^{a}
Average	39.00^{c}	43.00^{b}	45.20^{b}	50.45^{a}	44.41

Description: Different superscripts lowercase letters within the average columns represent significant difference. Superscripts with significantly different capital letters in the rows and columns of Feed A and Feed B represent the interaction of both factors.

The DMI value in Table 6 shows the DMI value of the interaction between the feed factors, i.e. Feed A and Feed B with the addition of DFM. The average value of DMI value of Feed B (61.43%) is greater than that of Feed A (29.60%). Dry materials are a reflection of the amount of carbohydrates contained in feed ingredients, because about 50-80% of the dry materials are composed of carbohydrate, so that the feed containing food substances and energy is relatively different, then the DMI is relatively different [11].

The value of OMI in Table 7 shows that the digestibility of the organic materials resulted from the interaction of the two different feed types, i.e. Feed A and Feed B, with DFM, had an effect on the OMI value. The average value of OMI of Feed B (62.00%) is greater than that of Feed A (26.82%), and then feed B has significant difference of influence.

The effect of the increased OMI is presumed as due to the DFM interaction with the feed where it affects the DMI. Therefore, if the DMI increases, then the DMI also increases [12]. The results show the increase of OMI value in line with the addition of DFM to 0.6% both on Feed A and Feed B.

4. Conclusion

The conclusion of the research result were shows the interaction DFM in the complete feed improve fermentability and digestibility values and DFM 0,6% shows the highest value.

Acknowledgments

We are grateful for partial support from Animal Nutrition and Feeding Laboratory, Universitas Padjadjaran. We also thank Salman Parisi and Tuti Yana for assistance.

References

[1] National Research Council 2006 Nutrient Requirement of Sheep National Academy Press (Washington DC)

[2] Ruminant Livestock Nutrition and Animal Feed Chemical Laboratory 2016 Proximate Analysis Result 2016 Animal Nutrition and Feeding Departement Faculty of Animal Husbandry Universitas Padjadjaran

[3] Conway PL, Gorbach SL and Goldin BR 1987 Survival of Lactic acid bacteria in the human stomach and adhesion to internal cell J. Dairy Sci **70**:1

[4] Association of Official Analytical Chemists 2000 Official Methods of Analysis of AOAC International 17th ed. Horwitz W (Gaithersburg MD)

[5] Tilley JMA and Terry RA 1963 A Two Stage Technique For The In Vitro, Digestion Of Forage Crops (London: British Grass)

[6] McDoughall EI 1949 Studies on Ruminant Saliva: the composition and output of sheep's saliva Institute of Animal Pathology University of Cambridge Biochemical Journal Vol 43(1)

[7] McDonald P, Hunderson AR and Heron SJE 2010 Animal Nutrition. 7nd ed. Cambrian Printers Lt (Aberystwyth)

[8] Parakkasi A 1999 Ilmu Nutrisi dan Makanan Ternak Ruminan (Jakarta: UI-Press)

[9] Yoon IK and Stern MD 1995 Influence of direct-fed microbial on ruminal microbial fermentation and performance of ruminants A Review. Asian Aust. J. Anim. Sci. **8 (6)**: 533 .

[10] Church DC 1974 Digestive Physiology and Nutrition of Ruminant Vol. 2 O&B Books.

[11] Abun 2005 Efek Fermentasi Ampas Umbi Garut (Maranta arudinacea LINN.) dengan Kapang Aspergillus niger terhadap Nilai Kecernaan Ransum Ayam Pedaging Artikel Ilmiah (Bandung: Universitas Padjadjaran)

[12] Fathul F and Wajizah S 2010 Penambahan Mikromineral Mn dan Cu dalam Ransum terhadap Aktivitas Biofermentasi Rumen Domba secara in vitro Jurnal Ilmu Ternak dan Veteriner **15**(1):9

IOP Conf. Series: Earth and Environmental Science **119** (2018) 012010 doi:10.1088/1755-1315/119/1/012010

Ruminal Ca and P Releases from Diets with Different Portion of the Sugarcane Bagasse

E. Pangestu[1], F. Wahyono[1], L.K. Nuswantara[1], J. Achmadi[1*]

[1]Faculty of Animal and Agricultural Sciences, Diponegoro University, Semarang, Indonesia

[*]Corresponding E-mail: jachmadi@undip.ac.id

Abstract

The in sacco technique was used to study the ruminal Ca and P releases from diets with different portion of sugarcane bagasse. Three diets containing 15, 25, and 35% of sugarcane bagase were tested their kinetic of ruminal Ca and P degradabilities. Two adult male sheep fitted with rumen cannula were used in the in sacco technique. In the in sacco experiment, feed samples were placed in the nylon bag and inserted into ruminal cannula for 0, 1, 3, 6, 12, 24, and 48 h. The kinetic of ruminal Ca and P degradabilities were focused on rapidly soluble fraction (fraction *a*), potentially degradable fraction (fraction *b*), and the degradation rate of fraction *b* (*c*). The data were tested using analyse of variance based on a completely randomized design. While the portion of sugarcane bagasse increased ($P<0.05$) fraction *a* of Ca diet, the portion *b* of P diet was decreased ($P<0.05$) by the portion of sugarcane bagasse.In conclusion, the effect of increasing portion of sugarcane bagasse in diet on ruminal release of Ca may be differed with that ofthe ruminal P release.

1. Introduction

The ruminal degradation of forage mineralshad been studied into some extent [1]; [2]; [3]; [4]; [5]; [6].Ruminal mineral availability from forages may be affected by the form of mineral and cell wall association [5]. Some part of mineral was bond in cell wall component, therefore the amount of ruminal mineral release from agricultural waste is more lower than its concentrationconsistently [7]. The mineral ruminal release from agricultural waste based diet remains to be clarified.

The use of sugarcane bagasse as ruminant feed is restricted because the cellulolisic component is tightly bond to its lignin [8]; [9]. It is postulated that the higher portion of sugarcane bagasse in diets may result in more lower the amount of mineral ruminal release. This study clarified the ruminal Ca and P releases from diets with different portion of sugarcane bagasse. This was accomplished using the in sacco technique.

2. Materialsand Methods

2.1. Diet sample and thein sacco technique.

Three diets with different portions of sugarcane bagasse were tested in this study (Table 1). The diets those contained sugarcane bagasse with portion of 15, 25, and 35%, respectively, were designed to be isoenergy and isonitrogenous. All samples of agricultural wastes were obtained from some local industries at Semarang city. Sugarcane bagasse was the waste product from local sugarcane mill.

Two male local thin tailed cross bred sheep with body weight average of 25 kg and aged at 24 months were used in this *in sacco* technique. Animals were fitted with ruminal cannulas with inside diameter of 3 cm.Sheep were housed in individual metabolic cages and fed on a diet containing 70% elephant grass, 9% rice bran, 11% coconut mill, 4% cassava waste, 5% sugarcane molasses, and 1% urea. The diet was offered at daily maintenance level and dringking water was available throughout experimental period.

The *in sacco* technique used nylon bags with dimension of 2 X 6 cm and mean pore size of 46 µm. Each sample of 3.00 g was placed in the bag which was attached with 0.25 m weighted chain. The bags were suspended in rumen of sheep via cannula 30 min after morning feeding. Incubation times were 0, 3, 6, 12, 24, and 48 h with three bags per incubation time. Five replicatesof each sample of diet were testedfor each each tested diet andincubation time. Zero-hour bags were not incubated in the rumen but were washed in the same manner as incubated bags.

After time of incubation bags were removed and immersed in ice water for 15 min, the bags were washed with demineralized water using a washing machine for 5 min. The bags and sample residues were then dried at 60°C for 48 h and sample residues were prepared for chemical analyses.

2.2. Chemical and statistical analysis.

A part of each sample residue was analysed for the content of dry matter(DM) using method of [10]. The other part of each sample residue was ashed at 550 °C for 4 h for determination of Ca and P contents. The ash then was solubilized with 10 ml of 3 N HCl and boiled gently for 10 min under a watch glass [3].The solutions were transferred to 25 ml volumetric flasks and brought to the volume with distilled water. Contents of Caand Pwere then determined using the atomic absorption spectrophotometer.

Theruminal DM, Ca, and P disappearances(expressed as % of the initial amount) were calculated based on the their content of sample residues. The degradability data were then calculated to obtain the kinetic of ruminal DM, Ca and P degradabilities according to the equation, $P=a+b(1-e^{-ct})$ [11]where P is the amount degraded at time t, a is the rapidly soluble fraction, b is the potentially degradable fraction, c is the rate of degradation of fraction b. The parameters of ruminal degradability kinetic were focused on rapidly soluble fraction (fraction a), potentially degradable fraction (fraction b), and the degradation rate of fraction b (c) for DM, Ca and P, respectively. The data were tested using analyse of variance based on a completely randomized design.

3. Results and Discussion

The ruminal release of mineral from forages were studied using the *in sacco* technique [1]; [2]; [3]; [4]; [5]; [6]. However, these studies reported only the amount of mineral release from roughages at given time of ruminal incubations, and the kinetic of ruminal mineral degradability was not discussed. Results of this study clarify the ruminal Ca and P releases from sugarcane bagasse based dietsby discussing their rapidly soluble fractions, potentially degradable fractions, and the degradation rates of potentially degradable fractions.

The rapidly soluble fraction of dietary DM were decrased ($P<0.05$) with the increasing portions of sugarcane bagasse (Table 2). The component of fraction a consists of soluble materials of cell content. The increasing portion of sugarcane bagasse increases contents of NDF and ADF, as the cell wall components, in diets (Table 1). The use of sugarcane bagasse in a bagasse based total mixed ration for goats is limited by their levels of fiber component [8]; [9].

IOP Conf. Series: Earth and Environmental Science **119** (2018) 012010 doi:10.1088/1755-1315/119/1/012010

Table 1. Ingredient and chemical composition of sugarcane bagasse based diets

Items	Portion of Sugarcane Bagasse		
	15%	25%	35%
Ingredient			
Rice bran	13	16	14
Coconut mill	10	10	10
Palmfrond mill	5	5	4
Wheat pollard	12	12	10
Ground nut shell	8	6	3
Coffee seed shell	10	5	3
Mize	2	2	5
Cassava waste	17	14	11
Sugarcane bagasse	15	25	35
Chemical composition			
Dry matter	90.18	90.09	89.88
Crude ash	7.43	7.73	7.72
Crude protein	12.00	12.09	12.35
Extract ether	4.04	4.06	3.95
Crude fiber	28.78	28.8	29.12
Nitrogen free extract	47.74	47.32	46.86
Total digestible nutrient[1]	60.65	60.63	60.30
Neutral detergent fiber	59	61.76	63.05
Acid detergent fiber	34.37	35.40	35.72
Calcium	0.34	0.32	0.30
Phosphorus	0.15	0.14	0.13

[1]Based on calculation

The potentially degradable fraction of diet was unaffected by the increasing portion of sugarcane bagasse (Table 2). The potentially degradable fraction is the slowly degraded part of feed in rumen. Table 2 shows that the effect of sugarcane bagasse portion on the degradation rates of potentially degradable fractions was non significant. This fact may be responsible for the effect of increasing portion of sugarcane bagasseon the potentially degradable fraction of diet.

Table 2 indicates that the rapidly soluble fraction of dietary Ca was increased ($P<0.05$) by the portion of sugarcane bagasse. However, the increasing portion of sugarcane bagasse in diet decreased ($P<0.05$) the potentially degradable fraction and the degradation rates of potentially degradable fraction of dietary Ca. Ruminal mineral availability of feed may be affected by the form of mineral and cell wall association [4]. It could be considered that amount of dietary Ca is more larger in cell content than that in cell wall component. However, the fractionation of dietary Ca in cell content and wall components remains to be elucidated.

IOP Conf. Series: Earth and Environmental Science **119** (2018) 012010 doi:10.1088/1755-1315/119/1/012010

The rapidly soluble fraction of dietary P was decreased (P<0.05) by the portion of sugarcane bagasse (Table 2). Whilst the portion of sugarcane bagasse increased (P<0.05) the potentially degradable fraction. In contrast with the dietary Ca, it could be considered that amount of dietary P is more larger in cell wall than that in cell content component. Although the information about fractionation of dietary P in cell content and wall components.

Table 2. Kinetic of ruminal degradability

Items		Portion of sugarcane bagasse 15%	25%	35%	SEM
DM	*A*	29.16[a]	28.5[a]	26.07[b]	0.81
	B	21.26	21.47	21.01	0.12
	C	7.47	5.61	6.86	0.47
	dT	41.88[a]	39.85[b]	38.21[c]	0.92
Ca	*a*	25.34[b]	28.39[a]	29.07[a]	0.99
	b	23.05[a]	16.33[b]	15.92[b]	2.00
	c	8.49[a]	8.49[b]	8.96[ab]	0.14
	dT	40.62[a]	38.21[c]	39.27[b]	0.60
P	*a*	26.76[a]	25.66[a]	17.67[b]	2.48
	b	22.12[b]	20.18[b]	25.98[a]	1.48
	c	7.91[b]	9.12[a]	7.64[b]	0.39
	dT	40.3[a]	38.69[b]	33.36[c]	1.82

[a,b,c] *P<0.05*
a : % of rapidly soluble fraction;
b : % of potentially degradable fraction;
c : degradation rates of potentially degradable fractionsin h;
dT : % of degradation theory. SEM: standard error of means.

The use of sugarcane bagasse as small ruminant feed had been clarified by some studies based on some parameters of protein and carbohydrate metabolisms[8]; [9]. Dietary Ca and P were known well involved partly in the nutrient metabolism. In this study, the sheep's ruminal Ca and P releases from sugarcane bagasse based dietwas elucidated. Such information may be useful for considering the supplementation of dietary mineral to meet the requirement of animal.

4. Conclusion

The increasing portion of sugarcane bagasse in diet increased ruminal Ca release and decreased ruminal P release. The mineral fractionation of sugarcane bagasse based dietremains to be clarified.

Acknowledgement

The study was partly funded by PNBP Fakultas Peternakan dan Pertanian Universitas Diponegoro Tahun Anggaran 2015.

References

[1] Emanuele, S.R. and C.R. Staples.1990. Ruminal release of minerals from six forage species. J. Anim. Sci. **68**: 2052.

[2] Ledoux, D.R. and F.A. Martz, 1990. Ruminal solubilization of selected macromineralsfrom forages and diets. J. Dairy Sci. **74**:1654.

[3] Emanuele, S.M., C.R. Staples, and C.J. Wilcox. 1991. Extent and site of mineral release from six forage species incubated in mobile dacron bags. J. Anim. Sci. **69**:801.

[4] Ibrahim, M.N.M., G. Zemmelink and S. Tamminga. 1998. Release of mineral elements from tropical feeds during degradation in the rumen. Asian-Aus. J. Anim. Sci. **11**: 530.

[5] Ibrahim, M.N.M. and G. Zemmelink. 1999. Effect of plant fibre on the solubility of mineral elements. Asian-Aus. J. Anim. Sci. **12**: 1277.

[6] Ceresnakova, Z., P. Flak, M. Chrenkova, M. Polacikova, M.A. Gralak. 2008. Ruminal phosphorus release from selected forages determined by the *in sacco* technique. Slovak J. Anim. Sci. **41**: 146.

[7] Pangestu, E. 2005. Pangestu E. 2005. Evaluasi serat dan suplementasi zink dalam ransum berbahan hasil samping industri pertanian pada ternak uuminansia. Fakultas Pascasarjana Institut Pertanian Bogor. Bogor (Disertasi).

[8] Ariyani, S.A., L.K. Nuswantara, E. Pangestu, F. Wahyono and J. Achmadi. 2014. Parameters of protein metabolism in goat fed on different portion of sugarcane bagasse. J. Indonesian Trop. Anim. Agric. **39**: 111.

[9] Suhada, A.T., L.K. Nuswantara, E. Pangestu, F. Wahyono and J. Achmadi. 2016. Effect of synchronization of carbohydrate and protein supply in the sugarcane bagasse based diet on microbial protein synthesus in sheep. J. Indonesian Trop. Anim. Agric. **41**: 135.

[10] AOAC. 1999. Official Methods of Analyse. 17th ed. AOAC Int., Arlington, V.A.

[11] Ørskov, E. R. and I. McDonald. 1979. The estimation of protein degradabality in the rumen from incubation measurements weight according to rate of passage. J. Agric. Sci. (Camb.) **92** : 499.

IOP Conf. Series: Earth and Environmental Science **119** (2018) 012011 doi:10.1088/1755-1315/119/1/012011

Utilization of Cacao Pod Husk Silage as Cattle Ration Mixture at Taluditi, Pohuwatu Regency, Gorontalo Province

M. Hidayat[*1], **H.Y. Pratama**[1], **E. Martono**[2]

[1] Faculty of Animal Science, Universitas Gadjah Mada, Yogyakarta, Indonesia
[2] Faculty of Agriculture, Universitas Gadjah Mada, Yogyakarta, Indonesia

*Corresponding e-mail: muhammad.hidayat93@mail.ugm.ac.id

Abstract

Cacao plantation produces cacao pod husks (CPHs) by-products during the harvest period. This research aimed to make benefits the CPH, preserved with silage technology as cattle ration mixture, investigate the adequacy of nutrient for the cattle after treated with addition of CPH silage, and investigate the quota of cattle treated with CPH silage addition. The research design was conducted by giving extension and field observation that was carried out at Taluditi on July to August 2015 during the activity of Student Community Service (KKN PPM UGM Unit Gorontalo 02). The secondary data was gathered from the Department of Agriculture and Plantation Pohuwatu Regency, and also supporting references. The number of respondents in each location was about 30 – 40 people. The research results showed that the silage preservation technology can be well received by the farmers. There was improvement of cattle ration nutrient supplemented with CPH compared to that of ration nutrient which was usually be used by the farmers or standard ration nutrient. The research also resulted in fresh CPH production 2,625.741 tons/year, CPH silage production 2,140.549 tons/year and load capacity ± 575 heads/day weight 250-300 kg.

1. Introduction

Taluditi District is an area with the largest production of cacao in Pohuwatu Regency, Gorontalo Province, with total plantation land 5,175 hectare out of 7,712.2 hectare or 67.1% of total cacao land at Pohuwato Regency [1]. The cacao plantation has potency to produce by-products in form of cacao pod husk (CPH) during the harvest time. Unutilized CPH is potential to be a carrier for cacao pest and disease. The field condition showed the pests and cacao disease at the cacao plantation in Taluditi was relatively high and can decrease the productivity of cacao. This occurrence appeared on the data in which the total production of cacao beans at Taluditi was 1,000,973 kg/year with production width/PW 3,750 hectare. Therefore, the average production of cacao bean was 266.9 kg/ha/year [1]. The value of cacao bean productivity at Taluditi was still lower than the average cacao bean production of Sulawesi island as much as 876 kg/ha/year or national production as much as 800 kg/ha/year [2].Based on that data, the utilization of CPH was necessary to be performed. One of the economical benefits of CP is by using it as feed material. Utilization of CPH as feed can be a solution to develop the animal production as the limited land for feed production, especially in Taluditi.

The abundant CPH production during the harvest time cannot be directly whole given to the animals. The CPH is not durable and rots quickly as it contains very high water. [3]. The CPH contains water content up to 85%and the dry matter (DM) is 15% [4]. Based on this condition, it needs a technology application to preserve the CPH as feed.

Some simple feed preservation technology had been studied in the universities and research institutions. One of simple feed preservation technologies is silage, at which the making process is called ensilage. Ensilage is the method of forage preservation through anaerob (no air) fermentation assisted with lactic acid bacteria [5]. The kind of lactic acid bacteria that often be used is *Lactobacillus plantarum* [3] and exctract of fermented grasses [5]. This preservation technology needs to be introduced to the farmers especially farmers at Taluditi.

So far, the education of feed preservation technology among the farmers at Taluditi, Pohuwatu Regency, Gorontalo Province is still minimal to perform. Based on the observation conducted on the Student Community Service (KKN PPM UGM) Unit Gorontalo 02, generally the farmers at Taluditi only acknowledge grasses and rice bran as feed and do not know alternative variation of other feed materials and also do not know how to store feed for long term as stock in the dry season.

2. Materials and Methods

2.1. Research Time and Site

This research was conducted for 2 months started on July 2015 to August 2015. The research was done at Taluditi, Pohuwatu Regency, Gorontalo Province during the activity of Student Community Service (KKN PPM UGM) Gorontalo Unit 02 year 2015.

2.2. Cacao Plantation Observation

The land observation was done to explore the cultivation system at the cacao plantation. The cultivation system covered data collection of harvesting time, cacao productivity, and condition of harvested cacao pod husks (CPH). The result of observation was then used as one of analysis basis for capacity of livestocks, extension, and CPH silage practice.

2.3. CPH Silage Simulation

The simulation was performed to give examples of the actual CPH silage showed to farmers during the extension and direct field practice. CPH was chopped into 5-10 cm length, then poured evenly with bacteria inoculants EM4+Molasses dosage 2% of DM CPH, dissolve EM4 with water with ration EM4+Molasses:water 1:10. The CPH added with EM4 was then fermented in the fermentation drum capacity 100 kg, and harvested 30 days after, at the extension and CPH silage practice.

2.4. Extension and CPH silage Practice

The extension was done in 6 villages at Taluditi that was based on the number of cacao production, which was at Panca Karsa 1, Panca Karsa 2, Makarti Jaya, Kalimas, Tirto Asri, and Puncak Jaya with number of respondents around 30 to 40 farmers for each village. The data obtained from the field observation and the secondary data from the Department of Agriculture and Plantation, and also related references were used as the basis for the extension and CPH silage practice material.

The field practice was performed by practice to make CPH silage with a simple principle that was carried out at three locations, but the participants of the practice still came from the six villages. It was based the geographic condition and to make the field practice effective, the three locations were:
Location 1 : Panca Karsa 2 Village (Participants from PancaKarsa 1 and PancaKarsa 2)
Location 2 : Makarti Jaya Village (Participants from Makarti Jaya village)
Location 3 : Kalimas Village (Participants from Kalimas, Tirto Asri, and PuncakJaya villages)

The fermentation was performed in the fermentation drum capacity 100 kg. The result of CPH silage practice was harvested at the 14th Day, and was reported its result that included organoleptic component such as texture, color, smell, and the existence of fungus. The harvest was carried out by farmers and to socialize the results.

2.5. *The Analysis of the improvement of Cattle Ration.*

Nutrient Supplemented with CPH SilageAddition of CPH feed as feed supplement for livestock was analyzed its improvement of ration nutrition compared to basal feed that consisted elephant grass *(Pennicetum purpureum)* only. Master Feed Formulawas made based on 2 kg of fermented CPH flour [6] converted to 10.2 kg silage/conventional fermented CPH, with calculation alike to equation (1).

$$\text{CPH Silage addition} = \frac{100}{16} x \frac{15}{18.4} x \text{ 2 kg CPH flour} \qquad (1)$$

annotation:

- 16: obtained from rendement of CPH fresh material into flour [6]
- 15: dry matter content (DM) of fresh CPH [4]
- 18.4: dry matter content (DM) of fermented CPH/silage [7]

The nutrient content of the elephant grass *(Pennicetum purpureum)* wasas much as DM 21%, protein 8.3%, Ca 0.53%, and P 0.29% [8]. The elephant grass feeding was based on the farmers sample at Taluditi who fed their cattle weighted $\pm$ 250-300 with fresh elephant grass as much as 30 kg, and the obtained grass DM cand be calculated with equation (2).

$$\text{Elephant grass DM} = \frac{21}{100} x \text{ 30 kg Fresh elephant grass} \qquad (2)$$

The content of protein, Ca, and P of fermented CPH/silage were 11.04%, 0.13%, and 0.14% [6], while the dry matter content (DM) of fermented CPH/silage was 18.4% [7]. Analysis was done by equality of the cattle weight about 250 to 300 kg with body weight gain 0.521 kg/day [6]. Further, the result obtained was compared to the cattle nutrient need, recommended for cattle weighted 300 kg with body weight gain 0.5 kg/day as much as 7000 gram DM, 679 gram protein, 19 gram Ca and 14 gram P every day [9].

2.6. Analysis of Capacity of Cattle Raising Treated with CPH Silage Supplementation

The CPH feeding as feed supplement was analyzed its capacity of cattle raising that can be accommodated each day. The analysis was based on the maximum quota of cattle that can be accommodated with CPH feed supplementation and performed by calculation according to supporting literature respectively started from Equation (3) to Equation (6).

$$\text{Whole Cacao pod husks Production} = \frac{100}{27{,}6} x \text{ Cacao Beans Production} \qquad (3)$$

$$\text{Fresh CPH Production} = \frac{72{,}4}{100} x \text{ Whole Cacao pod husks Production} \qquad (4)$$

$$\text{Silage CPH Production} = \frac{15}{18{,}4} x \text{ Fresh CPH Production} \qquad (5)$$

$$\text{Capacity of livestocks} = \frac{\text{Silage CPH Production}}{\text{10.2 kg } x \text{ 365 days}} \qquad (6)$$

Annotation:

- 27.6: obtained from the percentage of the cacao beans weight on the whole of the pod weight [6].
- 72.4: obtained from the percentage of CPH weight for the whole pod weight [6].
- 15: dry matter content (DM) of fresh CPH [4].
- 18.4: dry matter content (DM) of fermented CPH/silage [7].
- 10.2: the need of fermented CPH fed to the cattle, converted from 2 kg of CPH flour before [6].

-

3. Results and Discussion

3.1. *CPH Processing with Silage technology*

The product of CPH processing resulted in the research of feed technology development at Taluditi was the silage. The result of physical parameter of CPH silage practice during the 14 days fermentation is presented on Table 1.

Table 1. Organoleptic Parameter of CPH silage

Trial	Physical Parameter			
	Texture	Color	Smell	Fungus
Location 1	Tender and contains mucus	Yellowish brown	Sweet, sour and fresh	-
Location 2	Tender and contains mucus	Brown	Sweet, sour and fresh	Less
Location 3	Tender and contains mucus	Yeloowish brown	Sweet, sour and fresh	-
Simulation	Tender and contains mucus	Brown	Sweet, sour and fresh	Less

The reason to choose the silage product was based on the principle of its simple production and relatively easy to understand by the people at Taluditi who were newly introduced with innovation of feed processing technology. This method was considered as very suitable to apply as it was better if the people learned the simple and applicative technology before learning the advance technology which was its application was more complex.

The other factor that determined to apply the silage was due to the CPH production was not always every day as the harvest season for cacao at Taluditi was only twice a year (October to January and April to June). The abundant CPH production during the harvest time cannot be utilized all in one day by farmers, so it was necessary to do CPH preservation to create a regular cycle of CPH feeding each day. CPH should be preserved KBK to prevent from quick rottening as its water content is very high. It was confirmed that the CPH contains high water level (WL), which causes CPH to rot in three days and cannot be used as ruminants feed [3]. Based on this reason, CPH preservation was needed to make it possible to be used for a long time, but did not degrade its quality. One of the preservation that can be performed was by silage production.

The organoleptic parameter resulted in silage with texture contains a little bit mucus, color yellow to brown, smell sweet, sour, and fresh, and also found contamination of fungus on the surface. The quality of CPH silage can be identified by its organoleptic. The good quality is marked with the characteristic texture that contains a little bit mucus and does not clot, color yellowish brown, smell sour and fresh, and also there was no fungus contamination [3]. According to the literature, the result of CPH silage that was made was generally in quite good condition and the farmers are feasible to produce the silage in bulk.

Basically, the silage production only needs an air-tight room condition. However, the CPH silage making will be more quality by adding easy to dissolve carbohydrate and microbial innoculum. The implementation of extension was by using molasses as easy to dissolve carbohydrate and EM4 prebiotic as microbial innoculum. Kinds of easy to dissolve carbohydrates besides molasses are cassava flour, rice bran, and palm sugar. The addition of cassava flour as dissolved carbohydrate source was considered as sufficient for lactic acid bacteria growth in the silage to result in lactic acid that can decrease the pH of CPH silage [3]. The decrease of CPH silage pHcan preserve the CPH. The increase of the number of lactic acid bacteria in three days is needed to quicken the degradation of Ph caused by the lactic acid production in order to restrict the protein breakdown by inactivating the protease enzyme and inhibit the growth of undesireable anaerob microorganisms like *Clostridia* [10].

The consideration to add innoculant EM4 microbial with molasses as easy to dissolve carbohydrates was the application of effective and simple technology. It was due to the ease to obtain both ingredients. The analysis result in the field showed EM4 was available at the farming store in Pohuwatu, while molasses was available at Paguyaman Sugar Company at Boalemo which is only next to Pohuwato regency. In addition, the alternative for molasses was palm sugar as the local potency, especially as Taluditi produces many palm sugars. Thus, the CPH silage was expected can be continuously performed.

3.2. The analysis of the increase of cattle ration nutrient supplemented with CPH

The result of analysis of the increase of cattle ration nutrient supplemented with CPH can be seen in Table 2.

Table 2. The Comparation between Cattle Ration Nutrient

Kind of rations	Nutrient Contents			
	DM (gram)	Protein (gram)	Ca (gram)	P (gram)
Elephant grass	6,300	522.9	33.39	18.27
Elephant grass+CPH silage	8,175	729.9	35.83	20.90
Reccommended rations*	7,000	679	19	14

* [9]

Based on the results obtained, it was identified that the supplementation of CPH silage as cattle feed resulted in better ration nutrient compared to that of often be used by farmers at Taluditi who only used elephant grass. It was also identified that the CPH silage supplementation to the cattle rations had fulfilled the need of standard ration nutrients which contained minimum dry matter (DM) 7,000 gram, protein 679 gram, Ca 19 gram, and P 14 gram [9], while for the farmers rations, it still had not fulfilled the nutrient need especially dry matter and protein.

Good livestock productivity is always related to the good ration feeding. Indicator of good ration is the fulfillment of the livestock needs quantitatively from biomass aspect and qualitatively from nutrient aspect. Rations given to livestock are not always consumed at whole by the livestock. It related to the palatability characteristic of feed. Feed with high palatability will also increase the consumption. Fermented CPH/silage was identified to have high palatability than elephant grass, so it increased the consumption of feed DM [11]. The fulfillment of DM in the rations given was being the basis for other nutrients sufficiency. It was described further that the increase of DM consumption will be followed by the consumption of nutrient fractions contained in DM [12].

The CPH preservation with silage technique also gave benefits for the improvement of CPH quality. It is explained that the CPH fermentation can decrease the NDF and ADF [10]. It is caused by the activity of cellulolytic bacteria in the innoculum EM4 microbial that is able to break the cell cellulose lignin fraction. The cellulolytic bacteria during its process can grow in aciditic condition altogether with lactic acid in CPH fermentation process. It can be proven from a study by the degradation of NDF after the sixth day of fermentation from 58.1% into 54.9% DM with pH 4.94. In addition, in the further growth, the lactic acid bacteria obtain dissolved sugar from the result of cellulose reshuffle by cellulolytic bacteria [10].

3.3. Analysis of Capacity of Cattle Raising supplemented with CPH

The results obtained from the analysis of capacity of cattle rising supplemented with CPH can be seen on Table 3.

Table 3. Capacity of Cattle Rising

Data of Capacity of cattle rising	Amount
Production of cacao beans (kg)*	1,000,973
Production of cacao pods (kg)	3,626,714
Production of cacao pod husks (kg)	2,625,741
Production of fermented CPH (kg)	2,140,549
Fermented CPH Supllementation (kg)	10.2
Capacity of cattle supplemented/day (head)	$\pm$ 575

*[1]

According to the results presented on the Table 3 above, it showed the capacity of cattle rising (weight 250-300 kg) at Taluditi di Kecamatan Taluditi based on the production of CPH was $\pm$ 575 heads. This capacity will be reached if farmers can also fulfill the forage as much as 6.3 kg DM or 30 kg fresh elephant grass/head/day and after converted into 575 heads became 3.62 tons DM/17.25 tons of elephant grass every day (elephant grass DM 21%) [8]. It because the natural feed stuffs for ruminants are forages either grasses or leguminosae. This condition strong related to the process of fiber digestive in the rumen by microbial, and high fiber feed stuffs are forages. [13].

The CPH feed is complete feed or feed supplement. Basically, CPH is the feed with quite good fiber content. CPH contains crude fiber (CF) 30.45% to 31.22% [14] and cellulose lignin fraction/Neutral Detergent Fiber (NDF) 66.3% [15]. The contents of CF and NDF were nearly the same with grass elephant grasses CF which was 29.3 to 38.3% [16] and NDF as much as 64.22% [17]. Though CPH contains CF and NDF which nearly the same with CF and NDF of elephant grass as forage, the CPH feeding should be limited caused by anti nutrient substance in it. It is further explained that the anti nutrient substance/anti nutrition component in the cacao products is alkaloid theobromin, a heterocyclic substance contains nitrogen that can harm/obstruct the digestive process [18].The theobromin content can be minimized by the CPH fermentation. It is proven with a research that the CPH fermentation process for six days can decrease the theobromin level from 123.86 ppm to 102.79 ppm and states that this degradation is caused by the temperature increase during fermentation process. It is alike with the theobromin damage caused by the heat from the process of boiling or drying [10].

The 575 heads cattle that can be raised are a good stock for the cacao plantation-based livestock development at Taluditi and the effort to animal food products protein source. On the other hand, the CPH processing can be the pest's growth inhibitor and cacao plants disease as the characteristic of cacao husk is can be a carrier for pest and disease growth. The CPH silage feeding as feed supplement automatically can lower the production cost than if farmers choose other commercial feed supplement. It is due to the CPH condition at Taluditi is available for free as for todays not many farmers utilize it as feed.

4. Conclusion

CPH feed processing technology which is simply and suitable to apply in the farmers' community at Taluditi, Pohuwatu Regency, Gorontalo Province is by using silage. The utilization of CPH as cattle feed can be used as economical cacao's pest and disease inhibitor. The supplementation of CPH silage to the ration can fulfill the cattle nutrient need and provide the capacity of cattle rising at Taluditi ± 575 heads, thus it is expected to support the effort to realize the farming sovereignty in Indonesia and further to achieve the animal protein products sovereignty.

Acknowledgments

A sincere gratitude is extended to:
1. Allah SWT who gives the opportunity to the researcher to do a study and community service at Taluditi, Pohuwato Regency, Gorontalo Province.
2. Rector of Universitas Gadjah Mada Prof. Ir. Panut Mulyono, M.Eng., D.Eng. along with all academicians of UGM.
3. Fellows at KKN PPM UGM team Gorontalo 02 Unit year 2015
4. The mayor of Pohuwato Regency Syarif Mbwinga, S.Pd.I along with SKPD members.
5. Taluditi people who supported and made this program a great success

References

[1] The Department of Agriculture and Plantation Pohuwato Regency. 2015. Data of Development of Plantation Area

[2] Directorate General of Plantation.2014. Statistical data of Indonesian Plantation 2013-2015: Cacao.Available at http://ditjenbun.pertanian.go.id/ pdf. Accession date 5th Nov 2015.

[3] Zakariah, M.A., R. Utomo, and Z. Bachruddin. 2015. Pengaruh Inokulum Campuran *Lactobacillus plantarum* dan *Saccharomyces cerevisiae* terhadap Kualitas Organoleptik, Fisik, dan Kimia Silase Kulit Buah Kakao. Buletin Peternakan 39: 1-8.

[4] Tequia, A., H.N.L. Endeley, and A.C. Beynen. 2004. Broiler performance upon dietary substitution of cocoa husks for maize. Int. J. Poult. Sci. **3**: 779.

[5] Santoso, B., B.Tj. Hariadi, H. Manik and H. Abubakar. 2009. Kualitas Rumput Unggul Tropika Hasil Ensilase dengan Bakteri Asam Laktat dari Ekstrak Rumput Terfermentasi. Media Peternakan **32**: 137.

[6] Guntoro, S., Sriyanto, N. Suyasa, and M.R. Yasa. 2006. Pengaruh Pemberian Limbah Kakao Olahan terhadap Pertumbuhan Sapi Bali. Prosiding Seminar Nasional Teknologi Peternakan dan Veteriner page. 116-120.

[7] Anonymous. 2015. Cacao Pod Husk as Feed Alternative. Available at http://cybex.pertanian.go.id/materipenyuluhan/detail/783. Accession date 20th Okt 2015.

[8] Anggraeny, Y.N., U. Umiyasih, N.H. Krishna and L. Affandhy. 2006. Strategi Pemenuhan Gizi Melalui Optimalisasi Pemanfaatan Limbah untuk Pembesaran Sapi Potong Calon Induk. Prosiding Seminar Nasional Teknologi Peternakan dan Veteriner page. 82-87.

[9] Umiyasih, U., dan Y.N. Anggraeny. 2007. Technical Guidelines for balance ration Feed strategy for beef cattle. Indonesian Center for livestock Reseach and Development-Indonesian Agency for Agricultural Research and Development- Department of Agriculture. Jakarta.

[10] Wulandari, S. 2014. Peningkatan Nilai Cerna Serat dan Penurunan Theobromin Pod Kakaosebagai Bahan Baku Complete Feed pada Domba. Dissertation. Faculty of Animal Science UniversitasGadjah Mada. Yogyakarta.

[11] Suharto, M. 2004. Dukungan Teknologi Pakan Dalam Usaha Sapi Potong Berbasis Sumberdaya

Lokal. Prosiding Lokakarya Nasional Sapi Potong page. 14-21

[12] Suharti, S., D.A. Astuti, and E. Wina. 2009. Kecernaan Nutrien dan Performa Produksi Sapi Potong Peranakan Ongole (PO) yang Diberi Tepung Lerak *(Sapindus rarak)* dalam Ransum. JITV **14**: 200.

[13] Team of Laboratory of Feed Science and Technology IPB. 2012. Pengetahuan Bahan Makanan Ternak. CV Nuri Sejahtera.

[14] Munier, F.F. 2009. Potensi Ketersediaan Kulit Buah Kakao *(Theobroma cocoa l.)* sebagai Sumber Pakan Alternatif untuk Ternak Ruminansia di Daerah Istimewa Yogyakarta Prosiding Seminar Nasional Teknologi Peternakan dan Veteriner page. 752-759

[15] Buharman, B. Pemanfaatan Teknologi Pakan Berbahan Baku Lokal Mendukung Pengembangan Sapi Potong di Provinsi Sumatera Barat. WARTAZOA **21**:133.

[16] Hartadi, H., S. Reksohadiprodjo, S. Lebdosukojo, and A.D. Tillman. 1980. Table of Feedstuffs Composition for Indonesia. Utah:International Feedstuffs Institute Utah Agricultural Experiment Station,Utah State University Logan.

[17] Lubis, D., N.D. Purwantari, dan T. Manurung. 1999. Potensi Nutrisi Rumput Gajah dari Sistim Pertanaman Lorong dan Kapasitas Dukungnya untuk Sapi Perah Laktasi.Prosiding Seminar Nasional Teknologi Peternakan dan Veteriner page. 375-381.

[18] Mathius, I.W., and A.P. Sinurat. 2001. Pemanfaatan Bahan Pakan Inkonvensional untuk Ternak. Wartazoa **11**(2): 20.

The Effect of Herbaceous Legume of Feed in In-Vitro Digestibility

S. Ratnawaty[1*], Hartutik[2] and S. Chuzaemi[2]

[1]Agricultural Research Center of East Nusa Tenggara
[2]Faculty of Animal Husbandry, University of Brawijaya

*Corresponding E-mail: sophiaratnawaty@yahoo.com

Abstract

This study was carried out to evaluate in-vitro digestibility of herbal legumesin feed. The materials used in this study were three types of herbal legumes namely *Clitoria ternatea* Q5455 (CT Q5455), *Clitoria ternatea* cv. Milgarra (CT cv Milgarra) and *Stylosanthes seabrana (S. seabrana)*, The treatments were P0= 100% Grass; P1= 50% Grass+ 50% CT Q5455, P2= 50% Grass + 50% CT cv. Milgarra, P3 = 50% Grass + 50% S. seabrana. The result showed that the treatments had a significant effect (P <0.05). The highest dry matter (DM) digestibilitywas in P1 (60.35%) and P3 (60.22%). The DM digestibility of the highest raw materials was in CT cv. Milgarra (73.49%) and the lowest one was in *S. seabrana* (63.90%). The treatments had a very significant effect (P <0.01) on the organic matter (OM) Digestibility. The highest OM digestibility wasin P1 (63.04%) and P3 (61.89%). The highest value of OM digestibility of raw materials was in CT cv. Milgarra (73.90%) and the lowest one was in *S. seabrana* (63.85%). The treatmentshad a significant effect (P <0.05) on the crude protein (CP) digestibility. The average CP digestibility of feed was at the same value in all treatments but in CT Q5455 (67.25%). The treatmentshad a significant effect (P <0.05) on the total digestible nutrients (TDN). The highest TDN was in P1 (66.19%) and the lowest one was in P0 (51.38%). Average TDN of the highest raw material was in CT cv. Milgarra (77.59%) and the lowest was in *S. seabrana* (67.04%).

1. Introduction

Forage, which grows on natural meadows in Timor, is dominated by a type of native grass that has very low productivity and quality. One of the ways to increase the production and quality of natural grasslands is by introducing herbaceous legume, such as: (1) *Clitoria ternatea* Q5455 (CT Q5455); (2) *Clitoria ternatea* cv.Milgarra (CT cv. Milgarra); and (3) *Stylossanthes seabrana* (*S. seabrana*). Herbaceous legumes are selected because they have several conditions: (1)they are an annual plant capable of growing to the peak of the dry season; (2) they are able to provide forage as a source of protein during the dry season; and (3) they can improve soil fertility.

The results of Rubianti et al. (2009) showed that crude protein content (CP) of *C. ternatea* and *C. pascuorum* in the form of hay was 21.32% and 19.56%, respectively. This indicates that feeding of herbaceous legume in the form of hay will reduce protein deficiency, which always occurs during dry season and always results in changes of body weight of the livestock.

Basically, ruminants feed is divided into two parts, forage and concentrate. Forage feed is a class of feed ingredients that is a source of fiber for ruminants obtained in the form of grass, leguminous, and agricultural byproduct (Chuzaemi dan Hartutik, 1988). The important factor determining the success of ruminant rearing is the provision of forage. Fiber contained in forage is very useful in normalizing rumen function and can maintain milk fat content in dairy cattle

Feed containing herbaceous legume can improve the digestibility of dry matter and organic matter.Itis because tannins are not entirely binding proteins. There are still some proteins that can be used to the microbes activities in the rumen. Therefore, the increase of the microbes activities can improve the digestibility of dry matter and organic matter.

2. Material and Methods

The method used in this study was randomized block design with seven treatments, and rumen fluid taking was done three times as a group. Feed treatments were: (1) *CT Q5455*; (2) *CT cv.Milgarra*; (3) *S.seabrana*; (4) native grass; (5) native grass + *CT Q5455*; vi) native grass + *CT cv.Milgarra*; (6) native grass + *S. seabrana*.

The treatments applied in the research consisted of 50% grass and 50% herbal legume. The quality of the feed treatmentswas tested using proximate analysis and in vitro digestibility test. The in-vitro experiments used Vanille Meer's Tilley and Terry (1980) method. Samples were incubated for 96 hours then stopped by soaking syringe in ice waterthen centrifuged in a rate of 15,000 rpm and the residue was then filtered by Whatman 41 filter paper. For DM measurement, the residue was put into oven with temperature 105^0C, and then OM was measured by furnace with temperature 550^0C. The rumen fluid was extracted from the livestock having fistulae, swiftly filtered for fluid using muslin cloth that was trapped in two,and collected in a bottle, then inserted in an earlier flask filled with hot water. Rumen fluid was stored in thermos with temperature of 39^0C in order to be sent to the laboratory. McDougall's rumen and salivary liquids had a ratio of (1: 4). Both liquids had previously been mixed in a 1 liter Erlenmeyer tube bystirringit for 1 hourand flowed CO_2, pH 6.9 into it. Each polypropylene tube was filled with 50 ml of saliva mix made by McDougall and rumen fluid. The variables observed in this study were in vitro digestibility (DM, OM, CP). The data obtained were analyzed using various analysis, then tested using Duncan test if there was any difference among the feed treatments. The mathematical model was$Y_{ij} = \mu + \alpha_i + \beta_j + \epsilon_{ij}$. whereas$Y_{ij}$= Result, μ= median, α_i = treatment effect of herbaceous legume, β_j= Group effect-j, ϵ_{ij} = Experimental error treatment-i, group-j

3. Results and Discussion

The result of the proximate analysis can be seen in Table 1. The dry matter content was relatively the same among the four treatments. The average CP treatment content ranged from 11.74 to 12.86%, while treatment of P0 content of CP was 7.89%. This was consistent with higher content of ash compared to the other treatments. The ash content at P3 was lower than other treatments. This was because P3 was a treatment containing herbaceous legume type *S. seabrana* which has more ratio of stems than leaves. This condition resulted in higher crude fiber content in P3 than in other treatments.

The largest component of forage cell wall is composed of various structural carbohydrates depending on the type of forage, maturity, climate, fertilization, and other factors. Structural carbohydrate of plant is cell wall. This section is known as Neutral Detergent Fiber (NDF). NDF consists of cellulose, hemicellulose, lignin, and some pectins. Cellulose, hemicellulose, and pectin can be degraded by microbes, while lignin is difficult to degrade.

An Enzyme that plays a role in cellulose hydrolysis is called cellulase enzyme. Cellulase enzymes are produced by microbes in response to the presence of cellulose in their environment.

Van Soest (1967) stated that there was a correlation between cell wall digestion with cell wall content (in DM) in grass ($r = 0.46$) and legume ($r = 0.60$). NDF content is very influential in ruminant livestock's ability to consume the feed. Feed containing more than 50% of NDF will suppress the consumption rate of dry matter. Plant cell wall structure (NDF) containing lignin that varies among plants and their physiological age will affect the digestibility of rumen microbes. The lignin content that increases as the plant grows causes cellulose digestion to degrade, from potentially digestible cellulose fractions into lowdigestible cellulose fractions (Van Soest, 1994). The results of fiber component analysis on the four feed types used in the study are presented in Table 2.

Table. 1. Nutritional content of four types of feed treatment and herbaceous legume

Treatment	DM (%)	Nutrient (%DM)				
		OM (%)	Ash (%)	CP (%)	CF (%)	CF (%)
Legume						
CT Q5455	89.81	92.67	7.33	18.83	35.62	1.75
CT cv.Milgarra	91.96	87.49	12.51	17.49	32.59	2.16
S. seabrana	90.39	93.10	6.90	16.24	33.75	1.63
Feed						
P_0	89.67	85.44	14.56	7.89	32.60	1.60
P_1	92.06	89.41	10.59	11.74	31.19	2.65
P_2	87.87	89.66	10.34	12.86	33.64	2.38
P_3	88.95	94.54	5.48	11.18	37.26	1.55

Table 2. Results of Fiber component analysis of four types of feed treatment and herbaceous legume

Treatment	NDF (%)	ADF (%)	Hemi cellulose (%)	Cellulose (%)	Silica (%)	Lignin (%)
Legume						
CT Q5455	57.76	38.48	19.29	26.23	0.58	11.67
CT cv.Milgarra	57.44	33.50	23.94	22.16	0.67	10.67
S. seabrana	58.57	39.40	19.17	27.55	0.70	11.15
Feed						
P_0	58.94	44.04	14.90	30.74	0.92	12.38
P_1	47.54	36.42	11.11	26.89	0.02	9.52
P_2	48.88	38.33	10.56	28.46	0.12	9.74
P_3	48.45	38.80	9.65	28.71	0.13	9.96

Table 2 shows a four-feed fiber analysis of treatment, in which treatment P1, P2 and P3 had lower NDF, ADF, hemicellulose, and cellulose content than P0. This was consistent with its low lignin content. The higher NDF (cell wall) content of P0 was an indication of higher digestibility of the substances in the treatment of P1, P2 and P3.

Cellulose and hemicellulose are the fractions of structural carbohydrates which are the main components of plant cell walls. Fractions are often present in the form of binding to lignin making it difficult to digest by rumen microbes. Generally, forage in tropical areas has high levels of fiber content (NDF and ADF). SK content is the main factor causing low degradation of forage.

Fibre fraction contained in forage feed is not a problem in digestion process of ruminant livestock because, in rumen, there are microbes that can digest fiber,only if the fiber does not contain high lignin. Van Soest (1967) stated that lignin affects digestive process only if it is within the cell wall. This causes grass having low lignin content but high cell wall content to have lower degradation compared to legume having higher lignin but lower cell wall content.

The ability of microbes to degrade cellulose and other fiber fractions generates a source of energy because there are enzymes released during the fermentation process. Fiber fraction content will increase as the plant grows. The fraction varies among the species of forage which later will affect the amount of fibre that can be degraded on its nutrient content. The results of in vitro digestibility by DM, OM, CP in four types of treatment rations are presented in Table 3.

Table 3. Average digestibility of DM, OM, CP in feed treatment and herbaceous legume

Treatment	Nutrient Digestibility (%)			
	DM	OM	CP	TDN
Legume				
CT Q5455	70.96 ± 2.31^{c}	71.08 ± 3.05^{b}	67.25 ± 4.01^{bc}	74.63 ± 3.21^{b}
CT cv.Milgarra	73.49 ± 2.56^{c}	73.90 ± 2.94^{b}	61.25 ± 3.56^{bc}	77.59 ± 3.09^{b}
S. seabrana	63.90 ± 3.56^{c}	63.85 ± 5.78^{ab}	60.80 ± 3.49^{bc}	67.04 ± 6.07^{ab}
Feed				
P_0	41.69 ± 2.86^{a}	48.43 ± 3.96^{a}	38.65 ± 4.56^{a}	51.38 ± 4.16^{a}
P_1	60.35 ± 1.27^{bc}	63.04 ± 2.17^{ab}	59.64 ± 6.77^{b}	66.19 ± 2.28^{a}
P_2	54.36 ± 3.60^{ab}	56.27 ± 4.50^{ab}	58.78 ± 6.33^{b}	59.08 ± 5.14^{ab}
P_3	60.22 ± 6.37^{bc}	61.89 ± 6.83^{ab}	58.25 ± 4.38^{b}	64.99 ± 7.17^{ab}

Information:$^{a-c}$: Different superscripts in the same column show very significant differences (P <0.01)
Source: TDN: OM digestibility in vitro x 1.05 (Ibrahim, 1988)

The result showed that the treatment had significantly different effect (P <0.05) on the digestibility of in vitro DM of feed and was significantly different (P <0.01) from the digestibility of in vitro DM on leguminous type. The highest digestibility of In vitro DM feed was shown in P1 and P3. The highest in vitro DM digestibility was seen in CT cv. Milgarra and the lowest was in *S. seabrana*.

The treatment had significantly different effect (P <0.05) on in vitro digestibility of CP feed, and was significantly different (P <0.01) from in vitro digestibility of CP type of legume. The mean of in vitro digestibility of CP resulted in almost the same value in all feed treatments. In vitro digestibility CP was the highest legume type on CT Q5455.

The result showed that the treatment had significantly different effect (P <0.05) on the total digested nutrient (TDN) and was significantly different (P <0.01) from TDN type of leguminosa. The highest TDN feed rate was at P1 and the lowest was at P0. The highest TDN type of legume was found in *CT cv. Milgarra* and the lowest was in *S. seabrana*. At P1 treatment, feed had the highest dry matter digestibility, organic matter, crude protein, and total undigested nutrients. This was due to the low content of crude fiber in treatment P1 (Table 1). This indicates that the digestibility of a feed is affected by the crude fiber from the feed material.

According to Parakasi (1995), the higher the crude fiber of feed ingredients, the lower the digestibility. It is because the feed containing more lignin or silica content causes more energy loss through feces. Lubis (1963) stated that low crude fiber results in increasing gastrointestinal coefficients. This is due to the cell wall components of the plant (hemicellulose and lignin).

Furthermore Sutardi (1977) stated that the degradation of organic matter is closely related to thedegradation of dry matter because some dry materials consist of organic materials. Degradation of dry matter digestibility resulted in the decrease of organic matter digestibility and vice versa.

Feed containing herbaceous legume can improve the digestibility of dry matter and organic matterbecause it is assumed that tannins are not entirely binding proteins. There are still some other proteins that can be exploited in the rumen by microbes for their activity. With increasing activity of these microbes, it causes the increase of dry materials and digested organic material. In treatment P1, P2, and P3, protein digestibility was not different except for P0, this was because P1, P2, and P3 rumen microbes could degrade feed proteins fermentable in rumen, so that it can be used for microbial protein needs. In accordance with the opinion of Hubber and Kung (1981), feed protein degradation is determined by several things: (1) the ability of rumen microbes to degrade feed ingredients; (2) How long food proteins stay in the rumen; (3) protein source, (4) total consumption, (5) food particle size; and (6) microbial growth.

4. Conclusion

The feed treatment had a real effect on CP Digestibility. The mean of CP Digestibility of feed resulted in almost the same value in all feed treatments. The highest CP Digestibility value wasin *CT Q5455*. The feeding treatment had a significant effect (P5) on the total digested nutrient (TDN). The highest TDN feed rate was at P1 and the lowest was at P0. The highest TDN mean of the raw material was in *CT cv. Milgarra* and the lowest was in *S. seabrana*.

References

[1] Huber, J.T. and L. Kung, Jr. 1981. Protein and Non Protein Nitrogen Utilization in Dairy Cattle Science. J. Dairy Sci. 64:1170-1195.

[2] Ibrahim, M.N.M. 1988. Feeding Tables For Ruminants in Srilanka. First Edition. Kandy Offset Printers. Kandy.

[3] Lubis. D.A.1963. Animal Feed Science. Pembangunan Jakarta Press.

[4] Parakkasi, A., 1995. Nutrition Science and Ruminant Animal Feed, Universitas Indonesia Press, Jakarta

[5] Rubianti, A., P. Th. Fernandez., H.H.Marawali and E. Budisantoso. 2009. Digestibility of Dry Material and Hay Organic Material from *Clitoria ternatea* and *Centrosema pascuorum cv Cavalcade* on Weaned Balinese Cattle. Proceedings of National Seminar on AnimalHusbandry and Veterinary Technology, August 3-4, Bogor. Pg.177-181.

[6] Sutardi, T. 1977. Overview Ruminology. Forage Training for Animal Feed. Directorate General of Livestock – B.P.L.P.P. Agricultural Department, Jakarta

[7] Tilley, J.M.A. and R.A. Terry. 1963. A two stages tecnique for the in vitro digestion of forage crops. j. brit.grassland. Soc. 18:104-111.

[8] Van Soest, P.J. 1967. Development of a Comprehensive System of Feed Analysis and its Application to Forages. J. Anim. Sci. 26 : 119-128.

[9] Van Soest, P.J. 1994. Nutritional Ecology of The Ruminant. Second Edition. Comstock PublishingAssociates A Division of Cornell University Press. Ithaca and London.

[10] Van der Meer, J.M.1980. Determination of the In vitro Organic Matter Digestibility Coefficient of Feeds for Ruminant. Document Report no.67. IVVO. Lelystad

IOP Conf. Series: Earth and Environmental Science **119** (2018) 012013 doi:10.1088/1755-1315/119/1/012013

Calcium and Phosphor Status of Beef Cattle in Upland and Lowland of Jratunseluna River Basin in Central Java

Sutrisno[1], A. Subrata[1], Surahmanto[1], M. Christiyanto[1], Surono[1], J. Achmadi[1], F. Wahyono[1] and E. Pangestu[1]*

[1]Faculty of Animal and Agricultural Sciences, Diponegoro University, Semarang, Central Java, Indonesia.

*Corresponding E-mail:ekopangestu57@gmail.com

Abstract

The study was aimed to obtain information regarding feed given and mineralstatus (Ca, P) in fodder and beef cattle in Jratunseluna river basin. Feed and drinking water given by farmers identified for 14 days and extracted sampling for mineralanalysis, t-test was used to compare mineral status in upland and lowland. Results of the research showed that feed given by farmers were varying. The ratio of forage/concentrates in lowland and upland areas was different, i.e. 67: 33 and 30: 70, respectivelly. Ca content on forage given in upland areas ranged from 0.17 to 0.74%, and concentrates from 0.002 to 0.49%, while Ca content on forage given in lowland areas ranged from 0.33 to 0.52%, and concentrates ranged from 0.38 to 0.49%. P content on forage in upland areas ranged from 0.02 to 0.04%, concentrates ranged from 0.018 to 0.09%,while P content on forage in lowland areas ranged from 0.03 to 0.07%, and concentrates ranged from 0.04 to 0.07% . Ca and P consumption in upland areas were 301.06 and 54.73 g, and 391.92 and 65.70 g in lowland.Caand P content of beef cattle's hair in upland were 0.14 and 0.01%, while in lowland areas were 0.11 and 0.03%.It can be concluded that Ca and P intakeof beef cattle in Jratunseluna river basin were less and mineral status of Ca and P in marginal condition.

1. Introduction

Many river basin has not been handled properly, with the result the river basin areas already damage widespread and severe. Increasing number of population along with the increasing of development had resulted in the shifting pattern of land use, causing various problems, losing of fertile land, and soil contamination. Jratunseluna river basin is a river flow from Jrakah, Tuntang, Serang, Lusi, and Juwana, covers 12 administrative district area in Central Java Province[1]. Jratunseluna river basin area approximately cover 32,544.12 km^2 (the wide of Central Java around 3,25 million hectares). Livestock population in Central Java in year 2013 around 1,500,077 heads of beef cattle and 103,794 heads of dairy cow [2]. From livestock population above, 54% of beef cattle and 86% of dairy cow were raised in Jratunseluna river basin area. The carrying capacity of Jratunseluna river basin area only 32% of Central Java carrying capacity.The intensity of ruminant livestock, especially cows are supported by the availability of adequate feed. It is supposed to cause low productivity of ruminant livestock in that area[3].According to Jratunseluna geographic and the availability of inadequate feed will affect nutrient content, especially Ca and P of feed (grass and legume)given to livestock [4; 5]. The topography of Jratunseluna river basin area and the less vegetation that cover the land allow soil leaching which caused the less availability of nutrient or mineral for plant and causing diminish of the mineral on plant or grass, less mineral intake, and finally will decrease livestock productivity. It has been known that mineral is an essential nutrient on feed and metabolism process. Based on the problems above, study about mineral status (especially Ca and P) on beef cattle livestock in Jratunseluna area should be conducted.

2. Materials and Methods

The study was performed through a survey in Jratunseluna river basin area. The location was selected by stratified random sampling and represent beef cattle livestock condition. The location was selected based on Jratunseluna river basin topography and climate (micro climate), distinguished into upland and lowland area. On each upland and lowland area, selected 2 areas which had a lot of livestock population. The measured variable including total forage and feed given, mineral status (Ca and P) on land, water, feed, and livestock. Feed and water given by farmers were identified for 10 days and the sample was taken for Ca and P analysis. The land sample was taken in 4 areas around forage crops given to livestock for Ca and P analysis. The beef cattle's hair were taken after feed identification. Ca and P analysis was performed in Nutrition and Feed Science Laboratory, Faculty of Animal and Agricultural Faculty Science, Diponegoro University. Data obtained processed statistically using SPSS to determine mineral status of beef cattke, both in upland and lowland area. t-test was used to compare mineral status in upland and lowland area.

3. Results and Discussion

3.1. Composition and Mineral Content of Feed

Livestock in Jratunseluna river basin were given quite various feed, either fibrous feed like rice straw, corn straw, napier grass, *Arenga pinnata* waste, and native grass or other feed like tofu waste, wheat bran, rice bran, cassava waste, and concentrate feed (Table 1).

Table 1 showed that beef cattle in upland of Jratunseluna river basin got straw and grass as a feed with a huge portion compared to the concentrate, with the ratio of 67:33, while beef cattle in lowland of Jratunseluna river basin got concentrate much more than forage feed with forage, with the ratio of 29:71. The fibrous feed given to beef cattle appeared to be compatible with the potential of feed crops in the area, meaning that farmers used the feed crops in the form of straw given to beef cattle, as well as concentrated feed ingredients, utilizing the potential of agricultural products in each region. Tofu waste, rice bran, and *Arenga pinnata* waste often given to beef cattle in upland area, while cassava waste and rice bran as by-product of agricultural industry often given to beef cattle in lowland area. The interesting thing in lowland area of Jratunseluna river basin was the use of commercial concentrate, while in upland of Jratunseluna river basin, commercial concentrate never given to beef cattle. Concentrate feed ingredients provided were adjusted for availability in watershed area, with the exception of wheat bran. Field observations (Table 1) also showed that concentrate feed ingredients of protein source and mineral supplements were not given by farmers in both areas of Jratunseluna river basin.

Ca content in feed given to beef cattle in upland of Jratunseluna river basin for forage/fibrous feed ranged from 0.17 to 0.74%, concentrate ranged from 0.02 to 0.49%, while Ca conten of feed in lowland area for forage/fibrous feed ranged from 0.33 to 0.52%, and concentrate ranged from 0.38 to 0.49%. P content in feed in lowland area for forage/fibrous feed ranged from 0.03 to 0.07%, and concentrate ranged from 0.04 to 0.09%. P content in upland area for forage/fibrous feed ranged from 0.02 - 0.04% and in concentrate ranged from 0.02 to 0.09%. The data in Table 1 indicated that Ca content of feeds in upland were high compared to the requirement of beef cattle (napier grass, native grass, tofu waste, *Arenga pinnata* waste) and some were low (rice straw, rice bran, wheat bran).In lowland area, feedstuff with quite high Ca content were rice straw, napier grass, corn straw, cassava waste and commercial concentrate. P content in feedstuff and plants in upland and lowland areas were classified as 'border line' or low [7; 8]

Table 1. Composition and Content of Ca and P in Feed Consumed by Beef Cattle in Jratunseluna River Basin

1. Upland				2. Lowland			
3. Feed types	4. Proportion	5. Ca	6. P	7. Feed types	8. Proportion	9. Ca	10. P
11.	12. -------------(%) -------------			13.	14. -------------- (%) --------------		
15. Rice straw	16. 35	17. 0.19	18. 0.02	19. Rice straw	20. 26	21. 0.33	22. 0.03
23. Napier grass	24. 30	25. 0.49	26. 0.03	27. Corn straw	28. 2	29. 0.35	30. 0.03
31. Native grass	32. 2	33. 0.74	34. 0.04	35. Napier grass	36. 1	37. 0.52	38. 0.08
39. Tofu waste	40. 3	41. 0.49	42. 0.04	43. Cassava waste	44. 8	45. 0.49	46. 0.09
47. Wheat bran	48. 7	49. 0.04	50. 0.05	51. Concentrate	52. 39	53. 0.38	54. 0.04
55. Rice bran	56. 18	57. 0.002	58. 0.09	59. Rice bran	60. 24	61. 0.02	62. 0.07
63. Arenga pinnata waste	64. 5	65. 0.37	66. 0.02	67.	68.		
69. Forage/ Concentrate ratio	70.			72.	73.		
	71. 67:33				74. 29:71		

3.2. Consumption of Nutrients (dry matter, Ca, P)

Feed given to beef cattle in upland and lowland area of Jratunselunariver basin was insufficient to meet dry matter requirement of beef cattle (Table 2).

Ca consumption of beef cattle in upland and lowland area had already met the requirement of beef cattle (Table 2). Ca consumption of beef cattle in Jratunseluna river basin area was adequate, while dry matter consumption was less, because Ca content in feedstuff was high enough. The high proportion of feedstuff provision which contains high Ca caused Ca consumption also high. Ca consumption of beef cattle in upland area was different compared to downland area ($p<0.05$). The difference occured because the consumption of high Ca-containing materials in lowland area had more proportion and was derived from concentrate feed. Mean while, in upland area, it was derived from a fibrous feed (forage and straw).

P consumption of beef cattle in Jratunseluna river basin, both upland and lowland, had not met the requirement of beef cattle (Table 2). Most feedstuff given to beef cattle had a low P content, then the lack of dry matter consumption will cause P consumption did not meet the requirement of beef cattle.

3.3. Status of Ca and P in Beef Cattle's Hair

Calcium and P content in beef cattle's hair in Jratunseluna river basin were presented in Table 3. T-test results showed no difference in the status of the two minerals in beef cattle's hair between upland and lowland areas of Jratunseluna river basin. Ca content in beef cattle's hair was in marginal status if the content ranged from 0.1 to 1.85% [9] and $P<200$ ppm [10]. Thus, the status of Ca and P minerals of beef cattle in upland and lowland of Jratunseluna river basin areas were in defficient to marginal condition.

Ca consumption of beef cattle in both areas were sufficient, but the status of Ca in beef cattle's hair was in marginal condition. It was suspected that available Ca in feedstuff that can be absorbed bay beef cattle was low. Stated by [5], available Ca in forages feed in tropical regions are low, ranges from 40 to 50%. Feedstuff with low P content and low consumption of P would also certainly affect low status of P mineral in beef cattle (Table 2 and Table 3). The marginal status of Ca and P minerals in the body of beef cattle would certainly haad an effect on metabolic activity and performance of beef cattle.

Table 2. The Requirement and Consumption of Dry Matter and Mineral of Catle in Jratunseluna River Basin

75.	76. Upland	77. Lowland
78. Average body weight, kg	79. 361.0	80. 393.1
81. Requirement:	85.	89.
82. Dry matter, kg	86. 8.95	90. 9.13
83. Calcium (Ca), g	87. 17.7	91. 19.91
84. Phosphor (P), g	88. 14.91	92. 17.67
93. Consumption:	97.	101.
94. Dry matter, kg	98. 7.73	102. 8.89
95. Calcium (Ca), g	99. 20.1^{a}	103. 26.1^{b}
96. Phosphor (P), g	100. 4.36	104. 3.60

*) The different superscript in the column above shows the differences ($p < 0{,}05$)

Table 3. Status of Ca and P Minerals in Beef Cattle's Hair Cattle in Jratunseluna River Basin

105.	106. Upland	107. Lowland
108. Average body weight, kg	109. 361.0	110. 393.1
111. Calcium,%	112. 0.19	114. 0.11
Phosphor, %	113. 0.02	115. 0,01

4. Conclusion

It was concluded that nutrient intake of beef cattle in Jratunseluna river basin was less and mineral status of Ca and P was in marginal condition. Beef cattle in lowland area obtained better feed quality compared to beef cattle in upland area of Jratunseluna river basin.

References

[1] Departemen Permukiman dan Prasarana Wilayah. 2004. Proyek Induk Pengembangan Wilayah Sungai Jratunseluna. Semarang.

[2] Jawa Tengah In Figures 2015. BPS – Statistic of Central Java Province. Semarang

[3] Prawirodiputro, B.R. 2003. Sistem produksi hijauan pakan di lahan kering DAS Jratunseluna. JITV 8(3): 189-195.

[4] Rosero, O.R., R.E. Tucker, G.E. Mitchell Jr. And G.T. Schelling. 1980. Mineral utilization in sheep fed spring forages of different species, maturity and nitrogen fertility. J. Anim. Sci. **50**:128.

[5] McDowell, L.R., J.H. Conrad, G.L. Ellis, and J.K. Loosli.1983. Mineral of Grazing Ruminants in Tropical Region. Departement of Animal Science, Center for Tropical Agriculture, University of Florida. Gainsville.

[6] Pangestu , E. 1994. Suplementasi Mineral padaTernak Kambing di Bagian Hulu Daerah Aliran Sungai Serang. Fakultas PascaSarjana. Universitas Gadjah Mada. Yogyakarta. (Thesis).

[7] Kincaid, R. 1988. Macro Elements for Ruminant. In: D.C. Church (Ed.). The Ruminant Animal Digestive Physiology and Nutrition. A Reston Book. Prentice Hall. Englewood Cliffs. New Jersey.

[8] Underwood, E.J. and N.F. Suttle. 1999. The Mineral Nutrition of Livestock. 3rd Edition. CABI Publishing. New York.

[9] Puls, R. 1988. Mineral Levels in Animal Health Diagnostic Data. Published by Sherpa International.
Clearbook. British Columbia. Canada.

[10] Combs, D.K. 1987. Hair analysis as an indicator of minerals status of livestock. J. Anim. Sci. **65** : 1753.

Quality and Quantity of Sorghum Hydroponic Fodder from Different Varieties and Harvest Time

R. Chrisdiana[1]*

[1]The student of Master of Animal Sciences, Faculty of Animal and Agricultural Sciences, Diponegoro University, Semarang, Central Java Indonesia

*Corresponding E-mail: freya.rasheesa@gmail.com

Abstract

This experiment was designed to compare different varieties and harvest time of sorghum hydroponic fodder based on nutrient content and biomass production. Experimental design for fodder productivity was completely randomized design with 2 x 3 factorial, i.e., sorghum varieties (KD 4 and Super-1) and time of harvesting the sorghum hydroponic fodder (8,12 and 16 d). Total biomass and DM production, were affected significantly ($p<0.05$) on harvest time. Total biomass and nutrient content were increased in longer harvest time. The nutrient content were increased with decreasing total value of DM. Super-1 varieties produce larger biomass and nutrient content higher than KD4 ($p<0.05$). Based on sorghum hidroponic fodder quality and quantity, sorghum hidroponic fodder with Super-1 varieties harvested at 12 d had a good quality of fodder and it can be alternative of technology providing quality forage and land saving with a short time planting period and continous production.

1. Introduction

Forage is the main source of ruminant feed to supply the maintenance, production and reproduction requirements of livestock. Therefore, methods and technologies were required to improved productivity of forage like hydroponic system with continuously production without season-influenced and can be applied without requiring extensive land. Forage development technology that can be applied is hydroponic fodder system.

Hydroponic fodder system was hydroponics forage production with a short time harvest by using cereal plants. Moisture and nutrients were needed to enable germination and growth of plants. Germination process will produce green shoots and root-shaped mats to be harvested as animal feed [5]. The basic principle of a hydroponic fodder system is that cereal plants respond to water supplies and nutrients to germinate, grow, and then produce green plants in short time. The development of this system allows the production of fresh forages from grains produced with high yields of 15 to 30 cm with a production rate of 7 to 9 kg of fresh forage (depending on grain varieties) with 0.9-1.1 kg of dry matter of 1 kg of seeds within 6 to 8 days [7].

Various cereals can be used as animal feed, including barley, wheat, corn, and sorghum. Sorghum is an adaptive and suitable food crop developed in the tropics and efficient in producing high photosynthetic products. The optimum temperature for the growth of sorghum plants ranges from 21-35°C. Sorghum has advantages, including drought tolerance, high salt content, and extensive adaptability, and easier maintenance than sugar cane and other cereals. Sorghum has a greater biomass than maize [8]. Among the species of sorghum that can be developed as animal feed is Sorghum bicolor KD4 and Super-1 varieties.

The process of growing plants hydroponically will cause the chemical and structural changes that take place in the cereal grains through the germination process. Activation of enzymes in seeds during germination causes the hydrolysis of proteins, carbohydrates and fats to be simple components [6]. This hydrolysis increases the concentration of amino acids, soluble sugars and fatty acids in grains and produces sprout [4]. The growth or germination of the grains will affect the nutrient content by increasing fiber, fat, protein, but decreasing starch and total dry matter [7].

This study was designed to compare different varieties and harvest time of sorghum hydroponic fodder based on nutrient content and biomass production. The results of this study are expected to provide information about the technology of providing quality forage and land saving with a short time planting period.

2. Materials and Methods

2.1. Plants productivity Experiments

Six hundreds grams of Sorghum seeds were weighed and spread in 1 tray of 16 trays according to the treatment, then sorghum seeds soaked 20% sodium hypocloride solution 1 mL for 1 L of water for 20 minutes. The sorghum seed was washed again, then soaked with water for 24 hours. The water was drained and the seeds were spread on the tray. Sorghum seeds were sprayed two times a day until harvest time. Three replications of the experiments were observed for plant productivity including biomass, dry mater, crude protein production and subsequent proximate analysis. Harvested sorghum hydroponic fodder was dried in an oven at 60°C for 48 hours then ground for proximate analysis including Dry mater (DM), ash, crude protein (CP), extract ether (EE), crude fiber (CF), neutral detergent fiber (NDF), and acid detergent fiber (ADF). Proximate analysis using SNI (1995) method for DM, CF and Ash, AOAC (2005) method for CP, EE, ADF, and NDF.

2.2. Statistical Analysis

Experimental design for fodder productivity was completely randomized design with 2 x 3 factorial, i.e., sorghum varieties (KD4 and Super-1) and time of harvesting the sorghum hydroponic fodder (8,12 and 16 d). Data obtained were analyzed by using analysis of variance (ANOVA) and followed by Duncan's multiple range test.

3. Results and Discussion

The nutrient compositions of sorghum hydroponic fodder are presented in Table 1. Dry matter of the plant was reduced with a longer time of harvest, while CP, ash, NDF, and ADF were increased. In barley plant fodder, dry matter of the seeds is significantly reduced with increasing periods of 6-7 days. While CP, EE, ash, NDF, and ADF increased [7]. In hydroponic fodder plants, soaking seeds will increase the activity of enzymes that break down the seeds into simpler fractions such as starch into sugars, proteins into amino acids and fat into free fatty acids, while dry matter and total energy will decrease [10]. In a barley hydroponic, germination of barley plants have 18% loss in DM. CP, Ash, fiber content, NDF, and ADF increased in a 8 days of sprouting [9].

It has been known that loss of dry matter and total energy may be caused by oxidation of plants. During soaking and germination, the seeds lose dry matter because they use their own energy reserves for growth. Sprouts can regain their dry matter with effective mineral absorption and photosynthesis [10]. The lower dry matter value give the higher CP content. The increase in CP is probably due to the loss of carbohydrates in the dry matter because at the time of planting no source of nitrogen is externally added to the water used for irrigation [6].

The increase in NDF and ADF values occurs because of the synthesis and accumulation of lignin which usually occurs during the formation and thickening of secondary cell walls, therefore the rate of maturation and irrigation will affect the accumulation of lignin [3]. Furthermore, Dung et al [6] (2012) explains that the fiber content is mostly derived from root and seed husk due to the maturation process and endosperm formation. With the increasing value of ADF and NDF in plants, the organic matter content will decrease.

Table 1. Nutrient Composition of Sorghum Hydroponic Fodder

Composition	KD4			Super-1		
	8d	12d	16d	8d	12d	16d
	------------------------------			% ------------------------------		
DM	27.04	22.66	15.68	25.36	14.55	15.94
CP	13.94	18.04	18.68	17.27	26.94	26.68
EE	4.04	4.05	5.80	4.13	5.55	5.60
CF	15.17	16.06	19.66	25.55	31.62	36.57
Ash	2.25	3.25	3.69	1.56	2.10	2.41
NDF	22.57	35.70	39.04	23.18	32.68	33.41
ADF	8.99	10.9	15.36	9.96	17.65	18.26

Biomass Production of sorghum hydroponic fodder was significantly different ($p<0.05$) on harvest time and varieties (Table 2). Longer harvested time will produce higher biomass. At 16 d harvest time, biomass production tends to remain even decreased in Super-1 varieties. This is because plant growth depends on the availability of nutrients in the seeds. Without additional nutrients the plant will last only less than 16 days. As mentioned by Andriani and Isnaini [2], sorghum which grown with soil medium with age between 3-10 days after planting, its growth depends on nutrition and food reserves from seed. Shoots (coleoptile) appear on the soil after 3-4 days after planting, while the secondary roots will begin to growth 3-7 days after sprout.

Super-1 varieties produce higher biomass than KD4. Biomass production depend on plant species, varieties, and environmental conditions, 1 kg of grain will produce 15-25 kg of fodder with planting period of 8–10d [1]. The results and quality of the fodder are also influenced by many factors such as soaking time, seed quality, seed type and treatment, temperature, moisture, nutrient supply, and seed density [4]. Starch in cereal plants is a raw material that supports plant growth. Water Imbibition during the soaking process will activate enzymes in the seeds that hydrolyse starch into sugars that available for embryonic growth. Under conditions of optimum humidity, oxygen and warmth, the sugar will be used for cell wall synthesis and to provide energy for growth [5].

DM production of sorghum hydroponic fodder was significantly different ($p<0.05$) on harvest time but did not differ significantly in plant varieties (Table 2). This result is inverse if compare with biomass production. In this study, longer harvested time, biomass production plants will increase while dry matter production will decrease. Biomass production per kg of plants increased during the 6-7 day growing period caused by water absorption during seed germination resulting in a different percentage of DM [7]. Based on varieties, dry matter production is not significantly different between Super-1 and KD4. Yield and quality of the sprouts produced are influenced by many factors such as soaking time, seed quality, seed type and treatment, temperature, humidity, nutrient supply, and seed density to achieve maximum results [4].

Table 2. Biomass, DM and CP Production of Sorghum Hydroponic Fodder

Variables	Varieties	harvested time (d)			mean of sorghum varieties
		8	12	16	
Biomass (kg/m^2)	KD4	25.511	28.373	30.141	28.008^b
	Super-1	27.948	33.530	32.753	31.410^a
	mean of harvest time	26.730^b	30.951^a	31.447^a	
DM (kg/m^2)	KD4	6.893	6.439	4.725	6.019
	Super-1	7.070	4.877	5.229	5.974
	mean of harvest time	6.981^a	5.658^b	4.977^b	
CP (kg/m^2)	KD4	961	1.162	883	1.002^b
	Super-1	1.221	1.314	1.395	1.267^a
	mean of harvest time	1.091	1.238	1.139	

^{a,b}, means with different superscripts differ significantly (P<0.05).

In hydroponic fodder plants, soaking seeds will increase the activity of enzyme that breaks the seed into a simpler fraction. Dry matter and total energy will decrease due to plant oxidation [10]. When water imbibition occurs in the seed embryo, plant hormone (giberrelic acid) along with water will go to the aleuron layer. In the aleuron layer there will be DNA activation that converts amino acids into amylase enzymes. The amylase then goes to the endosperm layer to then catalyse the hydrolysis of starch into the glucose component. Glucose is then used for metabolic activity of newly grown plants. This oxidation will produce ATP and release CO_2 [6].

Crude protein production of fodder sorghum hydroponic value was significantly different ($p<0.05$) on plant varieties and did not differ significantly on harvest time. This explains that the increased CP percentage is proportional with DM percentage decrease, so the CP content does not actually occur in kilogram. Super-1 varieties CP (kg) is higher than KD4 because the content of CP seed in Super-1 varieties is higher than KD4. This result is consistent with statement that the percentage of proteins gives apparent increase as basically the total weight of the protein remains the same according to the dry matter [10]. The increase in the crude protein content derived from nitrate absorption, thus facilitating the metabolism of nitrogen compounds derived from carbohydrate reserves (Morgan et al., 1992) as cited by Saidi [9] (2015).

4. Conclusion

Based on sorghum hydroponic fodder quality and quantity studies, sorghum hydroponic fodder with Super-1 varieties harvested at 12 d had a good quality of fodder and it can be alternative of technology providing quality forage and land saving with a short time planting period and continuous production.

Acknowledgment

A sincere thanks was expressed to Animal Husbandry Training Center Batu for facilitating authors to conduct the experiment.

References

[1] Ghazi, N., Al-Karaki and M. Al-Hashimi. 2011. Greem Fodder Production and Water Use Efficiency of Some Forage Crops Under Hydroponic Conditions. ISRN Agronomy **1**. 15

[2] Andriani, A. and Isnaini M. 2012. Morfologi dan Fase Pertumbuhan Sorgum. Cereals Plant Research Institute.

[3] Carmi A, Aharoni Y, Edelstein M, Umiel N, Hagiladi A, Yosef E, Nikbachat M, Zenou A, & Miron J. 2006. Effects of Irrigation and Plant Density on Yield, Composition and In Vitro Digestibility of a New Forage Sorghum Variety, Tal at Two Maturity Stages. J. Anim. Feed Sci. Tech.**131**:121

[4] Chavan J and Kadam SS 1989 Nutritional Improvement of Cereal by Sprouting. Critical Rev. Food Sci. Nut. **28** (5) : 401.

[5] Cuddeford D. 1989. Hydroponic Grass In Practice **11**(5) 2

[6] Dung DD, Goodwin I R, Nolan JV 2010 Nutrient Content and *in sacco* Digestibility of Barley Grain and Sprouted Barley J. Anim. Vet. Advances **9** (19) 2485

[7] Fazaeli H, Golmohammadi HA,Tabatabayee SN, Tabrizi MA 2012 Productivity and Nutritive Value of Barley Green Fodder Yield in Hydroponic System. J World Appl Sci. **16** (4): 531.

[8] Rocateli AC, Raper RL, Balkcom KS, Arriaga FJ, Bransby DI 2012Biomass sorghum production and components under different irrigation/tillage systems for the southeastern US. Industrial Crops and Products **36** : 589.

[9] Saidi ARMA and Abo Omar J 2015 The Biological and Economical Feasibility of Feeding Barley Green Fodder to Lactating Awassi Ewes J. Anim. Sci., **5**, 99.

[10] Sneath R and McIntosh F 2003 Review of Hydroponic Fodder Production For Beef Cattle. Department of Primary Industries: Queensland Australia 84. McKeehen, pp: 54.

Development of Botanical Composition in Maribaya Pasture, Brebes, Central Java

N. Umami[1*], N. Ngadiyono [1], Panjono[1], N. Agus F[1], M. Shirothul H[1], I.G.S. Budisatria[1], Y. Hendrawati[2], I. Subroto[2]

[1]Faculty of Animal Science, Universitas Gadjah Mada, Yogyakarta
[2]Agency of Animal Science and Health of Brebes Regency, Brebes.

*Corresponding E-mail : nafiatul.umami@ugm.ac.id

Abstract

The research was aimed to observe the development of botanical composition in Maribaya pastures. The sampling method was cluster random sampling. The observed variables were the type of forages and the botanical composition in the pasture. Botanical composition was measured by using Line Intercept method and the production was measured by the estimation of botany production for each square meter using its dry matter measurement. The botani sampling was performed using square with size of 1x1 m^2. The observation was performed before the pasture made (at 2015) and after the pasture made (at 2017). Based on the research result, it was found that there was significant difference between the forage type in the pasture at 2015 and at 2017. It happens due to the adjustment for the Jabres cattle feed.

1. Introduction

Livestock activities are important part in supporting the implementation of integrated agricultural system in Indonesia. Among farmers, especially rural farmers, beef cattle is an important asset, cannot be separated from the agricultural system. One of the problems that often occur in the maintenance pattern of traditional beef cattle is the limited land for forage feed. Commonly, used forages come from the processed (productive) and unprocessed (unproductive) land. Meanwhile, most of the land is in the not productive land for plants. This phenomena happens in Maribaya village, Bumiayu District, Brebes Regency, Central Java.

Some breeders in the Maribaya region, Bumiayu Subdistrict, Brebes District, Central Java get forage for ruminants from farming areas as well as from agricultural remains and also in the watershed. The area around the river flow is the main place for the availability of forage sources of feed. For example is in the Yellow River.This area is the main place for farmers to take forage feed, but this area is also one of the forage sites that has been revegetated earlier. Therefore, the measurement of forage production and its quality is required to take a strategical step in utilizing the forage in order to increase the livestock population in the area. Observation of plants that can grow well in terms of quality and quantity is needed to know the development of pastures. Maribaya grazing paddock is a new pasture of grazing in Java island. It requires observation and maintenance with proper management. For this reason research about the development of Maribaya grazing is needed to make scientific studies related to the future development.

In order to create a standardized pasture, it is necessary to adjust the botanical composition that lives in the area. Since 2015, some efforts has been made to optimize the certain botanical composition as a source of forage feed. This study was conducted to evaluate the the botanical composition optimization results in Maribaya Pasture. The number of botanical compositions was observed before optimization (in 2015) and after optimization (in 2017).

2. Materials and Methods

This research was conducted in Maribaya village, Bumiayu District, Brebes Regency, Central Java. This research used 1x1 m square, lux meter, sickle, trash bag, sack, meter, hoe, sickle, scissor, wooden pin, rope, masking tape, newsprint and stationery.

The measurements of forage condition and other staple crops were performed using Line Intercept method. Observations on the number of botanical compositions were performed before optimization (in 2015) and after optimization (2017).

3. Results and Discussion

Plant species is one of the factors affecting plant growth. Forage feed physiologically has grown properties in specific areas corresponding to each plant. These properties must be considered in every forage planting and should be adapted to the quality of the forage adaptation which will ultimately impact the success rate of the forage development effort. The factors about the management of grazing areas should also be considered in the order to develop the herbs of feed in addition to growth factors. One of the core activities is the management of forages. Technical factors that need to be considered in the business of feed forage management in grazing area are soil fertility, physiological condition of forage feed which become marker if nutrient deficiency is needed, and the existence of water (Hasan, 2012).

Pastureland should be developed with a dry-resistant plant species, consisting of grasses and legumes. The botanical composition of Maribaya Pasture was shown in Table 1. In the beginning of Maribaya Pasture development, the botanical composition was consist of many grasses and legumes, either usefull or useless for livestock feed. The optimization of botanical composition was performed for 2 years in order to eliminate the useless plant and maximize the usefull plant growth.

Table 1. Botanical Composition in 2015 and 2017 of Maribaya Pasture, Brebes Regency

No.	Botani Type	Occurence in 2015	Occurence in 2017
1.	Elephant Grass	√	√
2.	King Grass	√	√
3.	Setaria sphacelata	√	√
4.	Alang-Alang	√	√
5.	Wangi Root	√	√
6.	Granjangan	√	√
7.	Carpet Grass	√	√
8.	Signal Grass		√
9.	Lamtoro		√
10.	Glirisidia		√
11.	Imperata		√
12.	Sentro		√

The types of grasses in Maribaya Pasture at 2015 are Elephant Grass, Grass King, Setaria sphacelata, Alang-Alang, Wangi Root, Granjangan, Grass Field, while the existing legume type was leguminous tree species and shrubs, legume *Gliricidea maculata* or *Leucaena leucocephala, Euchresta Horsfieldii* Benn, and *Gliricidia sepium*. And for the bush legume there are*stylosanthes, clitoria,* and *Mimosa pudica*.

Based on the identification results in the Maribaya Pasture at 2017, there is an increase in the number of grass species present in the area. It happens because there are some optimization effort. This area will be used for grazin. Some types of forage was found in the pasture.They are Grass Signal (*Brachiaria decumbens*), Lamtoro (*Leucaena leucocephala* L.), Local grass or field grass (*Axonopus compressus*), Glirisidia (*Gliricidia maculata*), Imperata (*Imperata cylindrica* (L.) Raeusch, and Sentro (*Centrosema pubescens*). Each type of forage has different appearance and characteristics.

Grass Signal (*Brachiaria decumbens*). The characteristic of this forage are long-spreading, longevity, height between 80 cm to 150 cm, leafy, short, dark green leaf texture, and grows well in rainfall conditions 1200 to 1750 mm/year.It has function as pasture grass and cut grass (Figure 1).

Figure 1. Grass Signal (*Brachiaria decumbens*)

Lamtoro (*Leucaena leucocephala* L.). This forage came from central america The characteristics of this forage are 2 to 10 meters tall, deep-rooted, has a double-pinnate leaf, dark gray-green leaf color, grows well in rainfall conditions ranging from 700 to1650 mm / year and altitude of place between 700 up to 1200 meter above sea surface. It can be protein sources for livestock and also can be used as fertilizer plants (Figure 2).

Figure 2. Lamtoro (*Leucaena leucocephala* L.)

Local grass or field grass (*Axonopus compressus*). It has shallow rooting characteristics, roots out of stolon books, long stolons and branches in all directions, leafy strands of the leaves, downy feathered and stemmed. The length of the leaves reaches 30 cm, while the width is 1.5 cm and has a compound interest in the form of grains (Figure 3).

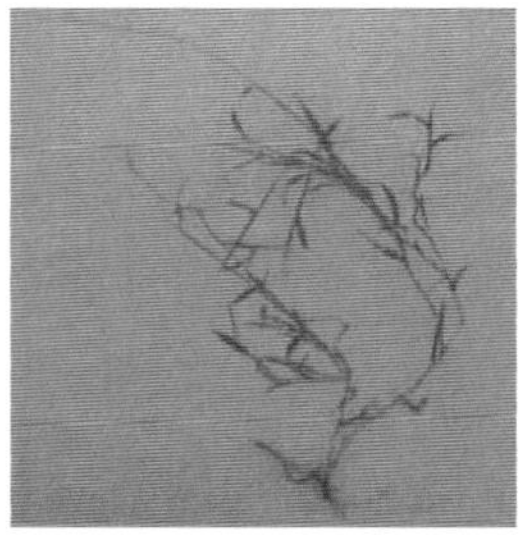

Figure 3. Carpet grass (Axonopus compressus)

Gliricidia (*Gliricidia maculata*). It has light brown or whitish brown trunks, stems grown, stem branches often come out at the bottom of the stem, the length of the petiole is 15 to 40 cm, a leaf stalk has 7 to 17 strands leaves with size 1x3 cm to 3x6 cm. The flowers are pale pink, medium size and the upper surface of the leaves are bright green, while the lower surface is pale green (Figure 4).

Figure 4. Glirisidia (*Gliricidia maculata*)

Imperata cylindrica (L.) Raeusch).It classified as annual grasses that favor sunlight. It has flammable parts on the ground and roots of rhizomes or rhizomes that spread widely below the soil surface. The breeding of this forage is through seeds and roots rhizomes and have allelopathy that can inhibit other nearby plants (Figure 5).

Figure 5. Alang-alang (Imperata cylindrica (L.) Raeusch)

Sentros (*Centrosema pubescens*). It has characteristics of growing, climbing or twisting, rather hairy and non-woody stems and adaptable to less fertile soils, acidic pH, inundated water or ugly drainage (Figure 6)

Figure 6. Sentro (Centrosema pubescens)

4. Conclusion

Based on the results of this study, it can be concluded that Maribaya Pasture has the potential of forage feed as a source of protein and fiber. There is a increase in the number of grass species that present in the area at 2015 and 2017.In 2017, the forages are Grass Signal (*Brachiaria decumbens*), Lamtoro (*Leucaena leucocephala* L.), Local Grass or Field Grass (*Axonopus compressus*), Glirisidia (*Gliricidia maculata),* Imperata (*Imperata cylindrica* (L.) Raeusch), and Sentro (Centrosema pubescens). Each type of forage has different appearance and characteristics. It means that the introduction of high quality forage in Maribaya Pasture is success.Further research is needed to optimize the potential of Maribaya Pasture.

Acknowledgment

The author gives thanks to Ministry of Research and Higher Education of Indonesian Republic which has fully funded this research through Universitas Gadjah Mada, Yogyakarta, in the form of *Penelitian Unggulan Perguruan Tinggi Grant* with contract number of 2470/UN1.P.III/DIT-LIT/LT/2017.

Reference

[1] Reksohadiprodjo, S. 1994. Produksi Hijauan Makanan Ternak. Badan Penerbitan Fakultas Ekonomi (BPFE) Universitas Gajah Mada, Yogyakarta.

[2] Sawen, D. dan M. Junaidi. 2011. Potensi Padang Penggembalaan Alam Pada Dua Kabupaten Di Provinsi Papua Barat. Seminar Nasional Teknologi Peternakan dan Veteriner.

[3] Susetyo, S. 1980. Padang Penggembalaan. Departemen Ilmu Makanan Ternak Fakultas Peternakan Institut Pertanian Bogor, Bogor.

[4] Williamson, G. dan W. J. A.Payne.1993. Pengantar Peternakan di Daerah Tropis.Gadjah Mada University Press.Yogyakarta.

In Vitro Ruminal Degradability of Soybean Meal Protein Protected with Natural Tannin

B.W.H.E Prasetiyono[1], A. Subrata[1], B.I.M. Tampoebolon[1], Surono[1] and Widiyanto[1]

[1]Faculty of Animal and Agricultural Sciences, Diponegoro University, Semarang, Central Java, Indonesia.

Corresponding E-mail: bambangwhep@ymail.com

Abstract

The influence of tannin from tea waste and *gambier* as natural tannin sources on ruminal protein degradability was studied in this investigation. The soybean meal was used as protein source in this investigation. There were three treatments in this investigation mainly without protection (NT); protection with tea waste (Tt); and protection with *gambier* (Tg). The measured parameters consisted of in vitro dry matter digestibility (IVDMD), in vitro organic matter digestibility (IVOMD), and the ruminal fermentation characteristics. Results of this experiment showed that protection with tannin from tea waste as well as *gambier* increased ($p<0.05$) the IVDMD (77.0 vs 85.10 and 86.20%) and IVOMD (75.70 vs 83.40 and 84.40%). The highest IVDMD and IVOMD ($p<0.05$) was found in Tg treatment group. The Ruminal Undegradable Protein (RUP) in Tt and Tg group was higher than that in NT group (66.29 and 69.20 vs 51.10%). The ruminal protozoa population decreased ($p<0.05$) as a result of tannin protection (6690 vs 3790 and 5480 microorganism/ml rumen fluid). The Tt treatment group showed the lowest ($p<0.05$) ruminal protozoa population. The ammonia concentration in Tt and Tg tend to be lower compared to NT (4.67 and 4.69 vs 5.00 mM). The soybean meal protein protection with tea waste and *gambier* tannin increased IVDMD, IVOMD, RUP. The ruminal protozoa population and ammonia concentration, on the other hand, were decreased by tannin protection from those tannin natural sources. The natural tannin from *gambier* was the most effective protection agent for soybean meal protein.

1. Introduction

Various protein supplementation techniques in ruminant ration formulations have been tested, but it was still necessary to improve efficiency, both in terms of process and increase of added value to meet high-quality proteins that can be absorbed in the small intestine. Selection of feed ingredients as a source of protein supplements is very necessary and important, especially related to material quality, procurement sustainability, and economic value. Feed ingredients of protein sources are still relatively expensive, so it is necessary to improve the efficiency of nutrients that have high biological value to increase the production of ruminant livestock. Soybean meals have a good biological value, and the crude protein with the content of about 41% can be used as a protein supplement. This is, however, still a high level of degradation in the rumen, which is about 50-80% [1]. Therefore, protection techniques were needed to decrease soybean meal protein degradation in the rumen [2], so the percentage of soybean meal protein intake and digestibility in the small intestine can be increased. Protection techniques in soybean meal protein with a natural materials have not been carried out, so that the research has been tested in vitro, via utilization of natural tannin extracted from tea waste of a beverage company of "PT Sosro Indonesia" and *gambier* (***Uncaria*** *gambir* ***Roxb***). This study aimed to evaluate the ruminal protein degradability of soybean meal protein products protected with natural tannins from tea waste or *gambier*.

2. Materials and Methods

2.1. Preparation of natural tannin extract and treatments

The tannin extraction from tea waste and *gambier* were studied by immersing them in a container using 95% alcohol [3,4] with a ratio between the ingredients extracted (tea waste and *gambier*) and alcohol ie 1: 3 for 12 hours. The obtained extract was filtered using a clean cloth, then evaporated until the extract solution was more concentrated and then dried in an oven at 40^0C to obtain tannin crystals. The concentration of condensed tannins was determined by using the drying extract oven. The condensed tannin concentration was calculated by the following equations:

(1) Tannin total = (weight of extract (crystal tannin) / sample weight) x 100%
(2) Tannin condensation to total tannin = (weight of sediment / weight of extract (tannin extract)) x 100%
(3) Tannin condensation= tannin total (%) x tannin condensation to total tannin (%)

The tannin concentration obtained from the extraction was used for soybean meal protection. The soybean meal protection was conducted by means of tannins from tea waste or *gambier* with 0.4% each (equivalent to 2.2 mg tannin / 0.55 g soybean meal) mixed on soybean meal. Then tannin extract was sprayed on soybean meal by using sprayer. The samples of soybean meal were divided into 3 treatment groups, namely Nt= the soybean meal without protection; Tt= protected soybean meal with tea waste; and Tg= protected soybean meal with *gambier*.

2.2. Chemical analyses

The rumen fermentation characteristic parameters were tested by in vitro fermentation technique according to Tilley and Terry [5]. The sources of inoculum were taken from the goat rumen that was cut in the slaughter house of Semarang, Central Java, Indonesia. The rumen fluid of the goat was taken and filtered and then inserted into the thermos, to be immediately taken to the laboratory for analysis.

A 3-g sample was put into the reaction tube and then 24 ml of inoculum liquid and 36 ml of McDougal buffer solution were added. Before being used McDougall solution was saturated with CO_2 so that the pH ranges from 6.8-6.9. Before being fermented the fermenter contents are also fed with CO_2 to create an anaerobic condition and then the fermenter is sealed with plastic rubber. Next they are incubated into shaker bath at temperature of 40^0C. The fermentation process is stopped with saturated HgCl2. The samples of the fermented rumen fluid were then analyzed for VFA, NH3, Protein total, Rumen Undegradable Protein (RUP) and Microbial Protein Synthesis [5]. In vitro dry matter digestibility (IVDMD) and In vitro organic matter digestibility (IVOMD) were also analized according to Tilley and Terry [5]. Ruminal protozoa population were calculated with method of Kurniawati and Supadmo [6]. Methan production was calculated according to Menke and Steingass [7]

2.3. Statistical analysis

The data were examined by Analyzed of Varians (ANOVA). The means differences between treatment were analyzed by Duncan's Multiple Range Tests with the general linear model procedures of SAS [8].

3. Results and Discussion

Several parameters were analyzed to show the influence of protein protection by tannin from tea waste and *gambier* on feed protein utility. Those parameters included the dry matter digestibility, and organic matter digestibility (IVDMD and IVOMD), ruminal ammonia concentration, VFA of rumen fluid, total protein production, Ruminal undegradable protein, protozoa population and microbial protein synthesis (Table 1).

Table 1. Characteristics of rumen fluid fermentation with respect to soybean meal protein protected with tannin from tea waste and *gambier*.

Parameters	Treatments			SEM	Significance
	NT	Tt	Tg		
VFA (mM)	110.01^{a}	91.70^{b}	93.30^{b}	3.64	*
Proportion of VFA (%):					
Acetate (C2)	68.10	66.60	67.00	0.29	NS
Propionate (C3)	19.10	19.59	18.90	0.07	NS
Butyrate (C4)	12.70	13.82	14.10	0.27	NS
C2:C3 ratio	3.59	3.40	3.49	0.02	NS
NH3 (mM)	5.00	4.67	4.69	0.05	NS
Protein Total (g)	86.00^{c}	123.10^{b}	147.50^{a}	1.66	*
RUP (%)	51.10^{b}	66.29^{a}	69.20^{a}	1.87	*
IVDMD (%)	77.00^{c}	85.10^{b}	86.20^{a}	0.14	*
IVOMD (%)	75.70^{c}	83.40^{b}	84.40^{a}	0.02	*
Microbial Protein Synthesis (mg/ml)	0.30	0.20	0.24	0.04	NS
Protozoa pupulation (microorganism/ml rumen fluid	6690^{a}	3790^{c}	5480^{b}	200.68	*
Methane (ml/ mg)	4.80^{a}	3.90^{b}	4.10^{b}	0.12	*

* $p<0.05$.
NS = non significant.
NT= Non Tannin.
Tt= Tannin from tea waste.
Tg= Tannin from *gambier*.
SEM = Standard Error of Means.

3.1. *In vitro ruminal fermentability*

In vitro dry matter and organic matter digestibility increased ($p<0.05$) due to protein protection with tannin from tea waste as well as *gambier*. The increasing of digestibility resulted from protein protection with tannin from *gambier* was higher than that resulted from protein protection with tannin from tea waste. The IVDMD and IVOMD in without protection treatment group, protection with tea waste tannin and protection with *gambier* tannin were 77.00 and 75.70; 85.10 and 83.40; 86.20 and 84.40, respectively. The increasing of IVDMD as well as IVOMD resulted from protein protection with tannin might occur because of the increasing of rumen bacteria population along with the decreasing ($p<0.05$) of protozoa population in tannin treatment groups (Tt and Tg) compared to Nt (6690 vs 3790 and 5480 microorganism/ml rumen fluid). The decreasing of protozoa population reduced the rumen bacteria predation by protozoa [9, 10] so that rumen bacteria population increased, including *cellulolytic* bacteria.

Fermentation constrain in soybean meal resulted from cell wall and seed shell especially the one containing fiber. The increase of bacteria population in the rumen would increase rumen microbial capacity to degrade the fiber, so that the IVDMD as well as IVOMD increased. The primary product of organic matter of soybean carbohydrate in the rumen was volatile fatty acids (VFA). The interesting phenomenon found was the significant decrease ($p<0.05$) of rumen VFA production in tannin treatment groups (Tt and Tg) compared to Nt (110.01 vs 91.70 and 93.30 mM) although its IVDMD and IVOMD increased compared to Nt. That case could occur because of the increasing of carbon skeleton use mainly an alpha keto acid (intermediary compound in VFA production) in Tt and Tg for microbial protein biosynthesis. That phenomenon was reflected in the increase ($p<0.05$) of total protein production in Tt and Tg treatment groups compared to Nt (123.10 and 147.50 vs 86.00 mg/g).

3.2. Energy efficiency of ruminal metabolism

Soybean protein protection by tannin from tea waste as well as *gambier* increased the ruminal energy metabolism efficiency. That case reflected on the decrease ($p<0.05$) of methane production with resulted from protein protection with tannin. The methane production in Tt and Tg treatment groups were lower ($p<0.05$) than that in Nt group (3.90 and 4.10 vs 4.80ml/mg) in (Tt and Tg vs Nt), respectively. The decrease of methane production in tannin protection treatment groups could occur due to using of hydrogen and de novo fatty acids biosynthesis which reflected in the tendency of butyric acids increase in Tt and Tg treatment groups compared to Nt treatment groups. This experiment result was in line with previous finding that tannin could decrease the gas production [11,12,13].

3.3. Efficiency of ruminal protein metabolism

The increase of IVOMD as result of soybean protein protection with tannin was not accompanied with the increase of protein degradability because of the decrease of soybean meal protein solubility due to the protein bounding by tannin. Table 1 showed that soybean meal protein protection with *gambier* tannin was more effective than that with tea waste. That was reflected by ruminal ammonia concentration [14] in Tt and Tg treatment groups which tent to be lower than that in Nt group although IVDMD and IVOMD in Tt and Tg treatment group were higher than that in Nt treatment group. That phenomenon also showed the increase of protein bypass due to bounding of that compound by tannin. The increase of that protein bypass was showed by the increase ($p<0.05$) of RUP in tannin treatment group (Tt and Tg) compared to Nt (51.10 vs 66.29 and 69.20%). The microbial protein synthesis in Tt and Tg treatment groups, on the one hand, were maintained because of the increase of an alpha keto acid availability as carbon skeleton for amino acid synthesis (reflected in the decrease of VFA production) so that the use of ammonia for microbial protein synthesis was more efficient. Portion of RUP, on the other hand, increased due to soybean meal protein protection by tannin [11]. The combination between the increasing of microbial protein synthesis efficiency and the increasing of RUP (bypass protein), resulted in the increase of ruminal protein metabolism efficiency [14, 15]. That result was showed by the increase of total protein production. The total protein production was the combination between bypass protein and microbial protein [16].

4. Conclusion

Soybean meal protein protection by tannin of tea waste as well as *gambier* improved the ruminal fermentability which reflected in the increase of IVDMD and IVOMD. Soybean meal protein protected by tannin increased the ruminal energy metabolism efficiency which reflected in the decrease of methane production. The use of tea waste and *gambier* as natural tannin sources for soybean meal protein protection increased the protein utility which reflected in the increase of RUP. The natural tannin from *gambier* was the most effective protection agent for soybean meal protein.

Acknowledgments

The authors would like to express their most warm gratitude to their students for their contribution in helping them conduct the laboratory research, especially Lintang, Bela, Ganang, Desti, and Mega. The authors would also like to extend their gratitude to Pak Mualimin, Faculty of Humanities, Diponegoro University, for editing the manuscript.

References

[1] Prasetiyono, B.W.H.E., Suryahadi, Toharmat T. and Syarief R. 2007. Strategi suplementasi protein ransum sapi potong berbasis jerami dan dedak padi. Med.Pet.**30** 207.

[2] Addisu, S. 2016. Effect of dietary tannin source feeds on ruminal fermentation and production of cattle; a review Online J. Anim. Feed Res. **6**: 45.

[3] Hussein, S.A. 2017. Utilization of tannins extract of acacia seyal bark (taleh) in tannage of leather. J. Chem.Eng.Process.Technol. **8.** 334.

[4] Moosophin, K., Wetthaisong T. 2010. Tannin extraction from mangosten peel for protein precipitation in wine. KKU Res. J. **15.** 377.

[5] Tilley, J.M., Terry R.S. 1969. A two stage technique for in vitro digestion of forage crops. J Br Grassland Society. **18**:104.

[6] Kurniawati, A. and Supadmo. 2010. Pengaruh tanin dan saponin pada protozoa dan aktivitas enzim CMC-Ase. Diktat Training on Ruminant Nutrition and Physiology, Universitas Gadjah Mada, Yogyakarta.

[7] Menke, K.H. and Steingass H. 1988. Estimation of the energetic feed value obtained from chemical analysis and in vitro gas production using rumen fluid. Anim.Res.Dev. **28** 7.

[8] SAS. 2009. SAS User's Guide. SAS Institute Inc., SAS Campus Drive, Cary, NC 27513.

[9] Baladi, R.M., Moghaddaszadeh-Ahrabi S. and Afrouziyeh M. 2014. Influence of the addition of different levels of tannin extracted from pomegranate pomace, on some nutritive value of soybean meal. European J. Ex. Biol, **4.** 148.

[10] Baldwin R.L., Allison M.J. 1983. Rumen metabolism. J.Anim Sci. **57.** 461.

[11] Dentinho M.T.P., Moreira O.C., Pereira M.S. and Bessa R.J.B. 2007 The use of a tannin crude extract from Cistus ladanifer L to protect soya-bean protein from degradation in the rumen Animal **1.** 645.

[12] Tavendale M.H., Meagher L.P., Pacheco D., Walker N., Attwood T., Sivakumaran S. 2005. Methane production from in vitro rumen incubation with Lotus pedunculatus and Medicago sativa, and effect of extractable condensed tannin fraction on methanogenesis. Anim.Feed Sci. Technol.**123** 403.

[13] Carulla J.E., Kreuzer M., Machmuller A. and Hess H.D. 2005. Supplementation of Acacia mearnsii tannins decreases methanogenesis and urinary nitrogen in forage-fed sheep. Aust. J.Agric.Res.**56** 961.

[14] Mendoza G.D., Britton R.A. and Stock R.A. 1993. Influence of ruminal protozoa on site and extent of starch digestion and ruminal fermentation. J. Anim Sci. **71** 1572.

[15] Khiaosa-ard R. and Zebeli Q. 2014. Meta-analysis of the effects of essential oils and their bioactive compounds on rumen fermentation characteristics and feed efficiency in ruminants. J. Anim. Sci.**91** 1819.

[16] Makkar H.P.S., Bummel M., and Becker K. 1995. In vitro effect on interaction between tannins and saponin and fate of tannins in the rumen. J.Sci.Food Agric. **69** 481.

In vitro Study of Noni Juice Extract Waste (*Morinda citrifolia* L.) and Pineapple Industrial Wastes (*Ananas comosus* L. Merr) as Energy Supplement in Dairy Goat Ration

D. Evvyernie[1*], A.S. Tjakradidjaja[1], I.G. Permana[1], T. Toharmat[1] and A. Insani[1]

Department of Nutrition and Feed Technology, Faculty of Animal Science, Bogor Agricultural University, Bogor, West Java, Indonesia

*Corresponding E-mail: erniedea8492@gmail.com

Abstract

The aim of the study was to evaluate the potency of noni juice extract waste (*Morinda citrifolia* L.) and pineapple industrial wastes (*Ananas comosus* L. Merr) as an energy supplement in dairy goat ration through in vitro study. This study used a complete randomized design with 5 treatments and 3 rumen fluid groups. The treatments were R0 as control (60% Napier grass (NG) + 40% concentrate), R1 (45% NG + 15% noni juice extract waste + 40% concentrate) + R2 (45% NG + 15% noni juice extract waste ammoniated + 40% concentrate), R3 (45% NG + 15% pineapple peel + 40% concentrate), and R4 (45% NG + 15% pineapple crown + 40% concentrate). The variables were totalbacterial population, protozoal population, fermentation characteristic (total VFA and NH_3 concentration), and digestibility (dry matter and organic matter).Data were analyzed with analysis of variance (ANOVA) and differences among treatments were determined by orthogonal contrast.The results showed that total VFA concentration was significant increased ($P<0.05$) by replacing 25% napier grass with noni juice extract waste (R1)or very significant increased($P<0.01$) by replacing the grass with pineapple peel (R3). The average increasing of total VFA concentration was 74% compared to control. As conclusions, 15% pineapple peel or 15% noni juice extract waste can use as an energy supplement by replacing 25% of napier grass in lactating dairy goat ration.

1. Introduction

Feed is one of important factors in livestock production. The quality and quantity of feed affect productivity in livestock. Lack of feed and its nutrient can cause livestock not producing optimally that can be impact on business farmers. Ruminants can digest forage as a main feed.[1] In dry season, ruminants facing a problem due to insufficient of forage or decrease of its quality.

Noni production (*Morinda citrifolia* Linn) in Indonesia reached 8.577.347 tons in 2014 [2]. Noni juice which is used as a traditional medicine for human, has a waste which called as noni juice extract waste (NJW) being disposed. NJW has been used as poultry feed, but still not many peoples use it as feed ruminant. [3] In formerly in vitro research, we found that total VFA production and NH_3 concentration linearly decreased when percentage of NJW linearly increased in ruminant ration due to of increasing the lignin content of NJW (24.51%). Lignin is one of the component in cell plants that has complicated degradation. The existence of lignin in feed can be a barrier in feed digestibility. Resolving lignin structure are required so that NJW can digest by ruminant. One of the methods to decrease the constraint such as ammonization may applicable to NJW. Ammoniation is chemically feed processing which used ammonia (NH_3) in the process for increasing feed digestibility and increase N (protein) concentration [4].In this present research we applied ammonization to NJW.

The others energy source come from pineapple industry [2]. In 2014, Indonesia produced 1,835,483 tons pineapple and it is increase annually. The majority of pineapple can be used as canned food, jam, syrup, and other. It can give more waste. Some of them already have been used as feed. In this present research we tried to use the peel and crown of pineapple as two kinds of energy source in ruminant ration due to high fiber but low lignin content.

The aim of this study was to evaluate the potency of noni juice extract waste (*Morinda citrifolia* L.) and pineapple industrial wastes (*Ananas comosus* L. Merr) as energy supplement in dairy goat ration through in vitro study.

2. Materials and Methods

2.1 Materials

The materials used in this research were noni juice extract waste (NJW), pineapple industrial waste (peel and crown) (PIW), and urea. The treatment rations that comprised of Napier grass, tempe waste, soybean meals, coconut meals, corn, $CaCO_3$, DCP, and premixwere composed according to NRC (2007) for lactating dairy goats which contained 12-17% crude protein and 53-66% TDN. Ration composition and nutrien content are placed in Table 1.

2.2 Methods

Noni juice extract waste (NJW) dried in the sun for 2 days. Then NJW had been dried again in oven 60 °C for 1 day. NJW which had dried in oven got weighed back then ground into flour. Some NJW flour been weight for ammoniation process. Ammoniation process used 3% urea from NJW total weight that being ammoniated. Pineapple industrial waste (PIW) consist of peel and crown then weighed, dried under the sun for 1 day weighed back. Both waste were then dried back with 60 °C oven for 1 day. After drying the peel and the crown of the pineapple are weighed back finally ground into flour.

This research used complete randomized design with 5 treatments and 3 rumen fluid groups, each group had 2 rumen fluid mixed from different dairy goats. The treatments were R0 as control (60% Napier grass (NG) + 40% concentrate), R1 (45% NG + 15% NJW + 40% concentrate) + R2 (45% NG + 15% NJW ammoniated + 40% concentrate), R3 (45% NG + 15% pineapple peel + 40% concentrate), and R4 (45% NG + 15% pineapple crown + 40% concentrate). The variables were totalbacterial population, protozoal population, fermentability (total VFA and NH_3 concentration), and digestibility (dry matter and organic matter). Data were analyzed with analysis of variance (ANOVA) and differences among treatments were determined by orthogonal contrast.

3. Results and Discussion

Feed undergoes a series of processes physically and chemically in the digestive tract before digest in the rumen[5]. In the rumen and reticulum, the microbes have roles to continue the fermentativedigestion of diets. Table 2 shows effect of substitution 25% napier grass with 15% noni juice extract waste, 15% noni juice extract waste ammoniation, 15% pineapple peel or 15% pineapple crown upon in vitro fermentation characteristic (NH_3 concentration and total VFA concentration), digestibility (dry matter and organic matter), and population of rumen microbes (bacteria and protozoa).

Table 1. Rations composition and nutrien content

Materials	R0	R1	R2	R3	R4
Ration composition (%)					
Napier grass	60	45	45	45	45
Tempe waste	1	1	1	5	4
Soybean meals	15.50	18.20	13.00	15.20	14.00
Coconut meals	7	5	6.2	5	4
Noni juice extract waste (NJW)	0	15	0	0	0
Ammoniated NJW	0	0	15	0	0
Pineapple industrial waste, peel	0	0	0	15	0
Pineapple industrial waste, crown	0	0	0	0	15
Corn	14	13	17	12	15
$CaCO_3$	0.3	0.3	0.3	0.3	0
DCP	1.2	1.5	1.5	1.5	2
Premix	1	1	1	1	1
Nutrien content (% DM)					
Dry matter[a] (DM)	91.13	91.03	92.10	90.37	91.18
Ash[a]	17.86	14.64	14.08	14.77	13.97
Ether Extract[a] (EE)	4.66	5.32	4.94	1.59	3.38
Crude Protein[a] (CP)	16.77	16.49	18.63	13.47	12.51
Crude Fiber[a] (CF)	25.78	25.38	26.67	25.75	25.40
Nitrogen Free Extract (NFE)	34.93	41.44	35.68	44.42	44.74
TDN[b]	51.78	54.56	54.94	53.94	54.48

[a]Proxymate analyses result fromLaboratory of Research Center for Biological Resources, IPB 2017; [b]TDN: *total digestible nutrients*; [b]calculation based on Hartadi *et al.* (1997) : 37.937 - 1.018 (CF) - 4.886 (EE) + 0.173 (NFE) + 1.042 (CP) + 0.015 $(CF)^2$ - 0.058 $(EE)^2$ + 0.008 (CF) (NFE) + 0.119 (EE) (NFE) + 0.038 (EE) (CP) + 0.003 $(EE)^2$ (CP).

There were no significant effect on NH_3 concentration of rations when 25% of Napier grass were substituted by NJW or PIW. However, the average ammonia (NH_3) concentrations were higher than the ideal concentration in rumen according to ranges from 6-12 mM [6]. This might caused by high content of soybean meal in the rationswhich categorized as rumen degradable protein (RDP) and also by addition of urea in ammoniation process.

Table 2. The effect of substitution 25% napier grass with noni juice extract waste or pineapple industrial waste upon in vitro fermentation characteristic, digestibility and population of rumen microbes

Variables	Treatment				
	R0	R1	R2	R3	R4
NH_3(mM)	15.13±2.20	14.99±2.00	15.74±1.43	12.41±2.20	13.76±1.92
VFA (mM)[1)]	95.77±13.10c	164.90±24.68b	110.96±25.96c	167.87±22.18a	114.70±37.76c
DMD (%)	52.62±8.56	58.85±4.67	58.53±4.92	59.77±6.44	58.45±7.16
OMD (%)	47.83±9.67	55.80±5.51	55.38±5.68	56.81±7.00	54.53±9.25
Bacterial Population (log CFU ml^{-1})	8.59±0.59	8.96±0.12	8.43±0.23	9.06±0.57	9.03±0.26
Protozoal Population (log cells ml^{-1})	4.66±0.03	4.69±0.08	4.74±0.02	4.71±0.04	4.62±0.09

1) The numbers followed by different letters in the same row show the signifficant difference (P< 0.05). R0= (60% Napier grass (NG) + 40% concentrate), R1= (45% NG + 15% noni juice extract waste (NJW) + 40% concentrate), R2= (45% NG + 15% NJW ammoniated + 40% concentrate), R3= (45% NG + 15% pineapple peel + 40% concentrate), and R4= (45% NG + 15% pineapple crown + 40% concentrate). NH_3= NH_3 concentration, VFA= total VFA concentration, DMD= dry matter digestibility, OMD= organic matter digestibility.

Substitution of 25% of Napier grass with 15% NJW or15% PIWin the rations showed significant effect on total VFA concentration (P<0.05). The average of total VFA concentrations were higher than the optimal range 70-150 mM [6], due to high content of readily available carbohydrate (RAC) such as corn in all rations.From the orthogonal contrast test, total VFA concentrations showed very significantly increased by addition of pineapple peel (P <0.01) and significantly increase by addition of NJW (P<0.05). The increasing of total VFA concentration in R1 and R3 achieved almost 74% compared to control. That increasing might caused by very high RDP content of soybean meal (18.2%) in R1 which supported the synthesis of microbial protein to digest NJW. The increasing of total VFA concentration in R3, might caused by high NFE as RAC sources of pineapple peel as well as corn. The lower total VFA concentration in R2, might caused by ammoniation process of NJW was not effectively done due to high content of lignin in NJW. In case of R4, total VFA concentration was low, it might caused by high availability of undegradable compound in pineapple crown. [7] There was relationship between the amount of soluble and degradable nitrogen or protein in the microbial protein synthesisas well as its rate of degradation, the amount of digestible organic matter fermented in rumen and alsoavailability ofcarbon chains (CHO's) to rumen microorganisms. Moreover, to increase more the microbial protein synthesis, both sources of nitrogen and degradable CHO's must be in a synchronized way.

The digestibility is an important factor in determining the value of feed[8]. In vitro digestibility is influenced by several factors such as mixing feed, composition feed, rumen fluids and materials, pH and temperature settings during fermentation, incubation length, sample particle size, and buffer solution [9].The founding results showed that the treatment rations not affected the dry matter as well as organic matter digestibility. The lowest value of both digestibility were showed in control (DMD 52.62% and OMD 47.83%), while similar value of both digestibility were achieved by all treatments (DMD 58.9% and OMD 55.63%). There is increasing the digestibility (12% in DMD and 16% in OMD) when substitution of 25% Napier grass were done with NJW and pineapple industry waste (PIW) in the rations. Availability of degradable protein such as soybean meal in the rations can support increasing the cellulolytical bacteria number, which in turn will digest NJW or PIW ration in longer time compare than control due to highly content of lignin or cellulose-lignin linkage.[10]Addition of RDP in diets containing non-fibre carbohydrates in Rusitec increased significantly apparent disappearance of diets, VFA total, acetate, acetate/propionate, methane, efficiency of microbial synthesis, *R. albus* and *R. flavefaciens*.

Rumen microbial activity determined the performance of the digestive system in ruminants. Microbes help ruminant in digesting feed up to 70-85% [11].The average of bacteria population as well as protozoa population in this present research showed slightly below than other study (9-10 log CFU ml^{-1} and 5 - 6 log CFU ml^{-1}, respectively) [6]. However, replacement type of fiber source (Napier grass with NJW or pineapple industry waste) in this research did not affect the rumen microbial population.Bacteria and protozoa population was not influenced by availability of NJW or PIW. In our previous research, we analyzed the phytochemical of NJW. NJW was not containsaponin and tannin, but contains high intensity of triterpenoid and flavonoid [3]. In medicinal herbs, saponin and tannin are usually function as anti-bacteria, and the triterpenoid and flavonoid function as antioxidant.

4. Conclussion

From in vitro investigation, total VFA concentration was significant increased ($P<0.05$) by replacing 25% napier grass with15% noni juice extract waste (R1) or very significant increased ($P<0.01$) by replacing the grass with15% pineapple peel (R3). The average increasing of total VFA concentration was 74% compared to control. Therefore, 15% pineapple peel or 15% noni juice extract waste can be used as an energy supplement by replacing 25% of napier grass in lactating dairy goat ration.

Acknowledgement

We would like to express our gratitude to Ministry of Research, Technology and Higher Education for research funding through PUPT-PSA Research Program, LPPM IPB 2017.

References

[1] Subekti E 2009Ketahanan Pakan Ternak Indonesia. Mediagro **5(2)** 63

[2] [KEMENTAN] Kementerian Pertanian 2015 *Statistik Produksi Holtikultura Tahun 2014*(Jakarta (ID): Direktorat Jenderal Holtikultura)

[3] Evvyernie D, Tjakradidjaja A S,Permana I G, Toharmat T, and Fakhirudin2017 *In vitro* evaluation of noni waste (*Morinda citrifolia* L.) as forage substituteinPEgoat diet. (Submitted paper for The 5thInternational Seminar ofAnimal Nutrition & Feed Science. Senggigi Lombok Indonesia 7- 9 November 2017)

[4] [DITJENNAK] Direktorat Jenderal Peternakan dan Kesehatan Hewan 2011 Pengolahan Bahan pakan Ternak*DITJENNAK* [internet] [Diunduh pada 2016 Agustus 14] Terdapat pada http://ditjennak.pertanian.go.id

[5] Setiyaningsih KD, Christiyanto M, and Sutarno 2012 Kecernaan bahan kering dan bahan organik secara i*n vitro* hijauan *Desmodium cinereum* pada berbagai dosis pupuk organik cair

dan jarak tanam J. Anim. Agriculture **1(2)**51

[6] McDonald P, Edward RA, Greenhalgh JFD, Morgan A, Sinclair L A, and Wilkinson RG. 2010. *Animal Nutrition* 7th Ed (England (UK): Pearson Hall)

[7] Andrade-Montemayor H, Gasca T G, and Kawas J 2009 Ruminal fermentation modification of protein and carbohydrate by means ofroasted and estimation of microbial protein synthesis. R. Bras. Zootec., v.38, p.277-291 (supl. special)

[8] Sutardi T 1979 Ketahanan protein bahan makanan terhadap degradasi olehmikroba rumen dan manfaatnya bagi peningkatan produktivitas ternak.*Prosiding Seminar Penelitian dan Pengembangan Peternakan.*(Bogor (ID):Lembaga Penelitian dan Pengembangan Pertanian).

[9] Selly 1994 Peningkatan kualitas pakan serat bermutu rendah dan amoniasi dan inokulan digesta rumen Skripsi Fakultas Peternakan Institut Pertanian Bogor Bogor

[10] Zhao X H, Gong J M, Zhou S,Fu C B, Liu C J, Xu L J, Pan K, and Qu M R 2015 Effects of degradable proteinand non-fibre carbohydrateson microbial growth andfermentation in the rumensimulating fermenter(Rusitec). Ital J Anim Sci vol.**14**: p.220.

[11] Thalib AJ, Bestari Y, Widiawati H, Hamid, and Suherman D 2000 Pengaruh perlakuan silase jerami padi dengan mikroba rumen kerbau terhadap daya cerna dan ekosistem rumen sapi. J Ilmu Ternak Vet. **5(1)** 1.

The Effect of Organic Phosphorus and Nitrogen Enriched Manure on Nutritive Value of Sweet Corn Stover

D. R. Lukiwati[1*], R. I. Pujaningsih[2] and R. Murwani[2]

[1]Agricultural Department, Faculty of Animal and Agricultural Sciences, Diponegoro University, Semarang, Indonesia.
[2]Animal Science Department, Faculty of Animal and Agricultural Sciences, Diponegoro University, Semarang, Indonesia.

[*]Corresponding E-mail: drlukiwati_07@yahoo.com

Abstract

The experiment aimed to evaluate the effect of some manure enriched with phosphorus (P) and nitrogen (N) organic ('manure plus') on crude protein and mineral production of sweet corn (*Zea mays saccharata*)and quality of fermented stover as livestock feed. A field experiment was conducted on a vertisol soil (low pH, nitrogen and low available Bray II extractable P). Randomized block design with 9 treatments in 3 replicates was used in this experiment. The treatments were T1(TSP), T2 (SA), T3 (TSP+SA), T4 (manure), T5 (manure+PR), T6 (manure+guano), T7 (manure+N-legume), T8 (manure+PR+N-legume), T9 (manure +guano+N-legume). Data were analyzed using analysis of variance (ANOVA) and the differences between treatment means were examined by Duncan Multiple Range Test (DMRT). Results of the experiment showed that the treatment significantly affected to the crude protein and calcium production of stover and nutrient concentration of fermented stover, but it is not affected to P production of stover. The result of DMRT showed that the effect of 'manure plus' was not significantly different on CP and Ca production of stover, mineral concentration, *in vitro* DMD and OMD of fermented stover, compared to inorganic fertilization. Conclusion, manure enriched with organic NP, resulted in similar on CP and Ca production of stover and nutrient concentration of fermented stover compared to inorganic fertilizer. Thus, organic-NP enriched manure could be an alternative and viable technology to utilize low grade of phosphate rock, guano and *Gliricidea sepium* to produce sweet corn in vertisol soil.

1. Introduction

Sweet corn (*Zea mays* var. *saccharata*) is utilized as raw materials for various uses with a majority for food, while corn stover and cornhusk are used for ruminants feed[1]. Sweet cornseed is used forfood ingredients while the remain i.e. corn stover and cornhusk are used for ruminants feed. Crop growth and yield in relation to application of different organic fertilizers has been widely reported [2]. Organic fertilizer from animal excreta or crop residue have been studied and shown to improve soil environment and fertility, activity of decomposer, promote crop yield, decrease green house gas emission and provide a carbon sink [3]. Organic fertilizer application is not as a substitute for inorganic fertilizers but as a complement because it contains macro nutrients, micro and organic acids and manure also acts as ameliorant. Application of organic fertilizers with the addition of inorganic fertilizers, increased soil fertility and crop production also saved the use of inorganic fertilizers [4]. This is in line with the Regulation of the Minister of Agriculture No.40/2007 that the purpose of organic material return or the provision of organic fertilizers equipped with inorganic fertilizers is to improve soil conditions and fertility.

Cattle manure as organic fertilizer is widely applied in Indonesia. The drawback is that it is relatively low in phosphorus, nitrogen, and potassium compared to commercial fertilizer. The composition of cattle manure differs between fresh and composted manure and varies with type of cattle, age, composition of feed, rations, type of bedding, manure storage and handling [5]. Cattle manure quality can be improved by enrichment utilizing organic P source such as rock phosphate or guano (bat faeces) and N source such as *Gliricidea sepium* [6, 7, 8]. Manure application on the crop used in the combination with NP organic, in comparison to inorganic fertilizer, reduces the stringent quality of environment protection requirements, and minimizes the negative impacts of manure on soil pollution. Its also changes the consideration of manure from a waste to a resource product [9] and residual effects of applied organic P enriched manure can be used for the next growing season [10].

The availability of the organic P from phosphate rock (PR) or guano, and the availability of organic N from *G. sepium* can be enhanced by composting it together with the cattle manure. The addition of decomposing microbes could aid the decomposition of organic matter during composting. Effective microorganisms (EM), which available in liquid form, has been introduced in 1991. The major groups of microorganisms contained in the EM include filamentous fungi, yeast, lactic acid bacteria, and other soil bacteria [11] which could aid the decomposition of organic matter during composting. The objective of this research was to evaluate the effect of mixed organic P (rock phosphate, guano) or organic N (*G. s*epium) with manure on crude protein and mineral production of sweet corn (*Zea mays saccharata*)and nutrient of fermented stover as livestock feed.

2. Materials and Methods

2.1. Composted Manure and Inorganic Fertilizers as Treatment and Experimental Design

The nine treatments were (1) TSP, application inorganic fertilizer on 66 kg P ha^{-1}; (2) sulphate ammonium, inorganic fertilizer on 200 kg N ha^{-1}; (3) sulphate ammonium + TSP, application of inorganic fertilizer NP (200 kg N ha^{-1} + 66 kg P ha^{-}); (4) manure, application of manure organic fertilizer on 20 t ha^{-1}; (5) manure+ PR, manure enriched with P-PR equivalen to 66 kg P ha^{-1}; (6) manure+guano, manure enriched with P-guano equivalent to 66 kg P ha^{-1}; (7) manure+ N-legume, application of manure enriched with N-*Gliricidea sepium* equivalent to 200 kg N ha^{-1}; (8) manure+P-PR+N-legume, manure organic fertilizer enriched with P-PR and N-*Gliricidea sepium* equivalent to 66 kg P ha^{-1} and 200 kg N ha^{-1}; (9) manure+guano+N-legume, application of manure organic fertilizer enriched with P-guano and N-*Gliricidea sepium* equivalent to 66 kg P ha^{-1} and 200 kg N ha^{-1}. Sulphate ammonium (SA), TSP, phosphate rock, guano, *Gliricidea sepium*, and manure were applied as the treatment, and KCl was added as basal fertilizer. Manure organic fertilizer made from cattle dung and urine mixed with waste of feed, and inoculated with biodecomposer (EM-4). After composting for one month only the manure could used as organic fertilizer, with or without enriched with organic-NP.

2.2. Study Area and Composted Manure Application

The experiment of organic and inorganic fertilizers treatment above for sweet corn was conducted on 350 m2 vertisol soil with low pH, P and N concentration in Sragen District, Central Java-Indonesia. The soil area was tilled and divided into 27experimental units and designed as randomized block design with nine treatments (T1 to T9) with three replicates. The size of each plot was 1.6 m x 2 m. In each plot a small holes was made for seed planting so that there were 20 holes per plot and each hole spaced 40cm x 40cm. Organic and inorganic fertilizer were evenly spread onto soil surface by hand and immediately incorporated by tillage before sowing. Tillage was done to 20 cm depth by shovel and followed by raking. Each plot received basal fertilizationof 125 kg K/ha as KCl [8], after which fertilization treatment was applied. The next day after the treatment two sweet corn seed (*Zea mays saccharata*) was planted into each holes. The sweet corn was harvested 70 days after planting, and analyzed for nutrient production (nitrogen and phosphorus) of stover and fermented stover. To measure nitrogen (N), phosphorus (P), and Caconcentration [12], samples were chopped, sub-sampled, and oven-dried at 70^{0}C for 48 h to constant weight. Samples were finally ground and analyzed for N and mineral Ca, P production of stover, mineral (Ca, P) and *in vitro* dry matter digestibility (DMD) and organic matter digestibility (OMD) of fermented stover.

2.3. Statistical Analyses

Crude protein and mineral (Ca, P) production of stover, nutrient concentration, IDMD and IOMDwere analyzed using one way analysis of variance (ANOVA). When significant differences among the treatments were found ($p<0.05$), the data were further tested using Duncan's Multiple Range Test (DMRT).

3. Results and Discussion

3.1. Chemical analysis

Chemical analysis result of soil and manure fertilizer showed that soil belong to deficient in N (0,17-0,20% N) and P (6,10-6,60 ppm P_2O_5). Manure analyses showed that concentration of N and P total (1,51%-1,75%) and (1,13-1,97%), respectively. Its manure quality was higher compared to Soelaeman (2008) 0,55% N dan 0,12% P total, respectively.This was affected by raw materials of organic fertilizers [13]. Manure made from cattle dung mix with urine and feed residue inoculated with effective microorganisms (EM-4). The major groups of microorganisms contained in the EM include filamentous fungi, yeast, lactic acid bacteria, and other soil bacteria [3, 11], which has the ability to increase decomposition process, mobilisation and mineralisation of nutrient [14]. Ratio of C/N of six manure treatment were 15,17; 14,66; 13,58; 12,83; 13,52 and 12,64%, respectively. According to [15], international quality standard of manure C/N ratio was below 20, and could be used for fertilization [16].

3.2. *Crude Protein and Mineral Production of Stover and Nutrient of Fermented Stover.*

Result showed that both of the CP and Ca production of stover were significantly influenced by the treatments. Phosphorus production of stover was not significantly influenced by the treatments.
Table 1 showed that CP production with SA(T2) was significantly higher compared to TSP (T1). Fertilization treatment of SA (T2) which KCl as basal fertilization had the high CP production of stover and that treatment contained nutrients sufficient enough to produce high CP. Nitrogen (SA) is very influential on plant growth [17], so that if there is a deficiency of N (T1) then the growth of theplant is inhibited and experienced chlorosis [18]. Availability of essential nutrients in adequate quantity and form enhanced protoplasmic development and cell proliferation. These culminated in proper crop growth and development [19].

Manure enriched with N, P or NP organic (T6, T7, T8, T9) showed that CP production of sweet corn stover (Table 1) and Ca production of stover (Table 2) were not different compared to inorganicfertilization of T2 (SA) and T3 (TSP+SA). This implies that these organic fertilizers contained adequate nutrients required for normal plant growth and development as contained in the NPK. Manure application on the crop used in the combination with organic NP, in comparison to inorganic fertilizer, reduces the stringent quality of environment protection requirements, minimizes the negative impacts of manure on soil pollution and changes the consideration of manure from a waste to a resource product [19]. The use of organic fertilizer and combination with inorganic (KCl)(T4, T5,T6, T7, T8, T9) showed similar effect on CP (Table 1) and Ca (Table 2) of stover compared to TSP+SA (T3). However, its not affected significantly on P production of stover (Table 2). That's mean application of various inorganic and organic fertilizers had equivalent ability on P production of stover. Cattle manure quality can be improved by enrichment utilizing organic P source such as phosphate rock or guano (bat faeces) and N source such as *Gliricidea sepium* [6, 7, 8]. Application of organic fertilizers with the addition of inorganic fertilizers, increased soil fertility and crop production also saved the use of inorganic fertilizers [4]. Manure increased the nutrient uptake of sweet corn stover when combined with inorganic fertilizers [20].

Table 1. Crude Protein Production of Sweet Corn Stover with Organic and Inorganic Fertilization

Treatments	Crude protein production kg/plot (3.2 m^2)
T1 (TSP)	380.50 ± 35.02 ^{b*}
T2 (SA)	631.62 ± 57.53 ^{a}
T3 (TSP+SA)	584.42 ± 52.36 ^{ab}
T4 (manure)	402.53 ± 28.75 ^{ab}
T5 (manure+P-PR)	445.84 ± 11.63 ^{ab}
T6 (manure+P-guano)	483.07 ± 34.03 ^{ab}
T7 (manure+N-legume)	422.10 ± 26.53 ^{ab}
T8 (manure+P-PR+N-legume)	415.39 ± 57.46 ^{ab}
T9 (manure+P-guano+N-legume)	486.75 ± 41.75 ^{ab}

*Significantly different (p<0.05)

Table 2. Mineral Production of Sweet Corn Stover with Organic and Inorganic Fertilization.

Treatments	Ca production (g/plot)	P production (g/plot)
T1. (TSP)	16.02 ± 2.05 b*	24.48 ± 1.47
T2. (SA)	25.11 ± 2.77 ab	35.23 ± 2.64
T3. (TSP+SA)	25.48 ± 3.12 ab	39.20 ± 4.67
T4. (Manure)	19.97 ± 0.98 ab	31.04 ± 1.64
T5. (Manure+P-PR)	20.11 ± 1.23 ab	31.20 ± 1.78
T6. (Manure+ P-guano)	24.20 ± 0.77 ab	32.79 ± 1.68
T7. (Manure+N-legum)	27.70 ± 2.36 ab	35.24 ± 1.11
T8. (Manure+P-PR+N-legum)	24.37 ± 2.25 ab	32.02 ± 4.44
T9. (Manure+P-guano+N-legum	29.32 ± 2.99 a	38.20 ± 3.56

* Significantly different (p< 0.05)

3.3. *Mineral concentration in vitro DMD and OMD of fermented stover*

Mineral concentration, *in vitro* DMD and OMD of fermented stover were significantly influenced by the treatments. Manure and organik N-P enriched manure (T4, T5,T6, T7, T8, T9) showed that Ca and P concentration of fermented stover (Table 3) and *in vitro* DMD and OMD of fermented stover (Table 4) were not significantly different compared to inorganic fertilization of T3 (TSP+SA). The carbonic acid and organic acid produced during the decomposition of organic matter solubilized insoluble phosphate in P organic (phosphate rock, guano), resulting in the release of phosphate and calsium into the solution [21]. That's mean application of some organic N-P enriched manure had equivalent ability on mineral concentration, *in vitro* DMD and OMD of fermented stover [22].

Table 3. Mineral Concentration of Fermented Sweet Corn Stover

Treatments	Ca concentration (%)	P concentration (%)
T1. (TSP)	0.41 ± 0.03 b*	0.27 ± 0.018 abc
T2. (SA)	0.58 ± 0.04 a	0.23 ± 0.006 bc
T3. (TSP+SA)	0.55 ± 0.04 ab	0.22 ± 0.009 c
T4. (Manure)	0.51 ± 0.02 ab	0.24 ± 0.013 abc
T5. (Manure+P-PR)	0.58 ± 0.05 ab	0.32 ± 0.012 a
T6. (Manure+P-guano)	0.43 ± 0.04 ab	0.25 ± 0.007 abc
T7. (Manure+N-legum)	0.51 ± 0.03 ab	0.31 ± 0.035 ab
T8. (Manure+P-PR+N-legum)	0.46 ± 0.04 ab	0.23 ± 0.012 abc
T9. (Manure+P-guano+P-legum	0.50 ± 0.07 ab	0.23 ± 0.002 abc

* Significantly different (p< 0.05)

Table 4. *In vitro* Dry Matter Digestibility and Organic Matter Digestibility of Fermented Stover

Treatments	DMD (%)	OMD (%)
1. TSP	64.68 ± 0.59 a	62,84 ± 0.61 a*
2. SA	64.10 ± 1.44 a	62,82 ± 1.49 a
3. TSP+SA	59.16 ± 2.17 ab	56.07 ± 2.16 ab
4. Manure	55.37 ± 0.59 b	53.46 ± 1.00 ab
5. Manure+P-PR	58.11 ± 2.00 ab	56.32 ± 2.37 ab
6. Manure+P-guano	55.23 ± 0.10 b	51.71 ± 1.18 b
7. Manure+N-gamal	61.19 ± 1.69 ab	59.89 ± 1.82 ab
8. Manure+P-PR+N-gamal	61.88 ± 1.54 ab	60.23 ± 1.72 ab
9. Manure+P-guano+N-gamal	63.51 ± 0.33 ab	62.20 ± 0.74 a

* Significantly different (p< 0.05)

4. Conclusion

Organic-N (N-legume) and organic-P (RP, guano) and organic-NP (manure+P-PR+N-legume; manure+P-guano+N-legume) enriched manure resulted in similaron CP and Ca production of stover, mineral concentration, *in vitro* DMD, and OMD of fermented stover compared to TSP+SA (T3). Application of organic NP enriched manurecould be a promising organic fertilizer to replace inorganic fertilizer (TSP, sulphate ammonium) to produce sweet corn in vertisol soil.

Acknowledgments

Thanks to Diponegoro University, Semarang – Indonesia (:Grant Competition Research Professorship Program 2016) for the support.

References

[1] Ranum P, Rosas JPP, and M.N.G-Casal MNG 2014 Global maize production, utilization, and consumption Ann.N.Y.Acad.Sci. 1312 105

[2] Akanbi WB, Ilupeju EAO, Togun AO, and Adeyeye SA 2010 Effect of Nigerian commercial organic fertilizers, compost and NPK on growth, shoot yield and nutritional quality of *Solanum macrocarpon*. International J. of Organic Agriculture Res. and Development. **1** (1) 135

[3] Leite MFA, Pan Y, Bloem J, Berge H, and Kuramaea EE 2017 Organic nitrogen rearranges both structure and activity of the soil-borne microbial seedbank *Sci Rep*. **7** 42634.

[4] Saha R, Saieed MAU, Chowdhury MAK 2013 Growth and Yield of Rice (*Oryza sativa*) as Influenced by Humic Acid and Poultry ManureUniversal J. Plant Sci. **1** 78

[5] Saskatchewan Soil Conservation Association 2000 *Solid Cattle Manure*. http://www.soilcc.ca /ggmp_fact_sheets /pdf/Cattle%20manure.pdf.

[6] Bah AR, and Rahman ZA 2001Gliricidea (*Gliricidea sepium*) Green Manure as a Potential Source of N for Maize Production in the Tropics. Optimizing Nitrogen Management in Food and Energy Production and Environmental Protection. Proceedings of the 2nd International Nitrogen Conference on Science and Policy The Scientific World (2001). **1** 90

[7] Shridhar MK, GO A, and OO A 2001 Alternate nitrogen amendments for organic fertilizers Scientific World Journal Dec 19 Suppl 2 142

[8] Lukiwati DR, Purbayanti ED, and Pujaningsih RI 2014 Sweet corn production and nutritive value of stover with manure enriched with rock phosphate fertilizer and biodecomposer. *J. of Agric. Scie. And Tech. A*.**410** 839

[9] Annicchiarico G, Caternolo G, Rossi E, and Martiniello P 2011 Effect of Manure vs. Fertilizer Inputs on Productivity of Forage Crop Models Int. J. Environ. Res. Public Health **8** 1893

[10] Lukiwati DR and Pujaningsih RI 2014 Efek Sisa Pupuk Kandang Diperkaya Fosfat Alam terhadap Produksi Jagung Manis dan Jerami di Lahan Kering J. Lahan Suboptimal **3** (2) 152

[11] Higa T and Parr JF 1994 Beneficial and Effective Microorganisms For a Sustainable Agriculture and Environment Int. Nature Farming Res. Center (INFRC). Atami, Japan.

[12] Islam AKMS, Kerven G, and Oweczkin. 1992. Methods of Plant Analysis. ACIAR 8904 IBSRAM QC

[13] Pardede ESbr, Mariati, and Sipayung R 2015 Pertumbuhan dan Produksi Tiga Varietas bawang Merah (*Allium ascalonicum L.*) Pada Pemberian Beberapa Jenis Pupuk Organik di Tanah Terkena Abu Vulkanik Sinabung J.online Agroekoteknologi **34** 1436

[14] Manoharachary C, Sridhar, Singh R, Adholeya A, Sauryanarayanan TS, Rawa S, and Johri BN 2005 Fungal biodiversity: Distribution, conservation and prospecting of fungi from India Current Science **89** (1) 58

[15] Ge B, McCartney D, and Zab J 2006Compost environmental protection standards in CanadaJ Eviron Eng Sci 5 221

[16] Edesi L, Jarvan M, Noormeths M, Lauringson E, Adamson A, and Akk E 2012 The importance of soil cattle manure application on soil microorganism inorganic and conventional cultivation. Acta Agric. Scandinavida. Section B – Soil & Plant Sci. **627** 583

[17] Abdissa Y, Tekallign T, and Pant LM 2011 Growth, bulb yield, and quality of onion (*Allium cepa* L.) as influenced by nitrogen and phosphorus fertilization on vertisol. I. growth attributes, biomass production and bulb yield. Afr J. Agric.Res. **614** 3252

[18] Suwandi GA, Sopha GA, and Yufdy MP 2015 Efektivitas pengelolaan pupuk organik, NPK dan pupuk hayati terhadap pertumbuhan dan hasil bawang merah. J. Hort. **253** 208

[19] Akanbi W B, Ilupeju E A O, Togun A O, and Adeyeye SA 2010 Response of Cassava and Soyabean grown sole and intercropped to inoculation of *Rhizobium* and Arbuscular Mycorrhiza FungiInternational J. of Organic Agriculture Res. and Development. **11** 135-149.

[20] Lukiwati DR, Surahmanto, and Kristanto BA 2010 Production and nutrient uptake improvement of sweet corn by rock phosphate combined with manure and mycorrhiza inoculation. Abstr.p.80. International Conference on Balanced Nutrient Management for Tropical Agriculture. Kuantan, Pahang. Malaysia, 12-16 April.

[21] Qureshi SA, Rajput A, Memon M, and Solangi MA. 2014 Nutrient composition of rock phosphate enriched compost from various organic wastes E3 J. Sci Res. **2** (3)

[22] Garray AH, Sollenberger LE, McDonald DC, Ruegsegger GJ, Kalmbacker, and Miskvy P 2004 Nitrogen fertilization and stocking rate affect stargrass pasture and cattle performance Crop Sci. **44** 1348

Observation of Muntingia Calabura's Leaf Extract as Feed Additive for Livestock Diet

R. I. Pujaningsih[1]*, B. Sulistiyanto[1] and S. Sumarsih[1]

[1]Faculty of Animal and Agricultural Sciences, Diponegoro University, Semarang, Central Java, Indonesia.

*Corresponding E-mail: retno.marwoto@gmail.com

Abstract

Using of synthetic antioxidants in feedstuffs continuously can cause negative effect for the livestock. This study observed the constituent compounds of cherry leaf powder using format method of descriptive qualitative. Comparative study was done between young and old leaves to identify the content of antioxidant and antimicrobial. Based on the results of phytochemical tests that have been done, old cherry leaves contain compounds of flavonoids more than young cherry leaves. From the results of this study can be concluded that the results of old cherry leaf isolation using soxhlet extraction has antibacterial power against *E. coli* bacteria, and *S. aureus* at concentration of 75% have greater inhibitory ability.

1. Introduction

One cause of rancidity in animal feedstuffs is the activity of enzymes in the tissues of foods containing fat and microbial activity. Symptoms of rancidity with the process of fat oxidation begins with the emergence of taste, flatness, oily nature, then change the taste and aroma. The next turns into a musty smell, and the last stage becomes rancid, damaging vitamin content and poisoning. That's why antioxidants are needed to prevent and inhibit the oxidation process.

Based on the type, antioxidant is divided into two groups, natural and synthetic antioxidants. Natural antioxidant derived from plants because they contain phenolic compounds, nitrogen [1]. The use of synthetic antioxidants began to be limited because the results of research resumed that synthetic antioxidants such as BHT (Butylated Hydroxy Toluene) be able to poison experimental animals and are carcinogenic. Therefore the food and medicine industries are switching to develop natural antioxidants and seeking new natural antioxidant sources [3]. One of plant that can be used is cherry leaf (*Muntingia calabura*). The first attempt to study the antibacterial activity of *M. calabura* was carried out by using the leaves and fruits collected in the State of Puebla and State of Veracruz, Mexico. The result showed both MEMCL (methanol extract of *M. calabura* leaves) and MEMCFr (methanol extract of *M. calabura* fruits) exhibited antibacterial activity against *E. coli* and *S. aureus* with the recorded minimum inhibitory concentration of 512 and 356 1024 µg/mL, and 128 and 256 µg/mL, respectively[5]. Utilization of cherry leaves as feed additive avoids competition with humans whom prefer to consume the fruit than the leaves. Based on the results of phytochemical tests that have been done, cherry leaves qualitatively contain flavonoid compounds, triterpenoids, tannins, saponins and steroids in accordance with the results of phytochemical tests according to Zakaria [2].

Flavonoids are phenol compounds that can change when added by alkaline compounds or ammonia. In nature, flavonoids are water-soluble compounds. Flavonoids are active compounds that can be used as antioxidants, antibacterial, and anti-inflammatory because it can inhibit the activity of disease-causing bacteria [6].

Isolation of flavonoids can be done with several methods of extraction. Each method has a deficiency and an advantage in producing cherry leaf extract. Research related to tea leaf aging explained that older leaves had higher mean total flavonoid values compared with young leaves ie 54.08 mg QE / g dry matter weight and 33.54 mg QE / g dry matter weight, respectively. Further informed that old avocado leaves contain higher levels of flavonoids [8]. The higher the level of flavonoids, the antioxidant potential will be higher. Flavonoids are a natural antioxidant and have biological activity, as an antioxidant that can inhibit various oxidation reactions, and capable of acting as a reducer of hydroxyl radicals, superoxide and peroxyl radicals. It can be assumed that the content of secondary metabolite compound of old cherry leaves has the ability as higher antioxidant than young leaves, so that antioxidant activity of old leaves is higher than the young leaves [2]. But the levels of these compounds in cherry leaves is not known quantitatively.

This study aims to observe the possibility of cherry leaf extract as feed additive that serves as antibacterial and antioxidant. Maceration, Soxhlet extraction and decoction were tested to get the best quality and quantity of cherry leaf extract. It is thought that older leaves contain more flavonoids that can function as organic feed additives.

2. Materials and Methods

Extraction is the separations of plants by using selective solvents through standard procedure. The purpose of all extractions is dissolved metabolites and leaving cellular cells insoluble (residue). The research was conducted in July - October 2017 at the laboratory of Feed Technology UNDIP and Biology laboratory of UNNES.

2.1. Materials

The cherry leaves used in this study are the leaves that are dark green and light green, still fresh, picked from cherry trees that grow in Tembalang District of Semarang. The leaves are then washed and dried with oven at 60° C for 24 hours. Next, it blend into powder. The solvents used were 96% ethanol, hexane, ether, petroleum ether, methyl chloride and alcohol.

2.2.Methods

This research is divided into 3 stages (sample preparation, extraction and data analyzing). The first stage is the making of cherry leaf simplicia powder. Cherry leaf simplicia powder used in this study is 300 grams for each of young and old leaves, that divided for 3 methods of extraction with the amount of each 100 grams. The second stage is extraction with maceration method, soxhlet extraction and decoction [7]. The data were analysed descriptively covering the physical quality of organoleptic from cherry leaf extract, the amount of rendement of each extraction method and the flavonoid content of cherry leaf extract [4] [5] [7].

2.2.1. Ekstraksi Maceration

One hundred grams of leaf simplicia powder cherry (each of young and old leaves) is put into vessel dark, then added 750 ml of solvent ethanol 96% and closed tightly as well protected from direct sunlight. Immersion process for 3 days while stirred every 8 hours. After 3 days, the mixture of simplicia and methanol was filtered so as to obtain the first macerate. The dregs were immersed with 250 ml of methanol for 1 day, filtered again and obtained a second macerate. The two macerates were applied overnight and then separated from the residue and concentrated using a rotary evaporator at a temperature of 40°C until a viscous ethanol extract was obtained. [9].

2.2.2. Soxhlet Extraction

Finely ground sample of each young and old leaves are placed in a porous bag made from filter paper which is place in thimble chamber of the Soxhlet apparatus. Extraction solvents is heated (70°C) in the bottom flask, vaporizes into the sample thimble, condenses in the condenser and drip back. As the liquid content reaches the siphon arm, the liquid contents emptied into the bottom flask again and the process is continued. Liquid extract that collected then concentrated by using rotary evaporator on temperature of 40°C until extract is obtained condensed ethanol [7].

2.2.3. Decoction Extraction

Simplicia cherry leaves of young and old leaves (each of 100 g) were extracted with water solvent at measured water temperature (96-98°C) for 30 minutes. The vessel is immersed in a boiling water bath. To obtain the condensed ethanol the rovap was used on temperature of 40°C upon collected liquid extract. [7]

2.2.4. Antimicroba test

The best extraction method then used to evaluate the antimicrobial activity of *E. coli* and *S. aureus*. The antimicrobial test was performed with an incubation time of 1 x 24 hour at a concentration of 75%.

2.3. Data collecting

The organoleptic parameters of the extract are descriptive, color, taste and smell [10]. Quantitative analysis of total flavonoids in cherry leaf extract was done by colorimetric method using aluminium chloride reagent [9]. Antimicrobial test was made for *E. coli* and *S. aureus.*

3. Results and Discussion

Many factors influence the quality of herbs and these include species variation, environmental conditions, and the time of harvesting, storage and processing. This study used the leaves of cherry (*Muntingia calabura*) due to the nature of the plant wherein leaves are the parts that are easy to collect in abundant throughout the year.

3.1. Organoleptic test

Old and young cherry leaf powder produced have a moisture content of 4.7% and 5%, respectively. Generally, water content of medicinal plant simplicia is maximal 10% [5]. Old and young cherry leaf simplicia obtained for 2.115 grams and 2.250 grams of 4.500 of cherry leaves, respectively. Organoleptic results of cherry leaf extract (*Muntingia calabura*) with maceration method, soxhlet extraction and decoction are the same. The texture is thick, dark green color and specific smell. [5][6][7].

3.2. Flavonoid

Studying from the total flavonoid content of all three extraction methods, the method that produced the highest total flavonoid percentage was the soxhlet extraction (Table 1). Based on the yield of cherry leaf extract, soxhlet extraction also yields more rendement than maceration method and decoction method, as well. This caused the total flavonoid content of soxhlet extraction is greater than the other two methods. In addition, the possibility of total flavonoids found in cherry leaves more easily soured with soxhlet extraction. In soxhlet extraction used less solvent than maceration method. So there is the possibility to reduce the compound that wasted within the solvent. Decoction is only suitable for extracting heat-stable compounds, hard plants materials (e.g. roots and barks) and usually resulted in more oil-soluble compounds compared to maceration and infusion [9].

Simplicia cherry leaf from old cherry leaf have a mean value of total flavonoids higher than young cherry leaf on all extraction methods. Factors affecting total levels of flavonoids in leaves are morphological and aged leaves, which will affect secondary metabolites and bioactive compounds produced [12]. The flavonoid content from decoction is the lowest because flavonoid compounds are not heat resistant and easily oxidized at high temperatures [2]. Flavonoids show different sensitivity in heat treatment depending on the structure. Regardless of its structure, flavonoids will be degraded at temperatures above 100° C. Flavonoids are heat sensitive due to hydroxyl groups and ketones, as well as unsaturated double bonds [11]. Table 1 showed the results of cherry (*Muntingia calabura*) leaf observation prior to become feed additive.

Table 1. Cherry (*Muntingia calabura*) leaf observation.

Observation	Old cherry leaf	Young cherry leaf
	--------------- % --------------	
water content	4,7	5
rendement[a] on maceration method	26,58	25,43
rendement[a] on soxhlet extraction	28,92	27,78
rendement[a] on decoction method	26,25	25,11
flavonoid[b] by maceration method	0,19	0,16
flavonoid[b] by soxhlet extraction	0,22	0,18
flavonoid[b] by decoction method	0,16	0,14

[a]Simplicia cherry leaf : 100 g
[b]The total flavonoid content is expressed as the number of grams of quercetin equivalent to each gram of the extract[7]

3.3. Antimicrobial test

The results of antimicrobial assays from cherry leaf extract with 1 x 24 hour incubation process can be explained in Table 2. Table 2 shows that extract from old cherry leaf produces higher inhibitory resistance to bacteria than the extract from young cherry leaf against *S. aureus* bacteria and bacteria *E. coli.*[5].

The old cherry leaf methanol fraction has higher concentration and stronger resistance to bacteria, whereas the young cherry leaf methanol fraction has lower concentration, indicating a small inhibitory power. This is due to the increasing levels of bioactive compounds that are increasingly bactericidal (microbial lethal agents), whereas lower levels are usually only bacteriostatic (agents that inhibit microbial growth, not microbial kill) [13].

Table 2. Inhibition's area of bacterial colony growth of *E.coli* and *S.aureus* after incubation for1 X 24 hours with soxhlet extract

Bacteria	Diameter (cm)	
	Old cherry leaf	Young cherry leaf
Escherichia coli	1,2	1,1
Salmonella aureus	1,4	1,1

4. Conclusion

Based on the data, it can be concluded that old cherry leaf will give better effect as antimicrobial and antioxidant at a concentration of 75% than young cherry leaf.

Acknowledgments

This research was supported by Faculty of Animal Agriculture, Diponegoro University. We are thankful to our colleagues (Tito, Ditto, Endah, Ade, Erry, Reni) who provided expertise that greatly assisted the research.

References

[1] Chiang C. J., Kadouh H., Zhou K. Q. 2013. Phenolic Com- pounds and Antioxidant Properties of Gooseberry as Af- fected by in Vitro Digestion," LWT-Food Science and Technology. **51** (2), 417-422. DOI: 10.1016/j.lwt.2012.11.014.

[2] Zakaria, Z.A. 2007. Free Radical Scavenging Activity of Some Plants Available in Malaysia. Iranian Journal Of Pharmacoglogy & Therapeutics. **6** :87-91.

[3] Takashi, Miyake and Takayumi Shibamoto. (1997). Antioxidant Activities of Natural Compound Found in Plants. J. Agric. Food. Chem. 45. 1819-1822.

[4] Zakaria Z. A., Jais A. M. M., Mastura M., Jusoh M. S. H., Mohamed A. M., Jamil M. N. S., Rofiee M. S., Sulaiman M. R. (2007). In vitro antistaphylococcal activity of the extracts of several neglected plants in Malaysia. J Pharmacol **3**: 428-431.

[5] Yasunaka, K., Abe, F., Nagayama, A., Okabe, H., Lozada-Perez, L., Lopez-Villafranco, E., Muniz, E.E., Aquilar, A., and Reyes-Chilpa R. (2005). Antibacterial activity of crude extracts from Mexican medicinal plants and purified coumarins and xanthones. J Ethnopharmacol **97**: 293-299.

[6] Binawati, D. K., dan Amilah, S. 2013. Effect of Cherry Leaf (*Muntingia calabura L.*) Bioinsecticides Extract Towards Mortality of Worm Soil (Agrotis ipsilon) and Armyworm (*Spodoptera exiqua*) on Plant Leek (Allium fistolum). Wahana, **61**(2):51-57.

[7] Harbone, J.B. 1998. Metode Fitokimia Penuntun Cara Modern Menganalisa Tumbuhan., ITB, Bandung.

[8] Felicia, N., I. Wayan, R.Widarta, Ni Luh, A.Y. 2017. Pengaruh Ketuaan Daun dan Metode Pengolahan terhadap Aktivitas Antioksidan dan Karakteristik Sensoris Teh Herbal Bubuk Daun Alpukat (Persea americana Mill.). Jurnal Ilmu dan Teknologi Pangan (*Itepa*). 5, no. 2, p. 85-94. ISSN 2527-8010.

[9] Azwanida, N. N. 2015. A Review on the Extraction Methods Use in Medicinal Plants, Principle, Strength and Limitation. Med Aromat Plants 4: 196. doi:10.4172/2167-0412.1000196.

[10] Depkes R. I., 2000. Parameter Standar Umum Ekstrak Tumbuhan Obat, Departemen Republik Indonesia, Jakarta

[11] Qiao, L, Y. Sun, R. Chen, Y. Fu, W. Zhang, X. Li, J. Chen, Y. Shen, and X. Ye. 2014. Sonochemical Effects on 14 Flavonoids Common in Citrus: Relation to Stability. PloS ONE **9** (2): e87766.

[12] Farhoosh, R., G. A. Golmovahhed, and M. H.H. Khodaparast. 2007. Antioxidant activity of various extracts of old tea leaves and black tea wastes (*Camellia sinensis* L.). Food Chemistry 100: 231 – 236.

[13] Volk, WA and Wheeler, MF. 1993. Mikrobiologi Dasar. Edisi V. Jilid 1. Diterjemahkan oleh Adisoemarto S. Jakarta: Erlangga.

Evaluation of Feed for Thin-Tailed Sheep Fattening with Supplemented Protected and Unprotected Aldehide

J. Riyanto[1]* and Sudibya[1]

Animal Husbandry Department of Agriculture Faculty, Sebelas Maret University, Surakarta, Central Java, Indonesia.

*Corresponding E-mail: jokoriyanto@staff.uns.ac.id

Abstract

The purpose of this study was to determine the effect of the use of soybean protection supplements in sheep ration in vivo in terms of consumption, digestibility, nutrient value index, and the digestible nutrients in the ration. Livestock used in this study were 15 heads of thintailed sheep male with an average body weight of 20.81 ± 1.40kg. The rations used in this study consisted of elephant grass, basal concentrate, soybean groats protected and without protected. The comparison between elephant grass and basalt concentrate is 30:70. Feed treatment in the form of supplementary concentrate from soybean groats ingredients without protection and protection. Protection of soybeans using 37% formaldehyde. The treatment given is P0 = 30% Elephant grass + 70% Basal concentrate, P1 = 30% Elephant grass + 60% Basal Concentrate + 10% soybeans groats without formaldehyde protection, and P2 = 30% Elephant grass + 60% Basal Concentrate + 10% soybeans groats formaldehyde protection. Supplementation of 10% soybean protected feeding in male thin tail sheep fattening ration had significant effect ($P < 0.05$) on crude protein digestibility, nutrient value index and digested crude protein. The use of 10% of soybean protected 37% formaldehyde protected soy by 1% of the dry weight of the concentrate in thin tail fattening rations could improve protein digestibility, nutrient value index and abrasive proteins that can be ingested in vivo.

1. Introduction

Fattening tailed sheep requires adequate nutrient intake to produce maximum weight gain. Protein is a nutrient that is needed for fattening. One of the feed ingredients that have high protein content comes from soybeans. Soy soybeans are soybean seeds obtained from the process of soybean seed peeling from pods then dried to obtain soybean fragments [1].

Soybeans are the ingredients of protein sources containing 35% protein content similar to whole soybeans [2]. On the other hand, during fermentation in the rumen, proteins are degraded and bio hydrogenated by the rumen microbes, so the utilization of proteins cannot be maximized. There is a need for protection against soybeans protection in order to avoid excessive degradation and bio hydrogenation so that it can go directly to the small intestine. One way of protection is done by using 37% formaldehyde. The use of 2% formaldehyde showed better results used as a protective substrate mixture of soybeans: lemuru fish oil ratio of 4: 1 compared to 2: 1 ratio [3].

Use of soybeans that are protected with formaldehyde can improve the performance of fattening beef cattle [1]. Through the method of protection with formaldehyde it is expected that protein in soybeans can be protected so that it can directly experience the enzymatic process in abomasum and intestine into amino acids [4]. Evaluation in vivo is a method of determining the digestibility of feed using thin-tailed sheep based on the digestion of feed ingredients that occur in the entire digestive tract, so that the value of the digestibility of the feed obtained close to the true value. Parameters that can be generated from the in vivo test can mainly be measured the level of consumption and digestibility of nutrient content of feed such as dry matter content, organic matter, protein, fat, crude fiber,extract material without nitrogen and Total Digestibility Nutrient (TDN) [5]. Based on the description above, this study was conducted to determine the effect of the use ofsoybean protected formaldehyde in thin tail fattening ration in terms of in vivo test parameters.

2. Material and Methods

This research was conducted at Boyolali, Central Java. Analysis of feed, waste of food waste and feces was carried out in the Laboratory of Nutrition and Feed Livestock, Animal Husbandry Program, Faculty of Agriculture Sebelas Maret University of Surakarta. Livestock used in this study were 15 male tailed sheep with an average body weight of 20.81 ± 1.40kg.

The research were used 15 thin tailed sheep maleindividual stage with a length of 110 cm and a width of 60 cm, made of bamboo and wood. Each cage is equipped with a permanent feeding and drinking place from a plastic bucket. At the time of data collection, the cage is equipped with a collection of feces made of nets and urine placed under the cage. Equipment used include sheep, feed weights, feces scales, feed grinders, feed mixers (feed mixers) and a set of detergent / van soest analysis tools, and tools for proximate analysis.

The rations used in this study consisted of elephant grass, basalt concentrate, soybean groats protected and without protection. Basal concentrate consists of palm cake, coffee skin, pollard, bran, mineral and molasses. The amount of feed given is 3% DM of body weight. The comparison between elephant grass and basalt concentrate is 30:70. Feed treatment in the form of supplementary concentrate from soybean ingredients without protection and protection. Protection of soybeans using 37% formaldehyde. Use of formaldehyde as much as 1% of the dry matter of the basal concentrate. Protection is done by means of water mixed with formaldehyde ratio of 10: 1. Formaldehyde solution was then sprayed on the soybean meal flour evenly and then watered for 12 hours then dried [3] and [6].

The nutritional needs of male thintailed sheep for fattening can be seen in Table 1, the composition of rations and nutrient content of the treatment ration can be seen in Table 2. This study used Randomized Block Design design with 3 treatments and 5 groups. When anova shows significant or significant differences, it is followed by a Duncan Multiple Range Test (DMRT) test. The treatment given was:

P0 = 30% Elephant Grass (EG) + 70% Basal Concentrate (BC)
P1 = 30% Elephant Grass (EG) + 60% Basal Concentrate (BC) + 10% soybeans groats without formaldehyde protection (SGUP)
P2 = 30% Elephant Grass (EG) + 60% Basal Concentrate (BC) + 10% soybeans groats formaldehyde protection

Table 1. The nutrient requirement of male thin tailed sheep for fattening

Requirement Nutrient	Content (%)
Crude Protein	8,93
Total Digestible Nutrient (TDN)	54,55
Calcium (Ca)	0,50
Phosphor (P)	0,35

Sources : Kearl (1982)

Table 2. Ration composition and nutrient content of treatment ration

Feed Ingredient	Treatment		
	P0 (EG 30% + BC70%)	P1 (EG 30% + BC 60% + SGUP 10%)	P2 (EG 30% + BC 60% + SGP 10%)
Elephant grass (EG)	30,00	30,00	30,00
Basal Concentrate (BC) :			
Bran	21,00	18,00	18,00
Pollard	19,00	16,00	16,00
Coffee skin	12,00	10,00	10,00
Palm cake	14,00	12,00	12,00
Mineral	1,00	1,00	1,00
Molasses	3,00	3,00	3,00
Soybean groats without protected (SGUP)	-	10	-
Soybean groats protected (SGP)	-	-	10
Total	100,00	100,00	100,00
Nutrient content (%)			
Crude protein	9.90	11.98	12.05
Crude Fat	4.53	5.54	5.49
Crude fiber	25.80	24.09	24.35
Ash	11.09	10.95	11.03
Extract material without nitrogen	44.74	43.53	43.57
Organic matter	84.91	85.05	85.37
Dry matter	89.19	89.67	88.88
Total Digestible Nutrient	63.72	65.38	62.48
NDF	65.77	60,53	59,90
ADF	26.15	23.50	22.83

This research was conducted through three stages, namely treatment feed adaptation, data collection and data analysis. The feed adaptation stage was conducted for two weeks according to [7]. The adaptation stage lasts for a minimum of 7 days. Feed is given three times a day, i.e., morning, noon and night. Feeding is done with a concentrate first and then elephant grass with a lag time of 1 hour. Provision of drinking water is ad libitum.

The data collection stage is conducted for 10 days. Collections carried out, among others, waste collection of fodder and feces. The remaining feed is accommodated and counted daily. Feed consumption is calculated by weighing the difference between the feed given and the remaining feed each day. Stool collection is done every day at the same time at 06.00 (1x24 hours). Taking feces for 10 days continuously. Feces collected whenever defecation cattle are then weighed and recorded the weight then taken 10% of the total stools to be dried in the sun. Sample of feed and feces then analyzed proximate and van soest in laboratory to know digestion.

The variables in this study include nutrient consumption and digestibility, nutrition value index, and nutrient can be digested.

1. Feed Consumption : a. Consumption of Dry matter (DM) (g / head / day), b. Consumption of Organic Matter (OM) (g / head / day), c. Consumption of Crude Protein (CP) (g / head / day)
2. Digestibility : a. Digestibility DM(%), b. Digestibility OM (%), c. digestibility CP (%)
3. Evaluation of nutrient index (NVI) : a. NVI DM (g / head / day), b. NVI OM (g / tail / day), c. NVI CP (g / head / day).
4. Evaluate the digestible nutrients in the ration : a. DM that can be digested (%), b. OM that can be digested (%), c. CP that can be digested (%)

3. Result and Discussion

Feed intake, digestibility, nutrient value index, and nutrient digestible ingredients of dried, organic, and crude protein rations containing protective soybean and non-in vitro protection in thin sheep can be seen in Table 3.

The result of analysis of variance of soybean protected soybean regimen had no significant effect (P> 0.05) on consumption of dry matter (DM). This indicates that the provision of soybean protected soybeans at the level of 10% of the total ration given had significantaffect the consumption of DM in livestock. Protection of soybeans with formaldehyde did not change the texture and smell of food when it is mixed homogeneously with concentrate so as not to reduce appetite in the cattle because basically when the cattle are hungry then the livestock will cope with the consumption of feed.

Protected soybeans also have the same palatability as control feeds when administered to livestock, so as not to disturb the appetite of livestock in consuming the feed and cause the consumption of BK to be the same between treatments. According [8] that one of the factors that influence feed consumption is palatability that is reflected by organoleptic on a feed ingredient such as appearance, smell and taste. The physical and chemical conditions of the feed are shown by the appearance of smells, flavors and textures that foster attraction and stimulate livestock to consume them. The results of this study are in line with the results of the research by [3] and [6], showing that the use of protected fish meal and soybean did not affect the consumption of DM in Simmental Ongole crossbred male. Rough protein digestion is the amount of crude protein from consumed rations that can be ingested in the digestive tract of livestock. The use of protected soybean soda has no effect on the digestibility of male thin-tailed sheep protein and has not been able to improve the digestibility of crude protein, this may be due to 2% less effective use of formaldehyde to rumen performance, the use of formaldehyde at 0.5 to 1% level effective for the performance of microbial in cattle rumen, so the use of 2% formaldehyde can not improve rumen performance, thus affecting the digestibility of the crude protein, which has no effect on the digestibility of crude protein [6]. Protected proteins are not able to be degraded in the rumen. It is thought that supplying protected soybeans can not be degraded by microbial rumen. In the small intestine too, the soybeans groats protection is still too strong for microbial rupture in the small intestine. It can also affect the rumen microbial activity, with the overcrowded protection of the rumen microbes unable to degrade the protein optimally, so the administration of soybean protected soy does not affect the digestibility of crude protein because the microbial can not break the soybean brewing protection so its activity can be disturbed [7]. Proteins degraded by microbial rumen into ammonia and will be utilized bacteria as a food source which will then be absorbed in the small intestine which will be utilized by livestock for metabolic processes, while non-degraded proteins will go directly into the small intestine and directly absorbed by bacteria.

Table 3. Average nutrient consumption of rations, digestibility, NVI and nutrients can bedigestedration for fattening thin sheep male tail in vivo.

Variable	Treatment		
	P0 (EG 30% + BC70%)	P1 (EG 30% + BC 60% + SGUP 10%)	P2 (EG 30% + BC 60% + SGP 10%)
	----------------	(g/head/day)	---------------------
Feed intake of Dry Matter	788.06±60.62	794.80±77.28	796.63± 38.88
Feed intake of Organic Matter	669.17±51.48	676.01±65.73	680.09±33.19
Feed intake of crude fiber	780.18±60.02	786.85±76.51	788.67±38.49
	---------------	(%)	--------------------
Dry matter digestibility	61.36± 0.88	62.41± 1.29	61.35± 1.58
Organic matter digestibility	67.03± 0.91	66.90± 1.01	67.40± 1.09
Crude fiber digestibility	72.23± 1.41^{a}	75.31± 1.58^{b}	75.15± 1.54^{b}
	----------------	(g/head/day)	---------------------
DM Nutrient Value Index	483.64±38.57	495.49±42.22	488.56±21.69
OM Nutrient Value Index	448.61±36.60	452.48±46.89	458.31±22.19
CP Nutrient Value Index	564.71±51.38^{a}	566.59±45.69^{b}	576.64±23.41^{b}
	--------------	(%)	------------------
Dried materials can be digested	54.73±0.78	55.66±1.15	54.72±1.41.
Organic materials can be digested	56.91± 0.77	56.81± 0.85	57.23± 0.93
Crude protein can be digested	7.16±0.14^{a}	8.64±0.19^{b}	8.81±0.19^{b}

^{ab}Different superscripts on the same line showed significantly different ($P < 0.05$). EG (elephant grass), BC (basal concentrate), SGUP (soybean groats unprotected), and SGP (soybean groats protected)

In line with research of [9] the average of crude protein digestibility in fish meal and soybean cultivation is protected by using formaldehyde respectively are 68.88% and 71.87%. Although the digestibility value is higher, the result of variance analysis in the study is not significant. The concentration of ammonia in rumen fluid depends on the digested protein and the quality of the feed protein. The more protein that is digested then the concentration of ammonia in the rumen is also higher. The results of research conducted by [3] concentration of NH_3 on fish meal, soybean cake, and protected palm oil meal showed unreal result. According to [10] that microbial protein biosynthesis reached its peak at ammonia concentrations in rumen fluid of about 10 mg% and an excess of 98.3% ammonia no longer stimulated microbial growth.

The productive function of feed can be measured by the nutritive value index which is the product of consumption with relative digestibility and predict the amount of ingested food intake [10]. The amount of Nutritive Value index of Dry Matter shows the amount of dry matter digested by livestock. Result of research on consumption of Dry Matter and Dry Matter Digestibilty showed no effect ($P > 0.05$). It is suspected that the value of NVI DM is significantly different due to the same amount of dry matter consumption. In line with the [6] study which stated that dry matter consumption significantly different from dry matter digestibility was not significant resulted in different NVI DM values was not real.

This shows that at the same level of digestibility the amount of consumption of dried ingested ingredients is not the same, the higher the consumption of feed the higher the amount of nutrients are also digested and vice versa. The organic matter NVI is closely related to the dry matter NVI, since the substances contained in the organic matter are contained also in the dry matter. The results showed that NVI OM is directly proportional to NVI DM. This is in line with research [6], which states that at the same level NVI OM is directly proportional to NVI DM.

This suggests that administration of soybean protected soybeans has not been able to increase the value of NVI OM in male thin tail sheep. NVI OM values are also influenced by the consumption of organic matter and the digestibility of organic matter. Result of research on consumption of BO and

Organic Matter Digestibilty showed no effect (P> 0.05). Riyanto *et al*[1] mentioned that nutrient digestibility figures indicate nutrients that can be utilized by livestock.

Digestible organic matter is the digestion of organic matter contained in a feed material that can be absorbed by the intestinal wall. The results showed that the treatment ration had no effect (P> 0.05) to the Digestible organic matter value. Meaning that with the addition of protective soybean broth did not increase or decrease the amount of digested organic material. Digestion is influenced by the rate of food travel in the digestive tract, the physical form or the size of the diet of the ration composer, the chemical composition of the ration and the influence of other food compounds [8].

The addition of protected soy broth in this study has not been able to affect the digestibility of organic matter. According to [7] the quality of feed ingredients is influenced by the composition of feed and will further affect digestibility, so if the quality of feed and composition of feeds [8] the digestibility of organic matter is the amount of food consumed by livestock, where the greater the amount consumed, the faster the travel rate of food in the digestive tract. According to Pujiastuti [1] and [3], the provision of soybeans, fish meal and soybean meal protected in Ongole crossbred cattle has not been able to improve the digestibility of organic matter. This shows that with the same dietary intake, the digested organic material will be the same, so it does not affect the digestibility of organic matter.

4. Conclusion

The use of 10% of soybean protected 37% formaldehyde protected soy by 1% of the dry weight of the concentrate in thin tail fattening rations can improve protein digestibility, nutrient value index and abrasive proteins that can be ingested in vivo.

Acknowledgment

Thank you to the Ministry of Research and Technology and higher education for the cost of this research in the first-year for Penelitian Unggulan Perguruan Tinggi (PUPT) scheme in 2017.

Reference

[1] Riyanto, J, E. Baliarti, T. Hartatik, D.T. Widayati and L. M. Yusiati. 2014. Supplementation Of Pufa Protected In Cattle Feed Based On Rumen Fermentation and Nutrient Digestibility Products By In Vitro Proceedings of International Conference the 16th Animal Science Congress of the Asian-Australasian Association of Animal Production Societies (AAAP). The theme of the 14th AAAP Animal Science Congress is "Sustainable Livestock production In thePerspective Food Security,Folicy,Genetic, Resources,And Climate Change" Faculty of Animal Sciences,
Gadjah Mada University, Yogyakarta, November, 10-14, 2014

[2] Riyanto, J., S. D. Widyawati, W. P. Suprayogi, A. P. Astuti danT. R. Setiadi. 2013. Degradasi secara in vivo tepung ikan, menir kedelai dan bungkil kelapa sawit terproteksi dalam ransum sapi potong. Dalam: Prosiding Seminar Nasional Peternakan Berkelanjutan 5. Jatinangor. Hal 247-
251.

[3] Riyanto, J, E. Baliarti, T. Hartatik, D.T. Widayati and L. M. Yusiati. 2015. Characteristics of polyunsaturated fatty acids and nutrient digestibility feed cattle of the fermented rumen fluid by one and two stage in vitro Proceedings of The 6th International Seminar on Tropical Animal Production (ISTAP 6) Faculty of Animal Sciences, Gadjah Mada University, Yogyakarta, October, 20-22, 2015.

[4] Kiernan, J. A. 2000. Formaldehyde, Formalin, Paraformaldehyde And Glutaraldehyde: What They Are And What They Do. Article published in Microscopy Today 00-1 pp. 8-12.

[5] Johnson, R. R. 1966. Techniques and procedures for in vitro and in vivo rumen studies. Journal Animal Science. 25: 855-875.

[6] Widyawati, S.D., W. Pratitis dan J. Riyanto. 2015. Suplementasi Asam Lemak Tak Jenuh (*Poly Unsaturated Fatty Acid=Pufa*) Berbasis Minyak Sawit Dan Minyak Ikan Lemuru Terproteksi Dalam Ransum Ditinjau Dari Performan Produksi Dan Kualitas Daging Domba. Proceeding of National Seminar on Zootechniques for Indogenous Resources Development. ISSN 978-602-097-243-5. ISAA Publication No. 1/2012. Faculty of Animal Agriculture Diponegoro University, Semarang, Indonesia.

[7] Tillman, A. D., H. Hartadi, S. Reksohadiprodjo, S. Prawirokusumo, dan S. Lebdosoekojo. 1998. Ilmu Makanan Ternak Dasar. Gadjah Mada University Press. Yogyakarta.

[8] Anggorodi. 1990. Ilmu Makanan Ternak Umum. P.T. Gramedia Pustaka Utama. Jakarta.

[9] Puastuti, W., I. W. Mathius dan D. Yulisiani . 2006. Bungkil kedelai terproteksi cairan batang pisang sebagai pakan imbuhan ternak domba : *in sacco* dan *in vivo* . Jurnal Ilmu Ternak dan Veteriner. 11 : 106-115.

[10] Soebarinoto, S. Chuzaemi dan Mashudi. 1991. Ilmu Gizi Ruminansia. Universitas Brawijaya. Malang.

Isolation of Rhizobium Bacteria from Forage Legumes for the Development of Ruminant Feed

E. Fuskhah[1]*, E.D. Purbajanti[1], S. Anwar[1]

[1]Faculty of Animal and Agricultural Sciences, Diponegoro University, Semarang, Central Java, Indonesia.

Corresponding E-mail: eny_fuskhah@yahoo.com

Abstract

The aimed of the study was to explore the presence of Rhizobium bacteria along the northern coast of Central Java, to develop a saline-resistant legumes. Rhizobium bacteria is a mutualistic bacterium capable of symbiosis with legumes so that legumes crop yields increase. The research begins with sampling of soil and root nodule of forage legumes along the Northern Coast of Central Java including Tegal, Pekalongan, Semarang, Demak, Pati. Soil samples were analysed for salinity, Total Dissolved Solids, and pH. Rhizobium bacteria were isolated from the acquired root nodule, then identified by biochemical test to ensure that the isolates obtained were Rhizobium bacteria. The results showed that the five districts/municipal sites sampled by the soil have very low salinity to very high levels. The highest level of soil salinity was found in Demak (Sayung) which has an electrical conductivity value (EC) of 17.77 mmhos/cm. The EC values of legumes overgrown soils showed a low salinity level while bare soils have high salinity levels. Feed crops legumes that could be found in the northern coast of Central Java were *Centrosema pubescens, Calopogonium mucunoides, Leucaena leucocephala,* and *Sesbania grandiflora.* The study obtained 6 kinds of isolates of rhizobium bacteria isolated from forage legumes, included 1) *Centrosema pubescens isolated from* Pekalongan, 2) *Centrosema pubescens isolated from* Tegal, 3) *Calopogonium mucunoides isolated from* Pekalongan, 4) *Leucaenaleucocephala* isolated from Tegal, 5) *Leucaena leucocephala* isolated from Semarang, 6) *Sesbania grandiflora* isolated from Tegal.

1. Introduction

Indonesia is the largest archipelagic country in the world with a total of about 13,667 islands. Between one island and another island is separated by the sea. Such conditions resulted in the Indonesian state having thousands of hectares of saline coastal land. The saline soil may also occur in dry areas where soil drainage is impaired, evaporation increases, and rainfall is low. Saline land in Indonesia which is thousands of hectares is a very potential land if managed properly, so it can be useful optimally. One of the alternative uses is to grow feed crops. As is known the availability of feed crops is declining, along with reduced land for feed crops and seasonal constraints. Reduced land for feed crops, among others, due to displace by food crops and settlements. Attention to the marginal lands such as saline land is now getting bigger. On the other hand, the need for animal protein is increasing as the population increases. Ruminant livestock producing meat and milk is a potential animal source of animal protein. Feed plants are an important factor for the development of ruminant livestock, as most of the feed for these animals is forage. For that reason, it is necessary to use the saline field to grow the feed crops. Efforts to grow feed crops and raise ruminant livestock in coastal areas are also an alternative of business diversification to improve the living standards of fishermen around the coast who are generally low income.

The salinity stress is one of the most influential factors in limiting agricultural production [1]. The term saline is used in conjunction with a soil having a conductivity of more than 4 mmhos/cm at 25°C, having the percentage of sodium exchangeable less than 15 and a pH less than 8.5. Water-soluble salts tend to accumulate in the top layer or ground surface. The process of soluble salt accumulation in the soil is called salinization. The accumulation of a number of salts such as chloride, sulphate, sodium chloride, calcium and magnesium salts are present in the saline soil. Sodium chloride (NaCl) is the dominant salt in saline soils in coastal areas [2]. The response of plants to the salinity or electrical conductivity (EC) levels expressed in mmhos/cm at 25°C is different. Tree legumes *Sesbania grandiflora* and *Leucaena leucocephala* exhibited higher salinity resistance than the cover crops legumes *Calopogonium mucunoides* and *Centrosema pubescens* [3]. Plant growing on saline soil is inhibited due to sodium ion poisoning (Na). The results of Sopandie [4] showed that with increasing NaCl concentration (up to 250 mM) will increase Na content in canopy and roots of barley and peanut plants. In addition, nutrient uptake is inhibited in saline soil. The addition of 2.0 mM calcium (Ca) will lower the Na content in the canopy of the plant. Calcium in saline conditions has contributed to the ion translocation regulation, thus reducing the toxicity of Na + ions while increasing K + ion uptake. In other words, the selectivity of K + / Na + in the membrane is influenced by calcium [4]. Fuskhah [5] showed that high salinity decreased nitrogenase activity of *Caliandra callothyrsus*. Efforts to increase the productivity of saline can be achieved through the cultivation of salt tolerant plants and reclamation

Legumes have mutualistic symbiotic ability with Rhizobium sp. bacteria that grow in the root region. The existence of this bacteria causes the formation of root nodules that are able to fix the free nitrogen from the air so as to supply the plant needs for nitrogen. The symbiotic result is expected to increase the production of forage plants. Fuskhah [6] demonstrated that the use of Rhizobium inoculum from 20-60 g / kg of seeds combined with phosphorus fertilization can increase the production of the dry matter forage Centrosema pubescens Benth. The ability to fix nitrogen can reduce the cost of purchasing artificial N-fertilizers, so the application of Rhizobium inoculation in legumes becomes very important to spur nitrogen fixation. For application in saline soils it is necessary to look for efforts to obtain saline-resistant Rhizobium isolates that can be applied to legumes in saline coastal areas. Based on these descriptions, then attempts to obtain saline-resistant Rhizobium strains are important to improve association success with forage legumes in coastal areas.

2. Materials and Methods

2.1. Analysis of Land Salinity of North Coast of Central Java

The objective of this study was to obtain information about salinity level of land area of North Coast of Central Java. This study used soil from North Coast of Central Java, plastic bags, chemicals and tools needed for pH analysis, total dissolved solids (TDS) and electrical conductivity (EC). The selection of soil sampling sites is based on the classification of climate according to Oldeman. Of the 11 districts along the North Coast of Central Java, most have a C climate of 7 districts. Climate D as many as 3 districts, and climate E as much as 2 districts. There are 1 districts with part of climate D and some having climate E. Thus, soil samples from C climate region were 3 districts, 1 location having D climate, and 1 location having E climate. Parameter observed is the pH, TDS, and EC of each soil sample.

2.2. *Isolation, and Identification of Rhizobium Bacteria from Forage Legumes*

The materials used in this study were root nodules of forage legumes from North Central Coast area of Central Java, 95% ethanol, 0.1% sublimate, distilled water, yeast mannitol agar medium (YMA), congo red (CR) bromine thymol blue (BTB), NaCl, tools for picking up root nodules, aluminium foil, rubber bands, cotton, cover paper, plastic, test tube, tweezers, erlenmeyer, stirring rod, petri dish, 'ose', oven, autoclave.

The sample of root nodules was taken from the North Coast of Central Java. The root aggregates were collected, then sterilization, isolation, identification according to [7].

1. Collection of root nodules
   The root nodules were collected by separating the root nodule and cutting 0.5 cm root on each side of the root nodules.
2. Sterilization of root nodule
   Root nodules was immersed in a 95% ethanol solution for 5-10 seconds, then immersed in a 0.1% sublimate solution for 3-5 minutes, then washed five times with sterile water.
3. Isolation of Rhizobium bacteria
   The sterile root was crushed with tweezers, then the root nodules were grown on a plate medium YMA + CR, incubated for 3-5 days at room temperature (28^0C) [7].
4. Identification of Rhizobium bacteria
   The colonies that grew apart and identified as Rhizobium bacteria were subsequently grown on the identification medium, using the inclined medium YMA + CR, YMA + BTB. Next, the bacteria were stained with Gram dye. All the acquired Rhizobium bacteria are collected. Pure culture obtained can then be stored as a Rhizobium culture stock, and may be used for further research.

3. Results and Discussion

3.1. Analysis of land salinity of north coast region of central java

The samples soil were obtained from Tegal, Pekalongan, Semarang, Demak, Pati during the dry season (August-September 2017). Based on the results of the survey and laboratory analysis, the results obtained as listed in Table 1.

Table 1 shows that from the five districts/municipal locations sampled the soils have very low to very high salinity levels. The salinity level is divided into 5 categories based on EC values in mmhos/cm that are very low (0-2), low (2 - 4), medium (4 - 8), high (8 - 16), and very high (> 16). The salinity level of the land usually changes according to the season. In the dry season, the level of soil salinity usually increases as the amount of water derived from rain is not enough to neutralize the amount of water lost by evapotranspiration, so that when water is evaporated into the atmosphere it causes the soluble salts to accumulate to the soil surface [8]. Causes the soil to be saline are :
(1) the soil has a salt-containing material [9], (2) sea water intrusion, the accumulation of salt from the irrigation water used or the groundwater motion reclaimed from the seabed [8], (3) high evapotranspiration rate with low rainfall so that minerals are not completely washed [10].

The highest level of soil salinity was achieved by Demak (Sayung) which has an electrical conductivity value (EC) of 17.77 mmhos / cm. EC values of legumes overgrown soils show a low salinity level while bare soils have high salinity levels. This shows that legumes are not able to live on soils that have high salinity levels. Feed crops legumes that can be found in the northern coast of Central Java are *Centrosema pubescens, Calopogonium mucunoides, Leucaena leucocephala*, and *Sesbania grandiflora*

Table 1. Soil pH, TDS, and EC Analysis Results of North Coast Region of Central Java

Java Location (Plants)	TDS (ppm)	EC (mmhos/cm)	pH
Tegal (*Centrosema*)	241	1.545	8.58
Tegal (*Leucaena*)	152	0.31	**6.93**
Tegal (*Sesbania*)	458	0.911	9.00
Pekalongan (*Calopogonium*)	372	0.729	6.46
Semarang (-)	7.83	3.15	6.08
Demak (-)	8840	17.77	7.39
Pati (-)	3.97	7.89	1950

Description: (-) no legume

3.2. Isolation and identification of Rhizobium bacteria

3.2.1. Collection of root nodules, based on the observation, it can be said that the nodule of the roots have a variety of shapes and sizes. Most root nodules in all species lie at the root of the root. In transverse slices, the root nodules appear pink. This shows that the root nodule is effective. In accordance with the opinion of [11], which stated that effective root nodules are pink, because they contain leghaemoglobin. A red pigment called leghaemoglobin is encountered in a root nodule between the bacterium and the membrane surrounding the membrane. The number of leghaemoglobin in the root nodule has a direct relationship with the fixed amount of nitrogen [12].

3.2.2. Isolation and identification of Rhizobium bacteria, effective root nodules, which are fresh, non-wrinkled and non-dried are taken for further sterilization and isolation. The results of laboratory analysis, obtained 6 kinds of isolates of rhizobium bacteria isolated from forage legumes were: 1) Centrosema pubescens isolated from Pekalongan, 2) Centrosema pubescens isolated from Tegal, 3) Calopogonium mucunoides isolated from Pekalongan, 4) Leucaena leucocephala isolated from Tegal, 5) Leucaena leucocephala isolated from Semarang, 6) Sesbania grandiflora isolated from Tegal. The growth of rhizobium bacteria as in Table 2.

Table 2. Growth of rhizobium bacteria

Various isolates	Density	YMA + CR	YMA + BTB
Centrosema isolated from Pekalongan	++	pink	yellow
Centrosema isolated from Tegal	+	pink	yellow
Calopogonium isolated fromPekalongan	++	pink	yellow
Leucaena isolated from Tegal	+++	pink	yellow
Leucaena isolated from Semarang	+	pink	yellow
Sesbania isolated from Tegal	++	pink	yellow

Description: + = grow solid, ++ = grows very solid. +++ = grow very dense

The obtained isolates were then tested biochemically. After incubation for 5 days, all of the isolates obtained grew well on YMA + CR medium, white colony color and medium color remained pink. The colony grown on the medium of YMA + BTB grows well, the medium color of green turns yellowish after Rhizobium grows. This indicates that all of the isolates obtained are included fast growing bacteria. Microbes have no real anatomical features, so bacterial identification is based on morphology, culture properties and biochemical properties. Morphology of microorganisms based on shape and size is not sufficient to identify. Other features such as staining properties, colony growth patterns, growth reactions to carbohydrates, and the use of amino acids are helpful in the identification of microbes [13].

4. Conclusion

The conclusions that can be drawn was that the five districts/municipal sites sampled by the soil have very low salinity to very high levels. The highest level of soil salinity is achieved by Demak (Sayung) which has an electrical conductivity value (EC) of 17,77 mmhos/cm. The EC values of legumes overgrown soils show a low salinity level while bare soils have high salinity levels. Feed crops legumes that can be found in the northern coast of Central Java were *Centrosema pubescens*, Calopogonium mucunoides, *Leucaena leucocephala*, and *Sesbania grandiflora*. The results of laboratory analysis, obtained 6 kinds of isolates of rhizobium bacteria isolated from forage legumes were: 1) *Centrosema pubescens* isolated from Pekalongan, 2) *Centrosema pubescens* isolated from Tegal, 3) *Calopogonium mucunoides* isolated from Pekalongan, 4) *Leucaena leucocephala* isolated from Tegal, 5) *Leucaena leucocephala* isolated from Semarang, 6) *Sesbania grandiflora* isolated from Tegal.

Acknowledgments

The authors would like to thank to the Ministry of Research, Technology and Higher Education of Indonesian Republic for the support of research funding. Acknowledgments were also conveyed to Widi Dwi Noviandi and Hafidz for assisting the research.

References

[1]Abolhasani, M., A. Lakzian, A. Tajabadipour, dan G. Haghnia. 2010. The study salt anddrought tolerance of Sinorhizobium bacteria to the adaptation to alkaline condition. Australian Journal of Basic and Applied Sciences, 4(5):882-886.

[2] Soepardi, G. 1983. Sifat dan Ciri Tanah. Fakultas Pertanian. Institut Pertanian Bogor. Bogor.

[3] Fuskhah, E, S. Anwar, E.D. Purbajanti, R.D. Soetrisno, S.P.S. Budhi, dan A. Maas. 2007. Eksplorasi dan seleksi ketahanan Rhizobium terhadap salinitas dan kemampuan berasosiasi dengan leguminosa pakan. Jurnal Pengembangan Peternakan Tropis. **32**(3) : 179.

[4]Sopandie, D. 1990. Studies on Plant Responses to Salt Stress. Disertasi PhD. Okayama Univ. Japan

[5]Fuskhah, E; Karno, dan F. Kusmiyati. 2003. Efek salinitas dan pemberian fosfor terhadap aktivitas enzim nitrogenase nodul akar Calliandra calothyrsus. Jurnal Pengembangan Peternakan Tropis Special Edition, Oktober 2003. ISSN 0410-6320.

[6]Fuskhah, E, E.D. Purbayanti, F. Kusmiyati, dan R.T. Mulatsih. 1997. Efek inokulasi Rhizobium Sp dan pemberian fosfor terhadap derajat katalisis enzim nitrogenase nodul akar Centrosema pubescens Benth. Majalah Penelitian. Lembaga Penelitian Universitas Diponegoro. IX(34): 19-25

[7]Vincent, J.M. 1970. A Manual for The Practical Study of The Root Nodule Bacteria. IBP Handbook No. 15, Blackwell Scientific Publ., Oxford.

[8]Tan, K. H. 1995. Dasar-Dasar Kimia Tanah. Gadjah Mada University Press, Yogyakarta (Diterjemahkan D. H. Goenadi).

[9]Buckman, H.O. dan N.C. Brady. 1982. Ilmu Tanah. Penerbit Bhratara Karya Aksara. Jakarta.

[10]Bintoro, M.H. 1983. Pengaruh NaCl terhadap pertumbuhan beberapa kultivar tomat. Bulletin Agribisnis XIV(1) : 13-35.

[11]Buchanan, R.E. and N.E. Gibbans. 1974. Bergey's Mannual of Determinate Bacteriology. Eight Edition. The William and Wilkins Company, Balltimore.

[12]Rao, N.S.S. 1994. Mikroorganisme Tanah dan Pertumbuhan Tanaman. Edisi kedua. Penerbit Univesitas Indonesia.

[13] Lay, B.W. 1994. Analisis Mikroba di Laboratorium. PT. Raja Grafindo Persada, Jakarta.

Commercial Cuts of Carcass of Thin-Tailed Lambs and Sheep Fed Complete Feed

N. Nurbaeti[1], C.M.S. Lestari[1]* and E. Purbowati[1]

[1]Faculty of Animal and Agricultural Sciences, Diponegoro University, Semarang, Central Java, Indonesia.

*Corresponding E-mail: cmslest@yahoo.co.id

Abstract

This research was conducted to examine the commercial cuts of thin tailed lambs and sheep fed complete feed. This study used 6 male thin tailed sheep aged ±11 months with average body weight of 23,01 ± 1,91 kg (CV 8,31%) and 6 male thin tailed lambs aged ±4 months with average body weight of 15,41 ± 2,11 kg (CV 13,72%). They were fed a complete feed as much as 3.5% from body weight. Fed contained 12% crude protein (CP) and 55% total digestible nutrients (TDN) and was given ad libitum. The animals were raised for 12 weeks and slaughtered being carcass. Commercial cuts were obtained from a half right body part of each sheep carcass and divided into eight commercial cuts, i.e., neck, shoulder, fore shank, breast, flank, rack, loin and leg. The collected data was analysed using t- test in 5% level. The results showed that slaughter weight, carcass weight and half carcass weight were significantly different ($P<0.05$) while the percentage of carcass weight and half carcass was not significantly different ($P>0.05$). All of the weight commercial cuts of thin tailed lambs and sheep were significantly different ($P<0.05$) but flank ($P>0.05$). The percentage of neck and leg were significantly different ($P<0.05$), while the other commercial cuts were not significantly different ($P<0.05$). Based on the result, it can be concluded that commercial cuts of sheep is better than lambs.

1. Introduction

Thin Tailed Sheep is a small ruminant that usually raised by local people. There are two kinds of feed for sheep that used to use which are grass and concentrate. Generally, sheep fed only grass could not achieve the requirements of production, therefore the concentrate being a feed addition with high nutrition. The concentrates utilization is used to improve the nutrient value and increase the energy and protein of feed [1]. In addition, the concentrates utilization also improves the dry matter feed digestibility, body weight gain, and feed efficiency. Growth can be manipulated with nutrients treatment by providing better nutrition [2] and good nutrition levels affects the body weight of sheep [3]. Different levels of nutrition will produce a carcass different composition [4]. Therefore, the concentrates contained complete nutrients and high protein is required.

Nowadays, the demand of lean meat is increasing. Lean meat can be obtained from lambs. The lambs has slow fat growth levels because feed nutrition given is still carrying out for bone growth, meat and internal organs requires. Bones is the skeleton body that grows and develops early also followed by the meat and fat growth [2]. The increased body weight and high carcass production depend on age of the animals [5]. According to the mature sheep, the bone growth rate were slowed then produced fat [6]. The variety composition of meat tendon depends on age as well [4]. Good slaughtered process is in lambs because the fat growth has optimum than sheep [7].

Commercial cuts are usually based on the consumer demand and to fulfil carcass selection of meat as desired [2]. The carcass commercial cuts of sheep divided into: shoulder, rib, loin, shank, breast, flank, leg and neck [8]. Every carcass commercial cuts of sheep have different proportions based on each carcass part. The purpose of this research is to examine commercial cuts of thin tailed lambs and sheep carcasses.

2. Materials and Methods

The materials used in this research were 6 thin tailed lambs aged ± 4 months with average body weight of 15.41 ± 2.11 kg (CV 13,72%) and 6 thin tailed sheep aged ± 11 months with average body weight of 23.01 ± 1.91 kg (CV 8,31%). They were raised intensively and fed complete feed that contained 12% CP and 55% TDN. The complete feed composed of 50% elephant grass and 50% concentrate that consisted of 50% elephant grass, 8% soybean meal, 9% rice bran, 2% mineral mix, 8% dried cassava, 18% wheat bran and 5% molasses. This research used randomized group design according to the age of the animals. The parameters observed in this study were dry matter intake, dry matter digestibility, feed conversion and body weight gain.

The research used independent sample comparison design method between sheep and lambs in commercial cuts. The sheep had 12 weeks periods treatment and fasted for 12 hours before slaughtering process. Afterwards, sheep's body were separated into carcass and non-carcass. Carcasses were weighed and split into 2 parts (right and left carcass) and stored in low room temperature for 10 hours. Each right carcasses were divided into eight parts according to the carcass commercial cuts that consisted of neck, shoulder, rack, loin, fore shank, breast, flank, leg [8]. The eight parts of carcass commercial cuts were weighed and analysed using T test significant in 5% level.

3. Results and Discussion

The average weight and percentage of sheep and lambs carcasses and commercial cuts are shown in Table 1. The results showed that all commercial cuts weight of thin tailed sheep and lambs carcass were significantly different ($P<0.05$) but flank ($P>0.05$). The percentage of commercial cuts of thin tailed sheep and lambs carcass on the neck and leg was significantly different ($P<0.05$), while the others were not significantly different ($P>0.05$).

The shoulder weight of sheep and lambs was significantly different ($P<0.05$). In the other hand, the shoulder weight of sheep and lambs was not significantly different ($P>0.05$). Shoulder is a mature part that has a large growth rate at the first [9]. The shoulder is a proponent of motion function. Shoulder percentage increased by the increasing of carcass weight [10].

The neck weight and percentage were significantly different ($P<0.05$). It indicated that neck weight in mature sheep was larger than lambs because bone growth in mature was optimal. The first growth rate section on lambs occurred slowly and became rapidly then. It became slowly and ended up reached mature [2]. According to [3] growth curve showed that sheep body weight was affected by age. Sheep growth was started from head and spreads to the body until foot tip to the upper body [4].

The shoulder weight of sheep and lambs was significantly different ($P<0.05$). In the other hand, the shoulder weight of sheep and lambs was not significantly different ($P>0.05$). Shoulder is a mature part that has a large growth rate at the first [9]. The shoulder is a proponent of motion function. Shoulder percentage increased by the increasing of carcass weight [10].

Table 1. Body Weight, Weight and Percentage of Carcass, Slaughter Carcasses and Commercial Cuts of Thin Tail Sheep and Lambs

Variables	Weight (g)		Percentage (%)	
	Sheep	Lambs	Sheep	Lambs
Body weight	22585.67[a]	16736.67[b]	-	-
Weight carcass	8934.00[a]	6110.97[b]	39.51[a]	36.41[a]
Slaughter weight of carcass	4472.75[a]	3060.81[b]	50.07[a]	50.09[a]
Commercial cuts :				
Neck	330.18[a]	196.62[b]	7.32[a]	6.41[b]
Shoulder	1185.12[a]	761.05[b]	26.48[a]	24.70[a]
Fore shank	245.67[a]	191.45[b]	5.54[a]	6.31[a]
Breast	356.83[a]	236.37[b]	7.97[a]	7.69[a]
Flank	69.82[a]	45.90[a]	1.57[a]	1.47[a]
Rack	374.03[a]	231.16[b]	8.29[a]	7.55[a]
Loin	486.84[a]	309.94[b]	10.78[a]	10.08[a]
Leg	1424.26[a]	1088.32[b]	32.05[a]	35.78[b]

Different letter in the same raw are significantly different (P<0.05)

The fore shank weight of sheep and lambs was significantly different (P<0.05) but fore shank (P>0.05). Fore shank functions are used to walk and move the body. Therefore, it becomes the earlier developed part [11]. The lamb's legs tend to have a greater proportion than other body parts [12]. Sari et al. [13] found that fore shank bone percentage has high proportion.

The breast weight of sheep and lambs was significantly different (P<0.05) but the percentage of breast (P>0.05). Breast functions are used to protect the lungs, heart and other important organs [9], therefore, the breast was included in the mature body part [14].

The rack weight of sheep and lambs was significantly different (P<0.05) but the rack percentage (P>0.05). Rack is one of carcass part that has a slow growth. The growth of the new rack part begins when other parts are declining growth [14]. The increasing fat and meat deposition will lead to the rack section when the body weight increase [15].

The flank weight and percentage of sheep and lambs were not significantly different (P<0.05). The flank average weight and percentage were 57.86 g and 1.52% respectively. Flank or groin is the smallest commercial cut because it has no bone. Flank function is used to hold stomach contents such as digestive tract and others internal organ [15].The digestive tract weight and size contents are related to the flank growth. Although the weight is increased, the flank weight is relative same. Therefore, the flank percentage is low when the weight increase [16].

The loin weight of sheep and lambs was significantly different (P<0.05) but the loin (P>0.05). Loin has a slow growth. Meat and fat accumulation occur in the loin portion [14]. Loin percentage increase according to the increase carcass weight [10].

The legs weight and percentage of sheep and lambs were significantly different (P<0.05). Leg grows early for walking and moving. According to [11], leg has a great bone growth rate. The leg percentage decrease while the carcass weight increase [10].

4. Conclusion

Based on the results, sheep carcass commercial cuts was larger than of lambs, but flank. Sheep neck percentage was larger than of lambs, while the legs percentage in lambs was greater than of sheep.

References

[1] Krisnan, R. 2013. Respon tiga rumpun kambing terhadap pemberian tambahan konsentrat. Seminar Nasional Teknologi Peternakan dan Veteriner, Sumatra Utara.

[2] Soeparno. 2015. Ilmu dan Teknologi Daging. Cetakan Ke-6. Gadjah Mada University Press, Yogyakarta.

[3] Owens, F. N., P. Dubeski and C. F. Hanson. 2014. Factors that alter the growth and development of ruminants. J. Anim. Sci. 1993. 71: 3138-3150

[4] Lawrie, R. A. 2003. Ilmu Daging. Edisi Ke-5. Universitass Indonesia Press, Jakarta.

[5] Speedy, A. W. 1980. Sheep Production. Longman, London.

[6] Hammond, J. Jr., Bowman, J. C. and Robinson, T.R. 1984. Hammond's Farm Animals. 5th ed. Butler and Tanner Ltd. London.

[7] Blakely, J. and D.H. Bade. 1994. Ilmu Peternakan. Edisi Keempat. Gadjah Mada University Press, Yogyakarta.

[8] Sunarlim, H dan H. Setiyanto. 2005. Potongan komersial karkas kambing kacang jantan dan domba lokal jantan terhadap komposisi fisik karkas, sifat dan nilai gizi daging. Seminar Nasional Teknologi Peternakan dan Veteriner.

[9] Frandson, R, D. 1986. Anatomi dan Fisiologi Ternak. Edisi Ke-4. Gadjah Mada University Press, Yogyakarta.

[10] Sugana, N., M. Duldjaman, A. Natasasmita dan Saparto. 1983. Paha dan lemusir domba local Priangan berdasarkan jenis kelamin dan pengelompokan bobot potong. Pertemuan Ilmiah Penelitian Ruminansia Kecil, 22-23 November 1983. Bogor.

[11] Berg, R, T and R. M. Butterfield. 1976. New Concepts of Cattle Growth. Sidney University Press, Sidney.

[12] Black, J. L. 1981. Sheep Production. Editor W. Haresign. Proc. 35th Easter School in Agric Sci. University Nottingham Butterworth, London.

[13] Sari, R. T., A. Purnomoadi dan R. Adiwinarti. 2016. Potongan komersial dan komponen karkas kambing Kacang jantan umur 1-1,5 tahun dengan pemeliharaan tradisional (Studi kasus di Kecamatan Wirosari Kabupaten Grobogan). Jurnal Animal Agriculture. 5(2): 11-16

[14] Sent, A, E., L. E. Walters and J. V. Whiteman. 1982. Performance and carcass characteristics of ram lambs slaughtered at different weight. J. Anim. Sci. 55 (6) : 1360-1368.

[14] Mawati, S., F. Warastuty dan A. Purnomoadi. 2004. Pengaruh pemberian ampas tahu terhadap potongan komersial karkas domba lokal jantan. J. Indon. Anim. Agric. 29 (3) : 172-176.

[15] Sonjaya, H. 2012. Dasar Fisiologi Ternak. IPB Press, Bogor.

[16] Hasnadi. 2014. Pertumbuhan relatif potongan komersial karkas domba Sungei Putih dan Lokal Sumatera yang mendapat pakan limbah kelapa sawit. J. Indon.Trop.Anim.Agric. 29 (3): 115-120.

The Effects of Manure at Saline Soil on Growth, Dry Matter Production and Crude Protein of *Sesbania grandiflora*

F. Kusmiyati[1]*, E. D. Purbajanti[1] and Surahmanto[1]

[1]Faculty of Animal and Agricultural Sciences, Diponegoro University, Semarang, Central Java, Indonesia.

*Corresponding E-mail: fkusmiyati@yahoo.co.id

Abstract

Saline soil is one of major abiotic stress in Indonesia. The research was conducted to evaluate growth, dry matter production and crude protein of *Sesbania grandiflora* at saline soil with different application dosage of manure. The research was done during June – December 2016at saline soil (EC = 4.1 dS/m) in Kaliori sub-district area, Rembang Regency, Central Java Province, Indonesia. The treatments were dosage of manure (0, 10 and 20 tonnes/ha). *Sesbania grandiflora* was planted as sole legume and inter-planted with *P. maximum.*Parameters measured were plant height, number of leaves, dry matter production and crude protein at first cutting and second cuttings. Results showed that plant growth was higher with application manure at saline soil. The effect of manure was obvious at second cutting. The highest dry matter production was obtained at application 20 ton/ha manure. Application manure at saline soil affected crude protein of *S. grandiflora* at first cutting. Manure was not affected crude protein *S. grandiflora* at second cutting. The conclusion was application of 20 ton/ha manure at saline soil (EC = 4.1 dS/m) increased growth, dry matter production and crude protein of *Sesbania grandiflora.*

1. Introduction

Saline soil is one of major abiotic stress in Indonesia. Excess salt, especially sodium chloride (NaCl) on the saline soil will cause ionic stress and osmotic stress [17]. Sodium (Na) and Chloride (Cl) are not essential for plant, therefore accumulation of these specific ions can be directly toxic for plants. The ability of plant to absorb water will reduce because of osmotic stress. Osmotic stress is similar to drought stress. Both ionic stress and osmotic stress will decrease plant growth. The effect of drought stress on *Oryza sativa* has been associated with the reduced chlorophyll content, number of tillers, stover yield and grain yield [8]. Plant ttolerance to soil salinity is varied. *Sesbania grandiflora* was the most tolerant crops on saline soil than *Leucaena leucocephala,Centrosema pubescens*, *Caliandra calothyrsus* and *Crotalaria juncea* in saline soil with electrical conductivity (EC) of 11 dS/m [5]. Biomass production of *Sesbania* has shown promisingly on moderately saline sodic soil among the forage species [9]. *Sesbania grandiflora* could be grown as sole crop or mixed cropping.

The negative effects of saline soil are intensified by the low levels of soil organic matter. Low organic matter at saline soil will reduce biological activity. The role of nitrogen becomes crucial at saline soil because of biological activity of saline soil is poor. A number of organic N-containing materials, such as animal manure, sewage sludge and plant residues are being used as soil amendments and a sources of N for improving the saline soil quality [18]. The research was conducted to evaluate growth, dry matter production and crude protein of *Sesbania grandiflora* at saline soil with different application dosage of manure.

2. Materials and Methods

2.1. Study area

The research was conducted at saline soil in the Kaliori sub-district area, Rembang Regency, Central Java Province, Indonesia (July– December 2016). Annual rainfall was 1140 mm year^{-1}. Soil pH, C-organic, total nitrogen and cation exchange capacity were 7.8 (alkaline), 1.08% (low), 0.13% (low) and 11,99 cmol/kg (low), respectively. Eelectrical conductivity (EC) was 4.1 dS/m. According to saline soil classification [1], saline soil at that area classify as moderately saline

2.2. Design and treatments

The experimental design was a randomized complete block design (RCBD) with three blocks. The treatments were dosage of manure (0, 10 ton/ha and 20 ton/ha). Cow dung was used as manure. *Sesbania grandiflora* was planted as sole legume and inter-planted with *P. maximum*. The size of each plot was 6 m x 7 m. As sole legume, *Sesbania grandiflora*was planted at 100 cm x 75 cm. In inter-planted, *S. grandiflora* was planted between *P. maximum* rows that planted at 100 cm x 75 cm. *S. grandiflora* and *P. maximum* were planted at the same time. The planting materials wereseed for *S. grandiflora* and tillers for *P. maximum.* Anorganic fertilizer was applied with recommended dosage of nitrogen (60 kg N/ha/cutting), phosphorus (150 kg P_2O_5/ha) and potassium (100 kg K_2O/ha). *S. grandiflora* was cut at 13 weeks after planting. The second cut was done 7 (seven) weeks after first cut. *S. grandiflora* was cut at 30 cm above soil.

Parameters measured were plant height, number of leaves, dry matter production and crude protein. One hundred grams of fresh forage yield was dried at 105°C until the weight of sample was constant to measure dry matter percentage. Crude protein and crude fibre was determined using the procedure of the Association of Official Analytical Chemists (AOAC, 1984). Data obtained were analyzed using analysis of variance, followed by Duncan Multiple Range
Test (DMRT) to compare the difference between treatments [12].

3. Results and Discussion

3.1. Plant Growth

Application manure on saline soil increased ($p < 0.05$) plant height and leaf number of *S. grandiflora* at first cutting and second cutting. Growth of S. grandiflora was significantly lower without manure (0 ton/ha). There were no significant difference ($p > 0.05$) between application 10 ton/ha and 20 ton/ha manure on growth of *S. grandiflora*. Growth of S. grandiflora when planted solely and when inter-planted with P. maximum were not differed ($p > 0.05$) at the same dosage of manure at saline soil (Table 1).

Plant growth was higher with application manure at saline soil. Manure can be function both as organic fertilizer and soil amendment. There are many beneficial effects of organic fertilizer as soil amendment such as soil biological and physical properties improvement and slow release of nutrients [3]. Organic amendments reduced the salt content at saline soil. This condition will improve the content of nutrient available at saline soil such as the availability of nitrogen (N), phosphor (P) and potassium (K) [16]. Potassium is one of essential nutrient for plant. The function of K is to maintain plant turgor pressure or osmotic pressure. Low osmotic pressure of saline soil with application of manure will increase absorption of water. Application manure at saline soil increased C-organic, organic matter, cation exchange capacity and potassium exchange [2].

Table 1. Plant height (cm) and number of leaves of *Sesbania grandiflora* at different dosage of manure

Manure		Plant heigh (cm)		Number of leaves	
		1st Cutting	2nd Cutting	1st Cutting	2nd Cutting
0 ton/ha	Sole legume	55.5^{b}	153.2^{b}	48.8^{c}	165.0^{b}
	Inter-planted	60.5^{b}	113.7^{b}	66.7^{bc}	143.3^{b}
10 ton/ha	Sole legume	105.6^{a}	233.7^{a}	139.8^{ab}	306.3^{a}
	Inter-planted	119.7^{a}	222.6^{a}	176.3^{a}	345.3^{a}
20 ton/ha	Sole legume	111.4^{a}	248.4^{a}	150.5^{a}	346.7^{a}
	Inter-planted	101.1^{a}	227.1^{a}	146.2^{ab}	293.0^{a}

Values within columns having the same superscript are not significantly different ($P<0.05$) using Duncan Multiple Range Test

3.2. Dry matter production

Dry matter production of *S. grandiflora* at saline was increased at treatment of 10 ton/ha and 20 ton/ha organic fertilizer both planted solely and inter-planted with P. maximum (Figure 1). An increase in DM production of S. grandiflora when planted solely at first cutting, from 0.97 ton/ha to 2.15 ton/ha and 2.38 ton/ha, as organic fertilizer increased from 0 ton/ha to 10 ton/ha and 20 ton/ha. The production of S. grandiflora when inter-planted with P. maximum were 1.08, 2.08 and 2.59 ton DM/ha for organic fertilizer application at 0, 10 and 20 ton/ha respectively. Dry matter production of S. grandiflora was 45 – 259 g/m^2 or 72.99 – 194.36 g/plant at first cutting. That production was higher than those reported by [15] whose reported dry matter production of S. grandiflora was 68 g/plant. Application of organic fertilizer increased significantly dry matter production of S. grandiflora at saline soil. The lowest dry matter production of *S. grandiflora* at first cutting was at no organic fertilizer treatment both planted solely and inter-planted with *P. maximum.*

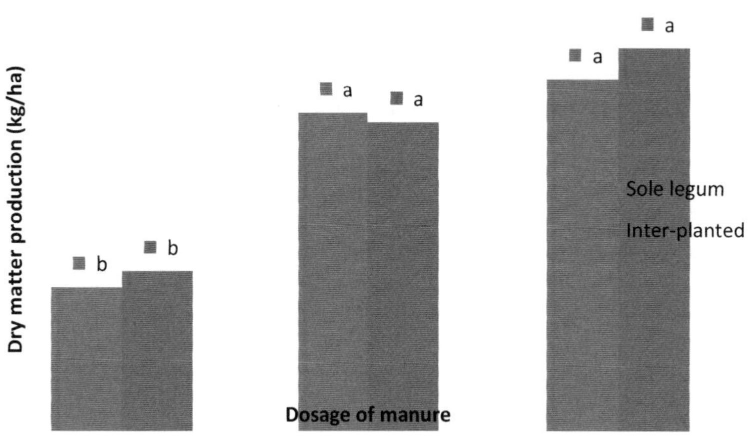

Figure 1. Dry matter production of *Sesbania grandiflora* at first cutting

The effect of manure was obvious at second cutting. The highest dry matter production was obtained at application 20 ton/ha manure (Figure 2). Manure as organic fertilizer release nutrient slowly, so the availability of nutrient increase time by time. Application of manure increased nitrogen availability. Total N at saline soil was 0.05 g/100 g soil. Application of manure at saline soil increased total N 60 to 90 % or 0.38 -0.55 g/100 g soil [19]. The influence of soil salinity on the decomposition of organic matter and nitrogen mineralization in animal manure amended soils. The amount of CO2-C released from animal manure amended soils was found to be increased at an increasing rate as incubation progressed [14]. Saline soil showed significantly ($P \leq 0.05$) low NH4 $^{+}$ -N content compared to the non-saline soil. The release of N-NH4 in the saline soil with the addition of manure is significantly higher than without manure (control treatment). The content of N-NH_4 in the control was 2 - 13 mg / kg of soil

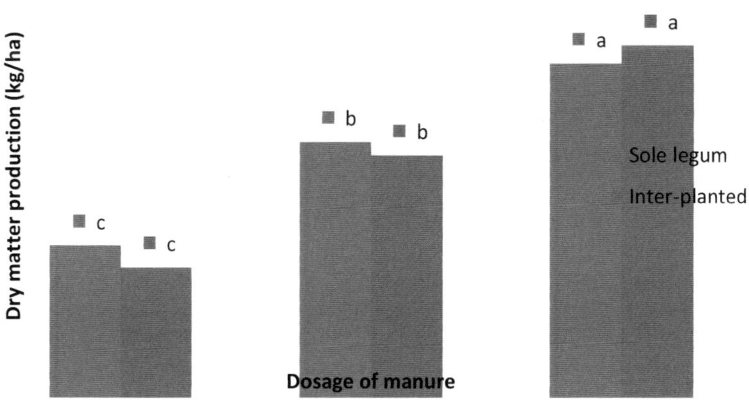

Figure 2. Dry matter production of *Sesbania grandiflora* at second cutting

3.3. Crude Protein

Application manure at saline soil affected significantly (P< 0.05) crude protein of *S. grandiflora* at first cutting. Manure was not affected crude protein *S. grandiflora* at second cutting (Table 2). Crude protein content of *S. grandiflora*was varied between 19.1 % - 25.6 % at first cutting, 23.3 % - 27.0 %. 2.3 % - 27.0 % at second cutting. These values were lower than the value of 33.78% reported by [6] for fresh leaf of *S. grandiflora* from Calcutta. However, that crude protein was higher than those reported by [7] and [11]. Crude protein recorded 23.3 % [7] and 23.1 % [11], respectively.

Application of manure at saline soil will decrease soil salinity. Application of manure to saline soil caused an increase in cation exchange capacity (CEC) greater adsorption of cations such as potassium, calcium and magnesium than sodium (Na^{+}). Conversely, Na+ leaching and a lower exchangeable sodium percentage (ESP) will decrease with addition of organic amendments [4]. Increased soil salinity at no manure treatment decreased crude protein of *S. grandiflora* at first cutting. Application manure did not affected crude protein at second cutting. Higher soil salinity affected several forage quality, such as organic matter (OM), crude protein (CP), neutral detergent fibre (NDF) and in vitro gas production. Generally, higher salinity level led to higher nutritional quality, although their significance varied amongst species and cuttings [10]. Iincreased electrical conductivity or salinity did not affected crude protein [13].

Table 2. Protein of leaves *Sesbania grandiflora* at different dosage of manure

Manure		Crude Protein (%)	
		1st Cutting	2nd Cutting
0 ton/ha	Sole legume	19.1^{c}	27.0
	Inter-planted	22.0^{b}	26.9
10 ton/ha	Sole legume	23.0^{a}	26.3
	Inter-planted	25.6^{a}	25.1
20 ton/ha	Sole legume	25.5^{a}	23.3
	Inter-planted	25.4^{a}	25.7

Values within columns having the same superscript are not significantly different ($P<0.05$) using Duncan Multiple Range Test

4. Conclusion

Application of 20 ton/ha manure at saline soil (EC = 4.1 dS/m) increased growth, dry matter production and crude protein of *Sesbania grandiflora.*

Acknowledgments

The authors would like to express gratitude to the Directorate General of Research and Community Service, Ministry of Research, Technology and Higher Education – Indonesia for the National Strategy Research funding.

References

[1] Abrol I P, Yadav J S V and Massaud F I 1988 Salt-Affected Soil and Their Management. *FAO*, Rome.

[2] Bharadwaj, A., V. Khandelwai, F. Choudhary dan A.K. Bhatin. 2011. Comparative studies of organic enrichers in the improvement of physic-chemical and microbiological characteristics of saline/usar soils. J. Chem. Pharm. Res. **3** 997

[3] Diacono M and Montemurro F 2015 Effectiveness of organic wastes as fertilizers and amendments in salt-affected soils. Agriculture **5**(2) 221

[4] Jalali, M. dan F. Ranjbar. 2009. Effects of sodic water on soil sodicity and nutrient leaching in poultry and sheep manure amended soils. Geoderma **153**. 194

[5] Kusmiyati F., Sumarsono and Karno. 2015. Seleksi legum pakan pada tanah salin berdasarkan karakter fisiologis dan kandungan mineral. *Procceding seminar Nasional* dan Lokakarya Teknologi dan Agribisnis Peternakan (Seri III), Fakultas Peternakan UNSOED 30 – 31 Mei 2015

[6] Nag, A. dan S. Matai. 2000. Fractionation of leaves and biochemical composition of the fractions. J. Trop. Agric Feed Sci. **28** 127.

[7] Nha, P.T., Nguyen Van Thu and T.R. Preston. 2008. Effects of different levels and sources of crude protein supplementation on feed intake, digestibility and nitrogen retention in swamp buffaloes compared to local cattle. Livestock Res. Rural dev. 20 :

[8] PurbajantiED, Kusmiyati F andFuskhah E 2017 Growth, yield, and physiological characters of three types of Indonesian rice under limited water supply. Asian J. Plant Sciences,**16** (2), 101.

[9] Qadir M, Tubeileh A, Akhtar J, Larbi A, Minhas P S and Khan M A 2008 Productivity enhancement of salt-affected environments through crop diversification. Land Degradation Development, **19** 429.

[10] Robinson, P.H., S.R. Grattan, G. Getachew, C.M. Grieve, J.A. Poss, D.l. Suarez dan S.E. Benes. 2004. Biomass accumulation and potential nutritive value of some forages irrigated with salin-sodic drainage water. Animal Feed Sci. Tech. **111** 175.

[11] Solorio-Sanches, F.J., Armandez-Yanez and J. Ku-Vera. 2000. Chemical composition and in vitro dry matter digestibility of some fodder tress from south-east Mexico. Livestock Res. Rural Dev. **12**(4)

[12] Steel R G D and Torrie J H 1991 Principles and Procedures of Statistics. McGraw Hill Inc

[13] Suyama, H., S.E. Benes, P.H. Robinson, G. Getachew , S.R. Grattan dan C.M. Grieve. 2007. Biomass yield and nutritional quality of forage species under irrigation under long term with saline-sodic drainage water : field evaluation. Anim. Feed Sci. Technol. **135** 329.

[14] Walpola B C and Arunakumara K K I U 2010 Effect of salt stress on decomposition of organic matter and nitrogen mineralization in animal manure amended soils. J. Agricultural Sci., **5** (1), 9

[15] Wandera F P, Dzowela B H and Karachi M K 1991 Production and nutritive value of browse species in semi arid Kenya. Tropical Grasslands **25** (4) 349.

[16] Wang L, Sun X, Li S, Zhang T, Zhang W 2014 Application of Organic Amendments to a Coastal Saline Soil in North China: Effects on Soil Physical and Chemical Properties and Tree Growth. *PLOS ONE*, 9(2) e89185.

[17] Xiong L, Schumaker K S and Zhu J K 2002 Cell signaling during cold, drough, and salt stress. Plant Cell 14 165.

[18] Yousif, A.M. and M.A. Abdalla. 2009. Variations in nitrogen mineralization from different manures in semi-arid tropics of Sudan with reference to salt-affected soils. International J. Agricultural Biol., **11** 515

[19] Zubair, M., F. Anwar, M. Ashraf, A. Ashraf dan S.A.S. Chatha. 2010. Effect of green and farmyard manure on carbohydrates dynamics of salt-affected soil. J. Soil Sci. Plant Nut. 6 1.

IOP Conf. Series: Earth and Environmental Science **119** (2018) 012024 doi:10.1088/1755-1315/119/1/012024

Feeding of Total Mixed Ration on the Productivity of Friesian Holstein Cross-Grade Cattle

Sunarso[1], M. Christiyanto[1], and L.K. Nuswantara[*1]

[1]Faculty of Animal and Agricultural Sciences, Diponegoro University, Semarang, Central Java, Indonesia.

[*]Corresponding E-mail: limbang.kn@gmail.com

Abstract

An experiment was conducted to evaluate feeding of total mixed ration (TMR) on the productivity of Friesian Holstein (FH) male cross-grade cattle at Semarang Municipality. TMR was a ration formulated with agricultural and agro-industrial by-product (no grass and/or green forage were used) to fulfilled the nutrient requirement of beef cattle. Total mixed ration were formulated on iso-energy of 66% of total digestible nutrients (TDN) and different level of crude protein (CP) content of 11%, 12%, 13%, and 14%. Twenty (20) heads of FH male cross-grade cattle with initial body weight of 292.40±33.06 kg were used in this experiment, and were arranged into 5 treatments T0, T1, T2, T3, and T4), and 4 replications. Data collected were analysed statistically using analyses of variance (Anova) based on the completely randomized design (CRD), then followed by Duncan's Multiple Range Test (DMRT) for different among treatments. Results of the experiment showed significantly different effect ($P<0.05$) on dry matter intake (DMI), CP intake, TDN intake, and average daily gain (ADG). There were no different effect ($P>0.05$) on feed conversion ratio (FCR), and feed efficiency. Others parameter showed that there were no significantly different ($P>0.05$) effect on the dry matter and organic matter digestibility in vitro, rumen ammonia concentration, and volatile fatty acid's rumen concentration. It was concluded that feeding TMR was potentially prospected for fattening of beef cattle, particularly as feeding strategy when there was no grass and/or green forage anymore.

1. Introduction

Poor growth rate has been established as a major disadvantage in beef cattle raising at the farmer, this was due to such factors as low genetic potential of the local cattle, improper management system, and lack of nutrients both in quality and quantity [1]. In the aspect of the availability of grass, the wet season is associated an abundance in grass. But, at the dry season there was no grass anymore or very limited production. Now, Java Island was very density island in Indonesia where many arable land area were already converted for housing and industrial development, this may cause the availability of grass or green forage for ruminant animal very limited. This situation indicates that the utilization of total mixed ration (TMR) by using agricultural waste and agroindustrial by-products ([2]; [3]; [4]; [5]; [6]) is prospective develop in order to increase beef cattle production and productivity in Indonesia particularly in Central Java. TMR [7] was a ration formulated with agricultural and agro-industrial by-product (no grass and/or green forage were used) to fulfil the nutrient requirement of beef cattle. Corbett et al. [8] stated that continuity of feed supplies for animals became reasonably assured and so there could now be continuity in the selection and breeding of improved livestock which, in turn, focused attention on methods of feeding that would enable expression of their production potential. This experiment was conducted to evaluate feeding of TMR on the productivity of Friesian Holstein (FH) male cross-grade cattle at Semarang Municipality.

2. Materials and Methods

2.1 Experiment 1 - Feeding of total mixed ration to Friesian Holstein (FH) cross grade cattle

Experiment 1 was conducted to evaluate the effect of TMR of iso-energy 66% TDN with different level of crude protein on the performance of FH cross grade cattle. Twenty male of the animals with initial body weight of 292.40±33.06 kg were used in this experiment, and were arranged into 5 treatments T0, T1, T2, T3, and T4, and 4 replications. All of the experimental cattle were fed 100 days, two times a day at 07.00 AM and 03.00 PM. Data collected such as follows: dry matter intake (DMI), crude protein intake (CPI), body weight gain (BWG), and feed conversion ratio (FCR). Completely randomized design (CRD) were used in this experiment, data collected were analysed using analysis of variance (Anova), and Duncan's multiple range test (DMRT) were used to test differences among treatments [9].Table 1 shows the experiment of total mixed ration treatment of iso-energy and different level of crude protein. As a control ration (T0), the experimental cattle was fed with the farmer ration consist of soya curd 51.0%, cassava 2.12%, rice straw 21.25%, rice bran 25.50%, and molasses 0.13% (dry matter basis), with nutrient content of 9.59% crude protein, 25.47% crude fibre, and 57.60% TDN. Feedstuffs and TMR were chemically analysed according of the procedures of [10].

Table 1. Total mixed ration composition (dry matter basis)

Feedstuff	T0	T1	T2 % DM	T3	T4
Palm oil meal	-	9.00	9.00	9.00	9.00
Yellow corn	-	10.00	10.00	10.00	10.00
Cassava	2.12	25.00	25.00	25.00	25.00
Rice bran	25.50	11.00	8.00	8.00	8.00
Kapok seed meal	-	13.00	15.00	15.00	14.00
Coffee hull	-	11.00	13.00	13.00	12.50
Palm oil	-	1.10	0.50	0.50	1.00
Copra meal	-	9.00	8.00	8.00	9.00
Urea	-	0.10	0.30	0.60	1.00
Molasses	0.13	10.00	10.00	10.00	10.00
Mineral mix	-	0.50	0.50	0.50	0.40
Salt	-	0.30	0.70	0.40	0.1
Soya Curd	51.00	-	-	-	-
Rice straw	21.25	-	-	-	-
Total	100.0	100.0	100.0	100.0	100.0
Nutrient Composition					
Crude Protein	9.59	11.24	12.12	12.98	13.99
Crude fibre	25.47	30.17	29.97	29.97	29.79
TDN	57.60	65.94	65.76	65.76	65.97

2.2 Experiment 2 - Digestibility and fermentability of TMR

Experiment 2 was aimed to evaluate the digestibility of TMR, and the in vitro fermentability of TMR. Data collected such as follows: dry matter digestibility, organic matter digestibility, rumen NH3 concentration, and total VFA's concentration in the rumen in vitro. Digestibility of DM and OM determination were based on the procedure of [11], and for NH3 and VFA's based on [12]. Data collected were analysed using analysis of variance (Anova), and Duncan's multiple range test (DMRT) were used to test differences among treatments [9].

3. Results and Discussion

Performance of experimental cattle after 100 days of feeding was shown in Table 2.

Table 2. Performance of beef cattle fed by experimental total mixed ration

Item	Treatments				
	T0	T1	T2	T3	T4
DMI, kg/h/d	11.07^{a}	6.24^{b}	5.92^{b}	6.37^{b}	6.09^{b}
CP intake, kg	0.98^{a}	0.70^{b}	0.72^{b}	0.83^{b}	0.80^{b}
ADG, kg/h/d	1.21^{a}	0.71^{c}	0.98^{b}	1.11^{a}	0.82^{bc}
FCR	9.24^{a}	8.75^{a}	6.34^{b}	5.76^{b}	7.75^{ab}

T_0: Farmer ration; T_1: T_2; T_3; T_4 (iso TDN ration66%; and 11; 12; 13; 14% CP). Number withdifferent letter at the same row was significantly difference (P<0.05)

Table 2 showing that after 100 days of feeding TMR there was different effect (P<0.05) on the DMI, CPI, ADG, and FCR of the experimental animal. On the DMI, compare to that farmer ration (T0) all the TMR (T1, T2, T3, and T4) significantly decreased the DMI from 11.07 kg/h/d (T0) vs 6.16kg/h/d, meanwhile, among TMR of 66% TDN there was no different effect on DMI, this may be due that farmer ration had lower energy (57.6% TDN) compare 66% TDN of TMR. The higher the energy content in the ration would be decreased the dry matter intake. The higher DMI of the farmer ration would consequently increase cost of production, and finally decreased their revenue. Although the CP content of T0 was lower than TMR of iso-energy, but, because of their DMI was significantly higher, then the total CPI was also higher compare to TMR. The higher the CPI would also increase the ADG of the experimental animal. Table 2 showing that the farmer cattle had an ADG significantly (P<0.05) higher compare to that T1, T2, and T4, and similar with T3, this might be correlated with their CPI was significantly higher. Protein intake was one of the nutrient which responsible to the ADG particularly at the growing animal when it was reflected by additional of meat as a protein biomass in the body. Although, all the TMR (T1, T2, T3, and T4) had ADG lower than farmer ration (T0), in fact, their DMI had lower than T0. So, the average of FCR (as calculated by DMI/ADG) had significantly lower than T0. It means that feeding with TMR more efficient compare to farmer ration, and could be concluded that feeding with TMR had positive and prospective aspect to substitute or to replace farmer ration because feeding T0 had higher cost and lower efficient, and practically would be used during dry season where there was not available grass or green forage, this will ensure the continuous cattle/meat production along year.

Results of Experiment 2 was shown in Table 3. Table 3 showing that there were no different effect of farmer ration (T0) with TMR of iso-energy 66% TDN and different level of CP. The DMD and OMD of all the treatment relatively similar among treatment and practically at the normal condition with generally OMD had higher DMD. In term of total VFA's there were no different effect among treatment, in fact all of the TMR produced the total VFA's of 128.33-165.00 mM higher than 80-160 mM the concentration needed for maximum microbial protein synthesis in the rumen.

Table 3. Digestibility and fermentability of experimental total mixed ration

Item	Treatments				
	T0	T1	T2	T3	T4
DM Digestibility, %	61.92	62.42	64.94	68.89	63.74
OM Digestibility, %	65.56	67.12	68.98	72.12	67.18
Total VFA's, mM	130.00	153.33	165.00	128.33	143.33
NH_3, mM	3.99	4.34	4.17	3.71	4.52

T_0: Farmer ration; T_1: T_2; T_3; T_4 (Iso TDN ration66%; and 11; 12; 13; 14% CP). Number with different letter at the same row was significantly difference (P<0.05)

The concentration of rumen NH3 were no different among treatment. Satteret al. [13] recommended that for maximum microbial protein biosynthesis required ration with ≥10% CP. Since all of the experimental ration content more than 10% CP, so, the concentration of rumen NH3 3.77-4.52 mM were relatively similar with 3.57-7.14 mM [8]or 5 mg% [12] for microbial protein synthesis [14]. Based on the results data from both total VFA's and rumen NH3 it could be concluded that all of the experimental ration had capacity to produce maximum microbial protein synthesis. Means that protein supply (from dietary protein and from microbial protein) for the experimental cattle could fulfilled the protein requirement for their maintenance and growth as reflected by their ADG. Increasing of CP may also cause on their nitrogen retention [3] which than reflected by their gain.

Generally, this experiment imply that feeding with TMR on iso-energy 66% TDN with different level of CP 11%, 12%, 13%, and 14% might be utilized by the farmers to replace their costly ration particularly at the dry season, in order to extend a continuous meat supply and/or for the upkeep of cattle production all year round.

4. Conclusion

Feeding of total mixed ration with iso-energy of 66% TDN, and different level of crude protein 11-14 % significantly affected on the dry matter intake, crude protein intake, energy intake, and average daily gain of the Friesian Holstein cross grade cattle. There were no different effect on feed conversion ratio (FCR), and feed efficiency. No different effect of total mixed ration fermentability *in vitro* on their dry matter digestibility, organic matter digestibility, rumen ammonia concentration, and volatile fatty acids concentration. It was concluded that feeding total mixed ration had no negative effect on the experimental animal, and was potentially prospected for fattening of beef cattle, particularly as feeding strategy when there was no grass and/or green forage anymore.

Acknowledgments

The authors thankto Aryani Nurul Hidayati and Alista Setyaningrum for assisting data collection and analysis during our experiment.

References

[1] Sunarso. 1993. The Utilization of *Setaria sphacelata* (SCHUM.) STAPF and C.E. HUBB. EX M.B. MOSS Silage by Sheep. PhD. Dissertation. University of the Philippines Los Banos.

[2] Hamidah, A., C.I. Sutrisno, Sunarso, M. Christiyanto, and L.K. Nuswantara. 2011. Performance of fat-tailed rams fed complete feed based oil palm fronds. Journal of the Indonesian Tropical Animal Agriculture **36** (3) 185-189

[3] Sunarso. 2012 The effect of King grass silage on the nitrogen balance and hematological profile of Ettawa grade male goat. International J. Sci. Engineering **3**(1) 13-16.

[4] Anggraeny, Y.N., H. Soetanto, Kusmartono, and Hartutik. 2015. Sinkronisasi suplai protein dan energi dalam rumen untuk meningkatkan efiseiensi pakan berkualitas rendah. Wartazoa 25 3 107-116.

[5] Yulistiani, D., Z.A. Jelan, J.B. Liang, H. Yaakub, and N. Abdullah. 2015. Effects of supplementation of Mulberry (*Morus alba*) foliage and urea-rice bran as fermentable energy and protein sources in sheep fed urea treated rice straw based diet. Asian Australian J. Anim. Sci. Vol **28**, No. 4: 494-501.

[6] Sani, F F, L K Nuswantara, E Pangestu, F Wahyono, and J Achmadi. 2016. Synchronization of carbohydrate and protein supply in the sugarcane bagasse based ration on in situ nutrient degradability J. Indonesian Trop. Anim. Agric. **41** (1) 28-36

[7] Hartadi, H., S. Reksohadiprodjo, dan A.D. Tillman. 1997. Tabel komposisi pakan untuk Indonesia. Cetakan ke-4. Gadjah Mada Press, Yogyakarta.

[8] Corbett, J.L., and M. Freer. 2003. Past and present definitions on the energy and protein requirements of ruminants Asian-Aust. J. Anim. Sci. **16 4** 609-624.

[9] Steel, R.G.D., and J.H. Torrie. 1980. Principles and Procedures of Statistics. McGraw Hill Book Co. Inc. New York.

[10] AOAC (Association of Official Analitycal Chemist). 2000. Official Methods of Analysis. 17th Ed. Association of Official Analitycal Chemist. Washington D.C.

[11] Tilley, J. M. A., and R.A. Terry. 1963. A two stage technique for in vitro digestion of forage crops. J. Br. Grassland Soc. **18** 104.

[12] Satter, L.D., and L.L. Slyter. 1974. Effect of ammonia concentration on rumen microbial protein production in vitro. Br. J. Nutr. **32** 199

[13] Satter, L.D. and R.E. Roffler. 1981. Influence of nitrogen and carbohydrate inputs on rumen fermentation. In: Haresign, W and D.J.A. Cole (Eds.): Recent developments in ruminants nutrition. 1st Publ. Butterworths, London.

[14] Seo, J.K., J. Yang, H.J. Kim, S.D. Upadhaya, W. M. Cho, and J.K. Ha. 2010. Effects of synchronization of carbohydrate and protein supply on ruminal fermentation, nitrogen metabolism and microbial protein synthesis in Holstein steers. Asian-Aust. J. Anim. Sci. **23** (11) 1455.

Effect of Chicken Bone Meal as Phosphorus Supplement on Blood Metabolites in Fattening Lambs

A. Pujiastuti[1*] and A. Muktiani[2]

[1]Master Program of Animal Science, Faculty of Animal and Agricultural Sciences, Diponegoro University, Semarang, Central Java, Indonesia

[2]Faculty of Animal and Agricultural Sciences, Diponegoro University, Semarang, Central Java, Indonesia

*Corresponding E-mail: asihpujiastuti13@gmail.com

Abstract

The aim of this study was to evaluate the effect of chicken bone meal (CBM) as phosphorus supplementon blood metabolites in fattening lambs. The experiment used 16 of 12 months old local male lambs with initial body weight 27.01 ± 1.51 kg. The experiment used a complete randomized design with 4 treatments and 4 replications. The treatments were T0 (basal ration = native grass + soybean curd waste), T1 (basal ration + 0.49% P Dicalcium phosphate), T2 (basal ration + 0.70% P CBM), T3 (basal ration + 1.39 % P CBM). The results indicated that CBM as phosphorus supplement was significantly different (P<0,05) on P intake, phosphorus and glucose serum and did not different significantly on dry matter intake and alkaline phosphatase activity. In conclusion, CBM is one of requirement organic phosphorus supplement which can be applied on ruminants.

1. Introduction

Forage is the main source of ruminant feed. Forage types that are mostly eaten by the ruminant animals are grass family *(Gramineae)* and legume *(Leguminoceae)*. The shortage of quality forage has become the limiting factor for development of ruminant farm resulting in the decreased ruminant production. Low productivity and nutritive value were the main problem in the ruminant farm. Native grass was one of forage that usually given in the ruminant farm. The only native grass consumption make a deficient on nutritive value especially minerals. Mineral requirement was needed in order to support optimal growth and production. Phosphorus was one of minerals suplement that needed by rumen and cattle microbes to maintain the integrity of membranes and cell walls, nucleic acid components and parts of high-energy molecules [18]. Sufficient phosphorus available in the rumen allows microbial rumen to ferment feed optimally.

Phosphorus is an important nutrient involved in most of the nutrient metabolism in the body. Phosphorus suplement source consist of anorganic an organic matter. An organic phosphorus usually used for example is Dicalsium phosphate [3]. The organic phosphorus usually used is ruminant bone meal. The problem is bone meal from ruminants is not recommended as a mineral supplement for ruminants as appropriate with the decision of the Minister of Agriculture (2002). Chicken bone meal was an alternative phosphorus source for mineral supplements.

Organic phosphorus derived from chicken bone flour has better utility. Its known from mineral solubility. The solubility of minerals from three bone meal such as chicken bone meal, goat bone meal and cow bone meal sequently are 19% chicken bone meal, 18.6% cow bone meal and 14.15% goat bone meal [16]. Based on the solubility, chicken bone meal can be used as an efficient mineral supplement.

Phosphorus is one of the minerals needed for ruminants especially lambs. P needed for bone growth and metabolism procces. Sufficienct P in diet was important to increase microorganism population and increase nutrient metabolism. Sufficient P in lambs can be described through the bloodmetabolites.

The aim of this study was to evaluate the effect of chicken bone meal (CBM) as phosphorus supplement on blood metabolites in fattening lambswith the purpose that chicken bone meal (CBM) is one of requirement organic phosphorus supplement which can be applied on ruminants.

2. Materials and Methods

2.1. Animals and Diet

The experiment used 16 of 12 months old local male lambs with initial live weight 27.01 ± 1.51 kg. The experimental basal diet (T0) was composedof native grass and soybean curd waste. The experimental treatments comprised three diets providing T1 (basal ration + 0.49% P Dicalcium phosphate), T2 (basal ration + 0.70% P CBM), T3 (basal ration + 1.39 % P CBM) (Table 1).
In allgroups, lambs were fed the corresponding diet in twoequal meals at 08.00 and 17.00 h and water were given ad libitum. The ration of native and soybean curd waste was 40:60. After 70-d of feeding treatment, blood serum was taken. The blood samples were withdrawn from jugular vein from each lambs by using a disposable sringe into 10 ml plain tubes.

2.2. Statistical Analysis

The experiment used a complete randomized design with 4 treatments and 4 replications.The data obtained were analyzed by variance analysis.

Table 1. Composition of basal diet

Item	T0	T1	T2	T3
Composition diet, % of dietary DM				
Native grass	40	40	40	40
Soybean curd waste	60	60	60	60
DCP		0.49		
Chicken bone meal			0.70	1.39
Nutrient composition				
Moisture (%)	12.77	12.77	12.77	12.77
Ash (%)	7.32	7.32	7.32	7.32
Crude Protein (%)	16.09	16.09	16.09	16.09
Extract Ether (%)	2.70	2.70	2.70	2.70
Crude Fiber (%)	35.48	35.48	35.48	35.48
Ca (%)	4.45	4.56	4.56	4.67
P (%)	0.56	0.65	0.61	0.65
TDN (%)	59.91	59.91	59.91	59.91

3. Results and discussion

The result of this study (Table 2), showed that dry matter (DM) intake not significantly different in each of treatment. The explanation of this phenomenon probably caused by the nutrient composition of diet such as energy and protein relatively was no different. Dry matter intake in each treatment was 2.46-2.54% of body weight (BW). DM requirement based on [12] was 3% of BW. This condition was assumed according to high moisture and crude fibre(CF) content (12.77 % dan 35.48%) of diet basal ration. High moisture and CF content will restrict the ability of consumption ration. This is due to limited rumen capacity and low rate of passage will reduce DM intake include other nutrients [10].

Feed intake was limited by energy requirements, which affects to DM intake. TDN on diet treatment was 59.91%. Energy requirement for lamb was 55-60% [14]. This condition causes lamb quit to consumtion of feed when energy requirement was fulfilled [13].

Based on Table 2, P intake between treatments showed significant differences. P intake of T1 and T3 (P <0.05) were higher than that in T2 and T0. The treatment on T1 and T3 was qualified the P requirement for lambs i.e. 2.39 g/d [12]. The results showed that adding CBM (1.39%) has been able to replace DCP (0.49%) as an organic phosphorus. The utilization of CBM has higher acceptability than DCP. It wasindicate from the higher P serum. P serum influenced by P intake [15]. Low P serum on T0 was assumed because of the low availability of organic P on the tissue [1].

This result was different from [17], where adding anorganic P (0.29%) dan organic P (0.35%) has unsignificant result in P intake (P>0.01). P intake anorganic dan organic was 1.34 (g/d) and 1.64 (g/d). The different was assumed because of the level of P addition and different source of organic P.

Fulfilled P in rumen was caused normal intake of P. Fulfilled of P in rumen supports rumen micro-organism fermented feed optimally. Organic feed material will be fermented by rumen micro-organism into microbial protein as a source of protein for livestock. Microbial rumen contributes 1/3-2/3 of the protein requirement of livestock.

3.1. P Serum

Phosphorus serum showed in every treatment was T0 (8.04 mg/dl), T1 (8.55 mg/dl), T2 (10.09 mg/dl) and T3 (11.21 mg/dl). P serum was higherin T3 treatment. This results was assumed from solubility factors as mention by [11]. One of the factors affecting P intake was solubility of P in the gastrointestinal tract. Increasing of solubility in gastrointestinal tract affect increasing of P absorption and it was associated with P intake.

Table 2. Phosphorus, Glucose and Alkaline phosphatase serum

Parameters	T0	T1	T2	T3
Dry matter intake (g/d)	664.24	670.09	684.54	677.84
P intake (g/d)	0.54^{c}	2.40^{a}	1.47^{b}	2.34^{a}
P Serum (mg/dl)	8.04^{c}	8.55^{bc}	10.09^{ab}	11.21^{a}
Glucose serum (mg/dl)	19.23^{b}	33.33^{ab}	30.03^{b}	46.38^{a}
Alkaline phosphatase (U/L)	63.87	79.39	81.00	83.87

Different superscripts on the same line show a significantly difference (P<0,05)

Phosphorus serum in all treatment was in a normal range. The normal range of P in lambs was 4-6 mg/dl [4]. Effect of CBM as an organic P supplement can increase P in serum (T2 and T3) higher than an anorganic P (DCP) in T1(P<0.05). This result was different from [17], where adding anorganic P (0.29%) dan organic P (0.35%) has unsignificant result in P serum (P>0,05). P serum anorganic dan organic was 6.16 (mg/dl) dan 7.20 (mg/dl). The different was assumed because of the level of P addition and different source of organic P. The normal P in the tissue shows supplement P to support metabolic activity on the tissues. This indicates that P was an essential metabolic in tissue [6].

3.2. Glucose Serum

The result of variance analysis showed that glucose serum was higherin T3 treatment (Table 2). Glucose serum of the study were below the normal range of 44-81.2 mg / dl [5], except T3 treatment.Phophorus supplementation in T3 treatment resulted normal glucose serum. Phosphorus intake was affected P concentration in the rumen. The sufficiency of P in the rumen helps fermentation carbohydrates. Glucose in ruminants derived from propionate (gluconeogenesis) was produced from the fermentation of fiber feed. Normal glucose serum especially in T3 treatment indicated energy from carbohydrate is fulfilled.

Phosphorus serum was affected glucose metabolism in blood. Glucose serum source was came from carbohydrates, glycogenic compounds that was produced from gluconeogenesis, and liver glycogen. Glucose serum was determined by the balance of glucose that enters the blood and which leaves the blood. Glucose serum was still to be secreted to supply the nutritional needs on various tissues. Glucose serum on normal limits, then most of the tissue will use glucose as an energy source.Increasing of glucose serum was followed by increase of carbohydrate absorption [7].

3.3. Alkaline Phosphatase Activity

Serum Alkaline phosphatase activityfrom the lowest was T0 (63.87 U/L), T1 (79.39 U/L), T2 (81.00 U/L) and T3 (83.87 U/L). P supplement on treatment did not influence to Alkaline phosphatase activity. The same compositionof energy and protein and P supplement were not significantly affected in alkaline phosphatase activity. The normal range of alkaline phosphatase activity is 63-387 U/L [8]. Indonesian local lambs have a range of alkaline phosphatase activity is 25.28-30.66 U/L [9].

Alkaline phosphatase activity is the total alkaline phosphatase released from tissue to blood. Alkaline phosphatase contains zinc used for enzyme activity. The physiological function of zinc in tissues or cells that are growing is to regulate protein synthesis [7].Increased alkaline phosphatase activity shows a high anabolic process in the tissues. This process can be either the synthesis of intra cellular materials or the growth of new cells in the tissue.

4. Conclusion

Effect of organic P supplements in diet was better than inorganic P and give the good effect on blood metabolites. Chicken bone meal (CBM) is one of requirement organic phosphorus supplement which can be applied on ruminants.

Acknowledgments

This research funding by Kemenristek Dikti for Master (S2) Program Scholarship.

References

[1] Braithwaite G D 1985 Endogenous faecal loss of phosphorus in growing lambs and the calculation of phosphorus requirements J. Agri. Sci.**105**:67

[2] Challa J and Braithwaite GD 1988Phosphorus and calcium metabolism in growing calves with special emphasis on phosphorus homoeostasis: 1. Studies of the effect of changes in the dietary J. Agric. Sci. (Camb.) **110**:573

[3] Dias RS, Lopez S, Patino RM, Silva, TS, Silva Filho JC, Vitti S, Pecanha MRS, Kebreab E and France J 2013 Calcium and phosphorus utilization in growing sheep supplemented with dicalcium phosphate J. Agric. Sci. **151**.424

[4]Engels, E A N 1981 Mineral status and profiles (blood, bone and milk) of the grazing ruminant with special reference to calcium, phosphorus and magnesium S. Afr. J. Anim. Sci. **11:** 171

[5] Fraser H, Mays A, Amstutz, Archibald J, Armour J, Blood D, Newberne P and Snoeyenbos G 1986 The Merck Veterinary Manual:Rahway, N. J. USA. pp 1-1677

[6] Georgievskii 1982 Mineral Nutrition of Animals Butterworths London Boston Sydney Durban Wellington Toronto

[7] Gropper S, Jack S, and James G Advanced Nutrition and Human Metabolism 2009 5th edWadsworth Cengage Learning

[8] Kaneko J, Harvey J and Brus M 1997 Clinical Biochemistry of Domestic Animals Academic Press, p: 932

[9] Mangkoewidjojo S and Smith 1998 Pemeliharaan, Pembiakan dan Penggunaan Hewan Percobaan di Daerah Tropik Universitas Indonesia Press

[10] McDonaldP, Edward RA, and Greenhagh JFD 1988 Animal Nutrition 2nd Ed. Longman London

[11] McDowell LR 1992 Minerals in animal and human nutrition Animal feeding and nutrition Academic Press San Diego

[12] National Research Council 1985 Nutrient Requirement of Lambs 6th Ed National Academy Press

[13] Parakkasi A 1999 Ilmu Nutrisi dan Makanan Ternak Ruminan Universitas Indonesia Press

[14] Ranjhan SK 1981 Animal Nutrition in Tropics 2nd Ed. Vilus Publishing House. Pvt Ltd, New Delhi

[15] Read, MVP, Engels EAN and Smith WA 1986 Phosphorus and the grazing ruminant. 4. Blood and faecal grab samples as indicators of the P status of cattle*S.* Afr. J. Anim. Sci.**16**: 18-22

[16] Sutardi 2011 Tingkat Kelarutan Mineral Tepung Tulang yang Berasal dari Tulang Berbagai Jenis Hewan Prosiding Seminar Nasional pp 134-141

[17] Shanklin RK 2001 Effect Of Form And Amount Of Phosphorus And PhytaseSupplementation On Phosphorus Utilization By Ruminants [Thesis] Faculty of the Virginia Polytechnic Institute and State University Blacksburg

[18] Underwood EJ and Suttle NF 1999 The Mineral Nutrition of Livestock (3rd ed) CABI Publishing New York.

IOP Conf. Series: Earth and Environmental Science **119** (2018) 012026 doi:10.1088/1755-1315/119/1/012026

Analysis of Non-Genetic Factors Influencing Reproductive Traits of Japanese Black Heifer

A. Setiaji[1,2,3*], and T. Oikawa[1,2]

[1]Faculty of Agriculture, University of the Ryukyus, Okinawa, Japan.
[2]United Graduate School of Agricultural Sciences, Kagoshima University, 1-21-24 Korimoto, Kagoshima 890-0065, Japan.
[3]Permanent address : Faculty of Animal and Agricultural Sciences, Diponegoro University, Semarang, Central Java, Indonesia.

*Coresponding e-mail: asepsetiaji93@gmail.com

Abstract

This study aimed was to identify non-genetic factors strongly associated with reproductive traits on Japanese Black heifer. Artificial insemination and calving records were analyzed to investigate non-genetic effect on reproductive performances. A total of 2220 records of heifer raised between 2005 and 2016 were utilized in this study. Studied traits were first service non return rate to 56 days (NRR), first service pregnancy rate (FPR), days from first to successful insemination (FSI), number of services per conception (NSC), age at first calving (AFC), and gestation length (GL). Test of significance for effects in the statistical model was performed using GLM procedure of SAS 9.3. The yearling trend was plotted on the adjusted mean of parameters, by the least square mean procedure. Means of NRR, FPR, FSI, NSC, AFC and GL were 72%, 53%, 52.71 days, 1.76, 760.71 days and 288.26 days, respectively. The effect of farm was significant (P<0.001) for FSI, AFC, and GL. The effects of age of heifer at first insemination was significant (P<0.001) for AFC. Month of insemination and sex of calf were significant (P<0.001) for GL. Compared with average value of reproductive traits, NSC and GL were generally within standard values for Japanese Black cattle, while AFC was slightly earlier. The result indicated that different management of farms strongly influenced reproductive traits of Japanese Black heifer.

1. Introduction

Production traits, especially carcass characteristics have been a primary object in Japanese Black breeding program. In recent decades, incorporating reproductive performances in breeding plan is a regular practice because of their influence on herd productivity [1]. Reproductive traits of heifers are traits measured relatively early in the life. Therefore, it can be effective to include it in breeding evaluation programs. Recently, [2] suggested that reproductive traits for heifer should be evaluated separately from reproductive traits for cows.

Non-genetic effects are factors contributing largely to reproductive performance of not only on cows but also on heifers. Most of the reproductive traits were significantly affected by environment effects. Curent studies have been reported in Simental cows [3], dairy cows [4;5], in Japanese Black cows reported recently by [6]. Hence, there should be sufficient improvement in these traits through better management. Environmental factors can be classified into factors with explicit effects, such as age, herd, year, season, parity and factors with embedded effects such as infectious diseases, endocrine fluctuation etc.[3]. Studies on reproductive performance and factors that affecting reproductive performance of Japanese Black heifer are scarce. The determined known effects can give information to solve fertility problem in the breeding program [7]. This study aimed to identify non-genetic factors strongly associated with reproductive traits on Japanese Black heifer.

2. Materials and Methods

Reproduction records of Japanese Black heifer were obtained from Artificial Insemination Center of Northern Okinawa. Data consisted of recorded artificial insemination and calving event. Six fertility traits were calculated for each heifer from artificial insemination and calving records: first service non return rate to 56 days (NRR), first service pregnancy rate (FPR), days from first to successful insemination (FSI), number of services per conception (NSC), age at first calving (AFC), and gestation length (GL).

2.1. Data set

The data set included records of heifer performance with following criteria: heifer that born in 2003 to 2015, first insemination of heifers occurred in 2005 to 2016, number of service less than 10 and farm with minimum five records. Heifer with incomplete records, embryo transfer donor or recipient, and had twin calves were eliminated to avoid errors of the reproductive recording scheme. After editing, records with 2220 heifers for NRR and FPR, and 2034 heifers for FSI, NSC, AFC and GL from 108 farms were available for analysis.

The following six reproductive traits were calculated: NRR was coded one when there was no subsequent insemination until 56 days after first insemination and zero otherwise. FPR was coded one when heifer become pregnant and calving in a time following the first insemination and zero otherwise. FSI was the number of days occurring between first and the insemination that resulted in conception, NSC was the number of insemination per gestation, AFC was the number of days between date of birth and date of first calving of a heifer, and GL was measured as an interval from the last insemination to the subsequent calving. To evaluate the effects of age, heifers were classified into three age classes: <16, 16 to 19 and >19 month [4]. All heifers in 108 farms were artificially inseminated by six AI technicians.

2.2. Statistical analysis

The data were analyzed using GLM procedure of SAS 9.3 software [8] to test significance of factors affecting reproductive performance. The linear model used for NRR, FPR, FSI and NSC was as follows:

$$y_{ijklmop} = F_i + Y_j + M_k + T_l + A_m + s_o + e_{ijklmop}$$

where: $y_{ijklmop}$ is the observation of NRR, FPR, FSI and NSC, F_i the ith fixed effect of farm, Y_j the jth fixed effect of year of insemination, M_k the kth fixed effect of month of insemination, T_l the lth fixed effect of AI technicians, A_m the mth fixed effect of age of heifer, s_o the oth random effect of sire, and $e_{ijklmop}$ the random residual of $y_{ijklmop}$.

The following model was used for AFC and GL:

$$y_{ijklmnop} = F_i + Y_j + M_k + T_l + A_m + X_n + s_o + e_{ijklmnop},$$

where: $y_{ijklmnop}$ is the observation of AFC and GL. F_i, Y_j, M_k, T_l, A_m, s_o, and $e_{ijklmnop}$ as described in the previous model, X_n the nth fixed effect of sex of calf. Trend was plotted on least square mean (LSM) of year effect for the traits in study by the GLM analysis.

3. Results and Discussion

Descriptive statistics for each trait are shown in Table 1.The average NRR and FPR of heifer during twelve years was 72% and 53%, respectively. Percentage of NRR following the first service was slightly lower than previous experiments. NRR of Canadian Holstein heifer showed 74% [9], and Holstein Friesian heifer in Netherlands showed 73.9% [10]. However, percentage of FPR was higher than reported in dairy heifer that was 50.7% [11]. Similar to the findings of the current study, FPR of

Japanese Black cows, reported 53.5 % [12]. FSI of Japanese Black heifer in Okinawa was 52.17. [6] reported that number of services per conception in Japanese Black cow was 1.41, less than average of the present study while the result of GL was consistence with their mean (288 days). The result of AFC in this study as 760.71 days, which are 20 days earlier than previous studies by [13] and [14], whereas based on records of Okinawa prefecture during 1990 to 2009, [15] reported those as 807.9 days.

Table 1. Descriptive Statistics for Parameters of Heifer Reproductive Traits.

Trait	No. of records	Mean	SD	Minimum	Maximum
NRR	2220	0.72	0.45	0	1
FPR	2220	0.53	0.49	0	1
FSI	2034	52.17	105.01	0	602
NSC	2034	1.76	1.20	1	10
AFC	2034	760.71	126.44	590	1298
GL	2034	288.26	4.97	263	303

The result of analysis of variance showed that there was few significant effects affecting NRR and FPR (Table 2) and NSC (Table 3) of Japanese Black heifer in Okinawa. In general, these was influenced by age, year and management of farm. Heifer was inseminated at older age for the first time had a slightly higher rate of success of the first insemination and smaller number of service [9]. Management level of farm had a significant effect on number of service in Holstein [16]. Furthermore, [17] concluded that environmental variations in that case, employment of an estrus detector and expertise of AI technician may play a greater part in success of first insemination.

Table 2. Analysis of Variance for Non-Return Rate (NRR) and First Service Pregnancy Rate (FPR).

Source	df	NRR				FPR			
		Sum of square	Mean square	F-value	*P*	Sum of square	Mean square	F-value	*P*
Farm	106	26.461261	0.249635	1.28	0.0307	28.941728	0.273035	1.13	0.1807
Year	11	2.005610	0.182328	0.94	0.5039	3.703450	0.336677	1.39	0.1701
Month	11	2.922537	0.265685	1.36	0.1831	2.793237	0.253931	1.05	0.3936
Technician	5	0.870125	0.174024	0.89	0.4843	1.923905	0.384781	1.59	0.1596
Age	2	0.475357	0.237679	1.22	0.2953	0.431953	0.215977	0.89	0.4098
Sire	120	27.325660	0.227714	1.17	0.1069	33.337967	0.277816	1.15	0.1362
Residuals	1963	382.22619	0.194715			475.057517	0.242006		

The present study reported that farm had a significant effect ($P<0.001$) on FSI, AFC and GL. The result indicated that different management of farms influenced heifer reproductive traits. According to [9] fertility of female will be lower in herds with a lower management. The effects of year was significant on FSI, whereas, the effects of month was not statistically significant. In contrast with the present study, effect of year and month on the FSI were found significant in Simmental [3]

The analysis of variance for AFC and GL are presented in Table 4. Different from the previous results, more effects were significant but month and sex of calf effects showed no significance in AFC. Significant effect of farm was reported on previous study of Japanese Black cows in Okinawa [15] but those study also reported a significant of year and sire effects. Among the factors affecting GL, the effect of farm, month and sex of calf were significant ($P<0.001$), slightly different with the result of [6], reported that year and month were significant in AFC and GL. Significant effect of sex of calf in the present result was consistent with previous study reported by [14] on Japanese Black cows. Heifer calves tend to shorten GL than bull calves, this difference may be caused by physiological reasons such as delayed of uterus recovery [18].

Table 3. Analysis of Variance for days from First to Successful Insemination (FSI) and Number of Services per Conception

Source	df	FSI				NSC			
		Sum of square	Mean square	F-value	*P*	Sum of square	Mean square	F-value	*P*
Farm	106	1 908 031	18000	1.78	<.0001	164.416762	1.551102	1.11	0.2073
Year	11	312 140	28376	2.80	0.0013	30.573922	2.779447	2.00	0.0254
Month	11	46 375	4216	0.42	0.9497	12.630278	1.148207	0.82	0.6155
Technician	5	146 842	29368	2.90	0.0130	11.491982	2.298396	1.65	0.1435
Age	2	51 264	25632	2.53	0.0799	5.949834	2.974917	2.14	0.1184
Sire	117	1 627 052	13906	1.37	0.0063	161.547551	1.380748	0.99	0.5096
Residuals	1780	18 032 206	10130			2478.925915	1.392655		

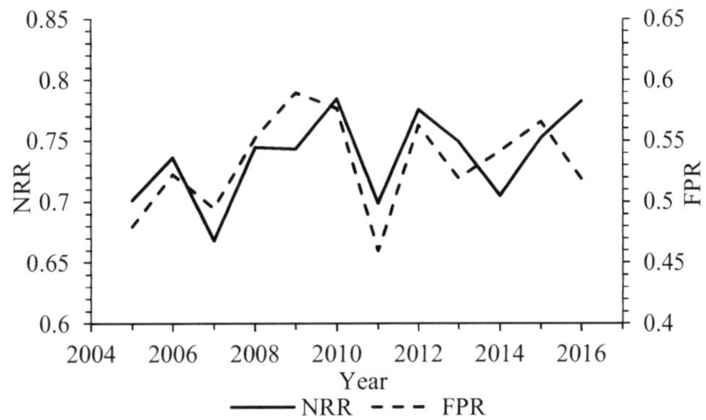

Figure 1. Yearly Trend of Non-Return Rate (NRR) and First Service Pregnancy Rate (FPR)

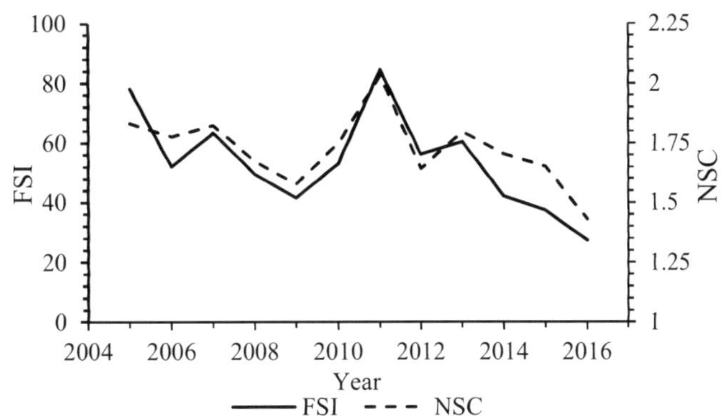

Figure 2. Yearly Trend of Days from First to Successful Insemination (FSI), and Number of Services per Conception (NSC)

IOP Conf. Series: Earth and Environmental Science **119** (2018) 012026 doi:10.1088/1755-1315/119/1/012026

Illustration of yearly trend of this study was based on an adjusted mean of the traits. Yearly trend of NRR and first FPR are illustrated in Figure 1. The lowest and highest LSM of NRR occurred in 2007 and 2010, respectively. NRR and FPR showed similar trend from 2005 to 2015, but in 2016 the trend of NRR was upward, while FPR was downward significantly. Figure 2 presents yearly trend of FSI and NSC. FSI tend to decrease from 2005 to 2016, as well as NSC. However, significant increase of FSI and NSC occurred in 2011.

Table 4. Analysis of Variance for Age at First Calving (AFC) and Gestation Length (GL)

Source	df	FSI				NSC			
		Sum of square	Mean square	F-value	*P*	Sum of square	Mean square	F-value	*P*
Farm	106	1 908 031	18000	1.78	<.0001	164.416762	1.551102	1.11	0.2073
Year	11	312 140	28376	2.80	0.0013	30.573922	2.779447	2.00	0.0254
Month	11	46 375	4216	0.42	0.9497	12.630278	1.148207	0.82	0.6155
Technician	5	146 842	29368	2.90	0.0130	11.491982	2.298396	1.65	0.1435
Age	2	51 264	25632	2.53	0.0799	5.949834	2.974917	2.14	0.1184
Sire	117	1 627 052	13906	1.37	0.0063	161.547551	1.380748	0.99	0.5096
Residuals	1780	18 032 206	10130			2478.925915	1.392655		

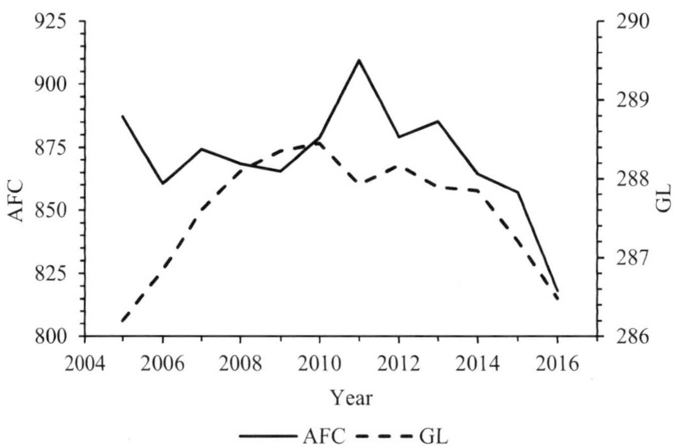

Figure 3. Yearly trend of age at first calving (AFC), and gestation length (GL)

Figure 3 presents yearly trend of AFC and GL. Trend of AFC showed some fluctuation, but significant decrease to the lowest at the last five years. The highest LSM of AFC was in 2011. GL showed significant upward trend in period 2004 to 2009, then downward toward shorter days at the last three years. The longest LSM of GL was in 2010, and the shortest LSM was in 2005. However the difference was two days.

4. Conclusion

Among mean values of the reproductive traits, NSC and GL were mostly within standard values for Japanese Black cattle, while AFC was slightly earlier than previous results. The result indicated that different management of farms strongly influenced reproductive traits of Japanese Black heifer.

Acknowledgments

The author thanks the staff of Artificial Insemination Center of Northern Okinawa for their collaboration on data inquiry and data collection

References

[1] Oyama K, Katsuta T, Anada K and Mukai F. 2004. Genetic parameters for reproductive performance of breeding cows and carcass traits of fattening animals in Japanese Black (Wagyu) cattle. Animal Science. **78**(2), 195.

[2] Abe H, Masuda Y and Suzuki M. 2009. Relationships between reproductive traits of heifers and cows and yield traits for Holsteins in Japan. J. Dairy Sci. **92,** 4055.

[3] Bolacali M and Öztürk Y. 2017. Effect of non-genetic factors on reproductive traits in Simmental cows reared in subtropical climate condition. J.Anim.Plant Sci. (in press).

[4] Liu Z, Jaitner J, Reinhardt F, Pasman E, Rensing S and Reents R. 2008. Genetic evaluation of fertility traits of dairy cattle using a multiple-trait animal model. J. Dairy Sci. **93,** 4333.

[5] Rzewuska, K and Srabel K. 2015. The effect of non-genetic factors on reproduction traits of primiparous Polish Holstein-Friesian cows. Anim. Sci. Pap. Rep. **33** (4), 347.

[6] Uchida H, Kobayasi J, Inoue T, Suzuki K, and Oikawa T.2002. Current level of reproductive performances in Japanese Black cows. Asian-Aust. J. Anim. Sci. **15**(8), 1098.

[7] Özbeyaz C, Kücük M and Colakoglu N. 1996. Fertility performance in Brown-Swiss herd at Malya state farm. Lalahan hay. Arast. Enst. Derg. **36**:1.

[8] SAS Institute Inc. 2011. *Base/STAT SAS 9.3 User`s Guide.* SAS Institute Inc., Cary, NC USA.

[9] Jamrozik J, Fatehi J, Kistemaker G.J and Schaeffer L.R. 2005. Estimation for Canadian Holstein female reproductive traits. J. Dairy Sci. **88,** 2199.

[10] Haer L.C.M. de, Jong G. de and Vessies P.J.A. 2013. Estimation of genetic parameters of fertility traits for Virgin heifers in the Netherland. Interbull Bulletin 47 p 142-146.

[11] Siddiqui M.A.R, Das, Z.C, Bhattacharjee J, Rahman M.M, Islam M.M., Haque M.A and Shamsuddin M. 2013. Factors affecting the first service conception rate of cows in smallholder dairy farms in Bangladesh. Reprod. Domest. Anim. **48**(3), 500.

[12] Sasaki Y, Uematsu M, Kitahara G, and Osawa T. 2016. Reproductive performance of Japanese Black cattle: Association with herd size, season, and parity in commercial scow-calf operations. Theriogenology. **86**(9), 2156.

[13] Bacó S, Harada H and Fukuhara R. 1998. Genetic relationships of body measurements at registration to a couple of reproductive traits in Japanese Black Cows. Anim. Sci. Technol. **69,** 1.

[14] Oyama K, Katsuta T, Anada K and Mukai F. 2002. Heritability and repeatability estimates for reproductive traits of Japanese Black Cows. Asian-Aust. J. Anim. Sci. **15,** 1680.

[15] Oikawa, T. 2016. Effect of heat stress on age at first calving of Japanese Black cows in Okinawa. Anim. Sci. J. **88**(3), 439.

[16] M'hamdi N, Aloulou R, Brar S.K, Bouallegue M and Hamouda B.M. 2010. Phenotypic and genetic parameters of reproductive traits in Tunisian Holstein cows. Biotech. Anim. Husbandry. **26**(5-6), 297.

[17] Bormann J.M, Totir L.R, Kachman S.D, Fernando R.L and Wilson D.E. 2006. Pregnancy rate and first-service conception rate in Angus heifers. J. Anim. Sci. **84** (8), 2022.

[18] McGuirk B.J, Going I, and Gilmour A.R. 1999. The genetic evaluation of UK Holstein Friesian sires for calving ease and related traits. Animal Science. **68**:413.

Hematologic Profile and Semen Quality of Male Timor Deer (*Rusa timorensis*) at Various Hierarchies

D. Samsudewa[1]*, S. S. Capitan[2], C. C. Sevilla[2], R.S.A. Vega[2] and P. P. Ocampo[3]

[1]Faculty of Animal and Agricultural Sciences, Diponegoro University, Semarang, Central Java, Indonesia
[2] Institute of Animal Science, College of Agriculture & Food Science, University of The Philippines Los Banos, Los Banos, Laguna, Philippines
[3]Institute of Biological Sciences, University of The Philippines Los Banos, Los Banos, Laguna, Philippines

Corresponding E-mail: daudreproduksi@gmail.com

Abstract

The aim of this research was to observe hematologic profile i.e. erythrocyte count, hemoglobin and hematocrit and semen quality, i.e. semen volume, sperm motility and sperm abnormality of α-male, β-male and subordinate male Timor deer raised under captivity. Twelve males (51 ± 6 months old; 68.29 ± 8.41kg body weight) at similar antler stages were use in this study. Before and after 43 days of establishment of dominance hierarchy blood were sampled after sedation for erythrocyte count, hemoglobin (mg/dL), and hematocrit (%). Likewise, semen was collected using electroejaculator and were analyzed for semen volume (ml), sperm motility (%) and sperm abnormality (%) to compare male deer at various heirarchies. Wilcoxon signed ranks test and Kruskal-Wallis H test of non-parametric analysis was done. Significant difference was tested with Mann-Whitney U test. The results showed that highest count of erythrocyte shown on α and β-male (1.60 million per μL). The highest increase in hematocrit was observed in β-male (5%) and then followed by S2-male (4%). S2-male had the highest increase in hemoglobin (0.13 g/dL). The highest increase in semen volume was observed in α –male (0.75 ml). Social stress affected negatively the sperm motility and abnormality ($P<0.05$). The highest decrease was observed in S2-male.

1. Introduction

The Timor deer (*Cervus timorensis* Blainville) is one of the medium size deer in Indonesia with 56.80 ±11.63 kg male body weight and 80.04 ±5.93 cm shoulder height [1]. Timor deer are currently being reared in captivity in Indonesia. But, increasing population of Timor deer in Indonesian captivity is still low. The captive farm in Kudus regency reported increasing population 0.5 head per year [1]. This research suggest that the productivity of Timor deer in captivity can be improved if some key aspects of management such as feeding, mating, aggressive male behavior and maternal care are addressed.

Aggressive male behavior was common in captive farm of Timor deer, to form dominance hierarchy. Aggressive male behavior during establishment of dominance hierarchy causes unnecessary stress i.e. partitioning nutrients towards fighting and wound healing. This social stress may lead to poor health of Timor deer and reproductive performance. Social stress has adverse effects on the reproduction system and health of both males and females [2]. Understanding and managing social stress of Timor deer will help establihsh the best management system esspecially in improving reproduction efficiency.

The research about understanding the effect of establishing dominance hierarchy of Timor deer on the reproduction was limited. Therefore this study was conducted to know the effect of dominance hierarchy on hematologic profile and semen quality of Timor deer.

2. Materials and Methods

2.1. Care & Management of Timor Deer

This research was conducted in H. Yusuf Wartono Timor deer captivity, Margorejo, Dawe, Kudus, Central Java, Indonesia. Blood and semen samples of 12 male Timor deer (51 ± 6 months old; 68.29 ± 8.41kg body weight) at similar antler stages were used in this research. Timor deer were placed in three communal stalls (23.5 x 21.5m). Four male Timor deer in every stall will form dominance hierarchy by fighting for about one week. Finally, Timor deer in every stall will form α-male (Superior male), β-male (Second rank male), subordinate-1 (S1) male (Third rank male) and subordinate-2 (S2) male (Fourth rank male). Hematologic profile and semen volume and quality were obtained before and at 43 days after establishment of dominance hierarchy. Seven kilogram napier grass (Pennisetum puprpureum) was fed for each individu per day to fullfil consumption. Water was prepared ad libitum.

2.2. Blood Collection and Assay

Blood samples were taken by jugular venepuncture from each male after ketamine and xylazine sedation. Blood analysis was done to obtain the hematologic picture (hematocrit, haemoglobin and erythrocyte).

Erythrocyte was counted using blood containing EDTA. Sipped the blood into a standard blood dilution pipette to the 0.5 mark, followed by methyl violet 2B diluents to the 1.01 mark. The final dilution was 1 : 200. Shaked the pipette briefly, place drops to the counting chamber of a neubauer haemocytometer. Let it settled for a few minutes, the red cell count was determined under high-dry magnification (40 x objectives and 10 x ocular). Erythrocytes appearing in the center and four corner squares were counted and the total multiplied by 10^4 to give erythrocyte count per microliter [3].

Haemoglobin was determined at the start by filling the Sahli tube with HCl 0.1 N up to 2 mark. Blood was drawn into a haemoglobin pippete up to the 20 mark and filled the Sahli tube. Blood was stirred for 3 minutes until hematin acid appeared. Placed the Sahli tube in comparator block. Water filled the Sahli tube and it was stirred until it obtained the same colour as the standard in the comparator block. Scale was read in Sahli tube [3].

Blood was collected in tubes with ethylenediamine tetraacetate (EDTA) and then filled in microcapillary tubes. Sealed the microcapillary tubes with wax and centrifuged at 16,000 rpm for 5 minutes and read using a standard microhematocrit graphic reader [3].

2.3. Semen Quality

Semen was collected from each male after ketamine and xylazine sedation with use of electroejaculator with 15 V and 500 mA. Increase of voltage was done gradually until 15 V [4]. Semen was analyzed macroscopically (volume) and microscopically (motility and sperm abnormality). Volume of semen was measured in ml by using tulip scale. Measurement range of tulip scale was 0-12 ml with the sensitivity of 0.1 ml.

Motility of spermatozoa was determined by the following steps: Mixed the semen of Timor deer by inverting vial 2-3 times. Placed a drop of the Timor deer semen and spread on the warm slide. Covered with cover slip. Examined the semen immediately for motility, using the low power objective of the microscope. Observe the edge of the semen to ascertain an approximation of the percentage of live and motil spermatozoa. Graded The semen based on the scale 0-100% [5].

Determination of spermatozoa abnormality was done using the following steps: Smear the semen on object glass and stained with eosin-negrosin. Examined the preparation under high objective (1000x) of the microscope. The sperms were observed for morphological abnormalities such as head abnormalities (giant, small, double etc.), middle piece abnormalities (double midpieces, swollen midpieces, coiled midpieces etc.) and tail abnormalities (coiled tails, absent tails, broken tails etc.). Minimum 200 sperms were counted using random fields on the different parts of the slide. Count the abnormality and convert it to percent (%) [5].

2.4. Data Analysis

Wilcoxon signed ranks test was used in comparing before and at 43 days after establishment of hierarchy for all parameters. The comparison of various hierarchy (Superior, 2nd, 3rd and 4rth rank) was accomplished using Kruskall-Wallis H Test for non-parametric analysis for this research. Mann whitney U test of non-parametric analysis was done when different rank were found significant [6].

3. Results and Discussion

3.1. Erythrocyte count

Average values of erythrocyte count of α, β, S1-male and S-2 male of Timor deer before and at 43 days after establishment of dominance hierarchy are shown in Table 1.

Table 1. Average values of erythrocyte count of male Timor deer at various hierarchies before and 43 days after establishment of dominance hierarchy

FACTOR	HIERARCHY			
	α-Male	β-Male	S1-Male	S2-Male
	..........Million per μL..........			
Before	5.42	5.66	6.85	6.61
After	7.77	7.05	5.57	6.68
Z	1.60	1.60	1.07	0.00

Wilcoxon signed ranks test for erythrocyte showed no significant difference (P≥0.05) in median values before and 43 days after establishment of dominance hierarchy for all hierarchy levels. The non-parametric Kruskal-Wallis H test showed no significant difference (P≥0.05) among the hierarchies before (χ^2 = 3.21) and 43 days after (χ^2 = 6.64) establishment of dominance hierarchy in terms of erythrocyte count.

The present study showed that establishment of dominance hierarchy did not influence the number of erythrocyte. This result contradicts the research on red deer stags that showed lowest red blood cell count (RBC) count during rutting period, since RBC count in this research (5.42 to 7.77 x 10^6/μL) still falls under the normal range (6.2 to 11.7 x 10^6/μL) [7]. However, these values are still low compared with those reported in Ragunan Zoo in Jakarta, 8.12 to 9.89 x 10^6/μL with an average of 9.01 x 10^6/μL ± 0.93 x 10^6/μL [8], and those found in male Rusa Deer (*Cervus timorensis russa*) in Thailand, 11.22 x 10^6 ±1.19 x 10^6/μL [9].

3.2. Hemoglobin

Average values of hemoglobin (Hb) of α, β, S1-male and S-2 male of Timor deer before and 43 days after establishment of dominance hierarchy are shown in Table 2.

Table 2. Average values of male Timor deer hemoglobin at various hierarchies before and 43 days after establishment of dominance hierarchy

FACTOR	HIERARCHY			
	α-Male	β-Male	S1-Male	S2-Male
	..........g/dL..........			
Before	11.80	11.80	12.07	12.61
After	11.60	11.40	11.13	13.00
Z	0.45	0.82	1.07	0.00

In this parameter, Wilcoxon signed ranks test for hemoglobin showed no significant difference ($P \geq 0.05$) in median values before and 43 days after establishment of dominance hierarchy groups for all male groups. Furthermore, the non-parametric Kruskal-Wallis H test showed no significant difference ($P \geq 0.05$) among the hierarchies both before (χ^2 = 1.56) and 43 days after (χ^2 = 2.61) establishment of dominance.

Hb in this study (11.60 to 13.00 g/dL) still falls below the nomal range (7.9 to 17.4 g/dL) [7], but is still low compared with those found in Ragunan Zoo, 14 to 17.6 g/dL with an average of 16.25 ± 1.63 g/dL [8], but comparable with that measured on Timor deer in Thailand, 12.5 ± 1.7 g/dL by [9].

3.3. Hematocrit

Average values hematocrit of α, β, S1-male and S-2 male of Timor deer before and 43 days after establishment of the hierarchy are shown in Table 3.

Table 3. Average values of male Timor deer hematocrit at various hierarchies before and 43 days after establishment of dominance hierarchy

FACTOR	HIERARCHY			
	α-Male	β-Male	S1-Male	S2-Male
	.... %			
Before	31.67	31.67	34.00	35.00
After	35.33	36.67	35.00	39.00
Z	1.63	1.60	0.82	1.63

Wilcoxon signed ranks test showed no significant difference ($P \geq 0.05$) in median values before and 43 days after establishment of dominance hierarchy for all male groups. The non-parametric Kruskal-Wallis H test showed no significant difference ($P \geq 0.05$) among the groups both before (χ^2 = 2.62) and 43 days after (χ^2 = 2.82) establishment of dominance hierarchy in terms of percent hematocrit.

Hematocrit in this study (31 to 39%) falls under the normal range (26 to 46%) [7], but is still low when compared with those reported in Ragunan Zoo, Jakarta (40.5 to 43.5%) [8], and that estimated on Timor deer in Thailand which were 41.75 ± 1.32% [9].

The results of the present work on the relationship between dominance hierarchy and hematologic profile varies. Although not statistically significant the highest increase of hematocrit after establishment of hierarchy was shown by β-male (5%) and followed by S2-male (4%). Hemoglobin, in S2-male showed the highest increase (0.13 g/dL). Both male groups had increased hematocrit and hemoglobin content which are probably due to stress related to dominance hierarchy establishment. Research on Roe deer showed that during the mating season stags need physical strength and endurance to be able to cope with the huge energy demands [10]. Further, they stated that since increased hematocrit improves oxygen supply to the muscles, it seems likely that during the rut, males with high testosterone level (the dominant male) will be able to endure more intense physical exercises during male-to-male combats. Furthermore, metabolic efficiency is needed to study for copulation with and in defense for females. The α-males had increased hematocrit (3.67 g/dL) and erythrocyte content (2.35 million x 10^6/μL). This condition indicates the fitness of α-male.

3.4. Semen volume

Average values of semen volume of α, β, S1-male and S-2 male of Timor deer before and 43 days after establishment of dominance hierarchy are shown in Table 4.

Table 4. Average values of male Timor deer semen volume at various hierarchies before and 43 days after establishment of dominance hierarchy

FACTOR	HIERARCHY			
	α-Male	β-Male	S1-Male	S2-Male
	....	.... ml	....	
Before	1.00	1.70	1.83	1.13
After	1.75	1.62	2.07	1.43
Z	1.07	0.00	0.27	1.07

Wilcoxon signed ranks test for semen volume showed no significant difference (P≥0.05) in median between before and 43 days after establishment of dominance hierarchy groups for all males. Kruskal-Wallis H test showed significant differences (P<0.05) among the groups before (χ^2 = 8.75) establishment of dominance hierarchy in terms of semen volume. On the other hand, no significant difference (P≥0.05) was found after (χ^2 = 0.64) the establishment of dominance hierarchy. The significant difference (P<0.05) before establishment of dominance was found in α-male compared with other males.

Alpha-male had the largest increase in semen volume (0.75 ml). Increasing semen volume in α-male is related with the rutting status of the animal. The α-males are the most active Timor deer in terms of mating. The semen volume non-rutting Eld deer stags was 2.11 ± 0.61 ml and during rutting Eld deer stags have a significant difference in ejaculate semen volume, around 3.49 ± 0.36 ml [11]. Semen volume of male Timor deer in this study (0.30 to 3.50 ml) was still within range compared with semen volume of Timor deer in Ragunan Zoo, Jakarta (2.06 ± 0.63 ml) [12].

3.5. Sperm motility

Average values of sperm motility of α, β, S1-male and S-2 male of Timor deer before and 43 days after establishment of dominance hierarchy are shown in Table 5.

Table 5. Average values of male Timor deer sperm motility at various hierarchies before and 43 days after establishment of dominance hierarchy

FACTOR	HIERARCHY			
	α-Male	β-Male	S1-Male	S2-Male
	....	.... %	....	
Before	70.00	69.17	70.00	75.00
After	60.83	46.67	46.67	33.33
Z	1.60	1.60	1.34	1.07

Wilcoxon signed ranks test showed no significant difference in median values before and after establishment of dominance hierarchy groups for all male groups. Kruskal-Wallis H test for showed no significant difference (P≥0.05) among the groups before (χ^2 = 1.77) and after (χ^2 = 0.12) establishment of dominance hierarchy in terms of sperm motility.

Establishment of dominance hierarchy among male groups did not affect motility of spermatozoa. Sperm motility of the present study (33.33 to 75.00%) is lower compared the sperm motility of Timor deer in Ragunan zoo (75.83 ± 3.76%) [12].

3.6. Sperm abnormality

Average values of sperm abnormality of α, β, S1-male and S-2 male of Timor deer before and 43 days after establishment of dominance hierarchy are shown in Table 6.

Table 6. Average values of male Timor deer sperm abnormality at various hierarchies before and after establishment (43 days) of dominance hierarchy

FACTOR	HIERARCHY			
	α-Male	β-Male	S1-Male	S2-Male
	....	.... %	....	....
Before	8.31	8.46	6.78	10.13
After	9.98	19.79	17.93	44.58
Z	0.00	1.60	1.60	1.07

Wilcoxon signed ranks test was showed no significant difference (P≥0.05) in median values before and 43 days after establishment of dominance hierarchy for all male groups. The non-parametric Kruskal-Wallis H test showed no significant difference (P≥0.05) among the groups before (χ^2 = 1.97) and 43 days after (χ^2 = 2.54) establishment of dominance.

Establishment of dominance hierarchy did not affect the proportion of sperm abnormality. Sperm abnormality of Timor deer in this research (8.31 to 44.58%) is higher than Timor deer in Ragunan zoo (7.31 ± 2.99%) [12]. Sperm abnormality found in this research includes sperms without tail, double head, double tail, small head and giant head. S2 male shown higher of sperm abnormality was connected with social stress and inhibition of mating for S2 male compared to other male. Eld's deer stags had lower number of normal sperm outside of rutting season (18.2 ± 5.9%) and stags in rutting season had the highest number of normal sperm (85.7 ± 5.4%) [11].

4. Conclusion

Dominance hierarchy was affected by decrease in sperm motility and increase in sperm abnormality esspecially to subordinate 2 (S2) male. Beta-male and S2-male had increased hematocrit and hemoglobin content which are probably due to stress affected by aggressive male behavior during dominance hierarchy establishment. The α-male is the fitness of male Timor deer compare with other hierarchy level.

Acknowledgments

Deepest thanks and appreciation is given to H. Yusuf Wartono the owner of Timor deer captive breeding for the permission and support. Thanks also to the Dean of the Faculty of Animal and Agricultural Sciences, Diponegoro University, Indonesia for allowing the conduct of this research.

References

[1] Samsudewa D, Handayani A and Aryadi B A 2016 Body Measurement of Timor Deer (*Rusa timorensis*) in Captivity of Central Java *Proc. The USR Internasional Seminar on Food Security* 2 pp 186-190

[2] Moberg P 1991 How Behavioral Stress Disrupts the Endocrine Control of Reproduction in

Domestic Animals J. Dairy Sci. **2**.304

[3] Praseno, K, Isroli and Susanti S 2007 Laboratory Guidelines: Basic of Animal Physiology. Laboratory of Physiology and Biochemistry (Semarang: Faculty of Animal Science Diponegoro University) p 17

[4] Samsudewa D, Sarjana T A, Lukman A, Sugiyanto E, Prasetyo I, Agustianto R, Jamaludin A and Saefulloh 2004 Effort of conservation germ plasm in Indonesia with electroejaculatore for stimulate ejaculation using ram models Proc. National Seminar in Ruminant p 28

[5] Capitan S S and Palad O A 1999 Manual for artificial breeding of farm animals (Los Baños: College of Agriculture University of the Philippines Los Baños) pp 27-29

[6] Santoso, S 2014 Statistik Nonparametrik (Jakarta: Elex Media Komputindo) p 23-25

[7] Boes K M 2010 Hematology of the cervids Schalm's veterinary 6th Ed. Weiss D J, Wardrop K J (Iowa : Willey-Blackwell) chapter 116 pp 918-926

[8] Salakij C, Salakij J, Kamthapanit C, Rochanaphat N and Nunklang G 1999 Hematological values of Rusa Deer (Cervus timorensis russa) in Thailand Kasetsart J. (Nat. Sci.) **33** 83

[9] Yusmin L M 1998 Morfology, karyotype and blood component of some species of deer in Indonesia (Bogor: Bogor Agriculture University) pp 67-75

[10] Blottner S, Hingst O and Meyer H H D 1996 Seasonal Spermatogenesis and Testosterone Production in Roe Deer (Capreolus capreolus) J. Reprod. Fertil. **108** 302

[11] Monfort S L, Brown J E, Bush M, Wood T C, Wemmer C, Vargas A, Williamson L R, Montali R J and Wildt D E 1993 Circannual inter-relationships among reproductive hormones, gross morphometry, behaviour, ejaculate characteristics and testicular histology in Eld's deer stags (Cervus eldi thamin). J. Reprod. Fertil. **98** 474

[12] Messang-Nalley W M 2006 Study on biology of reproduction and application of artificial insemination technology on Timor deer (Cervus timorensis) (Bogor: Bogor Agriculture University) pp 79-82

Estimation of 305 Day Milk Yield from Cumulative Monthly and Bimonthly Test Day Records in Indonesian Holstein Cattle

A.P. Rahayu[1,2], T. Hartatik[3], A. Purnomoadi[1] and E. Kurnianto[*]

[1]Faculty of Animal and Agricultural Sciences, Diponegoro University, Semarang, Central Java, Indonesia.
[2]Permanent Address: Agriculture, Fisheries and Food Center of Semarang Regency, Ungaran, Central Java, Indonesia.
[3]Faculty of Animal Sciences, Universitas Gadjah Mada, Yogyakarta, Indonesia.

[*]Corresponding E-mail: kurniantoedy17@gmail.com

Abstract

The aims of this study were to estimate 305 day first lactation milk yield of Indonesian Holstein cattle from cumulative monthly and bimonthly test day records and to analyze its accuracy.The first lactation records of 258 dairy cows from 2006 to 2014 consisted of 2571 monthly (MTDY) and 1281 bimonthly test day yield (BTDY) records were used. Milk yields were estimated by regression method. Correlation coefficients between actual and estimated milk yield by cumulative MTDY were 0.70, 0.78, 0.83, 0.86, 0.89, 0.92, 0.94 and 0.96 for 2-9 months, respectively, meanwhile by cumulative BTDY were 0.69, 0.81, 0.87 and 0.92 for 2, 4, 6 and 8 months, respectively. The accuracy of fitting regression models (R^2) increased with the increasing in the number of cumulative test day used. The used of 5 cumulative MTDY was considered sufficient for estimating 305 day first lactation milk yield with 80.6% accuracy and 7% error percentage of estimation. The estimated milk yield from MTDY was more accurate than BTDY by 1.1 to 2% less error percentage in the same time.

1. Introduction

The main need in evaluating the productivity of dairy cattle is the existence of milk yield data during one lactation period. In some countries that have advanced in the dairy industries like New Zealand, Australia and Canada, donations of recording program is large enough. For developing countries, alternatives are needed since facilities supporting the recording programs are lack. The alternative to replace daily records for selection purposes can be done based on a weekly, biweekly, monthly, or bimonthly records.

Previous studies showed the various methods that can be used to estimate complete milk production, including Test Day Models [1-4] and extending part lactation [5-6]. Test Day Records (TD) is a 24-hour total milk production record taken on certain test days only [7]. The former method requires TD for 1 lactation period (± 10 months). In fact, early estimates may be useful in a more efficient selection process because it can indirectly shorten the generation interval. The later method requires a sum of daily record yield during some months of lactation. Compared to the former method, the later method can predict milk yield earlier, but daily yield is more difficult to be recorded than

IOP Conf. Series: Earth and Environmental Science **119** (2018) 012028 doi:10.1088/1755-1315/119/1/012028

monthly. To further simplify and overcome the weaknesses of both methods, the experiment was conducted using monthly and bimonthly test day records on part lactation.

Other studies have also used TD on part lactation to predict complete milk yield, but with fairly complicated formulas, including the Bayes empirial method [8], Legendre polynomials [9] and artificial neural networks [10], so that they were rather difficult to be applied by farmers. Therefore, this study is expected to provide an appropriate estimation formula especially for small holder farming. The aims of this study were to estimate 305 day milk yield (dMY) of Indonesian Holstein cattle from cumulative monthly and bimonthly test day records and to analyze its accuracy.

2. Materials and Methods

2.1. *Materials*

Materials used in this study were the first lactation daily milk yield records of 258 Holstein-Friesian cows from 2006 to 2014 raised at Baturraden Dairy Cattle Breeding and Forage Centre (*Balai Besar Perbibitan Ternak Unggul dan Hijauan Pakan Ternak* / BBPTU–HPT Baturraden), Purwokerto, Central Java, Indonesia. The criteria of data used were 270 to 335 days of lactation length. Data used in this study was taken monthly after 8th days of calving as many as 2571 records and bimonthly as many as 1281 records.

2.2. *Methods*

Method used to estimate 305-d milk yield in this experiment was a simple linear regression analysis with one independent variable, i.e. cumulative test day yield [11]. The two cumulative monthly data was obtained by summing the first and second monthly test day milk yields (MTDY), as well as for the following cumulative months (3, 4, 5, 6, 7, 8 and 9). The two cumulative bimonthly data was obtained by summing the first and second bimonthly test day milk yields (BTDY), as well as for the following cumulative months.

Cumulative MTDY = 1st TD + 2nd TD + 3rd TD + ... + i^{th} TD

Cumulative BTDY = 2nd TD + 4th TD + 6th TD + ... + i^{th} TD

2.3. *Statistical Analysis*

For the calculation of regression values, the computer program SPSS 20 was applied using linear regression procedure by the following equation:

$$b = \frac{n\,(\sum X_{ci}\, X_{c10}) \quad (\sum X_{ci})(\sum X_{c10})}{n\left(\sum {X_{ci}}^2\right) \quad (\sum X_{ci})^2}$$

b = regression coefficient

X_{ci} = milk yield of the first i cumulative months

X_{c10}= milk yield of 10 cumulative months

n = number of cows

The equation of regression was:

$\mathbf{\hat{y}_{c10} = a + bX_{ci}}$

$\hat{y}_{c10}$= estimated milk yield of 10 cumulative months

a = intercept

b = regression coefficient

X_{ci}= milk yield of the first i cumulative months

The 305-d milk yield was estimated by the following formula:

$\mathbf{\hat{Y}e = \hat{y}_{c10} \times 30.5}$ (for monthly records)

$\mathbf{\hat{Y}e = \hat{y}_{c10} \times 30.5 \times 2}$ (for bimonthly records)

$\hat{Y}e$ = estimated 305-d milk yield during one lactation period
$\hat{y}_{c10}$ = estimated milk yield of 10 cumulative months by regression method
30.5 = assumption of average day in a month

Correlation to examine the relationship between estimated and actual milk yield were calculated as:

$$R=\frac{n(\sum \hat{Y}e.Y)-(\sum \hat{Y}e)(\sum Y)}{\sqrt{[n(\sum \hat{Y}e^2)-(\sum \hat{Y}e)^2][n(\sum \hat{Y}^2)-(\sum Y)^2]}}$$

R = correlation coefficient
$\hat{Y}e$= Estimated 305-dMY during one lactation period
Y = Actual milk yield during one lactation period

The accuracy of fitting regression models (R^2) were calculated by using the following formula:

$$R^2=\frac{\text{Sum of squares due to regression}}{\text{Total sum of squares}} \times 100\%$$

The error was estimated as a deviation of estimated milk yield from the actual milk yield. The absolute error was calculated without considering the positive or negative values.

$$E=\frac{|\hat{Y}ei-Yi|}{Yi} \times 100\%$$

E = Error percentage
$\hat{Y}e_i$= Estimated 305-dMY of the ith animal,
Y_i = Actual milk yield of the ith animal

3. Results and Discussion

The means of single test day milk yield were 17.2, 19.4, 18.3, 16.4, 15.1, 13.8, 12.8, 12.1, 11.4 and 10.5 for 1 to10 months, respectively (Figure 1). This result showed the peak of lactation occurs in the second month (acceleration phase) then decreased gradually until the dry period. The actual milk yield in dairy cows is a dynamic process of forming a curve. The curve at the beginning of lactation will increase toward lactation peak at 6 to 7 weeks after partus and then gradually decrease until the end of lactation [12]. The means of cumulative MTDY were 36.5, 54.8, 71.2, 86.3, 100.0, 112.8, 124.9 and 136.3 for 2 to 9 months, respectively, meanwhile the means of cumulative BTDY were 19.4, 35.8, 49.5 and 61.7 for 2, 4, 6 and 8 months, respectively (Table 1). These MTDY and BTDY were used to estimate the first ten months cumulative test day yield.

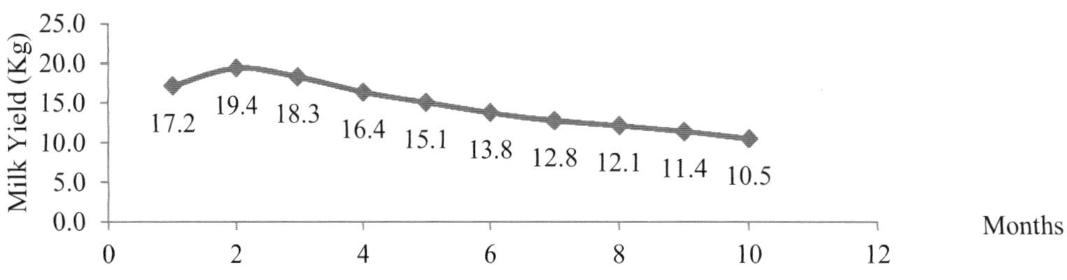

Figure 1. Observed Single Test Day Milk Yield

The actual and estimated 305-d milk yield for the cumulative monthly and bimonthly test day yield are presented in Table 2 and 3. The present mean±standard of deviation (SD) of 305-d actual yield was 4386.0±761.4 kg (Table 2). This present mean was higher than those reported by Rahayu *et al.* [12] in the same place, i.e. 4015.9 kg. The differences were in the number and criteria of the data being analyzed. Estimated 305-

dMY mean was very close to that observed for the both CMTDY and CBDTY (6.1 to 9.3 kg of differences).

Table 1. Observed Cumulative Test Day Milk Yield

	Number of cumulative months									
	1	2	3	4	5	6	7	8	9	10
CMTDY (kg)	17.15	36.54	54.84	71.21	86.27	100.04	112.80	124.94	136.34	146.45
CBTDY (kg)		19.39		35.75		49.52		61.66		71.77

CMTDY = average of cumulative monthly test day milk yield; CBTDY = average of cumulative bimonthly test day milk yield

Table 2. Statistic of The Actual and Estimated 305 Day Milk Yield Based on CMTDY

	Actual	Estimated milk yield based on number of cumulative months							
		2	3	4	5	6	7	8	9
Mean (kg)	4386.0	4393.0	4392.7	4394.1	4394.7	4393.9	4392.1	4394.5	4393.0
SD (kg)	761.44	564.53	625.27	664.68	689.30	713.11	731.84	748.39	760.05
CV (%)	17.4	12.85	14.23	15.13	15.68	16.23	16.66	17.03	17.30
Min. (kg)	2135.7	2886.4	2689.1	2627.9	2603.2	2527.4	2458.2	2394.2	2355.0
Max.(kg)	6314.5	6077.1	6205.8	6091.3	6231.6	6348.0	6233.6	6308.0	6373.0
A		57.093	42.061	30.687	22.683	14.710	7.098	0.430	-1.904
B		2.445	1.903	1.626	1.433	1.317	1.235	1.169	1.088
R		0.703	0.784	0.832	0.863	0.894	0.922	0.944	0.960
R^2 (%)		54.0	66.3	74.9	80.6	86.3	91.0	95.1	98.1
E (%)		9.9	8.7	7.9	7.0	6.4	5.4	4.6	4.0

The coefficients of variance (CV) for estimated 305-dMY based on CMTDY ranged from 12.85 to 17.30% (Table 2), meanwhile based on BMTDY ranged from 13.03 to 17.89% (Table 3), lower than reported by Atil [5] i.e. 26.94 to 42.10%. This lower CV indicated that cows at the BBPTU-HPT Baturraden have almost the same genetic qualities. In addition, the data used in this study is only the first lactation data with almost the same age (<36 months). Study by Khazaei and Nikosiar [2] using 1st to 8th lactation data showed a CV of 23.9%, showing that there was a large amount of diversity between milk yields of Holstein cows at different lactation stages. This was because the milk production curve generally follows a regular pattern in each lactation period. Milk production will continue to rise until it reaches its peak in the third to fourth lactation period (aged 6 to 8 years) due to an increase in the number and size of udder cells that are the places to synthesize milk. In subsequent periods, milk production began to decline gradually. CV, SD, the minimum and maximum values were closer to the actual value with the increasing number of TD used.

Correlation coefficients between actual and estimated milk yield by CMTDY were 0.70, 0.78, 0.83, 0.86, 0.89, 0.92, 0.94 and 0.96 for 2 to 9 months (Table 2), respectively, meanwhile by CBTDY were 0.69, 0.81, 0.87 and 0.92 for 2, 4, 6 and 8 months, respectively (Table 3). The correlation coefficient on Red Chittagong Cattle in Bangladesh using a similar method was slightly higher than that obtained in this study i.e. 0.84, 0.87, 0.90, 0.92, 0.95 and 0.99 for cumulative 2 to 7 months [11]. While the other study on Holstein in Egypt found the lower values, i.e. 0.14, 0.24, 0.36, 0.47, 0.62, 0.75, 0.87 and 0.96 for 2 to 9 months, respectively [5]. These differences might be caused by the difference of milk yield, weather, environment and genetic factors. Correlation coefficients (R) and the accuracies of milk yield estimation (R^2) were increasing along with the increasing number of TDs, otherwise, error percentages were decreasing (9.9 to 4% for CMTDY and 11 to 6.6% for CBTDY). It means that the more TDs used to estimate, the closer estimated to actual milk yield. The highest correlation achieved was in the 9th cumulative month as much as 0.96 by CMTDY and in the 8th cumulative month as much as 0.92 by CBTDY. The high correlation between actual and estimated 305-dMY indicated that part of lactation can be used for evaluating the milk producing ability in cows.

Table 3. Statistics of the Actual and Estimated 305 Day Milk Yield Based on CBTDY

	Actual	Estimated milk yield based on number of cumulative months			
		2	4	6	8
Mean (kg)	4386.0	4378.39	4378.84	4376.74	4379.89
SD (kg)	761.44	570.5	678.4	740.2	783.7
CV (%)	17.36	13.03	15.49	16.91	17.89
Min. (kg)	2135.7	2781.7	2721.2	2538.8	2282.4
Max.(kg)	6314.5	6197.3	6196.1	6441.4	6582.5
A		30.810	16.436	7.221	0.407
B		2.113	1.548	1.303	1.171
R		0.694	0.813	0.873	0.917
R^2 (%)		49.2	69.5	82.9	92.8
E (%)		11.0	9.2	7.9	6.6

Notes for Table 2 and Table 3: Mean = the average of 305 day milk yield; SD = standard deviation; CV = coefficient of variability; Min. = minimum value; Max.= maximum value; a = intercept; b = regression coefficient; R = Pearson correlation coefficient; R^2 = coefficient of determination; E = error percentage.

Error percentage indicated the deviation of estimation result with actual milk yield. Over-predictions were associated with cows with high yields at the initial stage of their lactation, which declined rapidly as the lactation progressed. As the rapid decline, the later stage of lactation was not properly accounted for. Under-predictions were associated with cows that are very persistent through their lactation. Therefore, the initial few TDs were inadequate in modelling the lactation curve of these cows [1]. The estimated milk yield from CMTDY was more accurate than CBTDY by 1.1 to 2% less error percentage in the same time. Study by Kokate *et al.* [4] based on fitting of stepwise multiple regressions, the regression equation with 3 variables (4, 6 and 10 months BTDY) was considered more appropriate for prediction of first lactation 305-dMY with 83% accuracy and 7.54% error.

The present results indicated that the first five months of lactation based on CMTDY ($\hat{Y}e = 22.683 + 1.433\ X$; R=0.86; R^2=80.6%; E=7.0%) and the first six months based on CBDTY ($\hat{Y}e = 7.221 + 1.303\ X$; R=0.87; R^2=82.9%; E=7.9%) were considered sufficient for estimate 305 day milk yield. As stated by Atil [5] that a minimum R^2 of 75% is accurate enough. Another statement by Olori [13] that R^2>70% implies a very good fit for a model, while R^2<40% should not be used for prediction. Atil [5] found that the seven months of cumulative daily records are sufficient of predictions 305 day milk yield (R^2= 75%), while, Abdel Glil *et al.* [14] found that the first four months are the best (R^2=0.91). Other study by Habib *et al.* [11] found the best correlations between single monthly test day records and the total records are 5th month of lactation, i.e. 0.81. If there is any situation when only single TD record is available, 5th month of lactation record would be the best test month to evaluate a cow. In another study, the use of 4 TDs can achieve a higher correlation and accuracy value of 0.95 and 90%, respectively, using artificial neural networks algorithm [10].

Selection on the basis of part records could be accurate enough to increase genetic progress by decrease the time required for progeny test, decrease the generation interval to increase the annual genetic gain, improve the accuracy of evaluation at farm level and useful for culling the low production cows to improve overall farm profitability [7, 15]. The use of simple regression coefficients make it easy to get equality of complete lactation milk yield. These benefits will be increasingly meaningful in productivity evaluation of local Holstein dairy cows at the small holder farms that most of them have not recorded their cattle productivity yet.

4. Conclusion

The accuracy of fitting regression models (R^2) increased with the increasing in the number of cumulative test day used. The used of 5 cumulative monthly test day yield (MTDY) was considered sufficient for estimating 305 day first lactation milk yield with 80.6% accuracy and 7% error percentage of estimation. The estimated milk yield from cumulative MTDY was more accurate than cumulative bimonthly test day yield (BTDY) by 1.1 to 2% less error percentage in the same time.

Acknowledgments

The authors would like to thank Head of BBPTU-HPT Baturraden for allowing the study and his staffs for gathering data. This study was fully financed by Indonesia Endowment Fund for Education through Indonesian Education Scholarship for Doctoral Programs contract number 20160222035511.

References

[1] Mrode RA, Swanson GJT and Lindberg CM. 2002. Ef ciency of part lactation test day records for genetic evaluations using xed and random regression models Anim. Sci. **74.** 189.

[2] Khazaei J and Nikosiar M. 2008. Approximating milk yield and milk fat and protein concentration of cows through the use of mathematical and artificial neural networks models Proc. World Conf. on Agricultural Information and IT (Atsugi*)* pp 91-105

[3] Cole J B, Null D J and De Vries A. 2011. Short communication: Best prediction of 305-day lactation yields with regional and seasonal effects J. Dairy Sci. **94.** 1601.

[4] Kokate LS, Singh A, Banu R, Gandhi RS, Chakravarty AK, Gupta AK and Sachdeva GK. 2014. Prediction of 305-day lactation milk yield based on bimonthly test day values in Karan Fries cattle Indian J. Anim. Res. **48.** 103.

[5] Atil H. 1999. Ratio and regression factors for predictinc 305 day production from part lactation milk records in a herd of Holstein Friesian Cattle Pak. J. Biol. Sci. **2.** 31.

[6] Rajeev, Singh R, Kumar R, Raja TV, Singh YP, Ali N and Maurya P. 2016. Factors affecting first lactation cumulative monthly milk yields in Frieswal cattle Indian J. Anim. Res. B **3259.** 1

[7] Swalve HH. 2000. Theoritical basis and computational methods for different test-day genetic evaluation methods J. Dairy Sci. **83.** 1115

[8] Pereira JAC, Suzuki M and Hagiya K. 2001. Prediction of 305 days milk production from early records in dairy cattle using an empirical bayes method Asian-Aust. J. Anim. Sci. **14.**1511.

[9] Ptak E, Satoła A and Czaja H. 2004. Prediction of 305-day lactation milk, fat and protein yields using Legendre polynomials and test-day yields from different parts of lactation Anim. Sci. Papers and Reports **22.**173.

IOP Conf. Series: Earth and Environmental Science **119** (2018) 012028 doi:10.1088/1755-1315/119/1/012028

[10] Gorgulu O. 2012. Prediction of 305-day milk yield in Brown Swiss cattle using artificial neural networks S. Afr. J. Anim Sci. **42.** 280.

[11] Habib MA, Bhuiyan AKFH and Amin MR. 2012. Estimation of regression factors for extending part lactation milk yield records of Red Chittagong Cattle (RCC) in Bangladesh IRJALS **1.**78.

[12] Rahayu A P, Kurnianto E and Johari S. 2015. Genetic gains of milk yield and milk composition as realized response to dairy cow selection in BBPTU-HPT Baturraden, Indonesia J. Indonesian Trop. Anim. Agric. **40.**79.

[13] Olori VE, Brotherstone S, Hill WG and McGuirk BJ. 1999. Fit of standard models of the lactation curve to weekly records of milk production of cows in a single herd Livest. Prod. Sci. **58.**55.

[14] Abdel Glil F, Omar AO and Salem AY. 1995. Ratio and regression factors for predicting 305-day production from part lactation milk records in Friesian cows in Egypt Egypt. J. Appl. Sci. **10.** 634.

[15] Njubi DM, Wakhungu JW and Badamana MS. 2010. Use of test-day records to predict first lactation 305-day milk yield using artificial neural network in Kenyan Holstein–Friesian dairy cows Trop. Anim. Health Prod. **42.** 639.

Influence of Meiotic Stages on Developmental Competence of Goat' Oocyte After Vitrification

S. Wahyuningsih[1*] and M. N. Ihsan[1]

[1]Animal Husbandry Faculty, Brawijaya University , Indonesia

[*]Corresponding E-mail : yuningyuning208@yahoo.com

Abstract

This objective of this research was to investigate effect of goat oocyte meiotic stages on developmental competence after cryopreservation. Ovaries were collected from slaugterhouse and oocytes was aspirated from2-6 mm of follicles. Oocyte with compacted cumulus cells and evenly granulated ooplasm were selected for this experiment. The lenght of in vitro maturation before vitrification was 8 or 22 h in IVM media TCM 199 + FCS 10 % + PMSG 10 IU + hCG 10 IU at 38.5 °C in a humidified atmosphere of 5 % CO_2 in air and were vitrified. After vitrification process, GVBD and MII oocyte were matured for 18 or 4 h to fullfill 26 h maturation requirement and then oocytes were subjected to IVF and culture. Cleavage and blastocyst formation rate were to asses their developmental competence. Cleavage rates were obtained for both GVBD (56.78 %) and MII (69.64 %) oocytes ($P<0.05$). Proportion of cleaved embryos from vitrified MII oocytes develop into blastocysts higher ($P<0.05$) than those from vitrified GVBD oocytes (10.25% vs 3.54%) repectively. Goat oocytes in different maturation stages response to vitrification differently and MII stages have better developmental competence than GVBD.

1. Introduction

Embryo production in vitro has received great attention in animal husbandry. Over the recent years, researchers have been trying to determine which conditions are needed during maturation, fertilization and development in vitro to enhance embryo production. The *in vitro* embryo production technique has greater potential of increasing the number of embryos produced for several studies such as embryo sexing, sperm injection, and nuclear transfer [1]. The embryo production in vitro from oocytes were obtained from abbatoir have potential for research and commercial use. If in vitro matured goat oocytes could be successfully cryopreserved, an embryo bank could be established and the techniques for advancing of genetic improvement could be promoted.

There are two techniques of cryopreservation, the first is slow rate freezing and and the second one is vitrification, but vitrification has been the favourite because of less damage. Recent progress in the development of protocols for cryopreservation indicated that vitrification in the ultra rapid cooling rate is a promising approach for oocyte cryopreservation. In general, the effectiveness of vitrification technique depends on several factors as the stage of oocyte , volume and cooling rate, cryoprotectant media, and the species, [2,3]. Until now cryopreserved oocytes remains lower than fresh oocytes in developmental capacity [4]. Several factors appear to be associate with the limited success including large size,low volume to surface ratio, high lipid contents and the nature of the plasma membrane of the oocyte [5]. However MII oocyte are high risk cryopreservation damage; such as disorganization of meiotic spindle , disruption ofmicrofilament ,and premature cortical granules release [6]. This research was conducted to investigate effect of goat oocyte meiotic stages on developmental competence after cryopreservation.

2. Materials and Methods

2.1. Preparation of oocytes

Ovaries were obtained from local slaugterhouse and brought to the laboratory in NaCl (0.9% (w/v) 30°C within 2 h. Cumulus oocyte complexes (COCs) were collected from follicular fluid aspirated through 18- G needle from follicles with diameter 3-6 mm.

2.2. Oocyte maturation and IVF

IVM process conducted based on [7]. After washing three times in TCM 199 (Gibco ,NY), Compacted cumulus cells oocytes and evenly granulated ooplasm were selected for experiments.nOocytes were matured in TCM 199 + 10 % FCS + PMSG 10 IU + hCG 10 IU under mineral oil and then incubation at 38.5 °C in 5% CO2. The lenght of maturation before vitrification was either 8 or 22 h. IVF was accomplised using frozen semen. Semen was rinse twice in Brackett Oliphant (BO) medium supplemented with caffeine-sodium benzoat (10 mM ,Sigma) to form a final concentration 2 x 10^{6} spermatozoa/ml.

2.3. Vitrification of oocytes

Oocytes were exposed to 15 % Ethylene Glycol (Sigma) and holding medium for about 60 s and then moved into 30 % EG+ 0.5 M Sucrose in a holding medium. Oocytes were loaded into straws and the straw was plunged directly into liquid nitrogen within 30 s beginning to concentrated vitrification solution.Warming procedure according to previous researcher [8]

2.4. Assesment of survival

Vitrified oocytes was evaluated by morphological examination. Oocyte with homogenous ooplasm, intact membrane and zona pellucida were classified as viable.

2.5. Developmental competence of vitrified oocytes

Oocytes which matured 8 or 22 h were vitrified. After warming, GVBD and MII oocytes were matured for additional 16 or 2 h, respectively, to fullfill 24 h maturation requirement. After the final maturation, all of oocytes were used to IVF and culture.

2.6. Statistical analysis

Data were pooled from 10 replicates. Data were analyzed by Chi square test. The rate of survival oocytes, fertilization and development stage were considered significant when $p<0.05$.

3. Result and Discussion

In this research, vitrification process reduced the developmental capacity of oocytes. Survival rate, cleavage rates and developmental competence of oocytes vitrified from MII groups were higher than GVBD ($P<0.05$).

Table 1. Effect of Meiotic Stage on Rate of Survival and Developmental Competence After Vitrification Process

Treatment	Matured oocyte	Number of Vitrified Oocytes	Number of Survived Oocytes	Cleaved oocytes(%)	Blastocytn(%)
GVBD	486	486	398	226(56.78)[a]	8(3.54 %)[a]
M II	488	488	448	312(69.64)[b]	32(10.25%)[b]
Control	482			418(86.72)[c]	154(36.84 %)[c]

Different superscript within columns indicate significant ($P< 0.05$)

The results showed that different status of meiotic stage would affects the ability of oocytes to developed after vitrification. Response of oocytes to vitrification differently at different stages of nuclear maturity. Oocytes must undergo not only nuclear maturation , but also cytoplasmic maturation to become delopmentally competent. There was many problem to continued development because of inadequate cytoplasmic maturation [6,9]. Oocyte in GVBD stage demonstrated more sensitive to cryopreserved compare to MII stage, but the reason for this problem is unknown [8]. There were many problem in cryopreservation of oocytes, including spindle disorganization, clumping of microtubules, fracture of zona pelucida, hardening of zona pelucida, increased polyploidy and decreased fertilization [6,10,11,12]. Previous researcher stated that among all of meiotic stage , MII stage is most resistant to cryopreservation [13,14].

4. Conclusion

Different meiotic stages of goat oocyte response to vitrification differently and MII stage have better developmental competence than GVBD.

Acknowledgments

The authors thank to BBIB Singosari for supply of goat frozen semen and Dr. Sasmito Djati for technical assisstance.

References

[1] Paramio, MT and Izquierdo, D. 2016. Recent advances in in vitro embryo production in small ruminants. Theriogenology **86** : 152.

[2] Arav A. 2014. Cryopreservation of oocytes and embryos. Theriogenology **81**:96.

[3] Menchaca A, Barreral N, dos Santos Neto PC, F Cuadro F, Crispo M. 2016. Advances and limitations of *in vitro* embryo production in sheep and goats. Anim. Reprod. **13(**3).273.

[4] Saragusty J and Arav A. 2011. Current progress in oocyte and embryo cryopreservation by slow freezing and vitrification *Reproduction* **141** : 1.

[5] Isachenko V, Soler C, Isachenko E, Perez-Sanchez F and Grishchenko V. 1998. Vitrification of immatured porcine oocytes : Effect of lipid droplets, temperature, cytoskeleton and addition and removal of cryoprotectants Cryobiology **36** :250.

[6] Gabriel PR, Tornes P, Fratto MC,Cisale H, Claver JA,Lombardo DM and Fischman ML. 2017. Effects of different sucrose concentrations on vitrified porcine pre antral follicles : qualitative and quantitative analysis. Cryobiology **76**: 1.

[7] Wahjuningsih S and Isnaini N. 2014. Supplementation of goat follicular fluid in TCM 199 based medium on in vitro matured oocyte. American Eurasian J of Suistanable Agriculture **8(7)** : 39.

[8] Men H, Monson RL, Rutledge JJ. 2002. Effect of meiotic stages and maturation protocols on bovine oocyte's resistance to cryopreservation Theriogenology **57**:1095.

[9] Leibfried-Rutledge ML, Florman HM, First NL. 1989. The molecular biology of mammalian oocyte maturation. In Schatten H, Schatten G, editors. The Molecular Biology of Fertilization. New York : Academic Press, 259-275.

[10] Wiesak T,Wasielak M, Zlotkowska A and Mileski R. 2017. Effect of vitrification on the zona pelucida hardening and follistatin and cathepsin, B genes expression and developmental competence of in vitro matured bovine oocyte. Cryobiology **76** : 18.

[11] Hochi S, Kimura K, Ito K,Hirabayashi M. 1997. Effect of nuclear stages during in vitro maturation on the survival of bovine oocytes following vitrification Theriogenology **47** : 345

[12] Smiljakovic T and Tomek W. 2006. Meiotic maturation and in vitro maturation on bovine oocytes. Biotechnology in Animal Husbandry **22(1-2)** : 29.

[13] Fuku E, Kojima T, Shioya Y, Marcus GJ, Downey BR. 1992. In vitro fertilization and development of frozen thawed bovine oocytes. Cryobiology **29:**485.

[14] Hyttel P, Vajta G and Callesen H. 2000. Vitrification of bovine oocytes with the open pulled straw method : Ultrastructural consequences Mol Reprod Dev 2000;**56**:80.

IOP Conf. Series: Earth and Environmental Science **119** (2018) 012030 doi:10.1088/1755-1315/119/1/012030

Identification of Pit-1 Gen Using PCR-RFLP of Padjadjaran Sheep and Evaluate of Growth Rate

S. B. K. Prajoga[1*], L. Andriani[1] and H. Subhandiana[2]

[1]Faculty of Animal Husbandry, Padjadjaran University. Jl. Raya Bandung – Sumedang Km 21 Jatinangor. Bandung.
[1]Faculty of Animal Husbandry, Padjadjaran University. Jl. Raya Bandung – Sumedang Km 21 Jatinangor.
Bandung
[2]Research Assistant of Animal Breeding and Biometric Laboratory, Animal Husbandry Faculty, Padjadjaran University, Bandung, Indonesia

[*]Corresponding E-mail: komarsri@yahoo.com

Abstract

The objectives of this research were to evaluate variation of Pit-1 gen of Padjadjaran Sheep and evaluate of their growth rate. The data comprised of 15 lambs female and 15 lambs male records of 0 – 12 mouths old lambs. Variation of Padjadjaran Sheep PIT-1 gen was analyzed using PCR-RFLP and used primer by 5'-GAGGGATAATTACAAATGGTCC-3' and 5'-TGTTAACAGCTGTGGGACACAC-3', length fragment analyse using HinfI restriction enzyme. Presence or absence of deference bands in a PCR fragment of the result can be distinguished by differences in the electrophoretic migration of the fragment. The result showed that the variation Pit1 gene showed that there was in the position 345 bp, 137 bp and 115 bp. In this experiment the difference migration did occur in all samples (monomorphic), that was amplificated with reverse and forward primer. The average male birth weight (BW), weaning weight (WN) and Yearling Weight (YW) was 2.29 ±0.42kg ; 8.96 ±1,89 kg and 30.12 ±5,65 kg. The average female birth weight (BW), weaning weight (WN) and Yearling Weight (YW) was 2.33 ±0.49 kg ; 8.23 ±1.99 kg and 26.11 ±5.50 kg. Correlation value between age and body weight was 0,99 for female and 0,97 for male. The highest body weight gain per month was 2.85 kg for female at the age of six months and 3.4 kg for male at the age of one year. The best equition of growth rate was $\hat{Y}$= 1,862 + 0,127X_1 0,076X_2

1. Introduction

Padjadjaran sheep is one of the local sheep, which were developed by researcher with the objective being to rapid growth as a source of meat. Characteristics of the Padjadjaran sheep have white fur, have big earlobe and a horned for male. Padjadjaran Sheep has variation sheep *mt*-DNA of deletion along 75 *bp* at 1447 *bp* position on the d-loop of *mt*-Chromosome. However, PIT-1 is one of the genes that are the family of growth genes in addition to IGF1, GH and GHRH. Although the criteria of performance body weight, body height and body length are variable that needs to be observed in the evaluation of livestock growth.

Growth is weight gain or body measurement gain until they reach adult size or per additional body mass per unit time. Specific growth occurs in young animals which is formed by the large bone, protein, and water networks. Growing character is important for a sheep breeding business whose final goal is meat production.

The study of growth can not be separated by the processes of various fields of physiology, biochemistry, genetic endocrinology and nutrition [1]. Growth starts from the front of the animal (anteriorly) to the back (posterior).

The appearance of a livestock is the result of a continuous growth process, throughout the life of the livestock. Body components have different growth rates [2]. Meanwhile, the size of the body surface and body parts have the utility to estimate body weight and carcass, as an image of the body shape of a livestock according to the character of his breed.

Shoulder height is usually used by researchers to know that a livestock grows perfectly, while the body length used to measure the growth of livestock can also be used as a quality guide of carcass.

Body weight of a livestock at a certain age can serve as a guide for growth in ages (0-12 months), and can also be correlated to form a growth curve. The growth rate is expressed as the weight gain or the length increase of a body part per unit time to adulthood.

Genomic molecular observation can be used as a tool to conduct selection by Marker Assistant Selection method (MAS), because it is expected to associate between gene marker with quantitative trait locus (QTL),

Genetic diversity is important in breeding programs for the genetic optimization of the acquisition of certain properties can be achieved when there is enough opportunity for the selection of genes or the desired properties.

Beside that, genetic diversity plays an important role in the survival of the population. The loss of genetic diversity can reduce the chance of survival of the population.

Through a complicated on differentiation phase of the anterior pituitary and the regulation of prolactin gene (PRL), growth hormone (Growth hormone-GH) and thyroid-stimulating hormone-β (TSH-β), then Gen sheep PIT-1 is recommended as a candidate gene to the nature of the production.

Somatic cells residing on white blood cells can be used as a source for the DNA analysis. In the microsatellite study to obtain superior genes for the purification of Padjadjaran sheep, so that the origin of the nation can be standardized based on Polymerase Chain Reaction (PCR) and PCR product was cut with restriction enzyme Hinf-1 and electrophoresed on agarose gel an stained with ethidium bromide (Etbr). The result showed that there were bandings in the difference position as Restriction Fragment Length Polymorphism (RFLP)

Therefore, based on the availability of the genetic resources, it is essential to form seeds of Padjadjaran sheep, which have criteria in accordance with the will of the breeder, for purposes of standardization and certification. The purpose of this study was to determine the variation of PIT-1 gene in blood samples and to take out Marker Assistant Selection in Padjadjaran sheep base on growth rate.

2. Materials and Methods

2.1. Material and Methods description:

2.1.1. Blood sampling as many as 30 samples, grouped into group female lamb and 15 male lamb.
2.1.2. Pre-treatment sample storage (using vacuum-tainer containing EDTA anticoagulant).
2.1.3. Isolation of DNA samples using the Genomic DNA Purification Kit (fermentage*)* [3]
2.1.4. PCR analysis for each of the isolated DNA with four pairs of primers using Tag Dream Green PCR Master Mix.

2.2. Electrophoresis of PCR Product

PCR product was cut with restriction enzyme Hinfl and electrophoresis purification used 1.5% agarose gel stained with ethidium bromide (Etbr) staining to see the pattern of band microsatellite, that was amplificated by primer.

2.3. Hardy Weinberg equilibrium

H-W equilibrium was calculated by Chi-quadrat [4]

$$X^2 = \sum \frac{(Oi-Ei)}{Ei}$$

Where:
X^2 = chi square value
O_i = value of observation
E_i = value of expectation

2.4. Growth rate evaluation

Evaluation of growth rate can be done by using correlation and regression

3. Results and Discussion

At the holding cell isolation method optimization with the Genomic DNA Purification Kit (fermentage), used as a comparison sample of whole blood and uppers (serum). It was found that isolation by using whole blood (mixture) at stage 3 there will be no separate phase 3 well (the upper aqueous phase), the middle (cell debris), and the lower (organic phase) as in isolation by using serum. solation using whole blood at stage 3 there will be only two phases, namely solid phase (gel-like) on the top and the organic phase. Because it was later used as an isolated sample is part of the sheep blood serum. Furthermore, by using the PR1 (primer 5'-GAGGGATAATTACAAATGGTCC-3' and 5'-TGTTAACAGCTGTGGGA CACAC-3').

The result showed that Fragment Length of PIT-1 gene was uniform on 345 *bp*, 137 *bp* and 115 *bp* after restricted by Hinft1 Enzyme. According to [5] that the PIT1 gene hasn't heterozygote allele in the population of Padjadjaran Sheep. The PIT-1 gene was not associated with body weight on difference age. It belongs to the gene group to encode the proteins involved with growth. Specific pituitary transcription in code by factor PIT-1 Code by PIT-1's gen [6].

The condition of data during the study can be said uniform because Variation Coefficient is still below 15%, however as it is known that growth will be grouped in a phase acceleration (self-accelerating phase) and then subsequently followed by a slow phase (self-inhibiting phase).

The point where the highest weight gain is the sixth month (2.83 kg), while the height increase shoulder also in the sixth month is 3.13 Cm and the longest increase is also in the sixth month is 2.47 Cm, usually at that time the animal enter puberty phase.

According to former researcher that growth is divided in every quarter, the first quarter of growth is 50 percent, in the second quarter it reaches 25 percent, while in the third quarter the remaining 25 percent . Sheep are said to be adults when they are 18 months old where the growth period is 100 percent and growth for up to 12 months is still included in the 75 percent growth period [3].

IOP Conf. Series: Earth and Environmental Science **119** (2018) 012030 doi:10.1088/1755-1315/119/1/012030

Table 1. Result of Electrophoresis After Cutting by Restricted Enzyme HinfI

No.	Region	Sample Code	Fragment Length as Restriction Result (*bp*)				
			62	115	137	283	345
1	Sempur	Blood sample ♂		√	√		√
2		Blood sample ♂		√	√		√
3		Blood sample ♂		√	√		√
4		Blood sample ♂		√	√		√
5		Blood sample ♂		√	√		√
6		Blood sample ♀		√	√		√
7		Blood sample ♂		√	√		√
8		Blood sample ♂		√	√		√
9		Blood sample ♂		√	√		√
10		Blood sample ♂		√	√		√
11		Blood sample ♂		√	√		√
12		Blood sample ♂		√	√		√
13		Blood sample ♂		√	√		√
14		Blood sample ♂		√	√		√
15		Blood sample ♂		√	√		√
16		Blood sample ♀		√	√		√
17		Blood sample ♀		√	√		√
18		Blood sample ♀		√	√		√
19		Blood sample ♀		√	√		√
20		Blood sample ♀		√	√		√
21		Blood sample ♀		√	√		√
22		Blood sample ♀		√	√		√
23		Blood sample ♀		√	√		√
24		Blood sample ♀		√	√		√
25		Blood sample ♀		√	√		√
26		Blood sample ♀		√	√		√
27		Blood sample ♀		√	√		√
28		Blood sample ♀		√	√		√
29		Blood sample ♀		√	√		√
30		Blood sample ♀		√	√		√

Table 2. Structure Data of Average Body Weight, Shoulder Height and Body Length

N	Age (month)	Average of Body Weight (kg)		Average of body Length (cm)		Average Height of Should (cm)	
		Female	Male	Female	Male	Female	Male
30	0	2.33	2.29				
30	1	5.19	6.67	33.47	39.07	37.53	39.07
30	2	7.06	8.19	41.43	43.33	41.13	43.33
30	3	8.23	8.96	42.47	44.53	42.47	44.53
30	4	8.71	9.80	44.33	45.47	44.33	45.47
30	5	9.83	10.59	46.33	46.27	46.33	46.27
30	6	12.66	12.74	49.47	48.57	49.47	48.57
30	7	14.92	15.51	51.30	50.00	51.30	50.00
30	8	17.10	18.26	52.53	51.53	52.53	51.53
30	9	19.40	20.79	53.47	52.80	53.47	52.80
30	10	21.53	23.65	54.73	54.13	54.73	54.13
30	11	23.65	26.72	55.93	55.20	55.93	55.20
30	12	26.11	30.12	57.53	57.13	57.53	57.13

4. Conclusion

Variation of reverse and forward primers, showed that migration is relatively the same for all the samples and estimated (based on comparison with a marker) that is the size of the fragments between 115 *bp*, 137 *bp* and 345 *bp*. PIT-1 gene can not be used as marker for Marker Assistant Selection (MAS). Pit1 gen has more functions as a coupling candidate (binding factor) in the process of transcription of protein biosynthetic from pituitary, formation of Growth Hormone, GHRH and IGF1. The highest body weight gain per month was 2.85 kg at the age of six month for female and 3.4 kg for male at the age of one year. The best eqution of growth rate was $\hat{Y}$= 1,862 + 0,127X_1 0,076X_2
The ram that is prepared for fattening can start from the age of 1 (one) year, and for the ewe at the age of 18 months can be mated to ready to become dam.

Acknowledgments

To Riset Dikti DP2M for supporting research project and also thank to Laboratory of Biochemical of Chemical Department, Faculty of Nature Science Padjadjaran University for supporting PIT-1 analyse.

References

[1] Owens K Q, Goodkand B, Tokach M, Kansas University , Swine Day . Report of Proogress, **695**: 58.

[2] Glodek H 1980 Populations Parameter. In: Tierzuchtungslehre, editor G. Comberg, Verlag Eugen Ulmer, Stutgart. Germany. pp 143-156.

[3] Sambrook J, Fritsch E F and Maniatis T 1989 Molecular Cloning: A Laboratory Manual 2nd ed, Cold spring Harbour, Cold Spring Laboratory Press, NY.

[4] Warwick R E, Astuti J M and Harjosubroto W 1990 Pemuliaan Ternak, Gajah Mada University Press. Yogyakarta.

[5] Bandiati SBK., Dedi Rahmat, Lovita Andriani, Nena Helmia, Dudi 2015 Identifikasi Gene Produksi Pada domba Padjadjaran. laporan Akhir . Academic Leadership Grand. University Padjadjaran. Jatinangor. West Java.

[6] Bandiati SBK 2006 Identification of Pit1 Gen of Cockerel Chicken and their Selection Respon of Haching weight. Int. Symposium Agriculture for Life, Life for Agriculture. Buchares Romania

IOP Conf. Series: Earth and Environmental Science **119** (2018) 012031 doi:10.1088/1755-1315/119/1/012031

Identification of Genetic on Blood Serum Protein of Prolific Ewes

Sutiyono[1*], Y. S. Ondho[1], E. T. Setiatin[1], Sutopo[1], A. N. Laily[1] D. E. Prasetyowati[1] and F. Noviani[1]

[1]Faculty Animal and Agricultural Sciences, Diponegoro University, Semarang, Indonesian

*Corresponding E-mail: barep.sutiyono@gmail.com

Abstract

The aim of the research was to identify the genetic specification of blood plasma protein in ewes that are prolific. The material of study of local sheep in Bawen and Jambu Sub-district of Semarang Regency is 132 which is determined by purposive sampling that have been give lambing three times. Ewes were divided into three groups that always has a single child (L1), ever had twins (L2) and twins more than two (LM2). Blood sampling was performed using dispossible syringe in jugular vein as much as 5 ml per ewe. Blood plasma was analyzed by Polyacrylamide Gel Electrophoresis-Thin Layer (PAGETLE) method in Biochemistry Laboratory of Veterinary Faculty of Gadjah Mada University. Data analysis is using descriptive statistics and the laws of equilibrium Hardy-Weberg. The research parameters were comparison type of ewes and frequency genetic of protein of blood serum. The results showed that the parent comparisons of L1, L2 and LM2 were 66 (50.00%), 49 (37.12%) and 17 (12.88%), respectively. The frequency genes haven a high propensity to relationship of prolificacy nature parent are Pal^2, Alb^B , CP^F, TF^B, PTF^S and Aml^B on pointes, 67.65, 55.88, 91.17, 70.59, 79.41 and 91.18%. Conclusion the mostly LM2 ewes have genotypes Pal^1Pal^2, Alb^BAlb^C, Cp^FCp^F, Tf^ATf^B, Ptf^SPtf^S and Aml^BAml^B whit frequency are 52.94%, 52.94%, 88.24, 47.06, 64.71 and 88.24% respectively.

1. Introduction

The right program to increasing the population and production of sheep in the future is to use prolific ewes parent. The use of prolific ewe parent will both speed up and simplify the production of prolific sheep breeds, as they are more effective and efficient. Prolific ewes are female sheep that have the ability to produce more than two sheep for each birth. The ability to produce lambs over two every birth is the fecundity gene expression of every ewes. According to [1] prolific sheep has the homozygote genotype of the dominant fecundity gene (FF), while those with twin and singular abilities are the genotypes of heterozygote (Ff) and homozygote recessive (ff). The prolificacy Indonesia of sheep occurs due to the mutation of genes B in the indigenous Java sheep [2]. The strategic program breeding was separated between of a group of sheep prolific and non-prolific [3]. The group of productive ewes mated with males of the productive offspring to get more types of prolific [4]. The selection of prolificacy trait in cattle showed a greater effect on the performance of litter size [5]. The problem in the selection of fertile ewe is the number very few and not all female have fecundity genes that can indicate to the litter size. The main factor influencing the nature of sheep prolificacy is genetic of fecundity, but for better performance in the system, it is necessary to consider the supporting factors. [6]. [7] stated that the appearance of fertile ewes can be seen from the maskers associated with the ease of pregnancy and birth process. The prolific ewes should be selected that have the capacity of the uterus to gestation twins, in order to have a survival after birth. [8]

Some researcher are explane that the results of electrophoresis can be used as a supporter of superior livestock selection. Electrophoresis results showed that considerable characteristic differences between the components of minor proteins of semen in cattle and buffalo. [9]. The method of electrophoresis examination of milk proteins can determine the phylogenetic origin of the type of livestock that the results are applied to the industry with respect to the control of the quality of commercial dairy products [10].

2. Material and Methods

The research was conducted by survey method. Sheep management is done traditionally by farmer breeding. Types of feed that is generally given were grass, peanuts or foliage with additional bran or not added. The range of coarse protein content in the feed given is 8.74-11.23% and energy metabolism 1797.00-1859.00 kcal. The research material is 132 local sheep owned by ranchers in Sub-districts of Bawen and Jambu, Semarang Regency. The materials are determined by purposive sampling based on birth type ie sheep that have given birth at least three births.

The number of 132 sheep by type of birth ever happened was grouped into three ie single, twin and triplet or more lamb are 66, 49 and 17 respectively. The three groups of ewes are, a. Always single child (L1), b. Once had twins (L2) and c. Once had more than two children (LM2) at their birth. Blood collected as much as 5 ml at 08.00-10.00 pm, through the yugolaris vein of each ewe. The blood samples are stored separately at 0-40C until the electrophoresis process. The electrophoresis of proteins of blood serum using PAGETLE (Polyacrilamide Thin Gel Electrophoresis) in Biochemistry Laboratory of the Faculty of Veterinary Medicine, University of Gadjah Mada Yogyakarta. Parameters of research were locus Per-albumin (P-Alb), albumin (Alb), ceruloplasmin (Crp), transferrin (Tfr), post-tranferrin (P-Tf) and amylase I (Aml). The homozygous genotype frequencies of 1st, 2nd and heterozygous calculated by the principle of Hardy-Weinberg equilibrium [11]. The genotype frequencies in each L1, L2 and LM2 sheep clusters were analyzed descriptively.

3. Result and Discussion

Number and percentage of L1, L2 and LM2 local ewes owned by small farmers, subdistric Bawen and Jambu was 66 (50.00%), 49 (37.12%) and 17 (12.88%) respectively. The comparison of the three groups of mothers is very contrast, and LM2 is the least. The factor that cause the number of LM2 ewes was little likely caused because the farmers do not choosing their sheep on based gen of prolific. According to [5] the selection of productive traits will increase prolific of mother sheep whose effect is on the number of litter size. The number of LM2 ewes is little can be caused by nutritional and body factors that do not support the birth of more than 2 litter sizes. According to [12] nutritional deficiencies in females caused endocrine and follicular disorders in the oogenesis process, so they can not be fertilized. Another factor is the premature death of children in the womb that causes reduced litter size that was born. The embryonic mortality in uteri of cattle repeat-breeder from 16 to 34 day, reached 51.7% [13].

Table 1 show that the genotype of six sheep blood serum proteins are very varied, which means every protein genotypes vary in L1, L2 and LM2 Ewes. The variation was directed to the frequency of homozygous genotype 1st and 2nd and heterozygote with range from 4.08 to 88.24%. The frequency of pre-albumin genotype from blood serum is presented in Table 1 and Figure 1-6.

Appearance of pre-albumin genotype on L1 and LM2 ewes showed a positive interaction between the Pal1 and Pal2 alleles so that the high value of 75.76 and 52.94% (Table 1 and Figure 1). Whereas in the case Ewes L2 the positive trend of frequency genotype from homozygote 1st heterozygote and homozygote 2nd with the highest score was 55.10%. Pre-albumin has an important physiological roles as a transporter of thyroxine and retinolbinding protein. Protein pre-albumin is synthesized in the body a low or decreased concentrations, because of the response of inflammation, stress and to nutritional conditions [14].

Table 1. Frekuensi of Genotype and gen Pre-albumin Albumin Ceruloplasmin Transferrine Post-transferrine and Amylase Protein of Serum Blood from L1, L2 and LM2 Ewes In Sub-district Bawen and Jambu.

Parameter	Classification genotype	Head % number	Birth Type		
			L1 (n=66)	L2 (n=49)	LM2 (n=17)
Pal	Pal^1Pal^1 : Pal^1Pal^2 : Pal^2Pal^2	(head)	3 : 50 : 13	5 : 17 : 27	1 : 9 : 7
		(%)	04.54:75.76:19.70	10.20:34.70:55.10	05.88:52.94:41.18
	Pal^1 : Pal^2	(number)	56 : 76	27 : 71	11 : 23
		(%)	42.42 : 57.58	27.55 : 72.45	32.35 : 67.65
Alb	Alb^BAlb^B: Alb^BAlb^C: Alb^CAlb^C	(head)	33 : 20 : 13	27 : 17: 5	5 : 9 : 3
		(%)	50.00:30.30: 19.70	55.10:34.70:10.20	29.41:52.94:17.65
	Alb^B: Alb^C	(number)	86 : 36	71 : 27	19 : 15
		(%)	65.15 : 34.85	72.45 : 27.55	55.88 : 44.12
Cp	Cp^FCp^F: Cp^FCp^S: Cp^SCp^S	(head)	30 : 23 : 13	30 : 12 : 7	15 : 1 : 1
		(%)	45.45:34.85:19.70	61.22:24.49:14.29	88.24:05.88:05.88
	Cp^F: Cp^S	(number)	83 : 49	72 : 26	31 : 3
		(%)	62.87 : 47.13	73.47 : 26.53	91.17 : 08.41
Tf	Tf^ATf^A: Tf^ATf^B: Tf^BTf^B	(head)	3 : 56 : 7	2 : 37 : 10	1 : 8 : 8
		(%)	04.55:84.85:10.60	04.08:75.51:20.41	05.88:47.06:47.06
	Tf^A: Tf^B	(number)	62 : 70	41 : 57	10 : 24
		(%)	46.97 : 53.03	41.84 : 58.16	29.41 : 70.59
PTf	Ptf^FPtf^F: Ptf^FPt^S: Ptf^SPtf^S	(head)	23 : 30 : 13	10 : 29 : 10	1; 5; 11
		(%)	34.85:45.45:19.70	20.41:50.18:20.41	05.88:29,41:64.71
	Ptf^F: Ptf^S	(number)	76 : 56	49 : 49	7 : 27
		(%)	57.58 : 42.42	50.00 : 50.00	20.59 : 79.41
Aml	Aml^BAml^B: AmI^BAmI^C: AmI^CAmI^C	(head)	36 : 20 : 10	34 : 10 : 4	15 : 1 : 1
		(%)	54.55:30.30:15.15	69.39:20.41:08.16	88.24:05.88:05.88
	Aml^B: AmI^C	(number)	82 : 40	78 : 18	31 : 3
		(%)	62.12 : 37.88	79.59 : 02.41	91.18 : 08.02

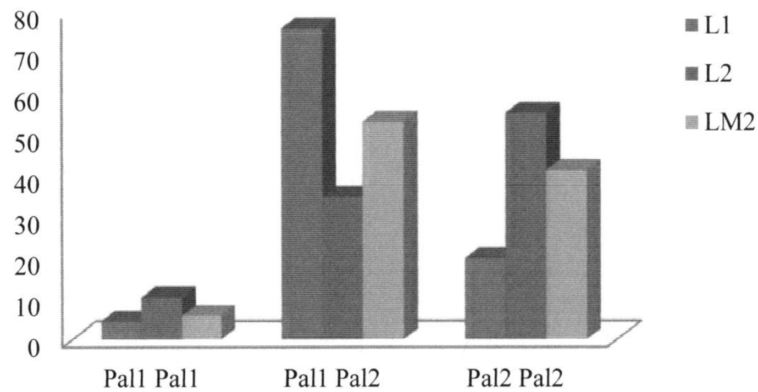

Figure 1. Genotype Frequency of Pre-albumin of Blood Serum On Ewes L1, L2 and LM2

The frequency of albumin genotype from blood serum is presented in Table 1 and Figure 2. The frequency of albumin genotypes on the L1 and L2 ewes shows decreased appearance from homozygote 1st, heterozygote and homozygote 2nd. Therefore, value of the highest frequencies in L1 and L2 ewes the genotype Alb^BAlb^B, which is 50.00 and 55.10%. While the genotype frequency at LM2 ewes are displays indications of interactions between alleles Alb^B and Alb^C, so that have the highest value 52.94% is Alb^BAlb^C of genotype. Albumin has a complex structure, which is responsible for a variety of biological functions ,and in case of decreased the albumin molecule is susceptible to modifications that may alter its biological activity [15]. The albumin genotype is responsible for the concentration of albumin protein that ideal for physiology of the livestock body. The results of the study indicated that the requirement of albumin protein in L1 and L2 ewes were fulfilled Alb^BAlb^B, while LM2 by genotype Alb^BAlb^C.

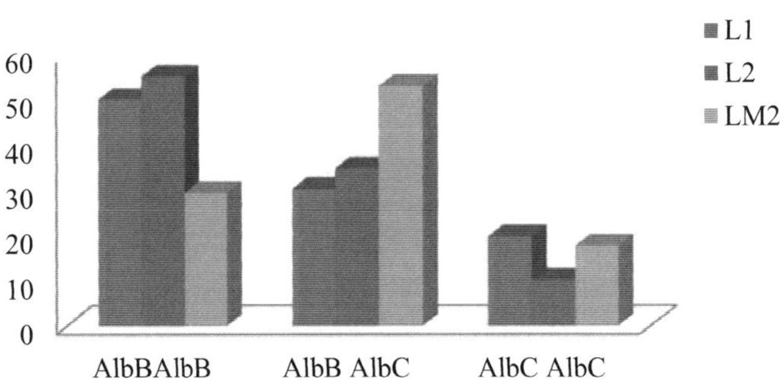

Figure 2. Genotype Frequency of Albumin Blood Serum Ewes L1, L2 and LM2

The frequency of Ceruloplasmin genotype from blood serum is presented in Table 1 and Figure 3. All frequencies in L1, L2 and LM2 ewes show a downward trend. Appearance of frequency genotype Cp^FCp^F, Cp^FCp^S and Cp^SCp^S on LM2 ewes were 88.24, 05.88 and 05.88%, respectively. Ceruloplasmin or caeruloplasmin is a ferroxidase enzyme whose work helps transport it in plasma with respect to transferrin controlled by the genotype CpCp. The plasma ceruloplasmin is increase highly significant ($P<0.001$) in uterine after mating, therefore fluctuations in levels of the above parameters point to their important role in the female reproductive system [16]. Physiology in the uterus after insemination is prepare pregnancy, that was the characterized by the growth of the uterine gland and produce the uterine milk. Therefore, the works of ceruloplasmin was greatly contributed to the occurrence of pregnancy.

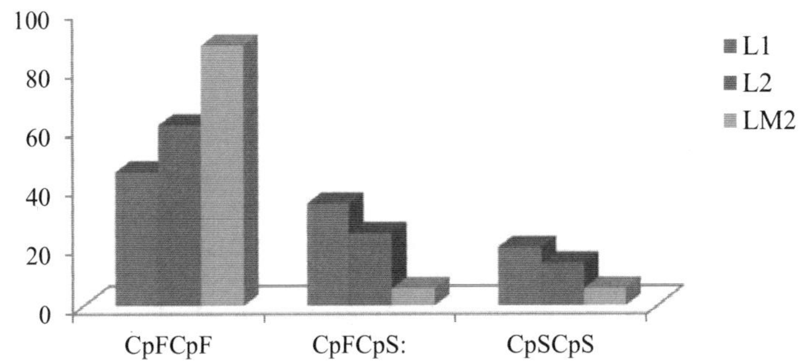

Figure 3. Genotype Frequency of Ceruloplasmin Blood Serum On Ewes L1, L2 and LM2.

The genotypes data of transferrin of blood serum are presented in Table 1 and Figure 4. Display data of frequency of transferrin genotype on figure 4 is shows that allele of Tf^A and Tf^B alleles have indicated events of positive interaction so that value of heterozygote (Tf^ATf^B) looks different and higher than homozygote (Tf^ATf^A and Tf^BTf^B). Frequency genotype Tf^ATf^A, Tf^FTf^S and Tf^BTf^B) on LM2 ewes are 05.88, 47.06 and 47.06% respectively. According to [18], the highest fertilization rate in Botosani Karakul sheep is found in Tf^C allele in combination with Tf^A and Tf^B alleles of 100%, while for the highest prolificacy rate is found in heterozygous allele that is on Tf^BTf^C genotypic. The heterozygous genotype of the ewes transferrin allele tends to have better production performance than homozygote so it can be used for the selection of genetically in order to increase the productivity of sheep [17].

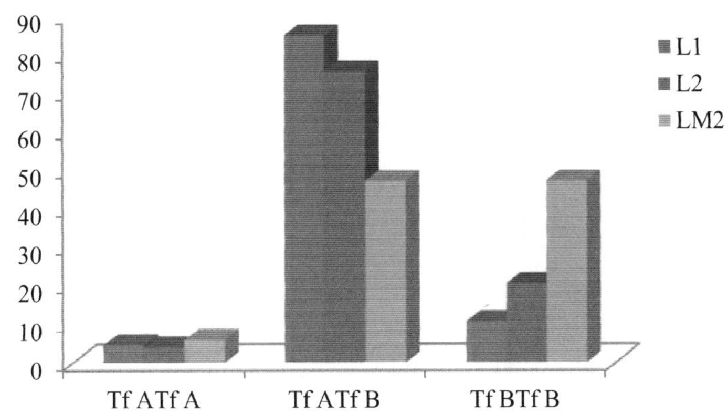

Figure 4. Genotype Frequence of Transferein Blood Serum On Ewes L1,L2 and LM2

The post-transferrin genotype in sheep serum Blood in Table 1 and Fig. 5 shows a positive interaction between alleles of Ptf^F and Ptf^S so that the Frequency of Ptf^FPtf^S on L1, and L2 ewes have the level high 45.45 and 50.18% respectively. Whereas in LM2 ewes showed a positive trend of genotype Ptf^FPtf^F, Ptf^FPtf^S, and Ptf^SPtf^S whose sequential frequency is 05.88, 29.41 and 64.71%. Post transferrin genotype of all groups of ewes have the level of high on the heterozygote, meaning they produce best protein levels in the process of reproducing local sheep in Bawen and Jambu Sub-districts.

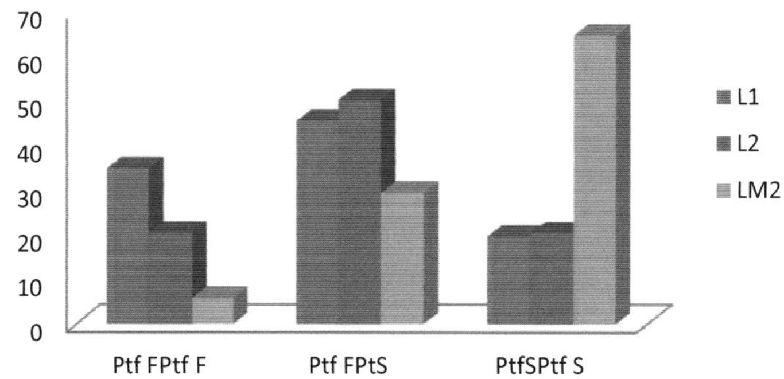

Figure 5. Genotype Frequency of Post-transferrin Blood Serum On Ewes L1, L2 and LM2.

The genotypes data of amylase of blood serum are presented in Table 1 and Figure 6. The all frequency of genotype from homozygote 1st, heterozygote and homozygote 2nd showed a decreasing trend on L1, L2 and LM2 ewes. Frequency genotype Aml^BAml^B, AmI^BAmI^C dan AmI^CAmI^C on LM2 ewes are 88.24, 05.88, 05.88% respectively. Amylase-I is an enzyme protein in the blood that are useful in increasing the rate of metabolism, which can used in the determination of gene loci of amylase through electrophoresis protein serum [19]. Amylase levels in the serum and urine is coupled with other laboratory tests such as liver enzymes, can be significant in important in predicting intra-abdominal injury [20].

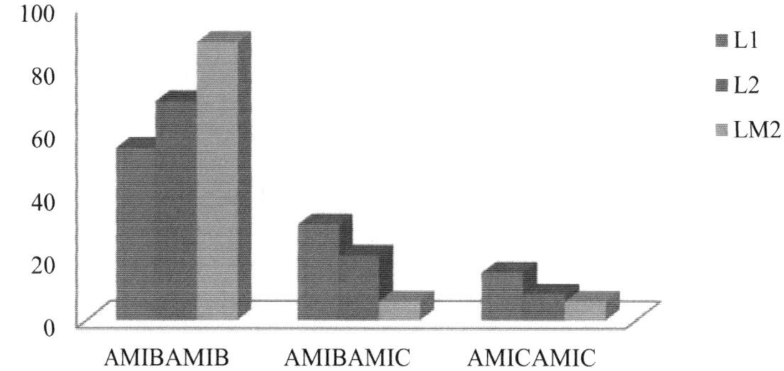

Figure 6. Genotype Frequency of Amylase I Blood Serum On Ewes L1, L2 and LM2

4. Conclusion

The mostly LM2 ewes have genotypes Pal^1Pal^2, Alb^BAlb^C, Cp^FCp^F, Tf^ATf^B, Ptf^SPtf^S and Aml^BAml^B whit frequency are 52.94%, 52.94%, 88.24, 47.06, 64.71 and 88.24% respectively.

Acknowledgements

This paper is part of a research project funded by the Competitive Grants Program of Higher Education Thank you for your belief in us carrying out research. Thanks to the Head of Department of Animal Husbandry and Fisheries Semarang Regency, with the permission so that we can do research smoothly in District Jambu and Bawen.

References

[1] Bradford GE. (1993). Small Ruminant breeding Strategy for Indonesian. Dalam Subandryo dan Gataby (Editor). Advences in small Ruminant Research in Indonesia PT. Gaya Teknik, Bogor.

[2] Davis, GH, SM Galloway, IK Ross, SM Gregan, J Ward, B V Nimbkar, PM Ghalsasi, P Mulsant, F Lecerf, JP Hanrahan, GE Bradford dan T Wilson. 2002. DNA Test in C. Mimbkar, G. D. Gray, Subandriyo, I. Inonunu, B. Tiesnamurti, E. Martyniuk, E. Eythorsdottir, Prolifics Sheep from Eight Cauntries Provide New Evidence on Origin of the Booroola *(FeeB)* Mutation. Biol. of Reprod. **66** (6) : 1869.

[3] Kumar S, AP.Kolte, and VK Singh. 2006. Twinning in Marwari and Bharat Merino ewes and its relationship Booroola *FecB* mutation. Indian J. Biotech. **5**: 482.

[4] Shelton, M, and T Willingham. 2002. Lamb mortality. Sheep and Goat Res. J. **17** (3): 15.

[5] Martin. P, J. Raoul1 and L. Bodin. 2014. Effects of the FecaL major gene in the Lacaune meat sheep population. Genet. Selec. Evol. **46** (1): 48.

[6] Riofrio, ELA, JBS Ferraz and E C. Mattos (2016). Influence of non-genetic factors on growth and reproductive traits of sheep Santa Inês in extensive systems. Livestock. Research. for Rural Development. 28 (7): *Article #121.* Retrieved January 14, 2017, from http://www.lrrd.org/lrrd28/7/agui28121.html

[7] Cemal, I dan O Karaca, 2007. Phenotypic and Genetic Parameters for Litter Size in Some Regional Synthetic Sheep Genotypes ; Evidence for a Major Gene Effect. J. of Biol. Sci. **7** (1) : 52.

[8] Echternkamp SE, RM Thallman, RA Cushman, MF Allan and KE Gregory. 2010. Increased calf production in cattle selected for twin ovulations. J. Anim. Sci. **85** (12): 3239.

[9] Banerjee AK and NC Ganguli. 1971 Differentiationof semen from Bos Bubalus anad Bos Taurus indicus By Starch Gel Electrophoresis. J. Anim Sci. **32** (6): 1212.

[10] Zagorchev L, M Dimitrova, M Odjakova D Teofanova and P Hritov. 2013. Electrophoretic characterization of milk protein from bulgarian rhodopean cattle. Bulgarian J. Agric. Sci. **19** (2):197.

[11] Warwick, EJ, JM Astuti dan W Hardjosoebroto. 1984. Pemuliaan Ternak. Gadjah Mada University Press, Yogyakarta.

[12] Boland, MP and Lonergan, P. 2003. Effects of Nutrition on Fertility in Dairy Cows. Adv. In Dairy Tech. **15**: 19.

[13] Hawk HW, JN Wiltbank, HE Kidder and LE Casida 1955. Embryonic Mortality Between 16

and 34 Days Post-Breeding in Cows of Low Fertility. Sammry. J. Dairy Sci. **38** (6): 673.

[14] Tothova C, O Nagy and G Kovac. 2016. Serum proteins and their diagnostic utility in veterinary. Vet. Med. **61**, (9): 475.

[15] Martinez RG, P Caraceni, M Bernardi, P Gines , V. Arroyo and R. Jalan. 2013. Albumin: Pathophysiologic basis of its role in the treatment of cirrhosis and its complications. Hepatology. **58** (5):_1836.

[16] Prabhu K, P. Kumar, SK Adiga, A Rao, A Lanka, and J Singh. 2009. Plasma protein thiols, ceruloplasmin, C-reactive protein and red blood cell acetylcholinesterase in patients undergoing intrauterine insemination. J. Hum. Reprod. Sci. **2**(1): 27.

[17] Yadav DK, S Taraphder, K C Dhara, S Batabyal, I Samanta and M Mitra. 2013. Association of transferrin polymorphism with different economic traits of Garole sheep. Glob. J. Anim.Sci. **1** (1) : 053.

[18] Hrinca GH, Groza M, Elena F, Padeanu I, Voia S, Ursu S, Chiorescu I (2008). Association of some biochemical-genetic markers with the production parameters of the Botosani Karakul ewes. Lucrari stiintifice Zootehnie si Biotehnologii, **41**(2): 751.

[19] Wyne, RZ, H Shelwyn and T Robert. 1995. Electrophoresis DNA: Allels of Blood Protein. Wallingford, United Kingdom.

[20] Kumar S, S Sagar, A Subramanian, V Albert, RM Pandey, and N Kapoor. 2012. Evaluation of amylase and lipase levels in blunt trauma abdomen patients. J. Emerg Trauma Shock. **5**(2):135.

Chromosome Characteristic of Peranakan Etawa (PE) Goat (*Capra hircus* Linn.) as Indonesian Local Breed

A. R. I. Putri[1]*, G. Ciptadi[2] and A. P. Warih[1]

[1]Faculty of Mathematic and Life science, MIPA-Biology, University of Brawijaya
[2]Faculty of Animal Husbandry, University of Brawijaya

*Corresponding E-mail: ardyah.putri@gmail.com

Abstract

Chromosome characteristics of Peranakan Etawa (PE) goat needs to be analyzed because information about Indonesian goat races is very limited. The purpose of this research was to determine the characteristics of PE goat chromosome as basic data as one of the genetic local resources. Blood was collected from pair of PE goat at Sumber Sekar Field Laboratory, Faculty of Animal Husbandry, Brawijaya University, Malang. Blood cultured using standard cytogenetic technique and stained with G-Banding. Observations being done in metaphase cells and analyzed using Genus Cytovision Image. Chromosomes arranged and numbered by standard goat karyotype. The result of this research showed that PE goat had number of chromosomes 2n=60, consisting of 29 pairs of autosome and a pair of sex chromosomes. Female goat had average of total length (TL) of autosome ranged from 47.91 μm±6.46 to 22.12 μm±3.33. TL of chromosome X are 45.96 μm±4,59 and 44.45 μm±3,96. Centromeric index (Ci) of chromosome X, 31,74 and 32,80. PE goat had average of TL of autosome ranged from 58.20μm±6.72 to 18.97μm±2.82. TL of chromosome X is 56,42μm±7,38 and Y chromosome is 15,80 μm±3,24. Ci in chromosome X and Y are 19.34 and 46.84. These results concluded that the total of goat chromosome was 60 with types of autosomal chromosomes were acrocentric as many as 58 chromosomes and pair of sex chromosomes XX and XY, X classified as subtelocentric and Y submetacentric.

1. Introduction

Peranakan Etawah (PE) breed of domestic goat (*Capra hircus*) is the one of the most popular goat for breeders in Indonesia and widely farmed for milk and meat [1]. PE goats originated from selection process undertaken in the District Kaligesing, around the hills Menoreh, Purworejo, Central Java. PE goats is often being cross breeds to improve productivity. Cross breeding is one of the ways to obtain the offspring that are considered have special advantages in accordance with the objectives of breeder [2]. PE goat originally a crossbreed between Kacang goat as indogenous breed with Etawah goat from India.

Analysis of chromosomes in goats in Indonesia, especially in PE goat as one of breeds from bovidae family has never been analyzed before. Normal domestic goat have 30 pairs of chromosome, 29 pairs of autosomal and pair of sex chromosome, XX for female/does and XY for male/buck [3]. Chromosome analysis is very important in the management and conservation of genetic resources in livestock [4]. The structure of the chromosomes can be seen in the metaphase stage and observed using certain staining techniques, one of them using Giemsa dye called G-Banding [5]. Chromosomes can be divided into several parts, namely centromere, telomere an darm chromosome called , p (short) arm and q (short) arms [6].

The purpose of this study is to determine the characteristics of each chromosome like structures, shapes, number, and idiogram of PE goats chromosomes as basic chromosome data. The advantage of this study is to get standard karyotyping of Peranakan Etawah goat that can be used as the first step for diagnosis of cytogenetic abnormalities in suspected cases and for the future is to prevent chromosome aberration for its offspring.

2. Material and Methods

2.1. Blood Collection and Cell Culture

Peripheral blood was collected from a pair of PE goat at Sumber Sekar Field Laboratory, Faculty of Animal husbandry, Brawijaya University, Malang. Blood culture applied according to standard cytogenetic techniques [7], medium RPMI enrichment with L-Glutamine, Fetal Bovine Serum US origin, phytohematoglutinin (PHA), HEPES, and Penicillin Streptomycin at 38 °C, 4% CO_2 for 72 hours. Colcemid (Gibco Karyomax) was added and incubated later. After incubation, it was centrifuged at 13000 rpm for 10 minutes and removed the supernatan. Hypotonic solution 0,075 M then added in the cell culture and incubated for 25 minutes then centrifuged at 13000 rpm for 10 minutes. Cold Carnoy's fixative solution was added as much as 10 ml per culture above the fortex and then centrifuged at 13000 rpm for 10 minutes, this process repeated three times until the pellet shows white clean color and cells was dropped onto clean microscopic slides.

2.2. G-Banding

Slide heated in an oven at 60°C for 18 hours before stained. Slides soaked in PBS containing 0,5% trypsin for 60 second and slides rinsed two times with PBS then stained using G-banding method with Leishmann dye:Gurr buffer with 1:3 ratio for 3,5 minutes. Slides was wind-dried and covered with cover glass. Chromosome spreading observed and photographed under Different Interfence Contrast (DIC) microscope with 100x magnification lens.

2.3. Chromosome Analysis

Observations and calculation of goat chromosomes being done in metaphases and analyzed Karyotyping was completed using Genus CytoVision Image software ver 4.5.1. Chromosomes were chosen for 25 spreading each. The chromosomes length measured by its short arm (p) and long arm (q). From both lengths we obtain and also centromic index (Ci). Chromosomes are arranged and numbered by size, from largest to smallest. Chromosomes types can be classified by the position of their centromer based [8]. Measurements were made using the software MicroMeasure ver 3.3 [9].

3. Result and Discussion

The result of this study showed the number of chromosomes in females Peranakan Etawah (PE) goat is consist 30 pairs of chromosomes (2n), consisting of 29 pairs of autosome and a pair of sex chromosomes. Female goat had average of total length (TL) of autosome ranged from 47.91 μm±6.46 to 22.12 μm±3.33. TL of chromosome X are 45.96 μm±4,59 and 44.45 μm±3,96. Whereas the male PE goat had average of TL of autosome ranged from 58.20μm±6.72 to 18.97μm±2.82. TL of chromosome X is 56,42μm±7,38 and Y chromosome is 15,80 μm±3,24.

Table 1. Lenght and Ci chromosome of female PE

No	TL[a] (μm)	Ci	No	TL[a] (μm)	Ci	No	TL[a] (μm)	Ci
1	44,30±5,72	-	21	34,39±2,70	-	41	27,98±0,74	-
2	47,91±6,46	-	22	34,80±6,29	-	42	28,17±1,66	-
3	45,83±5,65	-	23	34,28±3,02	-	43	24,70±2,69	-
4	46,46±7,00	-	24	35,84±3,86	-	44	24,56±2,63	-
5	45,37±5,01	-	25	34,05±0,96	-	45	24,85±4,63	-
6	42,13±7,67	-	26	34,90±2,60	-	46	23,43±3,14	-
7	42,31±5,98	-	27	35,71±1,38	-	47	23,82±1,45	-
8	42,68±6,22	-	28	35,03±2,39	-	48	26,17±2,06	-
9	41,09±6,17	-	29	28,99±3,68	-	49	25,12±3,66	-
10	42,67±6,79	-	30	31,39±5,08	-	50	25,02±3,51	-
11	41,28±2,85	-	31	31,41±3,56	-	51	25,67±2,87	-
12	38,60±4,98	-	32	30,19±3,16	-	52	26,21±3,81	-
13	39,69±6,37	-	33	30,65±1,86	-	53	23,09±2,63	-
14	39,23±5,86	-	34	28,91±1,09	-	54	24,37±1,69	-
15	34,64±3,29	-	35	29,19±4,65	-	55	25,59±7,66	-
16	38,04±7,96	-	36	29,59±3,16	-	56	22,12±3,33	-
17	36,74±3,57	-	37	29,73±2,97	-	57	22,52±1,35	-
18	35,44±5,98	-	38	29,72±4,32	-	58	23,28±1,59	-
19	37,05±4,48	-	39	27,90±1,41	-	59	45,96±4,59	31,74
20	35,36±4,96	-	40	29,63±2,41	-	60	44,45±3,96	32,80

[a]TL = total lenght

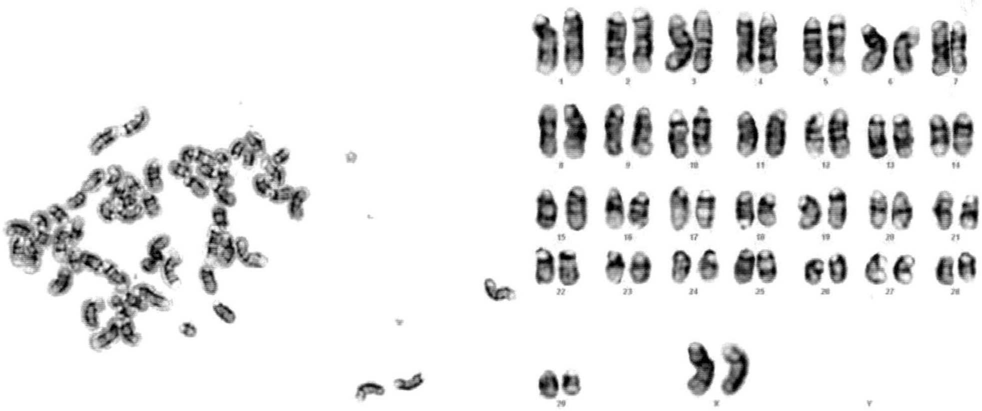

Figure 1. Female chromosome (Left) spreading, (Right) karyotyping

Table 2. Lenght and Ci chromosome of male PE goat

No	TL[a] (μm)	Ci	No	TL[a] (μm)	Ci	No	TL[a] (μm)	Ci
1	57,67±7,51	-	21	39,45±0,55	-	41	30,84±3,75	-
2	58,20±6,72	-	22	39,77±4,76	-	42	31,05±1,63	-
3	52,85±5,90	-	23	39,03±5,28	-	43	30,49±2,34	-
4	50,05±3,82	-	24	36,93±4,98	-	44	29,74±2,32	-
5	47,32±5,33	-	25	35,72±3,64	-	45	32,79±6,03	-
6	46,41±3,31	-	26	36,79±2,65	-	46	28,98±4,41	-
7	47,71±5,44	-	27	35,75±2,07	-	47	27,57±4,26	-
8	46,92±3,61	-	28	37,58±0,97	-	48	29,40±4,47	-
9	47,35±4,58	-	29	35,59±4,16	-	49	23,39±2,87	-
10	47,27±4,78	-	30	37,46±8,17	-	50	25,27±0,94	-
11	49,61±3,70	-	31	35,35±2,85	-	51	25,34±2,72	-
12	46,98±5,05	-	32	33,70±0,85	-	52	23,40±1,90	-
13	43,80±6,94	-	33	32,10±4,89	-	53	22,70±2,71	-
14	43,84±2,91	-	34	31,33±3,38	-	54	24,57±2,06	-
15	42,32±3,40	-	35	32,13±3,91	-	55	22,04±7,05	-
16	43,24±1,39	-	36	33,93±3,53	-	56	24,79±3,67	-
17	41,67±4,63	-	37	33,04±2,54	-	57	20,49±2,33	-
18	40,78±1,89	-	38	32,09±1,75	-	58	18,97±2,82	-
19	42,36±4,31	-	39	31,03±3,95	-	59	56,42±7,38	4,39
20	37,25±3,00	-	40	30,81±4,29	-	60	15,80±3,24	1,14

[a]TL = total lenght

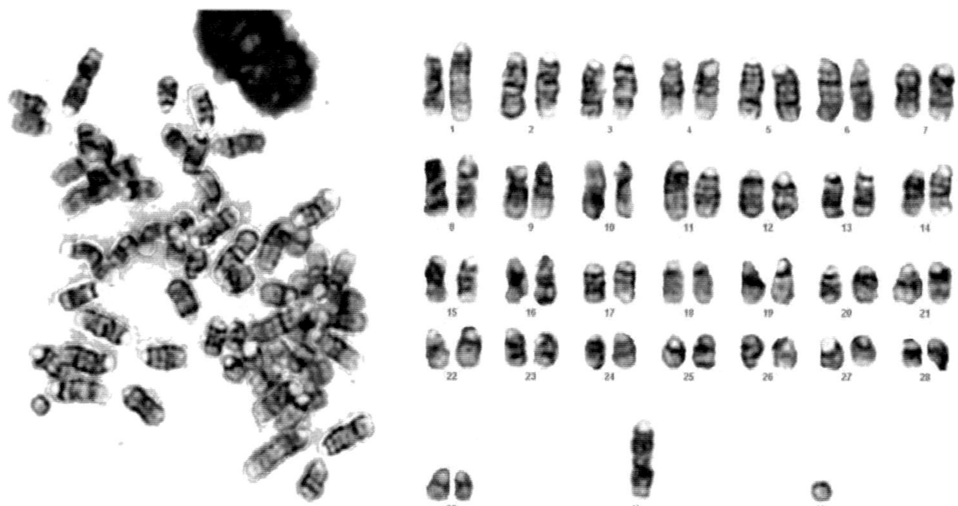

Figure 2. Male chromosome (Left) spreading and (Right) karyotyping

The characteristics of goat chromosomes under metaphase conditions may differ in size and shape, depending on the conditions of the mitotic stage in the sample preparation. Based on the centromere position of Ci calculation results to determine the chromosome types [8]. Centromeric index only found in sex chromosome of PE goat because all autosomes have short arms (p) that is very close to the centromere. Centromeric index from female PE goats are 31,74 and 32,80 and X chromosome from male PE goats 46.84 is indicates that the types is subtelocentric and Ci in chromosome Y from male PE goats is 19.34 called submetacentric.

4. Conclusion

The conclusion of this research is the total of goat chromosome is 60 with types of autosomal chromosomes were acrocentric as many as 58 chromosomes and pair of sex chromosomes XX and XY, X classified as subtelocentric and Y submetacentric. Female goat had average of total length (TL) of autosome ranged from 47.91 μm±6.46 to 22.12 μm±3.33. TL of chromosome X are 45.96 μm±4,59 and 44.45 μm±3,96. Male PE goat had average of TL of autosome ranged from 58.20μm±6.72 to 18.97μm±2.82. TL of chromosome X is 56,42μm±7,38 and Y chromosome is 15,80 μm±3,24.

Acknowledgement

This research financed by LPDP-RISPRO, Ministry of Finance of Republic of Indonesia No Contract. prj-1047/lpdp/2015. We thank to Helly Nurul Karima, Choirunil Chotimah, Reva Yuliasari, Rr. Fitria Dewi and Ali Harris for technical assistance in laboratory, Mr. Kusno and Mr. Sumali as technical assistance in field laboratory.

References

[1] Guillemot, E., Berland, H.M., Durand, V., Darre, E., Cribiu, E.P. 1991. Cytogenetic investigation in Saanen and Alpine artificial insemination bucks. Identification of a Robertsonian translocation. Genetics Selection Evolution, Biology Medical Central, 1991, 23 (5-6), pp.449-454

[2] Caraviello, D. Z. 2004. *Crossbreeding Dairy Cattle*. The Babcock Institute, Board of Regents of the University of Wisconsin-Madison. Reproduction and Genetics, No. 610.

[3] Santos, R.P., Affonso, P.R.A.M., Diniz, D., Medrado, A.S., Silva, K.M., Carneiro, P.I.S. 2013. Identification of Naturalized Goat Breeds Under Conservation from Northeastern Brazil Using Chromosomal Markers. Genetics and Molecular Res. Vol **12** (3): 3193

[4] Tiesnamurti, Bess. 2011. Rencana Aksi Global Sumber Daya Genetik Ternak dan Deklarasi Interlaken. Pusat Penelitian dan Pengembangan Peternakan. Bogor

[5] Moore, C.M. dan Robert G.B. 2001. Chromosome Preparation and Banding. Encyclopedia Of Life Sciences. Secondary Article. Nature Publishing Group

[6] Yuwono, Triwibowo. 2005. Biologi Molekuler. Erlangga. Jakarta. ISBN 9797811921

[7] Smith K. 2006. Basic Cytogenetic Techniques: Culturing, Slide Making, and G Banding Cytogenetic analysis and report were based on ISCN 2009. In: Celis, J.E. Cell Biology, a Laboratory Handbook Third Edition. Elsevier Academic Press. London, UK. Vol **3**:381.

[8] Levan, A., Fredga, K., dan Sandberg, A.A. 1964. Nomenclature For Centromeric Position On Chromosomes. Hereditas **52**: 201.

[9] Reeves, A. and J. Tear. 2000. MicroMeasure for Windows, version 3.3. Free program distributed by the authors over the Internet from http://www.colostate.edu/Depts/Biology/MicroMeasure

Effect of Age and Bulls on Fresh Semen Quality and Frozen Semen Production of Holstein Bulls in Indonesia

A. Argiris[1*], Y. S. Ondho[1], S. I. Santoso[1] and E. Kurnianto[1]

[1]Faculty of Animal and Agricultural Sciences, Diponegoro University, Semarang, Central Java,Indonesia.

*Corresponding E-mail: argi290582@yahoo.com

Abstract

Artificial Insemination is a compatible method of reproduction in an effort to increase dairy productivity. Artificial Insemination Center as a producer of frozen semen was required to maximize bulls in producing high quality frozen semen optimally. The purpose of this research was to determine effect of age and bulls on fresh semen quality and frozen semen production of Holstein bulls in Indonesia. The research was conducted at Lembang and Singosari AI Centers. The material used were 24.634 data of qualified fresh semen and frozen semen production from 81 Holstein Bulls aged 1-9 years that used as frozen semen producer in period of 2008 to 2016. The variables observed in this research were data of age of bulls, fresh semen volume (mL); sperm motility (%); mass movement; concentrations (million/mL) and frozen semen doses at each production at age of bulls. Nested design was applied to obtain and analyze data. Results showed that Age and bulls have significant effect (P<0,01) to volume, mass, motility, concentration and frozen semen production. Increasing the age of bulls resulted in increase semen volume until 7-year-old, while semen concentration decreased from 3 years old with increasing age. Frozen semen production, mass movement and motility shown the same relative value on 3-9 years old except on 1 to 2 years old had increase. Bulls would produce frozen semen optimally on 3-9 years old. Indeed, with knowledge of this factor, AI Centre might adapt management of AI bulls to improve semen production.

1. Introduction

Increased demand for milk was stimulating farmers to increase productivity of dairy cattle in order to supply market needs. Efforts to import dairy cows continuously will lead to dependency and depletion of foreign exchange, therefore another alternative was required by providing a source of high quality frozen semen that will be used in artificial insemination (AI). An increase in dairy productivity can be achieved through improved reproduction systems and genetic improvement.

Artificial Insemination is a compatible method of reproduction in an effort to increase dairy productivity. The success of AI implementation in the field was influenced by the quality of the frozen semen used [1]. Semen quality could be assessed based on its fresh semen characteristics through macroscopic or microscopic examination [2]. High quality frozen semen from a superior bull has a role in improving fertility and reproduction efficiency. Holstein bulls are "biological machine" having a productive age in producing frozen semen. Productive age is the period during the bull could maximize production optimally and efficiently in producing frozen semen. Lestari et al. [3] showed that age has a significant effect on fresh semen volume so that the age factor is very important in determining the quality and quantity of semen and can be used as part of the selection criteria for bull as frozen semen production.

Artificial insemination centre as a producer of frozen semen must be maximize bulls that capable to produce semen with good fertility so factors affecting semen production are very important such as effect of age [4]. Ismaya [5] stated that the quality of semen was influenced by several factor, those were: age of bulls, genetics, temperature and season, collecting and feeding frequency. The results of [6] showed that age had a significant effect on the volume of Simmental bulls in Austria.

2. Materials and Methods

2.1. Material and location

The research was conducted at Lembang and Singosari AI Centre. The material used were 24.634 data of qualified fresh semen and frozen semen production from 81 Holstein bulls aged 1-9 years that used as frozen semen producer in period of 2008 to 2016. The variables observed in this research were data age of bulls, fresh semen volume (mL); semen motility (%); mass movement; concentrations (million/mL) and frozen semen doses at each production age of bulls.

2.2. Statistical analysis

Data were analyzed with nested design, to calculate the value of production efficiency. Statistical model:

$$Y_{ijk} = \mu + A_i + B_{j(i)} + \varepsilon_{(ij)k}$$

Y_{ijkl} = The kth observation of jth age factor ith and bulls factor
μ = overall mean
A_i = effect of ith bull factor
$B_{j(i)}$ = effect of jth age factor of ith bull factor
$\varepsilon_{(ij)k}$ = effect error

3. Results and Discussion

Level of significance for bulls and bull's age class at 2 National AI Centre in Indonesia are presented in Table 1. As data of the 2 National AI centre had to be analyzed separately due to different data recording, comparison between centres are not feasible. Bulls and bull's age were highly significant for all traits in both AI centre.

Table 1. Level significance for age and bull effect at 2 national AI Centre

	Vol		Mass		Mot		Con		FS	
	1	2	1	2	1	2	1	2	1	2
Bulls	**	**	**	**	**	**	**	**	**	**
Bull's age class	**	**	**	**	**	**	**	**	**	**

1, Lembang AI Centre; 2, Singosari AI Centre; Vol, ejaculate volume; Mass, mass movement; Mot, motility; Con, sperm concentration; FS, total frozen semen.
**, (P<0.01)

a. Effect of Age and Bulls on Semen Volume of Holstein Bulls

The volume of fresh semen is one of the earliest standards in evaluating the quality of semen that can be seen directly on the tube scale of the collection. Data on the volume of fresh semen at 2 National AI center are presented in Table 2. The average volume of semen in Lembang and Singosari AI center had increased with age of bulls. The highest volumes of bulls achieved on 7 years old at Singosari and

9 years old in Lembang were in agreement with [6] stated that the highest ejaculate volume were achieved by bull being older than 72 months.

The results of the variance analysis showed a very significant difference (P <0.01) between the fresh semen volume of Holstein bulls at different ages (Table 1). It was accordance with research [6] which stated that the age had a significant influence on the volume of semen Simmental bulls in Austria.

Table 2. Volume of Fresh Semen at 2 National AI center

Singosari AI Centre			Lembang AI Centre		
Age	Mean	Noumber of Collection	Age	Mean	Noumber of Collection
1	-	-	1	5.41±1.70	95
2	6.18±1.73	1469	2	5.95±2.00	1139
3	6.86±1.84	2279	3	7.39±2.14	1186
4	7.41±2.02	3141	4	7.70±2.32	1908
5	7.72±2.21	2951	5	8.31±2.38	1319
6	8.09±2.27	2215	6	8.20±2.45	1143
7	8.41±2.30	1364	7	8.31±2.57	994
8	8.03±2.16	1278	8	8.49±2.29	741
9	7.95±2.26	1002	9	8.59±2.88	410
Total	7.54±2.19	15699	Total	7.74±2.47	8935

Total average of semen volume of Holstein bull per age at 2 National AI centers can be seen in Table 3. The volume of semen continues increased with the age of bulls until 7 years old and was relatively constant at the age of 9 years old. It were in accordance with the study of [5] which stated that the volume of fresh semen continuously enhanced with increasing age, but decreased for older bulls. Increasing volume with age was in accordance with previous studies [7] and [8] stated volume of fresh semen increased up to bulls aged 7 years old and relative constant until bulls aged 10 years. However, unlike [9], the highest semen volume was obtained in bulls aged 9 year-old, but not significant in bulls aged 5-9 years. The volume increased with the age of bulls due to the growing size of the larger testes where tubuliseminiferi growth up so that influenced increased production of spermatozoa cells [9]. The lowest semen volume in the bull group were <2 years old and increase with age. The volume of semen reach the top at bull's age 7 years (8.368 ± 2.41 mL) and after that time gradually decreased in older bulls. It was in accordance with the results of research [10] stated that the semen volume showed optimal results in the bulls age > 5 years and showed the lowest semen volume in age <2 years old. It was different from the study of [6] which obtained top semen volume at 6 year old of Simmental bulls in Austria.

b. Effect of Age and Bulls on Semen Mass Movement of Holstein Bull

The mass movement of semen is the simultaneous movement of spermatozoa that causes waves [11]. Semen mass movement was used as a selection in frozen semen production, where fresh semen must be has a minimum value 2+ to produce as frozen semen. Data on the mass movement of Holstein bulls per ages at Singosari AI center are showed in Table 4. Mass movement 2+ had highest percentage (97.6%) and mass movement 3+ had lowest percentage (0.5%). The highest percent value of 2+ mass movements were obtained in bulls aged 6 years (99%), while 1+ mass movement were mostly found in bull's age 2 year (3%). The percentage value of mass movement 1+ under 10% illustrates as good bulls management and semen handling, where the percentage of fresh cement unprocessed was less than 10%.

Table 3. Total average of semen volume of Holstein bull

Age	Mean	Noumber of collection
1	5.40±1.70	95
2	6.08±1.85	2608
3	7.04±1.96	3465
4	7.52±2.14	5049
5	7.90±2.28	4270
6	8.12±2.33	3358
7	8.36±2.41	2358
8	8.19±2.21	2019
9	8.13±2.47	1412
Total	7.61±2.29	24.634

Table 4. Semen mass movement at Singosari AI Center

			Ages								
			2	3	4	5	6	7	8	9	Total
Mass	1	Count	44	32	48	78	19	15	30	28	294
		% within age	3.0	1.4	1.5	2.6	0.9	1.1	2.4	2.8	1.9
	2	Count	1415	2234	3062	2857	2189	1312	1246	969	15284
		% within age	96.4	98.1	98.1	96.9	99.0	97.1	97.6	96.7	97.6
	3	Count	9	12	11	12	4	24	0	5	77
		% within age	0.6	0.5	0.4	0.4	0.2	1.8	0.0	0.5	0.5
Total		Count	1468	2278	3121	2947	2212	1351	1276	1002	15655
		% within age	100	100	100	100	100	100	100	100	100

Table 5. Semen Mass Movement at Lembang AI Center

			Ages									
			1	2	3	4	5	6	7	8	9	Total
Mass	1	Count	32	383	269	414	256	288	198	182	118	2140
		% within age	33.7	33.7	22.7	21.7	19.5	25.4	20	24.6	28.9	24
	2	Count	61	744	896	1474	1056	843	792	554	284	6704
		% within age	64.2	65.5	75.7	77.3	80.2	74.5	79.9	74.8	69.4	75.2
	3	Count	2	9	18	18	4	1	1	5	7	65
		% within age	2.1	0.8	1.5	0.9	0.3	0.1	0.1	0.7	1.7	0.7
Total		Count	95	1136	1183	1906	1316	1132	991	741	409	8909
		% within age	100	100	100	100	100	100	100	100	100	100

Based on data are presented in Table 5, mass movement 2+ in Lembang AI center had highest percentage (75.2%) and mass movement 3+ had the lowest percentage (0.7%). The highest percentage of mass movements 2+ were obtained in 5 year olds (80.2%), while mass movements 1+ were highest in bulls aged 1 and 2 years (33.7%). Hafez [12] stated that the normal mass movement of cement was 2+ to 3+. The percentage of mass movement 1+ more than 10% indicated bulls management and cement handling were non optimal, where the percentage of fresh cement than unprocessed greater than 10%.

Table 6. Average value mass movement of Holstein bulls

Age	Mean	Noumber of Collection
1	1.68±0.51	95
2	1.85±0.42	2604
3	1.92±0.30	3461
4	1.91±0.30	5027
5	1.92±0.30	4263
6	1.91±0.29	3344
7	1.92±0.31	2342
8	1.90±0.31	2017
9	1.90±0.32	1411
Total	1.91±0.32	24564

c. *Effect of Age and Bulls on Semen Motility of Holstein Bull*

Testing sperm motility is one of the important parameters that can be used as the basic information about the ability of spermatozoa in fertilization [14]. Data sperm motility of Holstein bulls at different ages are showed in Table 7. The value of semen motility at 2 national AI center had the lowest score (62.74%) in bull aged <2 years old at Lembang AI center and had the highest score (68.59%) in bulls aged 7 year old at Singosari AI center. The value of motility was increased with age. That was according to the study of [9] which stated that the increasing age had a significant effect on the sperm motility.

The result of variance analysis showed a significant difference ($P < 0.01$) motility of Holstein bulls at different ages (Table 1). This was subject to the research by [6] that stated age gave a real effect on the individual motility.

Total average motility semen of Holstein bull at 2 AI Center was presented in Table 8. Result of research got mean motility (66.02%) with range (62.74%-67.38%), consistent with [15] fresh semen had an average motility (48.7% -79.9). The motility value of the results of this study was still within the normal range in the opinion of [12] which stated that the motility of semen ranges from 40-75%.

Table 7. Semen Motility at 2 National AI center

Singosari AI Center			Lembang AI Center		
Age	Mean[1]	Noumber of collection	Age	Mean[1]	Noumber of Collection
1	-	-	1	62.74±0.17	95
2	64.91±0.11	1469	2	64.31±0.14	1135
3	67.13±0.09	2279	3	67.38±0.13	1183
4	67.97±0.09	3141	4	66.42±0.13	1906
5	64.88±0.12	2952	5	65.76±0.13	1316
6	66.57±0.09	2216	6	62.99±0.15	1132
7	68.59±0.08	1364	7	65.24±0.13	992
8	64.92±0.12	1278	8	64.56±0.15	741
9	65.69±0.11	1002	9	63.18±0.17	409
Total	66.44±0.10	15701	Total	65.27±0.14	8909

1. Motility (%)

Table 8. Total average motility semen of Holstein bull

Age	Mean (%)	Noumber of Collection
1	62.74±0.18	95
2	64.65±0.12	2604
3	67.21±0.10	3462
4	67.38±0.11	5047
5	65.15±0.12	4268
6	65.36±0.11	3348
7	67.18±0.10	2356
8	64.79±0.13	2019
9	64.96±0.13	1411
Total	66.02±0.12	24.610

d. Effect of Age and Bulls on Semen Concentration of Holstein Bull

Evaluation of semen concentration was used to determine the amount of diluent to be added in the frozen semen production process. The results of semen concentration of Holstein bulls at different ages are showed at Table 9. The highest semen concentration at Singosari AI center was obtained at bull aged 2 years old (1330.73 x 10^6), concentration decrease with increasing age, according to the study of [6] which stated that the concentration decreased with increasing age and reach optimum value for bull aged 18-20 months. While the highest cement concentration at Lembang AI center was obtained at bull aged 3 year old (1291,30 x 10^6) then decreased with increasing age.

Table 9. Descriptive Statistic for Semen Concentration

Singosari AI Center			Lembang AI Center		
Age	Mean[1]	Noumber of Collection	Age	Mean[1]	Noumber of Collection
1	-	-	1	949.89±448.46	95
2	1330.73±521.50	1469	2	1192.50±452.41	1139
3	1301.16±526.63	2279	3	1291.30±427.18	1186
4	1220.77±505.80	3141	4	1259.10±435.89	1908
5	1139.20±483.61	2952	5	1215.70±347.57	1319
6	1147.03±472.48	2216	6	1059.60±309.97	1143
7	1213.51±440.15	1364	7	1085.90±358.29	994
8	1019.80±457.00	1278	8	1125.60±434.07	741
9	1023.35±501.45	1002	9	1216.60±540.07	410

1. Semen Concentration (x10^6/mL)

The result of variance analysis showed that age had significant effect (P <0,01) to the semen concentration of Holstein bulls (Table 1). This is accordance with the study of [6] which stated that age had a significant effect (P <0.01) on semen concentration. Concentration of semen increases at age 1-3 years, and then decreases with age (Table 10). That was in accordance with [8] which stated that there was an increase in the concentration of semen up to the age 20-22 months then decreases with age. The lowest semen concentration (949.89 ± 448.4 x 10^6) was obtained in bulls aged <2 years old and the highest (1,297.8 ± 4.94 x 10^6) was obtained in bulls aged 3 year olds. This was different from [16] study which stated that bull's age 4 years old would have the highest concentration of semen.

Table 10. Semen concentration averages of Holstein bulls at different ages

Ages	Mean (10^6/mL)	Noumber of Collection
1	949.89±448.46	95
2	1270.40±497.18	2608
3	1297.80±494.80	3465
4	1235.30±480.89	5049
5	1162.80±447.41	4271
6	1117.30±426.20	3359
7	1159.70±412.41	2358
8	1058.60±451.50	2019
9	1079.50±520.23	1412
Total	1187.40±471.88	24.636

e. Effect of Age and Bulls on Frozen Semen Production of Holstein Bull

Total frozen semen results in the frozen semen production process were influenced by quality of fresh semen, diluent and production process. The average data on total frozen semen produced by the Holstein bulls at different ages are showed in Table 11. The highest average of frozen semen production was achieved at bulls aged 7 years (420 doses) in Singosari and bull aged 8 years (387 doses) in Lembang. This is accordance with [9] who stated that cement production increased significantly in bulls aged over 60 months. Total average production of frozen semen at singosari was higher (354 doses) than Lembang (321 doses).

Table 11. Descriptive Statistic for Frozen Semen Production

Singosari AI Centre			Lembang AI Centre		
Age	Mean	Noumber of Collection	Age	Mean	Noumber of Collection
1	-	-	1	203±96.54	63
2	319±172.33	960	2	272±114.85	742
3	355±180.12	1804	3	330±135.85	910
4	359±175.24	2672	4	330±145.24	1478
5	351±158.06	2157	5	313±125.77	1058
6	341±166.89	1805	6	294±123.65	838
7	420±194.22	1149	7	326±143.19	793
8	350±202.46	969	8	387±150.08	558
9	326±195.82	804	9	376±154.87	291
Total	354±178.86	12320	Total	321±139.21	6731

The results on Table 12 showed that the highest average frozen semen production at bull aged 7 years old (382 ± 181.14 straw) that was caused by bull aged 7 years old had highest average volume of fresh semen with other variable (semen concentration and motility) were quite large. This is accordance with [17] noted an increase in the amount of semen with age of the animal. The amount of semen volume, concentration and motility are factor that determines the quality and amount of frozen semen produced. The average of frozen semen production per collection increased at bull aged 1-3 years and reached a plateau at 3-9 years of age.

The determination of productive age was calculated based on the percentage of frozen semen production per collection per age that multiplied by average of frozen semen production per age are showed at Table 13. Result calculation of optimal production age shows that bulls age 3 to 9 year old had value of frozen semen productivity above the average total productivity, while 1-2 years old bulls had value of frozen semen productivity below total average, so it could be determined that bulls will be produce frozen semen optimally on 3-9 years old. Taylor et al. [7] observed increasing sperm production until 84 month of age and started to decrease after 112 month of age, while [18] reported that sperm production declined after 84 months of age. Indeed, with knowledge of this factor, AI Centre might adapt management of AI bulls to improve semen production.

Table 12. The average production of Holstein frozen semen per collection per ages

Age	Frozen semen/collection/age	
	Mean	Coeficient of Variation (%)
1	203±96.54	47.56
2	298±151.80	50.94
3	347±166.98	48.12
4	349±165.73	47.49
5	339±149.29	44.04
6	326±156.03	47.86
7	382±181.14	47.42
8	364±185.82	51.05
9	340±187.05	55.01
Total	343±166.67	48.59

Table 13. Calculation productive age of bull

Age	Noumber of Collection	Processing/collection/age		Frozen semen /collection/age	Productivity*
		Noumber of	%		
1	95	63	66.32	203	134
2	2608	1702	65.26	298	195
3	3465	2714	78.33	347	271
4	5049	4150	82.19	349	287
5	4271	3215	75.28	339	255
6	3359	2629	78.27	326	255
7	2358	1942	82.36	382	315
8	2019	1526	75.58	364	275
9	1412	1095	77.55	340	263
AVG	2737.33	2115.11	75.68	327	250

*, Productivity = (frozen semen/collection/age) x (% processing/collection/age)

4. Conclusion

The conclusions that can be drawn from this research are Age and bulls had significant effect (P<0,01) to semen volume, semen mass movement, semen motility, semen concentration and frozen semen production. Increasing the age of bulls result increasing semen volume until the 7 year-old, while the semen concentration decreases from 3 years old with increasing age. Frozen semen production, mass movement and motility show the same relative value in bull's age 3-9 years old except on bull age 1-2 years old has increase. Bulls would produce frozen semen optimally on 3-9 years old. Indeed, with knowledge of this factor, AI Centre might adapt management of AI bulls to improve semen production.

References

[1] Oliveira, L.Z, R.P. Arruda, A.F.C. Andreade, R.M. Santos, M.E.Beletti. 2012. Efect of sequence of insemination after simultaneous thawing of multiple semen straws on conception rate to timed AI in suckled multiparous Nelore cows. J.Theriogenology.**78**: 1800.

[2] Sumeidiana, I.S., Wuwuh and E. Mawarti. 2007. The volume of semen and concentration of Simmental, Limousin and Brahman at Ungaran Artificial Insemination Center. J. Indon. Trop. Anim. Agri. **32**(2): 131.

[3] Lestari, S., D.M. Saleh and Maidaswar. 2013. Profile of fresh semen quality of Limousin bulls with different ages at Lembang AI Center West Java. Jurnal Ilmu Peternakan. **1**(3): 1165.

[4] Karoui, S., D. Clara, S. Magdalena and R. Cue. 2011. Time trends, environmental factors and genetic basis of semen traits collected in Holstein bulls under commercial conditions. J. Anim. Reprod. Sci. **124**: 28.

[5] Ismaya. 2014. Artificial Insemination Biotechnology in Cattle and Buffaloes. Gadjah Mada University Press. Yogyakarta.

[6] Fuerst-Waltl, B., H. Schwarzenbacher, P. Christa and J. Solkner. 2006. Effects of age and environmental factors on semen production and semen quality of Austrian Simmental bulls. J.Anim. Reprod. Sci. **95**: 27.

[7] Taylor, J.F., B. Bean, C.E. Marshall and J.J. Sullivan. 1985. Genetic and environmental components of semen production traits of artificial insemination Holstein bulls. J. Dairy Sci. **68**: 2703.

[8] Mathevon, M., M.M. Buhr and J.C. Dekkers. 1998. Environmental, management and genetic factors affecting semen production in Holstein bulls. J. Dairy Sci. **81**(12): 3321.

[9] Brito, L.F.C., A.E.D.F. Silva, L.H. Rodrigues, F.V.Vieira, L.A.G. Deragon and J.P. Kastelic. 2002. Effect of environmental factor, age and genotype on sperm production and semen quality in Bos Indicus and Bos Taurus AI bulls in Brazil. J. Anim. Reprod. Sci. **70**: 181.

[10] Paldusova, M., T. Kopec, G. Chladek, M. Hasek, L.Machal and D. Falta. 2014. The effect of the Table environment and age on the semen production in the Czech Fleckvieh Bulls. Mandel. Net:178-182.

[11] Wahyuningsih, A. 2013. The effect of bull age and collection frekwency on the volume and motility of fresh semen Simmental bull in Lembang AI Center. J. Ilmiah Peternakan. **1**(3):947.

[12] Hafez, E.S.E. 1993. Reproduction in Farm Animals. 6th. Lea Febiger. Philadelphia. USA

[13] Makulska, J., Ch. Hagger, N. Kunzi and H.U. Kupferschmied. 1993. Genetic and environmental influences on semen traits in A.I. bulls. J. Reprod. Domest. Anim **28**: 279.

[14] Sarastina, T., G. Susilawati and G. Ciptahadi. 2012. Analysis of several parameters of sperm motility in various cattle using Computer Assisted Semen Analysis. J. Ternak Tropika **6**(2): 1.

[15] Stout, M.A. 2004. Comparison of Epididymal and Ejaculate Sperm Collected the Same Holstein Bulls. Desertation. University of Lousiana Lafayete

[16] Addass, P.A. 2011. Effect of age and body condition score on sperm production potential among some indigenous bull cattle in Mubi Adamawa State Nigeria. Agric. Biol. J. N. Am. **2**(2): 203.

[17] Foote, R.H.,G.E. Seidel, J.Hahn, W.E. Berndtson and G.H. Coulter. 1977. Seminal quality, spermatozoal output and testicular changes in growing Holstein bulls. J. Dairy Sci. **61**: 85.

[18] Everett, R.W and B. Bean. 1982. Environmentals influences on sperm output. J. Dairy Sci. **82**(65): 1303.

Semen Quality of Post-Thawed Local Ram's in Tris-Egg Yolk Extender with Different Glutathione Level

N. Solihati[1*], S.D. Rasad[1], R. Setiawan[1], E.N. Foziah[1], and E.T. Wigiyanti[1]

[1]Faculty of Animal Husbandry Universitas Padjadjaran, West Java, Indonesia.

*Corresponding E-mail: nurcholidah@unpad.ac.id

Abstract

The aims of this research were to find out the effect of glutathione level on semen quality of local ram and to find out the glutathione level that provide the best quality of local ram semen. This research use Completely Randomized Design (CRD) with five treatments of glutathione level (0 mM, 4 mM, 5 mM, 6 mM and 7mM) and was repeated for 10 times. Data were analyzed using analysis of varians (ANOVA) and differences between treatment was analyzed using Duncan test. The parameter were semen quality consist of motility, intact plasma membrane (IPM), abnormality, and recovery rate. Result of this research showed that glutathione level significantly ($p<0.05$) affect on motility, IPM, abnormality and recovery rate. Duncan test result showed that recovery rate from treatment of 5 mM glutathione level have significantly higher than other treatment, whereas for motility, IPM and abnormality were not significant different with 4 mM, and 6 mM. It is concluded that semen quality of local ram was affected by glutathione level, and the level of 5 mM in egg yolk tris extender provide the best quality of local ram semen.

1. Introduction

Local Sheep is sheep result of crossing or introduced from outside who had been bred in Indonesia until the fifth generation or more to adapt to the environment or local management [1].Local Ramhave been known to have qualified fresh semen qualities to be processed into chilled semen and frozen semen [2]. The process of freezing semen has been widely reported by researchers that it requires a cryoprotectant material to protect sperm from cold shock. Glycerol is one type of cryoprotectant that has been proven to provide protection against sperm so that the quality of sperm can meet the standards for use in artificial insemination (AI) program. Previous research on Local Ram has reported that the level of glycerol producing the best semen quality is 5% [3].

During the freezing process, sperm damage can occur due to the formation of ice crystals as well as damage to the membrane structure due to lipid peroxidation. It can occur during thawing, in the presence of extreme temperature and semen contact with oxygen, causes increase the metabolism and concentration of free radicals [4]. The effort to minimize lipid peroxidation that occurs in the freezing process of semen could be done by added an antioksidant, such as glutathione.Glutathione is the major non-protein thiol compound in mammalian cells. Glutathione participates in a number of cell functions including amino acid transport, DNA andprotein synthesis, reduction of disulfide bonds and protection against oxidative stress. Thesulphydryl groups of glutathione have been shown to confer protection against cell damage byoxidants, electrophiles and free radicals [5]. Based on the antioxidant properties that can neutralize free radicals, the addition of glutathione as a primary antioxidant can reduce the damage of the plasma membrane. The addition of glutathione in the yolk tris diluent is expected to reduce lipid peroxidation in the Local Ram cryopreservation process, but information on the optimum level is not yet available. This supports the authors to conduct a study about post-thawed Local Ram's semen quality in tris-egg yolk extender with different glutathione level. The objective of this study was to obtain the level of glutathione that produced the best quality of post-thawed Local Ram semen.

2. Materials and Methods

This research used two Local Ram weighing 45-50 kg and age around 24-36 months. Ram were fed with forage fed and additional feed given daily as much as ± 5 kg / head and ± 0.35 kg / head.

2.1. *Materials*

The research material to be used was semen produced from Local Ram that was collected for 10 times by the artificial vaginal method. Others main materials were glutathione, tris aminomethane, etc.

2.2. Semen Processing Methode

The semen diluent components used in this study consist of buffer, egg yolk and antibiotics. The diluent used was egg yolk tris. The diluted semen was then treated with glutathione levels, consisting of 0 mM (T0), 3 mM (T3), 4 mM (T4), 5 mM (T5), and 6 mM (T6). The freezing of semen begins with the sucking of semen that has been diluted into the straw by using an automatic suction pump, then the open straw end is emphasized on polyvinylalcohol adhesive powder. The straw used is ministraw (0.25 ml).

The next process was to insert a straw that has been sealed into a 5°C temperature cooler to be equilibrated for 2-4 hours. After the equilibration process the straw is arranged in a rack located approximately 10 - 15 cm above the liquid nitrogen surface (-130°C) for 15 minutes and subsequently inserted into the goblet which is directly dipped into the liquid nitrogen solution (temperature -196°C) contained in the container. Thawing is done on water with a temperature between 38 - 40°C

2.3. Research Design and Data Analysis

The study was designed with a complete randomized design (RAL) with five levels of glutathione treatment (0 mM, 3 mM, 4 mM, 5 mM, 6 mM)and 10 replications. Parameters observed included: fresh semen quality, post-equilibration and post thawed semen quality, consisting of motility, abnormality, intact plasma membrane, and recovery rate. The data were analyzed by analisys of variance and the differences between treatments were tested with Duncan's advanced test.

3. Results and Discussion

3.1. The effect of glutathione level on Motility

Motility of fresh, post equilibration and post thawed semen of local ram was showed in Table 1. This result showed that in post thawed motility, T5 (treatment of 5 mM glutathione) showed the highest results, whereas the lowest motility is in T0. Motility increases from T0 to T5, then falls back on T6.

The result of ANOVA showed that glutathione level treatment had significant effect ($P<0.05$) on motility. Further Duncan test results showed that the 5 mM level is significantly higher than 0 mM, 3 mM, but not significant different with level of 4 mM and 6 mM.

The results showed that the percentage of motility in all four treatments with the addition of glutathione in the tris-yolk was higher than the treatment without addition of glutathione. This is because diluents without glutathione get only protection from glycerol. Glycerol is an intracellular cryoprotectant that serves to protect spematozoa from the formation of ice crystals during freezing[6].

The addition of glycerol can not reduce the damage of the plasma membranes due to lipid peroxidation. It was reported that lipid peroxidation cause irreversible loss of motility, inhibition of respiration, leakage of intracellular enzymes, damage to spermDNA [7] and also deficiencies in oocyte penetration and sperm–oocyte fusion [8].The addition of glutathione serves to prevent lipid peroxidation of spermatozoa cell membrane during the freezing process of semen, thereby increasing the motility and integrity of the acrosome after thawing [9]. Glutathion has a positive effect on maintaining motility. Glutathione can protect sperm from oxidative stress to be caused by hydrogen peroxide in frozen semen [10]. Higher concentrations of glutathione may improve or maintain the morphological and sperm quality [11].

The results of this study indicate that the addition of glutathione less than 5 mM may be was less able to protect the sperm from reactive oxygen compounds so it has not been effective to keep sperm from lipid peroxidation. This is caused by an imbalance between the production of reactive oxygen compounds with existing antioxidants. The condition is called oxidative stress [12]. Similarly, the increase of reactive oxygen compounds besides damaging the sperm plasma membrane, also causes decreased motility and sperm capacitance [13].

In this research, the addition of glutathione ≥ 5 mM, does not produce high motility, even was decreased. This is may be due to the toxic effects of glutathione. Other researcher reported that the addition of excess glutathione levels can have negative effects or toxic effects that cause sperm death [14]. This toxic effect is reinforced that the effect of 5 mM glutathione on post thawed semen produces good motility, whereas 7 mM produces the lowest total mobility [15].

The addition of antioxidants in large quantities will further increase osmotic pressure dilution, plasma membrane damage and less well adapted by sperm adversely affect the on going process of sperm metabolism. In addition, glutamate content in glutathione can bind lipoprotein and lecithin to egg [16; 17]. This can decrease sperm motility due to impaired metabolic processes can lead to a lack of ATP production used spermatozoa to produce energy, while lipoprotein and lecithin are components of yolk that can protect sperm from the influence of cold shock.

The results of this study are consistent with other researcher that the addition of 5 mM glutathione resulted in the highest mobility and the lowest abnormality compared to the addition of 2 mM in Beetal, Black Bengal and Crossbred goats [18]. Similarly, study on sheep compared to 2 mM and 7 mM [15], while other reported that 6 mM glutathione levels resulted in optimal motility and viability in post thawed Boer goats [14].

Table 1. Effect of Glutathione Level on Sperm Motility

Treatment	Fresh Semen	Post-Equilibration	Post-Thawed
		.... %	
T0	91.18	70.26	34.61^{a}
T3	91.10	75.68	36.16^{a}
T4	91.58	78.87	39.38^{b}
T5	91.26	82.64	42.35^{b}
T6	91.23	78.48	36.50^{ab}

$^{a, b}$ Means with different superscripts within a column are significantly different (P<0.05).

3.2. The effect of glutathione level on Intact Plasma Membrane (IPM)

Based on Table 2, it can be seen that the average percentage of intact plasma membrane of Local Ram sperm post thawed, sorted from the highest to the lowest came from level 5 mM, 4 mM, 6 mM, 3 mM, and 0 mM. The result of ANOVA showed that the treatment of glutathione level had significant effect ($P<0.05$) on the integrity of plasma membrane of Local Ram sperm post thawed. Duncan's multiple-range test results showed that 5 mM was significantly higher ($P<0.05$) than 0 mM, 3 mM, but not significantly different with level of 4 mM and 6 mM. Giving 0 mM and 3 mM did not give significantly different results ($P>0.05$), the same result was also seen in the administration of 4 mM and 6 mM.

The results showed an increase in percentage of intact plasma membranes from 0 mM to 5 mM glutathione level, and 6 mM glutathione level was decreased but still above the average of 0 mM treatment, this is because increased levels of glutathione will keep and increase the morphology of sperm but in excess amount can have a negative effect. Provision of 5 mM glutathione markedly gives a higher percentage of intact plasma membrane than the other treatments, at this level glutathione is able to optimally bind hydroxyl-free radicals which are highly reactive and cause lipid peroxidation in cell plasma membranes so that the integrity of the plasma membrane can be maintained and the percentage which is obtained quite high, in addition to glutathione can also stabilize the membrane structure by eliminating or minimizing the formation of acyl peroxide in lipid peroxidation reactions [17]. At optimum levels, lipid peroxidation can be prevented because all the free radicals formed are bound and neutralized by glutathione, thus suppressing the oxidation of unsaturated fatty acids in the plasma membrane and the integrity of the plasma membrane can be maintained [19].

The addition of 0 mM, 3 mM, and 4 mM glutathione levels does not provide optimal results, this may be because glutathione levels have not been able to prevent the onset of lipid peroxidation reactions and may also be due to unbalanced glutathione available with the amount of free radicals produced, so that free radicals that form easily attack unsaturated fatty acids and lipid peroxidation occurs which ultimately damage the plasma membrane so that it is not intact anymore [17]. Lipid peroxidation is a very active process causing damage to the plasma membrane caused by loss of essential fatty acid membrane phospholipid constituents due to the bonding electron hydrogen unsaturated fatty acid constituent of plasma membrane. The addition of too little glutathione has not been able to capture all the free radicals produced, so some of the free radicals will bind to unsaturated fatty acids that ultimately damage the plasma membrane [20].

Result of this research showed that giving glutathione 6 mM showed a decrease in the percentage of intact plasma membranes. This decrease can be due to the addition of 6 mM glutathione is considered too excessive and cause negative effects for the sperm itself, this is in accordance with the opinion that the administration of too high level of glutathione will cause toxic and even caused death [21]. Inappropriate use of glutathione levels results in an increase in reactive oxygen species (ROS) during in vitro cultures that can cause sperm cell damage [22]. The addition of excessive antioxidants will further increase the osmotic pressure of the diluent solution and is less well adapted by spermatozoa, thus adversely affecting the ongoing process of sperm metabolism and can result in death [17].

Sperm plasma membranes are composed of proteins and phospholipids. The high ratio between unsaturated and saturated fatty acids in phospholpids and low cholesterol component causes the plasma membrane to have a high unstable hydrocarbon and thus become highly susceptible to lipid peroxidation. The presence of lipid peroxidation can damage the plasma membrane. The integrity of the plasma membrane is one of the most important microscopic evaluations of sperm to be performed because the plasma membrane serves to protect the organelles in it, able to regulate the substrate and electron traffic required for the biochemical processes of metabolism, so that metabolism can proceed well.

Sperm with intact plasma membrane is characterized by a circular or swollen tail due to its ability to retain fluid osmolarity in the cell, whereas sperm with incomplete plasma membrane is characterized by a straight tail due to inability to withstand the fluid entering the cell [23]. Sperm with intact plasma membranes are able to regulate incoming and outgoing traffic from all cell substrates and electrolytes needed in metabolic processes, so that metabolism can take place well.

Table 2. Effect of Glutathione Level on Intact Plasma Membrane

Treatment	Fresh semen	Post-Equilibration	Post-Thawed
	.... %		
T0	92.95	75.80	43.00^{a}
T3	92.65	80.05	45.05^{a}
T4	92.95	82.15	47.75^{b}
T5	93.05	84.95	50.95^{b}
T6	92.95	82.10	45.80^{ab}

$^{a, b}$ Means with different superscripts within a column are significantly different (P<0.05).

The freezing process of semen causes the sperm to be in much contact with the oxygen-rich outer air. This will speed up metabolism resulting in reactive oxygen species (ROS), ROS will attack sperm that rich in unsaturated fatty acids, this ROS attack can cause lipid peroxidation reactions that cause damage of sperm cells plasma membrane. The sperm plasma membrane damage will disrupt the entire biochemical process within the cell that will eventually lead to decreased motility, increased morphological damage and even cell death itself [21]. The reaction between free radicals and lipids, especially the unsaturated fatty acids that predominantly constitute cell plasma membranes, will lead to lipid peroxidation which may cause changes in spermatozoa plasma membrane composition [17].

3.3. The effect of glutathione level on Abnormality

In addition to freezing activity, secondary abnormalities may also occur due to lipid peroxidation [15]. Reactive oxygen compounds attack the sperm plasma membrane and damage their structures, resulting in high levels of oxidative stress in which the amount of reactive oxygen compounds more than the existing antioxidants, will increase the secondary abnormalities of sperm. The sperm can not defend itself by the attack and lose its cell integrity. This happens because the process of freezing and thawing sperm is an oxidative process in which in each process increases the level of reactive oxygen compounds that caused damage to sperm plasma membrane due to lipid peroxidation [24].

The result of ANOVA showed that glutathione level had significant (P<0.05) effect on abnormality. Result of Duncant test showed that 5 mM and 6 mM significantly lower than 0 mM and 3 mM but not different with 4 mM. Treatment 0 mM, 3 mM and 4 mM were not significantly different. Other researcher reported that the effect of glutathione on post thawed semen on acrosome integrity is still good at both 2 mM and 5 mM glutathione levels. Lipid peroxidation occurring by the attack of reactive oxygen to the plasma membrane causes damage to the integrity of the membrane, but has not yet increased the form of abnormality[15]. Lipid peroxidation is a reaction that causes structural damage and disruption of sperm metabolism resulting in sperm death. It is estimated that after the structure of the plasma membrane is damaged, the sperm cells have no longer ability to live, but the cell morphology is intact [16]. Result of this study also showed that percentage of sperm abnormality from all of treatment still in the normal value (< 20%).

3.4. The effect of glutathione level on Recovery Rate

Recovery Rate is the ability to restore sperma after freezing. Calculation of recovery rate is important to determine the percentage of sperm that can be recovered after freezing. It showed the success of the freezing process itself. This result showed that the average percentage of recovery rate were 37,97% (T0), 39,77% (T3), 43,06% (T4), 46,47% (T5), and 40,05% (T5). Result of ANOVA showed that glutathione level treatment had significant effect (P <0,05) on the recovery rate of local ram sperm post thawed. The results of Duncan's multiple-range test analysis showed that 5 mM significantly (P <0.05) higher compared with 0 mM, 3 mM, 4 mM and 6 mM, whereas between 0 mM, 3 mM, 4 mM and 6 mM were not significantly different.

Table 3. Effect of Glutathione Level on Sperm Abnormality

Treatment	Fresh Semen	Post-Equilibration	Post- Thawed
	.... %		
T0	0.6	1.75	2.65^a
T3	0.6	1.35	2.35^a
T4	0.6	1.25	2.20^{ab}
T5	0.6	1.00	1.70^b
T6	0.6	1.40	2.20^b

$^{a,\,b}$ Means with different superscripts within a column are significantly different (P<0.05).

Glutathione can increase motility, but the administration of glutathione in too high level will lead to toxic until death. This is evident from the research results obtained that is an increase in recovery rate from the level 0 mM of glutathione to 5 mM, whereas at 6 mM of glutathione level there is decrease again but still above the average of 0 mM treatment.

Provision of 5 mM glutathione showed a real recovery rate higher than other treatments with sperm recovered from freezing capacity of 46.77 ± 4.99%. The value of post thawed motility obtained from each treatment is related to the integrity of the plasma membrane. Provision of 5 mM glutathione is able to protect cells from damage caused by ROS attacks and able to neutralize free radical, prevent lipid peroxidation reactions, so that the integrity of the plasma membrane can be maintained. Sperm with intact plasma membranes capable of remodeling ATP in mitochondrial envelope activated by certain enzymes and produce energy used for sperm action, then motility can be maintainedand the value of recovery rate obtained can be greater[20].

Glutathione levels of 0 mM, 3 mM, and 4 mM have not provided optimal results, this may be because these glutathione levels have not been able to prevent the onset of lipid peroxidation reactions and also can not break the lipid peroxidation chain reaction and may also be caused by unbalanced glutathione available in the amount of free radicals produced. In cell metabolism balance between prooksidan and antioxidant is required and the balance is changed if there is an enormous increase of reactive oxygen species (ROS) production [17]. Excessive ROS can cause oxidative stress that eventually leads to lipid peroxidation reactions. This reaction can damage the plasma membrane causing the loss of enzymes needed for metabolism so that metabolism stops and energy is not produced, and then the next is permanent loss of sperm motility [20].

Giving glutathione 6 mM also has caused a decrease in motility, so the value of recovery rate obtained also decreased. This decrease can be due to the addition of 6 mM glutathione is considered too excessive or not appropriate so that the negative effects for the sperm itself. the negative effects can be caused by glutamate content in glutathione that binds lipoproteins to the yolk causing a lack of food supply for spermatozoa, resulting in reduced energy availability and ultimately causing death in spermatozoa, in addition to large amounts of antioxidant compounds can increase the osmotic pressure of diluent solution and less adaptable to spermatozoa so badly affect the ongoing process of spermatozoa metabolism [17].

4. Conclusion

Based on the results of research can be concluded that glutathione levels affect the quality of local ram semen and level of 5 mM glutathione in yolk tris diluent produces the best semen quality.

Acknowledgments

This research was supported by research grant "Penelitian Unggulan Perguruan Tinggi" (PUPT) from Ministry of Research, Technology and High Education Republic of Indonesia.

References

[1] Peraturan Menteri Pertanian. 2006. Sistem Perbibitan Ternak Nasional. www.bphn.go.id. (acccessed on 13 Oktober 2015, 8.15 PM).

[2] Solihati, N., S.D. Rasad, R.Setiawan, C. Alvionita. 2016. Quality and Viability of Javanese Local Ram Semen at Different Age. Proceeding International Seminar On Livestock Production and Veterinary Technology. 265-270.

[3] Solihati, N., S.D. Rasad, R. Setiawan, S. Nurjanah. 2016. Post-Thawed Semen Quality of West Java Local Ram at Different Level of Glycerol. Proceeding The 3rd Animal Production International Seminar and The 3rd ASEAN Regional Conference on Animal Production Enhancing Synergistic Roles of Stakeholders for Development of Sustainable Livestock Production 3rd APIS & 3rd ARCAP. 249-251.

[4] Rizal, M., M.R. Toelihere, T.L. Yusuf, B. Purwantara dan P. Situmorang. 2003. Kriopreservasi semen Domba Garut dalam pengencer Tris dengan konsentrasi laktosa yang berbeda. Media Kedokteran Hewan. 19 (2) : 79-83.

[5] Irvine DS. 1996. Gluthatione as a treatment for male infertility. Rev Reprod;1:6–12.

[6] Gazali, M dan S. N. Tambing. 2002. Kriopreservasi Sel Spermatozoa. ISSN 0854-8587, 9(1).

[7] White IG. Lipids and calcium uptake of sperm in relation to cold shock and preservation: a review. ReprodFert Dev 1993;5:639–58.

[8] Aitken RJ, Buckingham D, Harkiss D. Use of a xanthine oxidase free radical generating system toinvestigate the cytotoxic effects of reactive oxygen species on human spermatozoa. J Reprod Fertil 1993;**97**:441.

[9] Holt WV. 2000. Basic aspects of frozen storage of semen. Anim. Reprod. Sci. Aug. **18**;62(1-3):3-22.

[10] Bilodeau J. F., S. Blanchette, I. C. Gagnon, M. A. Sirad. 2001. Thiols Prevent H_2O_2-mediated Loss Sperm Motility in Cryopreserved Bull Semen. Theriogenology. **56**: 275.

[11] Raijmakers MT, Roelofs HM, Steegers EA, Steegers-Theunissen RRP, Mulder TP, Knapen MF, et al.2003. Glutathione and glutathione S-transferases A1-1 and P1-1 in seminal plasma may play a role in protectingagainst oxidative damage to spermatozoa. Fertil Steril;79:169–72.

[12] Tariq, M., M. S. Khan, M. G. Shah, A. R. Nisha, M. Umer, S. M. Hasan, A.Rahman, dan I.

Rabbani. 2015. Exogenous Antioxidant Inclusion During Semen Cryopreservation of Farm Animals. Journal of Chemical and Pharmaceutical Research. 7(3): 2273-2280, ISSN 0975-7384. 2274.

[13] Solouma, G. M. A. 2013. The Influence of Adding Glutathione on Semen Characteristics of Sohagi Rams. Egyptian Journal of Sheep and Goat Sciences. **8(2)**: 61.

[14] Amalia, F. R., Suyadi, dan A. Rachmawati. 2013. Pengaruh Glutathione terhadap Kualitas Semen Kambing Boer Post Thawing dalam Pengencer yang Mengandung Dimetylsulfoxide (DMSO). Universitas Brawijaya. Malang

[15] Silva, S. V., A. T. Soares, A. M. Batista, F. C. Almeida, J. F. Nunes, C. A.Peixoto, dan M. M. Guerra. 2011. In Vitro and In Vivo Evaluation of Ram Sperm Frozen in Tris Egg-Yolk and Supplemented with Superoxide Dismutase and Reduced Glutathione. Reproduction Domestic Animal.

[16] Hartono, M. 2008. Optimalisasi Penambahan Vitamin E dalam Pengencer Sitrat Kuning Telur untuk Mempertahankan Kualitas Semen Kambing Boer. J.Indon.Trop.Anim.Agric. **33**(1): 11.

[17] Gunawan I, D. N. D. I. Laksmi, dan I. G. N. B. Trilaksana. 2012. Efektivitas Penambahan B-Karoten dan Glutathione pada Bahan Pengencer terhadap Motilitas dan Daya Hidup Spermatozoa pada Semen Beku Sapi. Indonesia Medicus Veterinus 2012 **1**(3): 385.

[18] Sinha, M. P., A. K. Sinha, B. K. Singh, dan R. L. Prusad. 1996. The Effect of Glutathione on the Motility, Enzyme Leakage and Fertility of Frozen Goat Semen. Anim.Reprod.Sci. **41**: 237.

[19] Rizal dan Herdis. 2010. Peranan Antioksidan dalam Meningkatkan Kualitas Semen Beku. Wartazoa, Vol. 20 No. 3 th 2010.

[20] Werdhany, W.I. 1999. Efektivitas Penambahan α- Tokoferol di dalam Pengencer Tris dan Susu Skim terhadap Kualitas Semen Beku Kaming Peranakan Etawah.Thesis. Institut Pertanian Bogor. Bogor.59 -77

[21] Triwulaningsih E., P. Situmorang, T. Sugiarti, R. G. Sianturi, dan D. A.Kusumaningrum. 2003. Pengaruh Penambahan Glutathione pada MediumPengencer Sperma terhadap Kualitas Semen Cair (Chilled Semen). JITV.8(2).

[22] Syarifuddin, A., Desak, N. D. I. L dan Wayan, B. 2012. Efektivitas Penambahan Berbagai Konsentrasi Glutathione terhadap Daya Hidup dan Motilitas Spermatozoa Sapi Bali Post Thawing. Indonesia Medicus Veterinus. ISSN : 2301-7848: 173 – 185.

[23] Arsiwan, Saili, T., Baa, L.O., dan rahadi, D. 2014. Membran Plasma Utuh Spermatozoa Epididimis Kambing Peranakan Ettawa dalam Natrium Klorida dengan Konsentrasi Berbeda. JITRO **1** (1) : 83.

[24] Ansari, M. S. 2011. Antioxidant Fortification of Semen Extender to Improve Freezability and Fertility of Buffalo Bull Spermatozoa. Thesis Pir Mehr Ali Shah Arid University Rawalpindhi. Pakistan. 80-81.

IOP Conf. Series: Earth and Environmental Science **119** (2018) 012035 doi:10.1088/1755-1315/119/1/012035

Effect of Fermented Cocoa Pod Husk Level in Concentrate on Etawah Grade Semen Quality

F.F. Munier[1*]

[1]Assessment Institute of Agriculture Technology, Central Kalimantan, G. Obos Street Palangka Raya City, Central Kalimantan

*Corresponding E-mail: ffmunier@yahoo.com

Abstract

The purpose of study was to determine the effect of fermented CPH supplementation level in concentrate toward semen quality of Etawah Grade Buck (EGB). This research was conducted from August to October, 2011 at experiment pen and Feeds Science Laboratory, Faculty of Animal Science, Universitas Gadjah, Yogyakarta. Semen was collected from 4 heads of EGB with about 2 years old and weight approximately of 24 kg. Concentrate was main material of CPH fermentation with composition P0 0, P1 20%, P2 40% and P3 60% from total of ration. The composer was used by the *Aspergillus niger* BPT with density 9.1 x 10^7 of colony/g as many as 1% of CPH material (on dry matter). Semen collection was done every 1 time/week and to be done five times semen collection. An examination was consisting of volume, color, smell, consistency, and the pH. Meanwhile, for microscopic examination was motion of mass, individual, motility, viability of spermatozoa. pH of P0, P1, P2 and P3 were 6.95, 7.03, 7.50 and 7.79, respectively. Mass movement of P0, P1, P2 and P3 were +++, +++, ++ and ++, while viability of P0, P1, P2 and P3 were 79.57%, 79.47%, 59.97% and 51.13%, respectively. It was concluded that addition of CPH fermentation in concentrates up to 20% for of EGB has not affect to feed consumption but it has affect to decreasing of semen quality in both of microscopic and
microscopic.

1. Introduction

The low birth rate goat is usually caused malnutrition especially on ranch system. The malnutrition pregnant goat is giving birth to kid be weak and milk less available. This condition is generally causing death for kid. One of the objectives of this program was acceleration of increases in population through improving the reproductive system of goat. The goat farming is expected to make contributions of meat self-sufficient program because goat population in Indonesia is quite high. National data for goat population in Indonesia on 2016 about 19.608.181 heads, which it is the highest population in Central Java reached 4,140,130 heads and followed by East Java about 3,267,954 heads. While the population highest in the Kalimantan Island is North Kalimantan reached 160,230 heads and followed by East Kalimantan about 63,114 heads [1].

Increasing in population of the goat is through improving the reproductive system particularly the semen quality of semen. The good quality of semen has high fertilization when he has mating to ewe with high of pragnancy. One of the factors can affect on the goat semen quality with giving feed quality. The preparation of feed quality can use an agricultural waste that it is abundant available during harvest season. The cocoa pod husk (CPH) is an alternative as main matter for ruminants concentrate. The CPH is always available on farmers cocoa plantation wich it is can be harvested throughout the year [2].

The benefits value of CPH as a fodder can be increased by reducing of faction fiber like hemicellulose, cellulose and lignin [3]. CPH has low crude protein content but it has high crude fiber content about 9.14 and 35.74%, respectively, while content of faction fibers like neutral ditergent fiber (NDF) and acid ditergent fiber (ADF) are 58.78% and 47.04%, respectively [4], hemicellulose, cellulose and lignin are 6.01%, 35.33% and 38.78%, respevtively [5].

The decline of CPH fiber content faction is fermentation methods using microorganism in both of aerobic or anaerobic conditions. Feeding of concentrates to Etawah Grade Buck (EGB) can improve the semen quality. The semen quality is determined by the ejaculate volume, color, pH, consistency, viability, motility abnormality and concentration of spermatozoa [6]. The purpose of study was to determine the effect of fermented CPH supplementation level in concentrate toward semen quality of Etawah Grade Buck (EGB).

2. Materials and Methods

This research was conducted at pen experiment and Feeds Science Laboratory, The Faculty of Animal Science, Universitas Gadjah, Yogyakarta. Equipments were used in this research such as artificial vagina (minitub), microscope, object glasses, glass cover, test tube, pipet, neubauer haemocytometer improved, tissue paper, universal of pH indicator (merck), hand counter, refrigerator, ice container, thermometer, hummer mill. Materials were used EGB semen, sterile water warm, vaseline, eosin, physiological NaCl 0.3 %, alcohol 70 %, feed, vitamins and drugs.

Semen was collected from 4 heads of EGB with about 2 years old and weight approximately of 24 kg. The health reproductive condition of animals should be normal, before feeding trials were given worm drug. Concentrates were given from the CPH with a measure that was enumerated 1x5 cm, which it was fermented in aerobic condition. The composer was used by the *Aspergillus niger* BPT with density 9.1 x 10^7 of colony/g as many as 1% of CPH material (on dry matter). Concentrate was main material of Fermentation CPH with composition 0, 20% (124 g), 40% (287 g) and 60% 482 g) from tatal ration (Table 1).

The CPH was fermented in seven days. Drying of CPH after fermentation process was 3-5 days depend on irradiating the sun. CPH dry was grounded using hummer mill by the size of screen in 5 mm. Feed was given to animals in form of fresh forage for King grass (*Pennisetum hybrid*) and concentrate which it was the portion as follows:

Table 1. The Portion Feeding Trials

Feed materials	Dietary treatments			
	P_0	P_1	P_2	P_3
King grass				
(*Pennisetum hybrid*) (g)	3,350	2,605	1,860	1,116
Groats soybean (g)	90	26	13	6
Fermented CPH (g)	-	124	297	482

The total of concentrates protein in concentrate was 12%. Feed was given every day to animals of 4.0% from live weight on dry matter basis [7]. Concentrates were enriched by the addition of 2.0% minerals and 0.5% salt from total of concentrates material. Feeding trials were done twice a day, morning at 08.00 AM and 03.00 pm, started feeding of concentrates and to be continued by King grass chopping. Adaptation period of feed trial was done 15 days before started of the research. Water drinking was given unlimited (*ad-libitum*).

The semen collection was done using artificial vagina already cleaned, sterile and dried. Warm water with the temperature 45° C incorporated into the artificial vagina, then in the interior of the vagina artificial mouth was lubricated by vasseline. Teaser prepared was used to increasing the libido and get the good quality and tobe done false mounting [8]. Semen collection was done every 1 time/week and to be done five times semen collection. Semen collection was done started at 09.00 AM after feeding. The frequency of collection was three times for each individual treatment. Semen was collected to be sheltered in tube inside and put it in a ice container and taken to laboratory for next analysis. Variable was observed in microscopic examination. An examination was consisting of volume, color, smell, consistency, and the pH. Meanwhile, for microscopic examination was motion of masses, individual, motility, viability of spermatozoa according to procedures of [9].

Motility of spermatozoa or impulse spermatozoa was determined in the mass movement and the movement of individuals. Based on the results of the assessment of the mass movement of semen quality can be determined as follows: a) excellent (+++), which it is looks great waves, thick, dark, and many, like wisps of active black clouds move fast moving place; b) good (++), when visible waves of small, thin, sparse, less clear, and moving slowly; c) fair (+), if not visible waves but rather just individual movements active progressive; d) (N, necrospermia or 0) if only little or no movement of the individual.

Semen concentration was the number of spermatozoa per ml of semen. Directly calculation with hemocytometer that was by using a method of neubauer calculating room. The concentration of spermatozoa in the five chambered calculated according to a diagonal direction using a microscope with an enlargement 40x10. The formula to calculate the concentration of spermatozoa:

$$Y \times \frac{400}{80} \times \frac{200}{0.1} = Y \times 10{,}000 \text{ mm}^3 = Y \times 10{,}000{,}000/\text{ml}$$

Y	= Totally of observation spermatozoa in 5 big box
400	= Totally of small box in calculation room
80	= Totally of small box in 5 big box
200	= Dilution of 200 times
0.1	= Volume of calculation box (mm^3)

The percentage of life spermatozoa was a drop of a substance with eosin in one glass object sterile and one small drops sperm added and mixed evenly, to be observed under a microscope with 40x10 enlargement. Calculation of 100 to 200 semen cell and the best 500 semen cells [8] using hand counter. The life spermatozoa head was not absorb color but it was dead absorb a red color after that the percentage of the life spermatozoa to be determined.

The percentage of spermatozoa abnormally was after counting of spermatozoa concentration and this semen was used to calculate the percentage semen abnomally. Spermatozoa was calculated up to 200 cells on one or several views under the microscope 40x10 with enlargement, then to be determined the percentage of spermatozoa abnormally.

3. Results and Discussion

3.1. Feed Consumption

The total of feed consumption in the first day and eighth day of the feeding trial is presented in Table 2.

Table 2. Feed consumption

	Dietary treatments			
	P_0	P_1	P_2	P_3
Feed cunsumed* (daily):				
King grass (g)	2,727.47	2,366.45	1,654.90	1,116.00
Groats soybean (g)	90.00	26.00	13.00	6.00
Fermented CPH (g)	-	61.90	50.38	141.01
Feed cunsumed** (daily):				
King grass (g)				
Groats soybean (g)	3,069.66	2,538.79	1,825,91	1,116.00
Fermented CPH (g)	90.00	26.00	13.00	6.00
	-	80.15	197.71	298.39

* Feed collected in first day of feeding trial.
**Feed collected in eighth day of feeding trial.

Feeding trial of P0 as a comparison that is not granted an additional CPH fermentation in concentrate so that the portion of the King grass more as basal feed and fiber source. While the P1, P2 and P3 a bit more given the grass King because given additive of CPH fermentation as fiber source. The addition of groats soybeans were all treatments as the main protein source in the concentrate. According to [10, 11] groats soybeans contain crude protein reached 41.3-41.7%. The addition of groats soybeans in rations on this research can fulfill the standards of the crude protein needs of animal growing about 12%.

The high feed consumption on P0 was caused by feeding only from King grass and soybean groats. The results of observations during the research, soybean groats it was a favorite goat so that it can be spent in a short period of time that is around 5-10 minutes every time of feeding. The factors affect the feed consumption is the very nature of feed such as particles size, texture and scent. According to [12] that scent was an important factor of feed material. The animal can refused material feed given without feel because it was not fond of the scent. The other factors that were also affect of feed consumption like body sizes of animal, increasing feed consumption in line with increasing the size of animal [13].

Groats soybeans provided can be spent starting the first day until the eighth day collection of fodder though mixed with CPH fermentation, animal first was eat more groats soybeans and to be continued eating of CPH fermentation. This case is caused by the goats have selective properties in consuming the feed through the senses of smell so it takes longer for the adaptation of the new type feed is given. The selective nature of the goat in consuming the feed associated with the tendency of choosing a feed that is no fibrous. According to [14] that was the goats have the ability to choose the type of feed to be consumed and preferred types of feed without fibrous.

3.2. Characteristic of Semen Macroscopic

The test semen quality under macroscopic examination such as semen of volume, color, smell, consistency, and the degree of the acidity or pH. The result of the observation of macroscopic in of the characteristics of fresh EGB semen in this research is still categories good and in the normal range of especially on P0 and P1, while P2 and P3 relatively less than the normal range of such as are presented in Table 3.

Table 3. Average of macroscopic characteristic of EGB semen

Semen characteristic	Dietary treatments			
	P_0	P_1	P_2	P_3
Volume (ml)	0.90	0.87	0.43	0.37
Color	cream-yellowish	cream-yellowish	cream-yellowish	yellowish
Odor	specific	specific	specific	specific
Consistency	moderately viscous-viscous	moderately viscous-viscous	dilute-moderately viscous	dilute
pH	6.95	7.03	7.50	7.79

Analyzed in the Reproduction laboratory, Animal Science Faculty, UGM

The semen volume of EGB on P0 and P1 were 0.90 ml and 0.87 ml, respectively, these volume are still in the normal range, relatively the same with the volume of EGB semen was reported by [15] with average 0.95 ml. The volume of goat semen has varies every collecting with range of 0.5-1.5 ml [16]. Semen color is still in the normal range which it is creamy to yellowish on all feed treatment, but if the semen color is white to pale and it would affect on other semen characteristics. The semen consistency for P0 and P1 had moderately viscous to viscous, while P3 the semen quality was relatively low marked with dilute consistency. Color, consistency and concentration have links to each others. The semen color more faded, and the spermatozoa concentration become decrease and to be followed by more dilute of semen [17].

Semen pH of P0 and P1 were 6.95 and 7.03, respectively, semen pH in this research is still in the normal range to the EGB. According to [18], good quality of semen had a ph with a little acid from 7.0 and in average of 6.7. Semen pH in this research is closed to previous studies of semen pH for EGB. The results of previous studies reported that pH ranges about 6.85-7.13 for semen EGB [19, 15]. P2 and P3 had pH above the range of goat pH in both of 7.50 and 7.79 which is leads to the alkalis. The reaction of alkalis in spermatozoa was often associated with the low quality and low fertility [20].

The low semen quality in macroscopic in P2 and P3 had relation with feed consumption because during feed consumption was not optimal especially feed concentrates which was the more portion of CPH fermentation. P1 had low portion of CPH fermentation and higher portion of soybean groats in concentrate, while P0 had not give CPH fermentation and given higher portion of soybean groats. Groats soybean is a source of high protein that affects on the formation of the semen quality. The analysis of chemical composition was groats soybean having high crude protein content reached 33.07%, to be supported by organic matter 89.93% and crude fat 7.72%. The chemical composition of fresh semen plasma had high protein content about 214 mg/100 ml [21] or 4.59% as component in semen plasm.

3.3. Characteristic of Semen Microscopic

Microscopic examination of goat semen is mass motions, individual motions, concentration, motility, viabilitas and abnormality spermatozoa. Table 4 is show average of characteristic for EGB semen microscopic.

Table 4. Average of microscopic characteristic of EGB semen

Semen characteristic	Dietary treatments			
	P_0	P_1	P_2	P_3
Mass movement	+++	+++	++	++
Motility (%)	78.33	76.67	51.67	46.67
Concentration (ml)	2.40 x 10^9	2.24 x 10^9	1.01 x 10^9	0.96 x 10^9
Viability (%)	79.57	79.47	59.97	51.13
Abnormality (%)	5.59	5.93	10.63	16.77

Analyzed in the Reproduction laboratory, Animal Science Faculty, UGM, 2011

The mass movement of EGB semen of P0 about +++ and P1 +++, these treatment were includes the categories excellent marked with visible large waves, thick, dark, and many, like wisps of active black clouds move fast moving place. It is supported with the high motility of the sperm in P0 and P1 was 78.33% and 76.67%, this condition indicates that the individual movement of sperm is quite progressive. It is powered by a low sperm abnormality in P0, P1 and P2 about 5% 5.93% and 10.63%, respectively, but P3 had more over from standard recomendation reached 16.77%. The semen concentration in all the feed treatments was still in the normal range of 0.96-2.40 x 106/ml, but it is still lower than the research results [15], 2.94 x 106/ml. [22] recommend that the terms of semen deluted that it may had mass movement, individual movement more over 65%, abnormality spermatozoa in range 14-15%. While [8] reported that the minimum standards for the semen quality which can be used for artificial insemination (AI) was the minimum to contain 500 million cells/ml/mass movement/ejakulate, as well as 50% in borh of semen viability and motility. The characteristics of EGB semen on this research especially P0 P1 and P2 can be used to the AI program.

4. Conclusion

Based on the results of this research it can be concluded up as follows:

1. The addition of CPH fermentation in concentrates up to 20% for of EGB has not affect to the feed consumption.
2. The addition of CPH fermentation in concentrates up to 20% for of EGB has affect to decreasing of the semen quality in both of microscopic and microscopic.

Acknowledgements

The authors gratefully acknowledge to Mr. Wartono and Isnawan as Staff in Reproduction Laboratory, Animal Science Faculty, Universitas Gadjah Mada Yogyakarta for their colaboration and helpful during sperm collection and sperm quality analysis in Reproduction Laboratory.

References

[1]. Central Bureau of Statistics. The Goat Population for Province, 2009-2016.

[2]. Prabowo, A., E. Basri, Firdausil AB., B. Sudaryanto dan S. Bahri. 2004. Assessment of goat farming system in the community cocoa plantation in Lampung. Proceeding of National Seminar for Crop-Animal Systems Integration. Indonesian Center for Animal Research and Development in Colaboration with Assessment Institute for Agriculture Technology, Bali, Crop-Animal Systems Research Network (CASREN). pp. 375-380.

[3]. Munier, F.F., M. Takdir, M. Dewi, and Soeharsono. 2014. Body weight gain of local beef cattle fed on native grass supplemented with fermented cocoa pod husks. Proceedings of the 16th AAAP Animal Science Congress Vol. II 10-14 November 2014, Gadjah Mada University, Yogyakarta, Indonesia. pp. 255-258.

[4]. Alemawor, F., V.P. Dzogbefial, E.O.K. Oddoye and J.H. Oldham. 2009. Effect of *Pleurotus ostreatus* fermentation on cocoa pod husk composition: Influence of fermentation period and Mn^{2+} supplementation on the permentation process. African J. Biotechnol. **8** (9): 1950.

[5]. Laconi, E.B. 1998. Inproving of Pod Cocoa Husk Quality through Amoniation with urea and biofermentation use *Phanerochaete chrysosporium* and description into ruminants ration formulation. Dissertation. Graduate Program, Bogor Agriculture University, Bogor.

[6]. Husin, N., T. Suteky and Kususiyah. 2007. Quality test towards semen of Nubian Goat and its crossing (Nubian Goat X PE) and Boer Goat based on the storage time. Jurnal Sain Peternakan Indonesia. Vol. **2** (2): 57.

[7]. National Research Council. 1981. Nutrient Requiraments of Domestic Animal; Nutrient Requiraments of Goats. Phisiology of Digestion No. 15. National Academy of Science, Washington DC., USA.

[8]. Toelihere, M.R. 1981. Animal Reproduction Physiology. Angkasa Press, Bandung.

[9]. Hardijanto, T. Sarjito, T. Hernawati, T. Susilowati dan T.W. Suprayogi. 2008. Laboratory Prosedure in Fisiology and Reproduction Technology (AI). Veterinary Faculty. Airlangga Surabaya University. pp. 8-18.

[10]. Hartadi, H., S. Reksohadiprodjo dan A.D. Tillman. 2005. Table of Feed Composition for Indonesia. Fifth Edition. Gadjah Mada University Press, Yogyakarta.

[11]. Cordesse, R. 1990. Value of chickpea as animal feed. CIHEAM-Options Méditerranéennes. Série Séminaeres. 9: 127-131.

[12]. Preston, T.R. and R.A. Leng. 1987. Matching Ruminant Production Systems with Available Resources in The Tropic and Sub-Tropics. Penambul Books, Armidale, Australia.

[13]. Zain, M. 2009. Subtitution of natural grass with ammonition of cocoa pod husk in local sheep ration. Media Peternakan. **32** (1): 47.

[14]. Huston, J.E. 1998. Goats and Goat Nutrition. In: Livestock Feeds and Feeding. 4th Ed. Prentice-Hall, Inc. Simon and Schuster/A Viacom Company Upper Saddle River, New Jersey, USA. pp. 335-353.

[15]. Tambing, S.N, M. Gazali. dan B. Purwantara. 2001. Empowerment of Artificial Insemination Technology for Goat. Wartazoa. Vol.**11** (1): 1.

[16]. Devendra, C. dan M. Burns. 1994. The Production in The Tropics. Penerbit ITB Press, Bandung.

[17]. Mahmilia, F., M. Doloksaribu dan F.F. Pamungkas. 2006. Karakteristik semen kambing Boer. Prosiding Seminar Nasional Teknologi Peternakan dan Veteriner, Bogor 5-6 September 2006. Puslitbangnak, Bogor. Hal. 533-536.

[18]. Bearden, H.J. and J.W. Fuquay. 1984. Applied Animal Reproduction. 2nd Edition. Reston Publishing Company Inc. A Prentice Hall Company, Reston Virginia.

[19]. Tambing. S. N., Toelihere. M.R., Yusuf. T.L., dan I.K. Sutama. 2000. Effect of gliserol in tris deluent toward frozen semen quality of Etawah Great. J. Anim. Sci. Vet. Vol. **5** (2): 1.

[20]. Perry, E.J. 1969. The Artificial Insemination of Farm Animals. Oxford & IBH Publishing Co. Calcuta, Bombay, New Delhi.

[21]. Tambing. S. N., M.R. Toelihere, T.L. Yusuf, B. Purwantara, I.K. Sutama dan P.Z. Sitomorang. 2003. Effect of ejaculation frequency toward fresh semen characteristics and estrus ability of Saanen goat. J. Sain Vet. Vol. XXI (**2**): 57,

[22]. Hafez, E.S.E. 1987. Reproduction In Farm Animal. 4th Edition. Lea and Febiger. Philadelfia, USA.

The Different Calcium^{+2} Intensity Profile and Quality of Oocyte and Goat Sperms after Cryopreservation

G. Ciptadi[1]*, S. Rahayu[2], Fatchiyah[2], S. Wahyuningsih[1], A. Budiarto[1], M. Nasich[1], A.R.I. Putri[2], M. Mudawamah [3] and M.N. Ihsan[1]

[1]Faculty of Animal Husbandry, UB, Indonesia,
[2]Faculty of MIPA-Biology UB,
[3] Faculty of Animal Husbandry, UNISMA, Malang

*Corresponding E-mail: ciptadi@ub.ac.id

Abstract

This research aims were to study the effect of the oocyte and sperms cryopreservation of Indonesian local goat on the post-thawing quality and profile or characters of Calcium^{+2} intensity in relating with their fertility capacity. A study was conducted to test the freezing method and post-thawing viability both stock cells stored in the deep freezer and liquid nitrogen (-80°C of vs -196°C). A fertility test of sperms has been conducted through in vitro of sperm quality, while the oocytes cryopreserved test was done by in vitro maturation (IVM) rate (%). The profile of Calcium 2+ was performed and analysis by Confocal Laser Scanned Microscope (CLSM). The result showed that IVM rate of goat oocyte is considered lower when cryopreserved in –80°C than in –196°C. Meanwhile, sperm is considered having a good quality in 2 methods of cryopreservation with post-thawing motility > 40 % (SNI 2014). There is an important difference between Calcium intensity of fresh and post-thawing both for oocyte and spermatozoa. Calcium +2 profiles is varied individually on the peak of intensity, but it considered expressed the same profile of each fresh and post-thawing cell. In vitro fertilization test need to be performed to complete the viability and fertility competence of these sperm and oocyte freezing stocks.

1. Introduction

The induction of Ca^{+2} releases in fertilized M-II oocyte following fertilization by sperms was explained by the sperm receptor and the sperms factor theory. Sperm factor postulated as a cytoplasmic factor in the spermatozoa has been used to parthenogenetically activate mammalian oocytes. Spermatozoa release cytosolic substances into oocyte at fertilization and it induces Ca+2 oscillations [1,2,3]. The parthenogenesis activation of oocytes is an interesting and valid tool to investigate the comparative roles of paternal and maternal genomes in controlling early embryo development.

Repetitive Ca+2 rises at fertilization are necessary to accomplish degradation of Matured Promoting Factors (MPF). Calcium oscillation in oocytes is responsible for pronuclear formation. In general, egg activation is caused by a dramatic increase in intracellular Ca^{+2} concentrations common to every animal species including mammalian [4]. Mammalian oocytes at the time of fertilization occurred exhibit a series of transient increase in intracellular calcium ion (Ca^{+2}) concentration which is prerequisite and sufficient for oocyte activation. Immediately after the sperm attaches to the oocyte, increase in intracellular Ca+2 in the oocytes are firstly seen near the site of sperm-oocyte fusion.

Cytoplasmic factor in the spermatozoa have been used to activate mammalian M-II oocytes artificially. Ciptadi *et al* [5,6], reported research was being undertaken to characterize the phenotypic variation of intra cellular crude sperm extract (CSE) for improving activation rate of the M-II oocytes.

This research was attempted to study the characters of Ca^{++} intensity of goat sperm and oocytes after freezing by nonconventional method. It is well-established that calcium plays an important role in the capacitation and then fertilization. In this study, we would like to test whether goat fresh sperm have different calcium levels intensity than sperm from freezing (post thawing fertility

2. Material and Methods

2.1. In vitro Maturation (IVM) Preparation of Oocyte

The oocytes-cumulus cell complexes were isolated from goat antral follicles with the diameter range of 2 – 6 mm, cultured in bicarbonate-buffered medium 199 supplemented with 10 % FBS, 0.1 mg sodium pyruvate [7]. The culture was carried out in 5 % CO_2 in humidified air at 38°C for 48 h. Oocytes were completely denuded by pipetting after treated with 0.01 % hyaluronidase (Sigma). Only those with a compact, non-atretic cumulus and a homogeneous cytoplasm which were previously classified as Grades A and B [7,8], were selected. Oocytes emitting the 1st polar body were selected in this experiment.

2.2. Calcium Intensity Preparation and Analysis Sperm and Oocyte

Early sperm characters of Ca^{+2} Intensity was performed by Confocal Laser Scanning Microscope (CLSM) through Fluo-3 staining and analysis descriptively. Semen prior to freezing for research has previously been quality tested by SNI with a minimum of 70% of individual motility. Each ejaculate was cryopreserved in commercial diluter according to standard procedures. For each experiment, 0.25 ml straws (40x10 6 sperm/straw) were thawed in a 37°C water bath for 60 seconds. Freezing is done with conventional modified methods with the speed (-1oC /min) with the help of Mr. Frosty system ® and cryopreserved in – 45°C using deep freezer then immersed in liquid nitrogen (LN2) at a temperature of - 196°C, to do quality testing to post-thawing.

3. Result and Discussion

3.1. Different Calcium (Ca^{+2}) Intensity of Sperm and Oocytes

The result of this research showed that different characters of the Ca^{+2} intensity of sperm and oocytes after freezing by the modified conventional method (Figure 1). A freeze on modified simple conventional local goat semen showed good results and meet start to research and development of Artificial Insemination (AI) and its implementation [9]. The intensity of intra-cellular calcium spermatozoa with CLSM showed that there was no change in the pattern of fresh semen and results of modified conventional freezing, but different in the peak of Ca^{+2} intensity [9].

The oocytes intensity of Ca^{+2} showed different profile of each. M-II oocyte with no treatment showed strong enough intensity on the area of zona pellucid [10]. Meanwhile, the oocyte exposed with CSE, Ca+2 is well expressed spread out on cytoplasm. However, the responses to CSE supplementation are varied after 24 to 48 hour of culture (Fig. 1). It was assumed that if Calcium $^{+2}$

was expressed in the cytoplasm, it means that its exhibits increase in Calcium intracellular. Other researchers [2,4] mentioned that egg exhibits increase in Ca^{+2} is indications of egg activation. Several recent research results suggest that mammalian sperm factor would be introduced into the oocyte cytoplasm through sperm-egg cytoplasmic connection, initially distributing evenly throughout the cytoplasm [11].

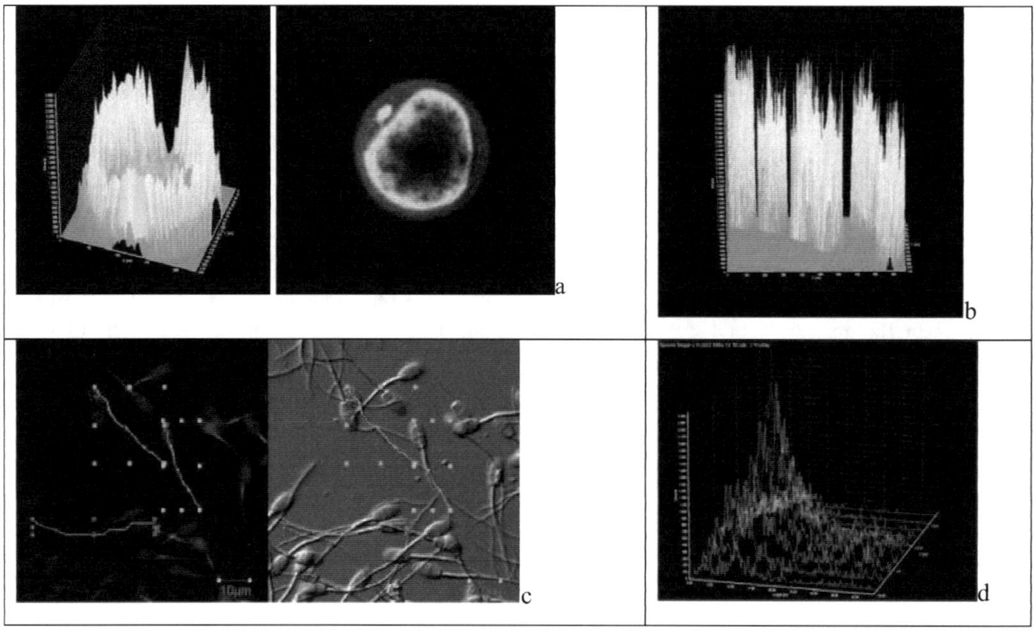

Figure 1. A comparison of Histogram profile of oocyte and sperms . A. Oocyte cultured IVM,48 hours of culture. Calcium intensity expressed mainly zona pellucida. Variation of oocyte response on different oocyte Calcium intensity [11]. B. . the intensity of intra-cellular calcium spermatozoa showed that there was no change in the pattern of fresh semen and results of modified conventional freezing, but different in the peak of Ca +2 intensity [9].

A number of experiments have attempted to study the molecular identity of sperm factor protein. Phospholipase C (PLC) zeta was identified as a strong candidate for the mammalian sperm factor[2]. The molecular weight of PLC zeta is approximately 74 kDa [12,13]. [14], confirmed that the most convincing evidence that PLC(z) is the mammalian sperm factor. Further work is required to clarify the component of sperm factor in this goat CSE.

3.2. The Ca^{+2} Profile of OOcyte activated (M-II)

In this research, we study the Ca^{+2} intensity of activated by crude sperm extract (CSE) to replace the sperm function. Activation of mammalian oocytes could be induced by a wide variety of chemical [15] and physical [16] stimulation that mimic the fertilization process by providing calcium signals to the MII arrested eggs by different mechanisms. The transfection method used in this research might be causing the lower result because of the improper method of activation Result of cleavage rate especially using CSE alone (2 %) was lower than achieved by [17] which resulted in approximately 30 % of oocyte cleaved using microinjection method of CSE (Table 1.). CSE may be a representation of the sperm factor complex that having a better influence in activation compared to CSE with 100 kD proteins specific. It is necessary that a complex of unknown composition of the active factor in CSE to be determined in next research.

The confirmation of Ca+2 intensity using fluo-3 through Confocal Laser Scanning Microscope showed the different profile of M-II oocyte activated using CSE. After 48 h activation of the oocyte, it was observed an increase in Ca^{+2} intensity only in the outer region of M-II activated by CSE alone) in contrary the increase of Ca^{+2} intracellular intensity of the oocytes was observed at all sites of M-II activated using chemical agent and CSE. Phenotype variation of Ca^{+2} intensity was varied large among slices observed (Fig.2). In mammalian eggs, Ca_2 oscillations associated with Ca_2 waves occur at fertilization and trigger egg activation [18,19]. Different cleavage rate resulted from activated M-II goat oocytes (Table 1.)

Table 1. Cleavage rate different treatment of parthenogenetic using CSE supplementation [5]

Activation method	Selected M-II Oocytes	Cleavage rate (%)
(1). Control (M-II) oocyte	40	(0.00)
(2). CSE 2.5 ug/ml	251	(36. 33)
(3). CSE of 100 kDa protein	240	(2.00)

The increase of Ca^{+2} intensity that equal to Ca^{+2} concentrations, might lead to egg activation. Oocytes activation was induced by chemical agent supplemented with CSE, with the evidence of rise Ca^{+2} intensity. This was considered to be sufficient factor for activation. [13] suggested that if a large quantity of protein with high ability for oocyte activation was extracted from spermatozoa, it might improve the rate of oocyte activation. Repetitive Ca^{2+} release from the endoplasmic reticulum is necessary for activation of mammalian eggs. Calcium influx during Ca^2 oscillations induced by injection of sperm extract into mouse eggs were investigated by Mn2 quenching of intracellular Fura-2 after adding Mn2 to external medium [20]. A qualitative image analysis of Ca^{+2} intensity done by histogram and line series. [21] reported that Calcium regu-lation by sperm from sires of low fertility appears to be deficient because their post-thaw relative intracellular cal-cium level is higher than it is in bulls of good fertility. Cryopreser-vation is known to disrupt the sperm plasma membrane and it in-duces premature capacitation of a sperm subpopulation, which may be a result of the increased internal calcium levels after thawing [22] . Figure 2 showed the variation of Ca^{+2} intensity in different slice of cells by CLSM.

Simon et al., (2000) [22] reported that sperm intracellular calcium level is correlated with in vivo fertility, with access to intracellular calcium levels in frozen-thawed sperm from bulls of varying degrees of fertility. The ultra-structural changes affect the ability of cells to control calcium flux across their plasma membranes. The different profile of ca intensity has been observed between fresh and post-thawing sperms.

The result showed that different diluter is influenced on viability motility and abnormalities of goat semen. Pre-freezing qualities of viability and motilities were 88.67+4.16 % and 66.33 +1.53 % respectively. Meanwhile, post-thawing qualities are considered as good as standard qualities at least more than 40 % (56.0+5.6 %). The different diluents were influenced highly significant ($P<0.01$).The best diluents ration is 1:4 (v/v) for both final sperms stocked. However freezing sperm conserved in -196°C is better than –45°C (i.e. motility 39.3.94 % vs. 56.0 + 5.6 %). Meanwhile, the sperm characters of two local breed showed the important variation of Ca+2 intensity, with the length of region measurement of 39.06 + 4.595 and 32.696 + 9.011 um each. It was concluded that this conventional method of the freezing system was considered as a feasible method for goat semen in the reason for practical purposes of Artificial Insemination.

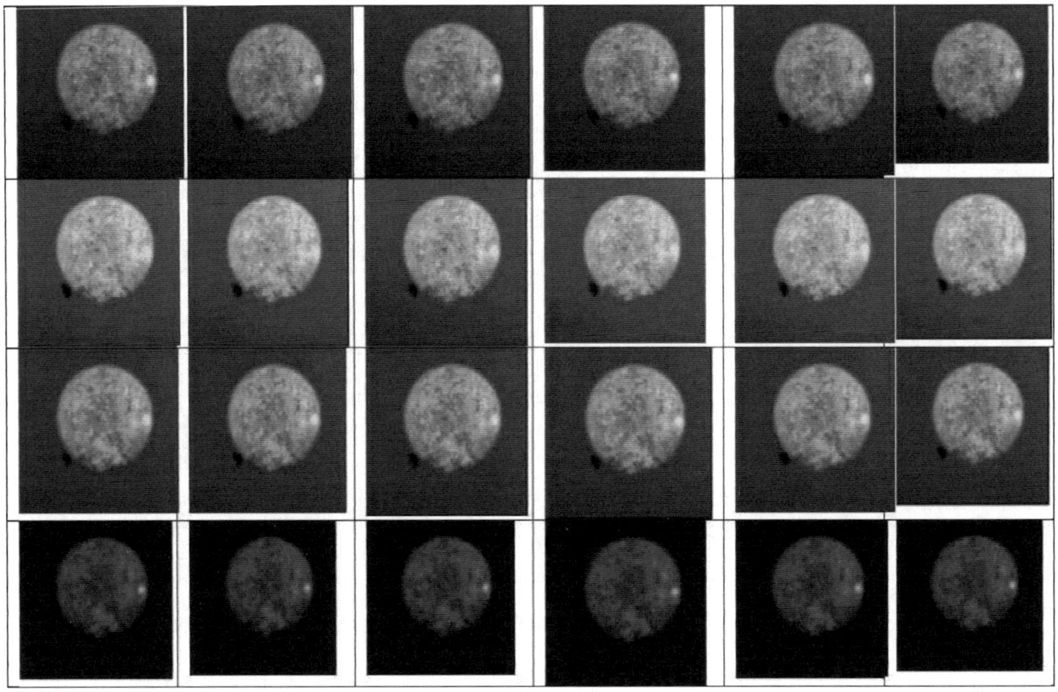

Figure 2. Different Calsium intersity base on the slice of oocyte performed by CLSM.

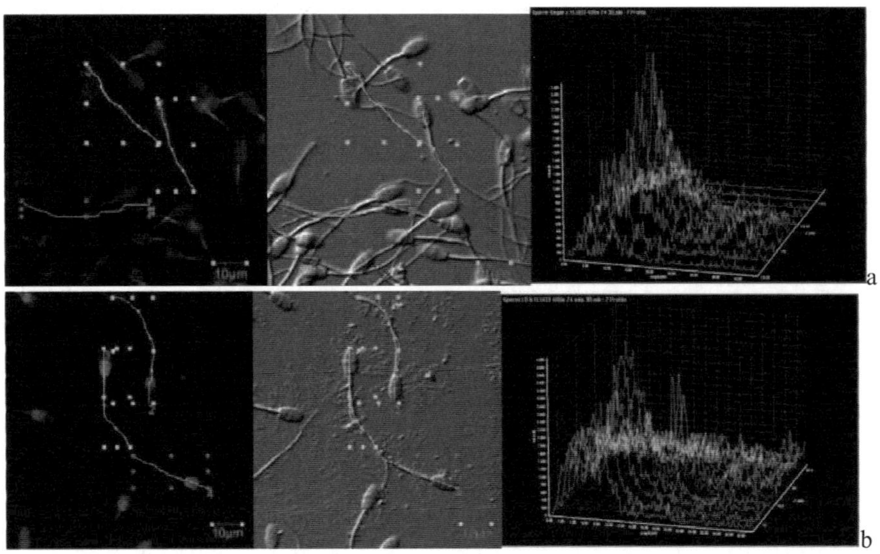

Figure 3. Different sperm Calsium Intensity: a. Fresh spermatozoa fresh and post thawing. Different profilws were observed mainly in peak of calcium intensity.Different profile of Calcium intensity of local goat sperm. It was analyses by Confocal Laser Scanning Microscope CLSM : 400, Z.4 , 3 D. [6,9]

4. Conclusion

The increase of Ca^{+2} intensity was considered to be sufficient indicator activation of M-II oocyte is occurred. The supplementation of CSE could improve the induction of M-II oocyte activation. The calcium intensity is varied more and higher in fresh sperm than freezing sperm. The sperm frozen by nonconventional could be used for the application of AI. The increase of Ca+2 intensity was considered to be sufficient indicator that showed activation of M-II oocyte occurs. This method needs to be enhanced using diluents and cryoprotectant more appropriate.

Acknowledgments

Research was supported by grant RISET BERBASIS *KOMPETENSI 2016-2017,* from Ministry of Research, Technology and Higher Education Education Republic of Indonesia. Especially thanks to Dr. Tanaka Hozumi, Jichi Medical School, Japan. for the helpful assisted cell cultured and assistance of micromanipulation, and thank to my students teamwork: Eva Ari Wahyuni, Helly, Ifa, Soepriandodo and Setyawati for very good support and collaboration

References

[1] R.E.Fissore R.E., Gordo A.C., Wu H. 1998. Activation of development in mammals: is there a role for sperm cytosolic factor?. Theriogenology **49**: 43.

[2] Oda. 2006. Mammalian Sperm Factor and Phospholipase C Zeta. J. Mamm. Ova Res. **23**: 2 .

[3] H. Wu., C.L. He and R.A. Fissore. 1987. Injection of a porcine triggers calcium oscilations in mouse oocytes and bovine eggs. Mol. Reprod.Dev, **46**: 176.

[4] S. Miyazaki, and M. Ito. Calcium Signals for egg activation in mammals. 2006. J. Pharmacol. Sci. **100** : 545.

[5] Ciptadi, G. S. Rahayu and B.Siswanto· 2012. The Phenotype Character Of Ca^{+2} Intensity M-Ii Oocyte Activated By Chemical And Crude Sperm Extract Stimulation. Paper WASET Seminar International. Paris. France.

[6] Ciptadi, G[1], A.Budiarto[1], S. Rahayu, [2], S. Wahjuningsih[1], M. Nasich,[1] M.N. Ihsan[1] , Fatchiyah[2] , A.M. Munazaroh[2] , and Mudawamah. 2015. The Quality And Ca $^{+2}$ Intensity Characters Of Local Indonesian Goat Sperm After Freezing By Nonconventional Methods Of Cryopreservation, Proc Seminar International AAAP, Fukuoka, Japan, 2015.

[7] H.Tanaka, 2001. Reproductive biology and biotechnology. JICA, Japan International Cooperation Agency, Indonesia.

[8] G.Neglia, Gasparrini B, Caracciolo di Brienza V, Di Palo R, Campanile G. and G.A. Presicce. 2003. . Bovine and buffalo in vitro embryo production using oocytes derived from abattoir ovaries or collected by transvaginal follicle aspiration. Theriogenology 2003;**59**:1123.

[9] G. Ciptadi[1]. M.N. Ihsan[1] S.Rahayu[2] , S. Wahjuningsih[1], M. Mudawamah[3] , A.M. Munazaroh[1] , And R.P Putra[1]. Post Thawing Sperm Quality And Ca $^{+2}$ Intensity Characters Of Local Goat Sperm After Freezing By Simple Method Using Deep Freezing **.** Journal of RLJS. Vol. 4 No.3 (accepted)

[10] Bailey JL, Buhr MM. Cryopreservation alters the Ca2 flux of bovine spermatozoa.Can J Anim Sci.1994;**74**:45.

[11] N. Ogonuki., Sankai T, Yagami,K., Shikano, T. Oda , S.,, Miyazaki , S., and Ogura A. 2001. Activity of a sperm-borne oocyte activating factor in spermatozoa and spermatogenetic cells from cynomolgus monkey and its localization after activation. J. Biol. Repro., 65:

[12] D. Matsuura, W. Heifeng and T. Terada. 2006. Effect of miniature pig sperm extract (CSE) and their treatment on porcine or bovine oocyte activation. J. Mamm. Ova Research. **23**: 52 .

[13] D. Matsura and T Maeda. 2006. Effect of sperm extract and its molecular weight fraction on oocyte activation in miniature pig spermatozoa. J. mamm Ova Research. **23**: 122 .

[14] S. Fujinomoto, N. Yoshida N, Fukui T, Amanai M, Isobe T, Itagaki C., Izumi T and Perry AC. 2004. Mammalian phospolipase Czeta induce oocyte activation from the sperm perinuclear matrix. J. Dev. Biol, **274**: 370 .

[15] D.Kline, JT., Kline . 1992. Repetitive calcium transients and the role of calcium in exocytosis and cell cycle activation in the mouse egg. Dev Biol 1992;**149**:80.
351-357.

[16] JP..Ozil . 1990. The parthenogenetic development of rabbit oocytes after repetitive pulsatile electrical stimulation. Development 1990;**109**:117.

[17] K. Okada, , T. Miyano and M. Miyake. 2004. Activation and development of pig oocytes after microinjection of Ceude Sperm Extract. J. Mammal Ova Research. **21**: 134 .

[18] S. Miyazaki S, Shirakawa H, Nakada K, Honda Y. Essential role of the inositol 1,4,5-trisphosphate receptor/Ca2_ release channel in Ca2_ waves and Ca2_ oscillations at fertilization of mammalian eggs. *Dev Biol* 1993; **158**: 62.

[19] S. A. Stricker . Comparative biology of calcium signaling during fertilization and egg activation in animals. *Dev Biol* 1999; **211**: 157.

[21] Cormier N, Sirard MA, Bailey JL. 1997. Premature capacitation of bovine sper-matozoa is initiated by cryopreservation.J Androl.1997;**18**:461.

[22] Simon Collin, Marc-Andre´ Sirard, Maurice Dufour, And Janice L. Bailey. .2000. Sperm Calcium Levels and Chlortetracycline Fluorescence Patterns are Related to the In Vivo Fertility of Cryopreserved Bovine Semen .Journal of Andrology, Vol. 21, No. 6, November/December 2000. CopyrightAmerican Society of Andrology. J Androl,**21**:938.

IOP Conf. Series: Earth and Environmental Science **119** (2018) 012037 doi:10.1088/1755-1315/119/1/012037

The Productivity and Natural Increase of Swamp Buffalo in District Malang

A. Budiarto[1*] and G. Ciptadi[2]

[1]Faculty Animal Husbandry University of Brawijaya
[2]Research group resources

[*]Corresponding E-mail: budiartoagus57@gmail.com

Abstract

The purpose of this research was to collect the basic information needed to develop a sustainable breeding program, which includes the potential for production and reproduction of buffaloes. This research was conducted on swamp buffalo in Malang Regency East Java. The research method used was survey method. Primary data was obtained from direct observation on 323 tails owned by 98 breeders. Variables observed were population growth and reproductive performance. The data obtained were analyzed descriptively. The result showed that the initial population study of swamp Buffaloes in Malang as many as 1155 with male and female ratio 1: 2. The ratios of male and female swamp Buffalo 20 percent male and 80 percent female. Overall, the buffalo reproduction performance was still low. Service per conception 2.06 ± 0.88 ; Anestrus Postpartum 7.46 ± 3.83 months ;calving interval distance 17.82 ± 4.86 months; 20.43 % birth rate and 4.33% mortality rate of Natural Increase (NI) population was about 16,1%. In conclusion, the value of NI of swamp Buffalo in Malang Regency is still low. To increase buffalo productivity, buffalo breeding program is continuously based on reproduction control.

1. Introduction

One of the fundamental problems of livestock productivity is the tendency of declining quality and quantity of local livestock. The government's efforts in recent years to achieve self-sufficiency in meat face quite serious obstacles such as the continued decline in the population and quality of local livestock, especially large ruminants that have a significant contribution to the provision of meat from livestock. Cattle and buffaloes are two types of large ruminant livestock are very strategic in the fulfillment of the need for meat. However, not only cattle are reported to be degrading in quality and quantity but the buffalo population also continues to experience a large drop of 3% per year [1,2].

To ensure the success of the meat self-sufficiency program, real efforts are needed to improve the population, quality and productivity of cattle and buffalo. The main problem in the current buffalo livestock is the very sharp trend of population decline, estimated at 3% in recent years. The physiological condition of buffaloes requiring specific microclimate conditions, such as the need for water puddles, makes the arena of buffalo maintenance increasingly pressured by rapid population growth, especially in Java. Limited availability of males in small environments also allows for inbreeding of family marriages that may reduce genetic quality. Reported the occurrence of phenotypic deviations of swamp buffalo in East Java, for example on the island of Kangean [3].

In East Java, cows has been known as part of the cultural social life of farmers. Potential production and reproduction of Buffalo have not received special attention by farmers, given its main function is as a working cattle. Other functions as a producer of meat, for example, has not been considered its potential. Efforts to improve genetic quality, feed quality, maintenance management, disease prevention and improvement of reproductive performance are expected to increase the productivity of buffalo cattle significantly, especially to the contribution in the provision of meat from livestock.

Various ways have been done in the business of preservation and improvement of the quality of large ruminant livestock in the main areas of livestock production in East Java, especially cattle and buffalo. However, there is still considerable attention from researchers and governments on other types of livestock, such as buffalo in some of the main areas of buffalo production with a high population. In certain areas, especially in wet paddy fields or wetlands in various regions of Indonesia, buffaloes still have a large role as cattle, however, the potential for production and reproduction of buffalo associated with other benefits, such as meat, has not received much attention. So far there is not enough data to describe the potential of this type of buffalo. Problems related to the productivity and reproducibility (natural increase) of buffalo are still rarely analyzed from the genetic aspect, but it is important to avoid some defective or defective aberrations or disabilities. The identification of livestock is still based on exterior features of great variability and low accuracy.

Base on the aim of the research is basic information through a comprehensive identification primarily of its reproduction and production potential. The data is very needed in the preparation of livestock breeding program strategies in a region and will enable the recognition of potential buffalo farms that exist today, based on a natural increase

2. Materials and Methods

The material used in this research was 323 swamp Buffalo owned by farmers. Tools used were measuring stick, tape measure, and questionnaire. The research method used surveys. The statistically vital production performance (body length, height body, chest girth) was taken from buffaloes each 10% of the tailed population in the district of Malang. The reproduction performance data (S / C, APP, calving interval) were based on interviews with buffalo breeders in each location. The data in the tabulation and analyzed descriptively.

3. Results and Discussion

From the composition of the population consisting of adult, young, easier in interpreting the condition of the population in the field. However, it will give more meaning if the composition of cattle has been grouped by age. Buffalo is classified as slow breeding, this is due to the long pregnancy period, and the emergence of long after birth (APP), Service per Conception (S / C) is high, impacts long days open (DO). From the variable, it can be said that the distance of birth (birth distance) buffalo long. The natural increase (NI) of buffalo can be calculated if the data of reproduction efficiency and mortality rate are known. Natural increase is the natural population growth calculated within the period of one year of breeding. From the results of the study obtained buffalo population structure in the survey area and reproduction data as shown in Table 1.

Table 1. The Structure of The Buffalo Population in Malang District

Sex		Adults (> 24 months)	Young (< 24 months)	Total
Male	(tail)	38	58	96
	(%)	11.76	17.96	29.72
Female	(tail)	158	69	227
	(%)	48.92	22.37	70.29
Total	(tail)	196	127	323
	(%)	60.68	39.32	100.00

From Table 1 above showed that the percentage of adult females is 48.92% of the population, this percentage is large enough to produce offspring. But of that percentage, the number of an adult female of productive age (24 -48 months) is only 18% of the population or 36% of the number of adult females. This affects the low percentage of births of 20.43% of the population, and the high female buffaloes who have a mutation out of 21%. The buffalo reproduction data and mutation are presented in Table 2.

Table 2. Data reproduction, mutation, NI

Variable	Average ± Stdev
The first age of mating (months)	29.17 ± 13.77
Age at first birth (month)	40.33 ± 12.83
S / C	2.06 ± 0.88
Mating after birth (month)	7.46 ± 3.83
Age weaned (month)	9.78 ± 4.19
Pregnant period (month)	10.94 ± 2.29
Birth distance (month)	17.82 ± 4.86
Output mutation (%)	21.36
Input mutation (%)	19.20
Birth (% of population)	20.43
Death (% of population)	4.33
Natural Increase (%)	16.1

The reproduction capacity of buffalo mud from the results of this study is still not much changed. Buffaloes (swamps and rivers) have a very high birth weight age and long birth intervals due to seasonally-dependent marriages. Male buffaloes will have an adult at the age of 2 years, while young buffaloes begin to experience estrus at the age of 2 - 2.5 years. The main factor causing the low productivity of buffalo mud is the pattern of maintenance of adult female buffalo has not been intensive, no recording, and facts in the field sometimes cycles estrus silent heat buffalo (silent heat). So that reproduction management effort becomes inefficient. Internal factors such as (the first of mating, mating after birth, etc) as shown in Table 2, more dominantly affect the increase in productivity compared with external factors such as quality and quantity of feed. Based on the percentage parameters of productive females; birth rate and mortality rate of swamp buffalo population in Malang regency, the value of natural increase is 16.1%. The direct marriage system with unselected males (quality and pedigree), has a little contribution as a cause of inbreeding.

The body size of swamp buffalo produced from this study as presented in table 3, and when compared to the size of the requirements set in the Indonesian National Standard (SNI 7706.1: 2011) in table 4, then the mud buffalo in Malang Regency for all good body sizes both adult and young buffaloes are still above the quantitative requirements of mud buffalo seeds. Based on data if seen from the performance of swamp buffalo in Malang, has a larger grade, and this is solely the influence of environmental factors is the feed consumed.

Table 3. Data of Vital Statistic Size of Swamp Buffalo Result of Research

	Male		Female	
	Adults	Young	Adults	Young
HB (cm)	122,08 ± 8,37	120,66 ± 8,92	123,22 ± 7,33	120.09 ± 7,45
BL (cm)	130,87 ± 6,90	128,72 ± 7,50	130,20 ± 7,77	127.40 ± 8,98
CG (cm)	188,87 ± 9,04	179,97 ± 12,18	186,08 ± 16,86	178.35 ± 13,87

Note. HB: height body; BL ; body length; CG : chest girth

Table 4. Data of Statistical Vital Size of Swamp Buffalo Standard SNI [4]

	Male		Female	
	Adults	Young	Adults	Young
HB (cm)	120	110	115	105
BL (cm)	125	110	120	105
CG (cm)	190	180	170	160

4. Conclusion

Additional buffalo population of the amount to be born (a natural increase) in the past year is still low, indicating the slow swamp buffalo breeding. The appearance of buffalo reproduction obtained is still below the ideal because the pattern of adult female maintenance is still very simple.

Acknowledgments

This field action research was supported by the grant of PUTP Program from Directorate General of Higher Education (DIKTI), Department of Research, Technology, Higher Education, Republic Of Indonesia, 2017. Contract No: 063/SP2H/LT/DRPM/IV/2017.

References

[1] Anonymous 2010 Polulasi kerbau turun 3 % JPNN-ekonomi, 5 Nopember 2010

[2] Anonimus 2005 Buku statistik Peternakan Direktorat Bina Perbibitan Jakarta

[3] Ciptadi G, V M A Nurgiartiningsih, Marjuki S, Wahjuningsih and Moh N Ihsan 2015 Early Investigation Of Genetic And Phenotypic Abnormalities Of Artificial Insemination Sire: A Case Of Inbreeding And Inter-Breed Crossing Of Ruminant Abstract International Basic Sci. UB- 2015

[4] BSN. ICS 65.020.30. 2011 Bibit kernau bagian 1: lumpur SNI. 7706.1.2011. Gd Manggala bhakti Wanabakti Blok IV Lt 3,4,7,10 Email Dokinfo@bsn.go.id , www.bsn.go.id

The Utilization of Sago Waste as Cattle Feed

B. M. W. Tiro[1]*, P. A. Beding[1], and Y. Baliadi[1]

[1]Assesment Institute for Agriculture Technology (AIAT) of Papua.

*Coresponding E-mail: batsebatiro68@gmail.com

Abstract

This study aimed was to evaluate nutrition value of sago waste and its effect on cattle performance.The collected data were analyzed using analysis of variance. The results of the study showed that of the utilization of sago waste had a positive effect on average daily gain (ADG), where with 2% sago waste of body weight (P2 treatment) gave the highest ADG 0.43 ± 0.02 kg/h/day and cattle which consumed only forage without sago waste (P0) gave the lowest ADG 0.26 ± 0.04 kg/h/day. Statistical analysis showed that the addition of sago waste significantly affected the ADG ($P<0.05$). The consumption of dry matter (DM) and crude protein (CP) also increased with the supplementation of the sago waste, where the highest consumption of DM was on the treatment P2 (5.09 ± 1.27 kg/day), and the lowest on the treatment P0 (4.25 ± 1.69 kg/day), while consumption of CP was highest at treatment P2 (0.37 ± 0.09 kg/day), and the lowest on the treatment P3 (0.34 ± 0.06 kg/day), while the feed conversionshowed the lowest level on the treatment P2 (12.01 ± 3.35) and highest on the treatment P0 (18.10 ± 7.39). However, supplementation of sago waste were not affect CP consumption ($P>0.05$), but significant affect($P<0.05$) DM consumption and feed conversion. Based on the results of this study it can be concluded that the sago waste as local resources have the potential to be used as a source of energy of feed supplement to beef cattle.

1. Introduction

Cattle productions in Papua are promising since it supported by ample land and diversity of local feed resources. Besides beef cattle has contribute both as a source of income for farmers and as a source of protein to meet the nutritional needs and educate the community. Beef cattle population in Papua Province during the last five years (2011-2015) has increased despite relatively small i.e. 5.82%/year. By the 2015 population of beef cattle was 100.542 and scattered in various counties and cities. This population has not been able to meet the demand for meat is increasing so it still must be supplied from outside Papua. Beef production in Papua on 2015 was reaching 3,068,235 kg compared to the previous year increased namely 2,711,011 kg (13.18%). However this meat production is still not meet demand for beef so that it still has to be supplied from outside. Yet to satisfy the needs of beef is possible because even though the population increased but increased beef cattle livestock has not been offset by the rate of reproduction and production of beef. In General, the system of maintenance of cattle in Papua is still traditional in nature, where only removed and feed still depends on the availability of nature certainly will affect the productivity of cattle produced. In addition breeders have not mastered the technology of feed appropriately so that the feed source of untapped optimally to support the growth of cattle. Utilization of local resources optimally is one of the strategic steps in an attempt to achieve the efficiency effort ruminant cattle production. Sago (*Metroxylon sp.*) many scattered in Eastern Indonesia, and about 90% of the area of sago in Indonesia occur in New Guinea [1]. Jayapura Regency is one of the areas that are quite widely spread plant Sago. Sago is a staple food communities who forever are still more consumed in the form of a papeda, but some are in the form of dry sago and sago cake. As the biggest sago producers in Indonesia of course waste sago is quite abundant in Papua.

Lees sago is waste that comes from the Sago flour processing, where in the process the obtained flour sago and sago dregs with 1:6, where comparisons of 1 sago can produce 220 kg flour Sago. This means that the potential of the dregs of sago generated considerable i.e. 1,320 kg/tree that consists of a mix of fiber and starch that does not extracted [2]. During this time, waste of untapped and sago are left piled up in the place of processing sago so that potentially cause environmental pollution.

In Papua, the sago waste utilization to feed cattle, pigs and beef cattle is still limited in certain location that near to site where farmers maintain to process sago flour. Ways of utilization of sago waste for cattle were took it from the place of processing and give it directly to livestock without controlling the amount of his deed and let cattle consume directly on site processing of Sago. However, according to the observations of farmers, cattle that consume it in processing site showed better performance. Nutrient content of sago waste, in this case protein low roughness ranges from 2.30 - 3.36%, yet starch dregs in sago is high i.e. 52.98% [3]. This allows the sago waste to be utilized as one of the alternative sources of energy to feed beef cattle. Several studies report that the utilization of sago waste as animal food to chickens and pigs can reduce uses of some other feed ingredients such as corn and rice bran. However it is not known yet the extent to which the grant of the waste sago is able to repair or improve the productivity of the cattle, so that needs to be examined his deed as cow feed. This study aims to find out the influence of utilization of waste of sago added body weight against the beef cattle.

2. Materials and Methods

The study was carried out in the village Kwadeware, District Waibu, Jayapura Regency is one of the area of the development of beef cattle and also the distribution of sago large enough. The study was carried out on farm research, in the land of farmers by involving farmer groups, related institutions, extension officers, and researchers began planning to implementation.

Use as many as 16 head of Bali cattle (aged 18-24 months) belong to the breeder. The equipment used is the scales, livestock scales feed, feed the elephant grass forage/field grass/corn straw, etc., sago and salt. The enclosure used is individual models equipped dining and drinking spot. The feed materials used will be analyzed proximate[4]. The study used a randomized block design (RBD) consisting of 4 treatments and each treatment was repeated 4 times. The treatments were: P0: 100% forage, P1: P0 + 1% sago waste of body weight, P2: P0 + 2% sago waste of body weight and P3: P0 + 3% sago waste of body weight. Variable observed include: feed consumption, body weight increase of cattle, and feed conversion. The data obtained were analyzed using Analysis of Variance, to see the difference between the treatments continued with the Duncan Multiple Range Test (DMRT).

3. Results and Discussion

3.1. An Overview of The Study

At the beginning of the implementation of the study all the cattle body weights were originally weighed and given a vitamin B complex and anthelmintic (Mectyson), then the cattle were randomized based on body weight to enter into treatment. Before the data retrieval, preceded with a period of preliminary for two weeks. This is intended so that cattle can adjust with the condition system maintenance (enclosure) and feed to be used, given the cattle used for these removable in the Pastorate and unfamiliar with the feed to be used (sago waste). The content of the dry matter (DM), Crude Protein (CP) and Crude Fiber (CF) feed sees on Table 1.

In terms of quantity, the sago waste enough available for use as animal feed primarily on flour producers areas sago, but in terms of quality, the sago waste has low nutritional value particularly low protein levels, although the core rate pretty high. The chemical composition of sago waste can be seen in Table 1.

Table 1. Chemical Composition of Material Feed (Sago Waste and Forage)

Material Feed	Dry Matter (%)	The levels of (% DM Basis)			
		Crude Protein	Crude Fiber	BETN	TDN
Sago waste	43,32	2,1	9,99	71,3	50,1
Forage	26,57	8,25	20,00	1,23	59,18

Source : Beef Cattle Research Station Laboratory, 2016.

The data in Table 1 show that the nutrient content of sago waste in this CP is very low, only 2.1%, so that the sago waste is used more as a feed source of energy. CP levels low cause suboptimal utilization of sago waste as animal feed. Further processing with a touch of technology is needed to improve the nutritional value of sago waste especially protein content of which through the process of fermentation. In the process of fermentation of carbohydrate availability can be used as an energy source for mold to grow. Fermentation is in principle able to enable the growth and metabolism of microorganisms that can enhance cerna and resulting aromas and flavors preferred cattle. Some research results reported, fermented sago waste can increase the content of research results at [5], retrieved content CP fermentation increased to sago waste 12.29 – 13.99%, [6] using Aspergillus niger fermentation processes in the womb at sago waste increased to 7.04%; [7] reporting the content of CP fermented after the sago waste increased to 4.3% while through the process of amoniasi increased to 4.1%. Obstetrician by-product CF sago obtained relatively low (9.9%), is lower than any other analysis results.

3.2. Feed Consumption

Observation of the feed consumption, average daily gain (ADG) and feed conversion is seen on Table 2. At the beginning of the activities, cattle fed on the sago waste not everything can directly consume the sago waste, first they just kissed sago waste given but there is also a livestock directly consume sago waste awarded. But on the second day of all animal can already consume sago waste awarded. The data in Table 2 show that the consumption of DM feed also looks likely to rise with the addition of sago waste. Total consumption of DM on the control treatment (P0) of 4.25 ± 1.69 kg/head/day; P1 of 4.86 ± 2.64 kg/head/day, P2 of 5.09 ± 1.27 kg/head/day and P3 amounted to 4.97 ± 0.91 kg/head/day. The relults of statistical analysis showed that there was an effect of treatment on feed consumption, were DM consumption between P0 there was a significant difference ($P<0.05$) with treatment of P1, P2 and P3. While between the treatment of P1,P2 and P3 are not significant. Supplementation with a visible sago waste, decline in forage consumption DM treatment P1, P2 and P3 (Table 2). This is due to the cattle have first consumes the waste sago then provided forage for animal, so that will reduce the consumption of forage.

Dry matter needs feed recommendations according to [8], in which the bull with the weight of the body of 135 kg and 180 kg with the added weight of 0.2 – 0.4 kg, consumption DM minimum range 3.5 – 4.6 kg/day, then consume the feed on this study DM already are above the recommended minimum DM needs. Compared with the results of previous research, consumption DM feed obtained in this study were lower than the results of the research [9] who reported consumption of DM Bali cattle Wallaby that consume feed basal supplements (energy sources: protein, 30:70 ratio) is 6.79 ± 0.53 kg/day. So are the results of the study [10], which reported that the consumption of a Bali cows DM consume additional forage and feed the tofu waste and liquid waste made from cassava is ranged 6.12 – 6.92 kg/day. However, the consumption of feed on the results of the study the DM is higher than the results of the research of [11] who reported consumption of a Bali cows DM consume grass fields with palm waste supplements range 2.93 – 4.35 kg/day. Likewise research results [12], where consumption of cow feed DM Aceh cows who consume feed forage and concentrates ranging 2.72 – 4.45 kg/day.

Table2. Average of Body Weight, Final Weight, ADG, Consumption of Feed DM, CP, TDN andFeed Conversion.

Description	P0	P1	P2	P3
Avarage of initial weight (kg)	138.83	139.07	131.03	126.40
Average of final weight (kg)	162.23	173.60	169.53	163.80
ADG (kg/head/day)	0.26 ± 0.04a	0.38 ± 0.01b	0.43 ± 0.02b	0.37 ± 0.02b
Consumption of DM (kg/head/day)	4.25 ± 1.69a	4.86 ± 2.64b	5.09 ± 1.27b	4.97 ± 0.91b
- Forage	4.25	4.16	4.16	3.81
- Sago waste	-	0.70	0.93	1.16
Consumption of CP(kg/head/day)	0.35 ± 0.14a	0.36 ± 0.19a	0.37 ± 0.09a	0.34 ± 0.06a
- Forage	0.35	0.34	0.34	0.31
- Sago waste	-	0.02	0.03	0.03
Consumption of TDN(kg/head/day)	2.21 ± 0.08a	2.51 ± 1.36a	2.63 ± 0.65a	2.56 ± 0.47a
- Forage	2.21	2.16	2.16	1.98
- Sago waste	-	0.35	0.47	0.58
Feed Convertion	18.10 ± 7.39a	12.53 ± 6.28b	12.01 ± 3.35b	13.51 ± 1.93b

Description: different Superscript on the same line showed a significant difference ($P < 0.05$). P0: forage (100%); P1: (P0 + 1% sago waste of body weight); P2: (P0 + 2% sago waste of body weight) P3: (P0 + 3%sago waste of body weight)

The existence of these feed consumption difference DM allegedly caused by the feed material is consumed, the level of palatability and age or body weight cattle. Beef cattle ration consumption affected the size of the body, the environment as well as the condition of the feed and forage consumption is correlated with in vivo digestibility of dry ingredients and organic materials [13]. Further described [14] that is usually a high digestibility gives a high consumption. DM feed consumption is determined by body size, range of rations, age, and condition. The dried material consumption ration usually decreases with increasing content of feed substances that can digest.

The factors that affect the rate of consumption of feed are 1). Factors include power feed, digestibility and palatability, and 2). Factors for livestock, which includes age, nation, gender and health conditions of livestock[15]. Further [8], explained that the palatability of the feed is one of the factors that affect the amount of consumption of livestock feed and the ability to consume DM contained in feed and relates to the physical capacity of the stomach as well as the condition of the gastrointestinal tract. High low feed consumption in ruminant livestock are strongly influenced by environmental factors as well as the feed material. Fast feed rate of motion is the latest in the rumen will have an effect on consumption of feed. The limitation of the capacity of the digestive tract almost always pressing the feed consumption, so that the incoming feed is always lower than the supposed [16]. Low feed kecernaannya will longer be in the digestive tract so that it will slow down the rate of feed in rumen and will cause a decrease in the consumption of feed.

Average consumption CP on P0 treatment amounting to 0.35 ± 0.14 kg/head/day; P1 of 0.36 ± 0.19 kg/head/day, P2 of 0.37 ± 0.09 kg/head/day and P3 of 0.34 ± 0.06 kg/head/day. The results of the statistical analysis, showed the presence of sago waste supplementation as influential feed not significant effect on CP consumption (P>0.05). Nevertheless empirically the average of CP consumption on treatment of P2 (2% sago waste of body weight) is higher and the lowest at the treatment of P3 (3% sago waste of body weight). This suggests that the higher the addition of sago waste as beef cattle feed will reduce consumption caused by nutrient content of CP in this case CP sago waste low so even though the DM consumption is higher than P0 (forage treatment 100%) CP consumption is lower. When compared with the recommendations of CP[8], consumption at cattle beef 135 kg and 180 kg have to add with 0.2 – 0.4 kg consumption CP range0.34 – 0.48 kg/day, then consumption CP study on feed is in compliance with the needs of CP are recommended.

Therefore consumption of CP can be said already suffice the needs of a Bali cows is currently growing. Increased consumption of CP along with increased consumption of feed, where DM consumption increases with the presence of suplmentasi sago waste, raising also the consumption of CP are contained within the feed. However granting excessive sago waste (3% of body weight) turns out to lower the consumption of CP because the low content of waste at sago. Consumption of CP feed affected by the increase of the weight of cattle, the amount of feed consumed as well as the quality of the feed they are given [8].

AverageTDN consumption for the treatment of P0 was 2.21 ± 0.88 kg/head/day, P1 was 2.51 ± 1.36 kg/head/day, P2 was 2.63 ± 0.65 kg/head/day and P3 was 2.56 ± 0.47 kg/head/day. The results of the statistical analysis, showed that the use of sago waste showed not significant influence on the consumption of TDN (P >0.05). Empirically, there is an increase in the consumption of waste supplementation with the TDN Sago. Increased consumption of TDN in line with consumption of its DM, because the consumption of nutrients is affected by the consumption of the nutrient content of feed and DM. The increasing consumption of BK then it will result in a rise in consumption of other nutrients.

3.3. Average Daily Weight Gain (ADG)

The data in Table 2 shows that, the highest ADG average was in treatment P2 (2% added of sago waste) and the lowest was on treatment P0 (without sago waste). Although in the first month ADGbeef cattle is very low (<0.2 kg/head/day), but in the next month there is a significant increase in body weight. While the treatment of P3 (3%of sago waste) gave ADG lower than P1 (1% of sago waste) and P2 (2% of sago waste). The low ADG in the treatment of P3 although the total consumption of high DM but low consumption of CP due to low content of CP sago waste. This shows that the provision of sago dregs for beef cattle ranges from 1 - 2% of body weight. The result of statistical analysis showed that the mean of ADG of cow was influenced by treatment, where between treatment P0 there was a significant difference (P<0,05) to treatment of P1, P2 and P3. While between treatments P1, P2 and P3 differ not significant. Body weight description and ADG of male Balinese cattle during observation are shown in Figures 1 and 2.

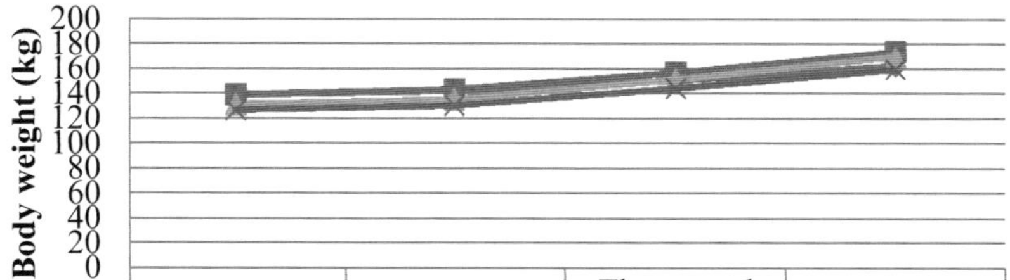

	Initial weight	First month	The second month	Third month
P0	138.83	141.97	150.20	162.23
P1	139.07	143.33	156.77	173.60
P2	131.03	135.10	151.20	169.53
P3	126.40	130.17	144.10	159.33

Figure 1. Graph of Male Balinese Cow Body Weight for 3 Months

In Figure 1, it can be seen that the mean body weight at treatment P1 is higher than other treatment because the initial weight average is also higher. While on treatment P2, mean body weight was higher than P0 treatment even though the mean body weight at treatment P0 was higher. This is thought to be caused by higher consumed feed and also with the addition of sago dregs in the feed can improve the weight of livestock than those who only consume forage.

Figure 1 also shows that in the first month of observation, the average increase in animal body weight is still very low (3.14 - 4.26 kg), this is because in the first month the livestock is still in adaptation with new maintenance and feed system. Maintenance systems are usually removed, where the cattle are allowed to find their own feed is on this livestock cattle so that suspected cattle stress with a new maintenance and feed system. But in the second month, the average increase in body weight is quite high (8.23 - 16.0 kg) as well as in the third month the average increase in body weight reached 12.03 - 18.43 kg. The highest body weight increase in livestock treated with P2 (forage 100% + sago waste 2% of body weight) reached 18.43 kg, then treatment P1 (forage 100% + sago waste 2% of body weight) reached 16.83 kg; the treatment of P3 (forage 100% + sago waste 3% of body weight) reached 15.23 and the lowest in the treatment of P0 (forage 100%) ie 12.03 kg. This illustrates that the sago waste as one of the locally available resources has the potential to be used as beef cattle feed.

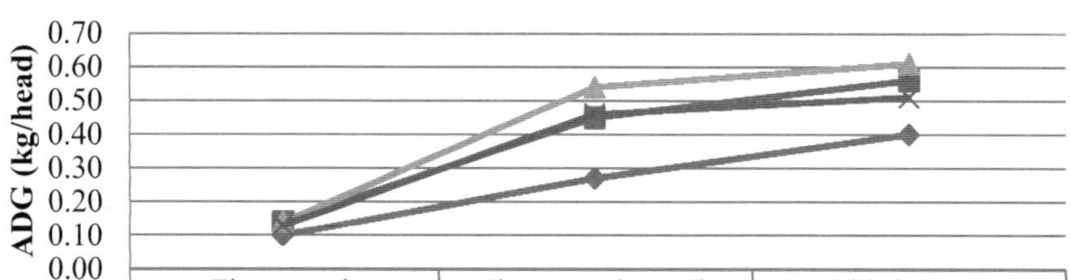

	First month	The second month	Third month
P0	0.10	0.27	0.40
P1	0.14	0.45	0.56
P2	0.14	0.54	0.61
P3	0.13	0.46	0.51

Figure2. Graph ofADG of BaliCows

In Figure 2, it is seen that the highest treatment ADG of P2 (forage 100% + sago waste 2% of body weight), then P1 and P3, while the lowest ADG at P0 (forage treatment 100%). Look also that ADG in the first month of observation is still low, then rise in second and third. This is because entering the second month cattle is familiar with system maintenance and new feed so that a positive effect against rising ADG. Views of the percentage increase in livestock that consume between ADG sago waste than just consume the forage, in treatment P1 occur an increase of 49.35% higher than P0, P2 treatment treatment increased by 67.53% compared to P0 and P3 treatment treatment increased by 42.86% compared to treatment of P0. Sago waste contain high starch so that it is used as a feed source of energy to a positive effect against beef cattle body weights of cattle, while protein needs would be sure of forage.

3.4. Feed Conversion

Feed conversion was one of the benchmarks for assessing the level of efficiency of the use of body weight added against rations of livestock. The data in Table 2 show that the feed conversion numbers on treatment of P0 of 18.10 ± 7.39, P1 of 12.53 ± 6.28, P2 of 12.01 ± 3.35, and P3 of 13.51 ± 1.93. The results of statistical analysis show supplementation sago waste as influential feed real ($P < 0.05$) against the feed conversion numbers. The lowest feed conversion figures are on treatment P2 and highest at treatment P0, this illustrates that livestock in treatment of P2 is more efficient in utilizing the feed. Low feed conversion treatment rataan P2 caused though the consumption of high pakannya but the resulting ADG also high so as to form a 1 kg ADG require less feed. High feed conversion on average P0 treatment due to the high consumption of feed is not offset by a high mean ADG to form ADG 1 kg need more feed so that it can be said that the feed consumed less efficient. The feed conversion figures obtained in this study are still lower than the results of the research [9],are reporting the number of conversion feed Balicows which consume fodder basal supplements (energy sources: protein, 30:70 ratio) was 39.9 ± 4.5; but higher than the results of the research of [17], where the male cow feed conversion of Aceh that consume feed forage and concentrates – 4.60 9.55. The difference of this feed conversion figures alleged to be caused by the feed materials consumed (quality and quantity), livestock conditions and environmental factors. However, the rate of conversion of feed obtained in this study (treatment of P1 and P2) is still appropriate, where according to [16],a good feed conversion figures for beef is 8.56 – 13.29. Feed conversion was influenced by the condition of livestock, power cerna, gender, nation, the quality and quantity of feed and also environmental factors. States that the use of the feed efficiency is influenced by several factors including the ability of cattle in the feed material, digesting the sufficiency substance staple feed for life, growth, body functions as well as the type of feed to be consumed [18].

4. Conclusion

The content of nutrient sago waste is very low especially CP, however, the core content is high enough so that it could potentially be utilized as energy sources in beef cattle feed. The addition of sago waste as the most efficient beef cattle feed on treatment of P2 (P0 + sago waste 2%of body weight), seen from the aspect of improved ADG 67.53% higher than P0 (forage treatment 100%), increased consumption of nutrients (DM, CP or TDN) and numbers the lowest feed conversion (12.01).

Acknowledgements

Credits given to Agricultural R & D Agency SMARTD through research cooperation KKP3SL who had financed the activities of this study, as well as Jayapura District Department of agriculture which has been cooperation in the activities of this study.

References

[1] Pusat Penelitian dan Pengembangan Perkebunan 2010 Warta Penelitian dan Pengembangan Tanaman Industri Volume 16 Nomor 2 Agustus 2010 http:///perkebunan.litbang.deptan.go.id/wpcontentuploads/2010/perkebunan_warta_2010.pdf

[2] Rumalatu F J 1981 Distribusi dan Potensi Pati dari Batang beberapa Jenis Sagu (*Metroxylon sp*) di Daerah Seram Barat Fakultas Pertanian Universitas Ambon

[3] Ralahalu T N 1998 Pengaruh tingkat penggunaan ampas sagu yang difermentasi dengan Aspergillus niger dalam ransum pada pertumbuhan babi selama periode pertumbuhan Tesis Program Pascasarjana Institut Pertanian Bogor

[4] AOAC 2005*Official Methods of Analysis. Association of Official Analytical Chemists* Editor: Horwitz W And G W Latimer Jr. Pubished by AOAC International 18th Edition: United States America

[5] Martaguri, Mirnawati dan H. Muis 2011 Peningkatan kualitas ampas sagu melalui fermentasi sebagai bahan pakan ternak. Jurnal Peternakan **8** (1): 38.

[6] Tampoebolon B I M 2009 Kajian perbedaan aras dan lama fermentasi ampas sagu dengan *Aspergillus niger* terhadap kandungan protein kasar dan serat Prosiding Seminar Nasional Kebangkitan Peternakan: Semarang

[7] Sangadji I 2009 Mengoptimalkan Pemanfaatan Ampas Sagu sebagai Pakan Ruminansia Melalui Biofermentasi dengan Jamur Tiram (*Pleurotus ostreatus*) dan Amoniasi Disertasi Program Studi Ilmu Ternak Institut Pertanian Bogor

[8] Parakkasi A 1999 Ilmu Nutrisi dan Makanan Ternak Ruminansia Universitas Indonesia: Jakarta

[9] Widyaningrum Y, D Pamungkas, M Kote dan Mariyono 2013 Pertumbuhan sapi Bali jantan pada agroekosistem lahan kering iklim kering Nusa Tenggara Timur Prosiding Seminar Nasional Teknologi Peternakan dan Veteriner

[10] Kurnia Sari D W, M H Astuti dan L Sinta Asi 2016 Pengaruh pakan tambahan berupa ampas tahu dan limbah bioetanol berbahan singkong (*Manihot utilissima*) terhadap penampilan sapi Bali (*Bos sondaicus*). Buletin Peternakan **40**(2) 107.

[11] Nanda D D, A. Purnomoadi dan L K Nuswantara 2014 Penampilan produksi sapi bali yang diberi pakan dengan berbagai level pelepah sawit. Agromedia **32** (2): 52.

[12] Usman Y, E M Sari dan N Fadilla 2013 Evaluasi pertambahan bobot badan sapi Aceh jantan yang diberi imbangan antara hijauan dan konsentrat di Balai Pembibitan Ternak Unggul Indrapuri Jurnal Agripet **13**(2)

[13] Minson D J 1990 Forage in Ruminant Nutrition Academic Press INC: California

[14] Davies Z S, D Mason, A E Brooks, G W Griffitth, R J Merry, and M K Theodorou 1990 An automated system for measuring gas production from forage inoculated with rumen fluid and its use in determining the effect of enzymes on grass silage J. Anim Feed Sci. Techno. **83.** 205

[15] Yudith T A 2010 Pemanfaatan pelapah sawit dan hasil ikutan industri kelapa sawit terhadap pertumbuhan sapi Peranakan Simmental fase pertumbuhan Departemen Pendidikan Fakultas PeternakanUniversitas Sumatera Utara

[16] Church D C and W G Pond 1988 Basic Animal Nutrition and Feeding John Wiley and Sons New York

[17] Siregar S B 1994 Ransum Ternak Ruminansia Penebar Swadaya Jakarta

[18] Campbell J R, Kenealy MD and Campbell K L 2006 Animal Sciences 4th Edition McGraw Hill New York

The Effect of Different Location of Muscle on Quality of Frozen Simmental Ongole Grade Male Meat

D. Triasih[1*], D. Krisdiani[1], J. Riyanto[1], W. Pratitis[1], and S. D. Widyawati[1]

[1]Animal Husbandry Department, Faculty of Agriculture, Sebelas Maret University.

*Coresponding E-mail : triasihdyah@gmail.com

Abstract

The aim of this research was to identify the influence of different types of muscle on the characteristics physical quality frozen meat of Simmental Ongole Crossbreed Male frozen meat. The research had been conducted at the Laboratory of Meat Technology and Processing and Laboratory of Nutritional Biochemistry, Faculty of Animal Science, University GadjahMada, Yogyakarta. The physical quality with 4 levels treatments, the name was *Biceps femoris* (BF), *Longissimus dorsi* (LD), *Triceps brachii* (TB), and *Pectoralis profundus* (PP). The chemical quality with 3 levels treatments, the name was *Biceps femoris* (BF), *Longissimus dorsi* (LD), and *Triceps brachii* (TB). The research used Completely Randomized Design with 5 replications for each treatment. The variables of the physical quality test were pH, tenderness, cooking loss, and water-holding capacity. The chemical quality test were water content, protein content, fat content, and cholesterol content. The result of the physical quality test showed that the different types of muscle were significantly influence the pH value (P<0,01), also the influence the tenderness and cooking loss (P<0,05), but there was no significant different on water-holding capacity. The chemical quality test showed that the different types of muscle significant influenced on protein content and fat content (P<0,01). They were significant different (P<0,05) on water content, and there was significant effect (P<0,05) on cholesterol. In conclusion, the BF have high value of pH, cooking loss, water content, protein content, and cholesterol than other muscle, but have low value of tenderness and fat content.

1. Introduction

Meat was origin food of livestock that high nutrient content for human consumption. Quality of the meat depends in the cow muscle. There was difference of anatomical, location, and muscle function given effected for meat produced. Active muscle yield meat has colored dark, water holding capacity, and high pH. Texture and size active muscle fibers more rough and more than the less active so muscle yield meat with low tenderness [1]. Quantity and size fibers each variety muscle bundles between muscle tissue to equal cattle and the equal muscle in different cattle. The thickness muscle fibers, muscle tissue, and connective fibers affected delighted consumer for meat, especially closely related with tenderness [2]. Muscle used as a determination quality meat was muscle contain direction obvious muscle fiber and proportion was large enough [3]. *Biceps femoris* (BF) muscle, *Longissimus dorsi* (LD) muscle, Triceps brachii (TB) muscle, and *Pectoralis profundus* (PP) muscle was category muscle cow muscle often used as sample determination meat quality and more savour to consumers [4].

Freezing was good method for curing meat and meat process. Process freezing has no effect for qualitative properties and organoleptic including color, flavor, and juice meat after cooking but frozen storage can affect degradation power received smell and flavor. The nutritional value meat was relatively unchanged during freezing and frozen storage for a limited time. Frozen meat product was one of preserving meat products to maintain the shelf life, because it inhibits the process of meat damage [3]. Based on that reason, this research was conducted to evaluate difference physic and chemical quality to location frozen meat muscle Simmental Ongole crossbreed male.

2. Materials and Methods

The research was conducted in Meat Processing Technology Laboratory and Nutritional Biochemistry Laboratory, Faculty Animal Husbandry GadjahMada University Yogyakarta. The samples meat source from frozen meat Simmental Ongole Crossbreed male.

The chemical quality content of frozen meat is known to use Food Scan type 78810 foss electric A /S 69 DK Siangerupgade Denmark. Place of research in Laboratory of Animal Technology and Products Faculty of Animal Husbandry Universitas GadjahMada Yogyakarta. Meat cholesterol levels are known using the Libberman Burchad method.

Statistic analysis research used for physical quality was Completely Randomized Design (CRD). Unindirectional pattern with 4 levels treatments and chemical quality was Completely Randomized Design (CRD) unidirectional with 3 levels treatments each level using 5 replications.

a. Physical quality of frozen meat
   P1: *Biceps femoris* (BF) (Muscle Silverslide Meat)
   P2: *Longissimusdorsi* (LD)(MuscleShortloin Meat)
   P3: *Triceps brachii* (TB) Muscle (Chuck Meat)
   P4: *Pectoralisprofundus* (PP) Muscle (Brisket point and Meat)
b. Chemical quality of frozen meat
   P1: *Biceps femoris* (BF) (Muscle Silverslide Meat)
   P2: *Longissimusdorsi* (LD) (Muscle Shortloin Meat)
   P3: *Triceps brachii* (TB) Muscle (Chuck Meat)

The variable observed physical quality research were pH, tenderness, cooking less, and water holding capacity [3]. Meat used has been stored in the freezer for 24 months and experienced re-freezer 2 time. Thawing was carried out by taken out the frozen meat from freezer (-18C), placed it in the– refrigerator for ±24 hours with temperatureof 3 - 5°C , and then placed the meat atdhe room temperature 27 – 27,5°C until the frozen meat become fresh meat and ready or the further analysis BF, LD, TB, and PP muscle preparations appropriate with analysis which will be implemented). The pH was measured using pH water . Tenderness level meat was measured using Warner-Bratzler Shear Force tool. Count cooking loss using formulations was percentage difference weight before cooking and after cooking to the weight before cooking. The water binding method used was the Hamm method.The variable observed chemical quality research were moisture content, protein content, fat content, and cholesterol level. Calculation of water content appropriate with AOAC, protein content using Kjeldahl method appropriate with AOAC, Determination of fat content by Soxhlet method [4] and cholesterol level of Lieberman-Burchad method.

Result data physical and chemical quality test were analyzed by using variance analysis (ANOVA). If significant, further Duncan Test will be performed [5].

3. Result and Discussion

The result variance analysis for physical quality test showed that difference location muscle had significant effect for pH ($P<0,01$), tenderness and cooking loss ($P<0,05$) but no significant effect for water holding capacity (Table 1).

Table 1. Physical Quality of Various Kinds of Frozen Meat Muscles

Variabel	Various Muscles			
	P1	P2	P3	P4
pH	6.55^{A}	6.46^{C}	6.51^{B}	6.52^{B}
Tenderness (kg/cm^2)	6.81^{b}	7.37^{a}	7.41^{a}	7.53^{a}
Cooking loss (%)	34.23^{a}	29.56^{b}	32.94^{a}	33.60^{a}
Water Holding Capacity (%)	5.60	5.32	9.10	6.72

^{A,B,C} Different superscripts on the same line showed highly significant (P<0,01).
^{a,b,c} Different superscripts on the same showed significant (P<0,05).
P1: *Biceps femoris* (BF), P2: *Longissimus dorsi* (LD), P3: *Triceps brachii* (TB) and P4: Otot *Pectoralis profundus* (PP)

3.1. pH

The result variance analysis showed that P1 (BF) treatment had significant different pH value (P<0,01) compared with P4 (PP), P3 (TB), and P2 (LD) treatments. The active muscle (BF) generally had lower pH value than passive muscle (LD, TB, and PP). This was because the active muscle contains more fat and high levels protein solubility [6] and passive muscle contains less fat and air, but higher protein content [7]. It was different from this research, pH for BF muscle was higher than LD, TB, and PP muscle.BF muscle had higher content marbling fat that low fat level and high protein compared with other muscle [8]. High content marbling fat can inhibit damage effect freezing period by reducing occurrence meat fluid drip and protect protein structure meat. Reduced fluid drip meat came out with hydrophilic proteins effect reduce amino acid content and caused occurrence increase pH value because amount released N and loss water so meat had alkaline character [9], this was different in acidic meat. Frozen meat for a long time caused high drip fluid exudation meat during time. When fluid drip meat came out and hydrophilic protein also take part came out so hydrophilic protein meat was reduced. Protein hydrophilic came out contain was hydrophilic amino acid residues [10], this results meat was acidic character.

Based on Duncan's results showed that P4 (PP) and P3 (TB) treatments were not different. This was expected because TB and PP muscle based location had function muscle almost same. So with effect old freezing causing maximum damage for both muscles. Changed microstructure meat occur during storage in a long time, effect occurrence high protein denaturation, so that hydrophilic group was reduced maximal[11]. Maximal damage can affect meat microstructure experienced same change, so fluid meat out at drip contain hydrophilic protein with same content.

Based result from different Duncan Test showed that P2 (LD) treatment was different significantly (P<0,01) compared with treatment P3 (TB) and P4 (PP). LD muscle lower pH was lower than TB muscle and PP [1]. This was accordance with this research, LD was lower pH than TB and PP. Passive muscle had low protein levels and high fat levels compared active muscle [12]. Passive muscle damage level protein structure was higher because freezing of the old, so causing drip flesh out. Hydrophilic proteins, which comes out containing hydrophilic amino acid residues [10].

Research using frozen meat has been stored during 3 – 12 days, result pH value in LD muscle was range 6,34 – 6,45 and BF muscle was range 6,15 – 5,43 [13], but this research result value pH in LD muscle was range 6,46 and BF muscle was range 6,5 in meat frozen have been stored during 24 months.

3.2. Tenderness

The result variance analysis showed that difference location muscle had significant effect (P<0,05) for tenderness level. P1 (BF) Treatment has a more tender level of tenderness than P2 (LD), P3 (TB), and P4 (PP) treatment. The level tenderness meat generally varies in muscle type.

Tenderness can decrease from pelvic extremity in the SM muscle, almost similarly to the BF and ST muscles, and increase from center both extremity LD. It was different in this research, BF muscle more tender than other muscles. BF muscle tender higher was suspected because it contains higher marbling fat and experience a long freezing (24 months). Meat has a high marbling fat can resist the occurrence of large ice crystals from the liquid of the meat inside, inhibits occurance freezer burn and occurrence muscle stretching or reducing contraction miosin actin, so resulting in increased sarcomere myofibril length [3]. It was different in the muscle LD, TB, and PP with micro structure less fat effect high protein structure damage with long freezing time. Mikro structure damage cause protein degradation impact occurrence ofmiosin actin overlaps, so can't bind water any more properly and cause muscles contract and muscle become tough [8].

Duncan test showed that P2 (LD), P3 (TB), and P4 (TB) treatments tenderness was not significantly different. This was presumably because marbling fat contained in the muscle was not as high in the BF muscle so that with the effect of the old freezing effect damage to the micro structure of the maximum meat in each muscle. Damage micro structure meat caused weakening hydrogen bonding system, so result disintegration three dimensional structure proteins and causing protein aggregation. Damage to the microstructure of the meat results in weakening the hydrogen bonding system, resulting in the disintegration of the three-dimensional structure of proteins and causing protein aggregation. Consequently protein tissue can't bind water[14]. It was caused damage to the microstructure of the meat with changes in the same muscle location, so level contraction miosin actin muscles was same. Contraction meat causing ice was perfectly formed in the protein matrix, so maximum freezer burn. At the time thawing effect water content in meat was reduced, so meat with content marbling fat was almost same level tenderness same. Frozen meat has been stored during 6 months in Ongole Grade Cattle, LD muscle portion it was tenderness range 1,91 – 3,17 kg/cm^2[15] and frozen meat have been stored during 14 days result level tenderness 5,47 - 16,88 kg/cm^2[16]. This research frozen meat has been stored during 24 months in the Ongole Grade Cattle BF, LD, TB, and PP muscle had level tenderness range 6,81 – 7,53 kg/cm^2. Level tenderness different caused because Duration of freezing in each research [16].

3.3. Cooking Loss

The result variance analysis showed that difference muscle location had significant effect (P<0,05) for cooking loss value. P2 (LD) treatment was significantly different (P<0,05) than cooking loss value frozen meat treatments P1 (BF). P4 (PP), and P3 (TB). LD muscle has lower water cooking loss than BF, TB, and PP muscle. LD muscle has capability higher binding water than BF, LD, and PP muscles [17]. This was accordance with this research that state LD muscle had LD muscle a low cooking loss value compared with other muscles, because it had the capability higher than other muscles.

Based on Duncan's different test results showed that treatment P1 (BF), P3 (TB), and P2 (PP) did not different significantly in cooking loss value. This is same with result frozen meat LD muscle and BF muscle with stored 6 days in temperature -18°C has value cooking loss was not significantly different. The microstructure meat was damaged by the effect old freezing, the damage muscle micro structure was maximal [18]. Function time and temperature greatly affect level damage protein structure meat because amount drip increases with increasing storage time [3]. Besides same time and temperature setting factors during boiling cooking loss test causing occurrence perfectly protein denaturation, so maybe liquid comes out same. Liquid meat comes out same cause cooking loss value not different, boiling by using temperature 30 and 40°C myofibril protein coagulated , temperature 55°C myofibril protein experienced perfect denaturation and temperature 60°C sarcoplasmic protein experienced perfect denaturation.

Cooking loss value LD muscle frozen meat Ongole Grade Cattle which has been stored for 2 weeks range 30,24 – 36,16%. Cooking loss value was higher than cooking loss value was higher than cooking loss value LD muscle Simmental Ongole Crossbreed Cattle has been stored for 24 months was range 29,56% [19]. BF muscle this research was cooking loss value range 34,23% higher than with cooking loss BF muscle in bull has been stored for 12 days range between 28,30 – 35,09%. Difference cooking loss value was caused length storage time and species cattle [3].

3.4. Water Holding Capacity

The result variance analysis showed that water holding capacity had no significantly for muscle location. General water holding capacity active muscle water holding capacity was lower than passive muscle (LD) because BF muscle many activities [20]. Presence amount positive charges released and negative charges surplus effect rejection myofilaments and giving more room to water molecule, so water holding capacity decreases [3]. The research was different from this research. This research was not different various location muscle. Cause not difference water holding capacity expected length freezing time effect structure damage in tendon with appearance extracellular fluid and formation ice crystals [3]. Formation ice crystals longer gets bigger consequently make the amount of ice crystal various muscles maximal, so causing distortion, damage brain fibers, and sarcolemma. The high ionic strength extracellular fluid also effect denaturation some muscle proteins, so increase displacement of a number of weakly bound water extracellular spaces and muscle fibers fail to reabsorb all water was translocated or exited during freezing process. The amount water bonded was rather weak as the second layer water molecules against hydrophilic group and weakly bound free water molecules between protein molecules exiting maximally. During freezing process with a long time full damage, so release water was weak and free in microstructure meat was also maximal when it occur on the pressing water holding capacity test. It enable water holding capacity value various muscles was not different.

The mean water holding capacity various muscles frozen meat Simmental Ongole Crossbreed Cattle treatments P1 (BF), P2 (LD), P3 (TB), and P4 (PP) were 5.60%, 5.32%, 9.10% , and 6.72% higher than the water holding capacity frozen meat Ongole Grade Cattle portion LD muscle range 2,49% - 10,22% [18]. This difference was caused difference time freezing and species cattle [3].

The result of variance analysis for physical quality test showed that the difference of muscle location had significant effect on protein content, fat content (P <0,01), water content and cholesterol (P <0,05) (Table 2).

Table 2. Chemical quality of various kinds of frozen meat muscles

Variabel	Various Muscles		
	P1	P2	P3
Water Content (%)	71.11^{a}	71.19^{a}	70.51^{c}
Protein Content (%)	24.94^{A}	22.01^{C}	23.28^{B}
Fat Content (%)	2.55^{C}	5.13^{B}	10.61^{A}
Cholesterol Content (mg/100gram)	18.99^{a}	14.62^{b}	14.01^{b}

^{A,B,C} Different superscripts on the same line show very different very real (P<0,01).
^{a,b,c} Different superscripts on the same line show a real different (P<0,05).
P1: Biceps femoris (BF), P2: Longissimus dorsi (LD), P3: Triceps brachii (TB) and P4: Otot Pectoralis profundus (PP)

3.5. Water Content

The location muscle significantly effect for water content of frozen meat. Different Duncan test result showed significantly different mean (P<0,05) was TB muscle higher than BF and LD muscle data. Level fat in each muscle can be different, because active muscle was more used to move.

Proportion high fat content can decrease water content the meat, so muscle have fat content will have low water content. It was contrast with this research, that semi-active muscle (TB) had a lower water content than less active muscle (LD), so suspected a long freeze (24 months) in meat. The old freezing result heating process in analysis water content, fat contained in it dissolves, so fluid secreted by BF muscle more and fat content will be less, then the water content in more BF muscle. This was contrast with passive TB and LD muscles a lower water content, because passive meat has more fat than BF muscle. Accumulation intramuscular fat content can stretch tissue structure meat, so cause a lot of water dissolved [3].

3.6. Protein Content

The location muscles had a significant effect on the protein content frozen meat (P<0,01). The result from Duncan's different test showed that protein content was significantly different (P<0,01) in each muscle location. The BF muscle has highest protein content compared with TB and LD muscle. It was because BF muscle was active muscle than TB and LD muscle. The increase myofibrils react contribution protein, so increasing protein levels. The more active muscle has a higher myofibril structure, so muscle myofibril protein increases [3].

3.7. Fat Content

The location muscle had very significant effect P (P<0,01) in the fat content of each muscle location. The highest fat content found in TB muscle. In muscle LD and BF have lower fat content than TB muscle. Fat content in each type muscle can be different. Passive muscle proportion fat can be higher than active muscle, because active muscle was more used to move [8]. This was contradict with this research which state that the semi active muscle (TB) has a higher fat content than the passive muscle (LD). The difference was possible because muscles used have been frozen a long time, so the meat experienced maximum freezer bun. Maximum freezer burn effect formation condensed layers of the vein tissue near the surface, it can prevent water entry from below [8]. This results formation large crystals. Large ice crystals can cause distortion and damage muscle fibers, so occur change meat structure. Changing on the microstructure of meat can cause abridgment muscle fibers in the LD, so fat will be exudation out when thawing and cooking with the meat fluid. It was possible that LD fat content was lower than TB. There was very close relationship between water content with fat content in beef from various muscle locations. This was in accordance with research that water content LD content This is in accordance with research that water content LD muscle was significantly different with BF muscle and inversely proportional with fat content [3].

3.8. Cholesterol Content

The muscle location affect effect each quality of the beef cholesterol content. The BF muscle had highest levels cholesterol compared with LD and TB muscle. This was suspected because different amount red muscle fibers and different amount nucleus cell in muscle. Active muscle (BF) contains more red muscle fibers, so amount of cell nucleus was also more than the passive muscle (LD and TB) [8]. Different muscle fibers effect differences amount of marbling fat, because the passive muscle there was lower marbling fat, so it tends to contain lower cholesterol levels than muscles that have more marbling fat. Cholesterol content of fresh meat reached 19.04-39,88 mg / 100g [18]. In this research the cholesterol content reached 14.01-18,99 mg /100g. It showed the results of cholesterol this research was below the value of cholesterol range fresh meat. The difference in the results of this research was suspected because effect of long freezing and heating processes during analysis resulting in dissolved marbling fat, resulting in lower cholesterol levels. Muscles was contain few marbling fat had lower cholesterol levels.

4. Conclusion

Differences muscle location in meat of the Simmental Ongole Crossbreed when viewed from the characteristics of physical quality and chemical quality showed that Bicep femoris muscle has pH, cooking loss value, water content, protein content, and cholesterol content was the highest among was other muscles, but it has the lowest level of tenderness and fat content.

Reference

[1] Riyanto J 1998 Physical Characteristic Quality Meat Ongole Grade Cattle Male on Various Muscle Padjajran University, Bandung

[2] Sulaksana I 1999 Effect of Storage Temperature on Physical Histologis Beef Characteristic in The Rigor Mortis Phase Institute Pertanian Bogor

[3] Soeparno 2009 Science and Meat Technology 5th ed. University GadjahMada Press Yogyakarta

[4] AOAC 1984 Official Method of Analysis 14th ed. Pub Arlington Virginia

[5] Gazpers V 1991 Experimental Design Method 2nd ed. CV Armico Bandung

[6] Zhang S X, Farauk M M, Young Q A, Wellicko K J, and Padmore C. J. 2005 Function stability of frozen normal and high pH beef Food Sci. **69** 766

[7] Winarso D 2003 Physical characteristic change due to difference of age, muscle type, cooking time, and temperature combination in native chicken meat J. Indonesian Tropical Animal Agriculture **28**(13) 119

[8] Lawrie R A 2003 Meat Science 5th ed, ed Parakkasi Universitas Indonesia Press Jakarta

[9] Djazuli N, Wahyuni N, Monintja D, and Purbayanto A 2009 Modification technology of surimi processing through utilization "by-catch" of shrimp net in arafura sea J. Processing of Fishery Products of Indonesia **7**(1) 17

[10]Sumardjo D 2008 Handbook of Medical Students Lecture and Programa Strata I Faculty of Biosekta EGC Press Jakarta

[11] Syartiwidya 2003 Study of Texture and Microstucture Changes of Fish Nugget During Processing and Storage Institute Pertanian Bogor

[12] Nuraini and Harapin H H 2006 Characteristic of beef quality of Peranakan Ongole from Longissimus dorsi and Gastrocnemius J. IlmiahIlmu – Ilmu Peternakan **9**(4) 250.

[13] Rahim S 2009 Effect muscle type and storage of beef quality J. Ilmu – IlmuPeternakan**12** (2) 67.

[14] Alfrianto W and Evi L 1991 Preservation and Fish Processing. Kanisius Publisher Yogyakarta

[15] Jamhari 2000 The change physical and organoleptical characteristics of beef meat during frozen storage. Buletin Peternakan B. Peternakan **24** (1) 44 .

[16] Utami R U, Jamhari, and Rusman 2006 Effect of thawing methods on physical quality and microstructure of frozen meat of mature male ongole grade B. Peternakan **30** (3) 147.

[17] Ali H M 2010 Effect of Effect of Muscle Types and Liquid Smoke Levels To Water Holding Capacity and power disconnect of Bali PrerigorBeefHassanudin University Makasar

[18] Rianto E, Rahmawati M F, and Purnomoadi A 2010 Physical Characteristics of Ongole Bulls' Meat at Various Body Weigh Seminar NasionalTeknologi Peternakan dan Veteriner 301

[19] Widati A S 2008 The Influence of Aging Period, Freezing Temperature and Packaging Material on Frozen Beef Chemical Quality J. Ilmu dan Teknologi Hasil Ternak **3**(2) 39.

[20] Mulyono A *2011* Effect Use of Fish Flour and Protected Soya Salts in Rations on The Quality of Meat Longissimus Dorsi (LD) chemical Simmental Ongole (SIMPO) crossbreed cattle Beef Sebelas Maret University Surakarta

Milk Production, Physiological Condition and Performance of Etawa Crossbreed Goats Feed by Ration Supplemented with Mangosteen Peel Flour

Dzarnisa[1*], D. Rachmadi [1], A. Azhar[2], R. Fakhrur Riza[1] and A. Hidayati [1]

[1]Faculty of Agriculture, Syiah Kuala University, Darussalam, Banda Aceh.
[2]Faculty of Veterinary Medicine, Syiah Kuala University, Darussalam Banda Aceh.

[*]Corresponding E-mail: dzarnisa@yahoo.com

Abstract

Study on the effect of the addition of mangosteen (*Garcinia mangostana L.*) peel flour on physiological condition and performance of Etawa crossbreed goats was done. This was to grant the use of mangosteen peel flour that rich of antioxidants and has variety good benefits for health as feed additive for cattle. This study used a Complete Randomized Block Design consisting of 4 treatment groups and 4 replications each. Subjects were 16 female Etawa crossbreed goats randomly designed into treatments group based on lactation periods. Subjects were feed with traditional rations (control, A), traditional rations and 2.5% mangosteen peel flour (B), tradition rations and 5% mangosteen peel flour (C), and traditional rations and 7,5 % mangosteen peel flour (D). Data on performance (milk production) and physiological condition (respiratory frequency, rectal temperature, and heart rate) obtained were analyzed using analysis of variance (ANOVA). The results showed that the addition of mangosteen peel flour as food additive in the rations resulted in variations in the milk production, physiological condition (rectal temperature, heart rate and respiration frequency) and performances (daily weigh gain, food consumption, ration conversion and breast volume) of Etawa crossbreed goats, but significant effect was only observed in the respiration frequency. The addition of 2.5% mangosteen peel flour in the ration caused the best, expected effects on milk production physiological condition and performance of Etawa crossbreed goats.

1. Introduction

The demand of animal products has progressively increased from year to year in line with increased numbers of population, better earnings and education. The increased demand animal products such as milk and meats can be fulfilled by raising bifunctional ruminants such as dairy goats.

The raise of dairy goats is an alternative for diversification program from non-cattle dairy animals. Among dairy goats, Etawa crossbreed goat has potential to develop for profitable entrepreneurial business. This goat is widely distributed in Indonesia, highly adaptive to the soil surface in various regions,. Raising management of the goat is relatively easy, cheap and does not require extensive land. Reproduction of the animal is also shorter than that of dairy cattle [1].

Efforts to increase milk production and quality could be done by improving genetic quality and raising management as well as by feeding appropriate, good rations. The use of agricultural wastes that contain medically beneficial biological compounds as feed additive might bring some benefits for animal production.

Mangosteen peels are among antioxidant-rich byproducts can be used for this purpose. The peels consist of xanthone that has been known as potential antioxidant, anti-proliferative, and antimicrobial agent that is absence in other fruits. Xanthone compounds include mangostin,

IOP Conf. Series: Earth and Environmental Science **119** (2018) 012040 doi:10.1088/1755-1315/119/1/012040

mangostenol A, mangostinon A, mangostinon B, trapezifolixanthone, tovophyllin B, alfa mangostin, beta mangostin, garcinon B, mangostanol, flavonoid epicatechin and gartanin [2].

The addition of mangosteen peels that contain biological compounds beneficial for health in the rations might provide antioxidant required for blocking deleterious effect of free radicals produced in metabolisms, support cell regeneration/replacement, and trigger cell growth. It is expected that these beneficial effects also occur in udder cells and bring good influence to physiological condition and performance of dairy goats. This study was done to investigate the influence of provision of mangosteen peel flour in rations on milk production, physiological condition and performance of Etawa crossbred goats.

2. Materials and Method

This study was done in the UD. Atjeh Livestock located in the Limpok Village, Aceh Besar from March 18 upto May 18, 2016. The study used Complete Randomized Block Design consisting of 4 treatment groups and 4 replications each. Here, 16 female Etawa crossbreed goats used and designed into treatment groups based on their lactation periods. Feed treatments used were the additions of mangosteen flour of 0% (A, control), 2.5% (B), 5% (C), and 7.5% (D). Parameters observed were goat milk production, physiological condition (respiration frequency, hearth rate, and rectal temperature) and performance data obtained was analyzed using analysis of variance (ANOVA) and Duncan Multiple Range Test [3].

3. Results and Discussion

3.1. Milk Production

Milk production of Etawa crossbreed goats given several amounts of mangosteen peel flour is presented in Figure 1.

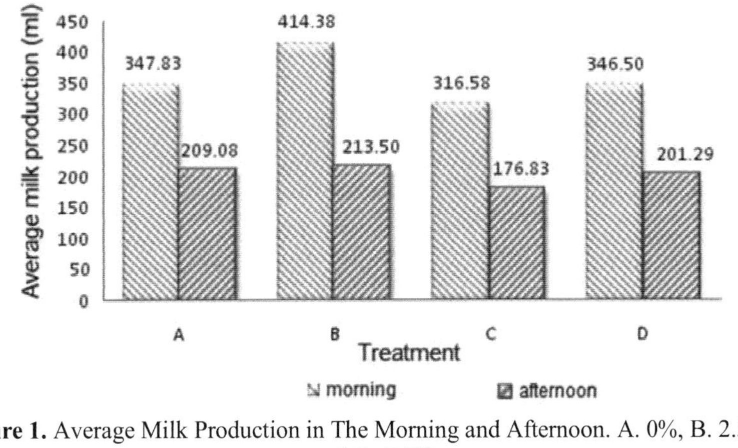

Figure 1. Average Milk Production in The Morning and Afternoon. A. 0%, B. 2.5%, C. 5.0% and D. 7.5%.

As illustrated in Figure 1, the goats feed with ration added with 0% (A), 2.5% (B), 5.0% (C) and 7.5% (D) mangosteen peel flour produced more milk in the morning than in the afternoon. The highest morning and afternoon milk productions were obtained from Etawa crossbred goats given 2.5% and 5.0% mangosteen peel flour, respectively. There were slight variations in morning and afternoon milk production of Etawa crossbreed goats among treatments, but the difference was insignificant (P>0.05). In other word, the addition of mangosteen peel flour 2.5-7.5% did not significantly influenced milk production of Etawa crossbreed goats.

It is predicted that antioxidant and antibacterial effects of additive compound in the mangosteen peel flour were able to restore structure of intestinal villi function in absorption process. Antibacterial effect of herbal plants is able to suppress the growth of pathogenic bacteria in the intestine [4]. Based on those statements Etawa crossbreed goats given ration contain 2.5% mangosteen peel flour showed high milk production due to increased food consumption. The addition of higher mangosteen peel flour might result in lower food consumption due to unbalanced nutrient compositions or poor palatability, causing low milk production. Good quality food is important for production of good quality milk. Stimulation of additive food might act the promoter of performance efficiency where additive food brings some changes in the digestive tracts, resulting in better growth and food efficiency [5].

3.2. Physiological Condition

a. Respiration Frequency

Results showed that Etawa crossbred goats given 2.5-7.5% mangosteen peel flour had slightly higher morning, afternoon and evening respiration rates than untreated goat controls. The variations in respiration frequencies observed, however, were still under normal limits because according to [6] normal range of respiration frequency of goats was 26-54 times per minute. The highest morning respiration rate (47.64 ± 1.92 times/minute) was found among goats received 7.5% mangosteen peel flour, but the rate was not significantly different from those observed in other treatment groups. In the afternoon, goats received mangosteen peel flour had significantly higher respiration rates than control, and the highest respiration rates was found in the goats given 7,5% mangosteen peel flour. In the evening, respiration frequency of goats received 2.5% and. 7.5% mangosteen peel flour had significantly higher (P<0.01) than those given no or 5% mangosteen peel flour. The highest respiration frequency in the evening was also found in Etawa crossbred goats given 7.5% mangosteen peel flour.

Some factors might contribute to respiration frequency of animals are physiological response to the changes in environmental temperature and humidity. Goat respiration frequency increases if environmental temperature raises [6].

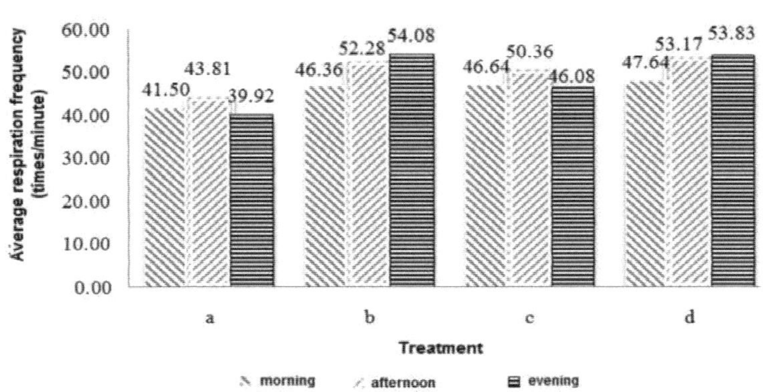

Figure 2. Average Respiration Frequency of Etawa Crossbred Goats Given Different Amount of Mangosteen Peel Flour. a. 0%, b. 2.5%,c. 5.0% amid d. 7.5%.

b. Rectal Temperature

Rectal temperature of Etawa crossbred goats received different doses of mangosteen peel flour is presented in Figure 3. There were no significant differences (P>0.05) in morning, afternoon and evening rectal temperature among treatment groups. Rectal temperatures observed belonged normal rectal temperature of goats reported by [7] [8], 38.5-40°C and 38.5 – 39.7°C, respectively.

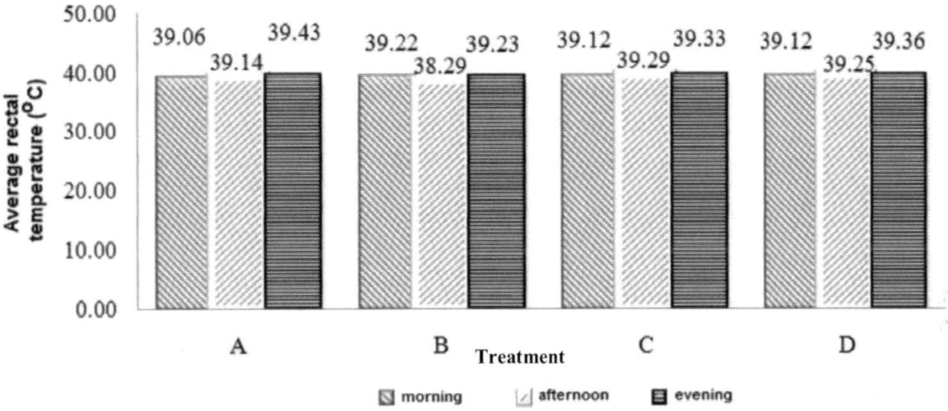

Figure 3. Average Rectal Temperature of Etawa Crossbred Goats Given Certain Amounts Mangosteen Peel Flour. A. 0%, B. 2.5%, C 5.0%, and D. 7.5%.

c. Heart Rate

Results of this study indicated that average heart beat frequency of Etawa crossbred goats in the morning, afternoon and evening (Figure 4) belonged to normal range reported [6] for goats, i.e. 70 – 135 times/min. That heart rate of goats in the morning ranged from 70-95 times/min [9].

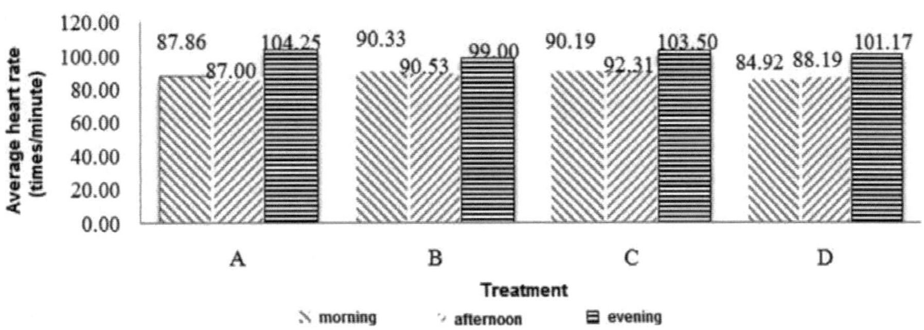

Figure 4. Average Heart Rate of Etawa Crossbred Goats Given Certain Amounts Mangosteen Peel Flour. A. 0%, B. 2.5%, C 5.0%, and D. 7.5%.

As illustrated in Figure 4, average goat heart rates in the evening were higher than those in the morning and afternoon. This might be related to the lower environmental temperature. In lower environmental temperature animals perform many activities to stabilize their body temperature. According [10], over activities might increase in body temperature, followed increased respiration frequency, heart rate and pulse.

3.3. Goat Performance

a. Weight Gain

Effect of mangosteen peel flour addition on average daily weight gain o Etawa crossbred goats are presented in Figure 2 below. Figure 2 showed that the addition of 2.5% and 5% mangosteen peel flour resulted in the highest and the lowest average daily weight gain of Etawa crossbreed goats (97.5 g - 37.5 g). Goats feed with ration supplemented with 5% mangosteen peel flour achieved average daily weight gain significantly lower ($P<0.05$) than those given control rations or ration added with 2.5% and 7.5%mangosteen peel flour.

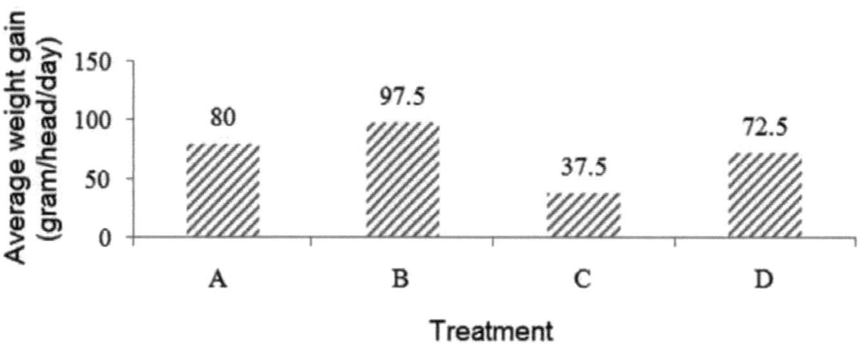

Figure 5. Average Daily Weight Gain of Etawa Crossbred Goats Given Certain Amounts Mangosteen Peel Flour. A. 0%, B. 2.5%, C 5.0%, and D. 7.5%.

The differences of weight gain achieved due mangosteen peel flour substitution in rations of Etawa crossbreed goats were assumed related to the changes in ration nutrient composition due to the increased dosage of mangosteen peel flour added. Appropriate dose of mangosteen peel flour might keep balanced feed nutrients and increase feed palatability. Mangosteen peel contains antocyanin, tannin and phenolic acids that might influence ration taste, resulting different food composition among goats in treatment groups. Antocyanin was not only antioxidant agent but also water soluble pigment responsible for the different colors of plant such as red, blue, purple and yellow. The higher concurrence of tannin might delay microbial digestion of ration by making a complex with protein and carbohydrate [11].

b. Feed Consumption

Feed consumption is the amount of food ate in certain period of time. Effect of mangosteen peel flour addition on food consumption of Etawa crossbreed goats can be seen in Figure 3.

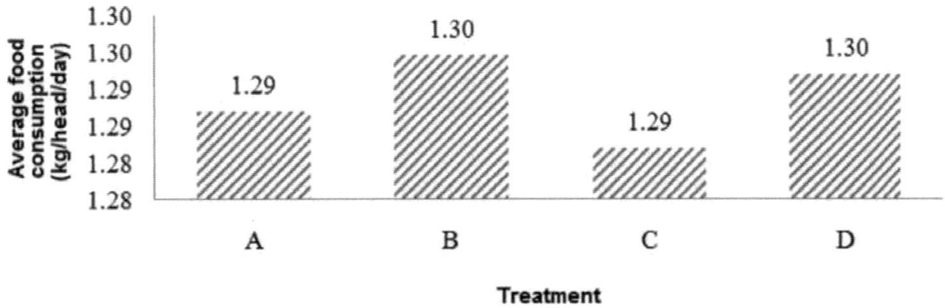

Figure 6. Average Feedd Consumption of Etawa Crossbred Goats Given Certain Amounts Mangosteen Peel Flour. A. 0%, B. 2.5%, C 5.0%, and D. 7.5%.

According to Sumoprastoso [12] the rate of food consumption and the capability to consume food of animals vary according to the type of food and animals body weight. Low nutrient rations are consumed more than high nutrient rations. Ruminants have normal condition (healthy and non-lactating) consume certain amounts of food to meet the need for basal metabolism and production. Nutrients absorbed are then metabolized to produce energy and to synthesize other nutrients. The amounts of food consumed by animals vary according to species [13].

c. Ration Conversion

Ration conversion represents the ratio between the amount of food consumed by animals within certain period of time and the yielded products. The amount of food given influences *Feed Conversion Ratio* (FCR), the relationship between total feed consumed and body growth. Low conversion values manifest low feed consumed by animals to produce meat 1 kg [14].

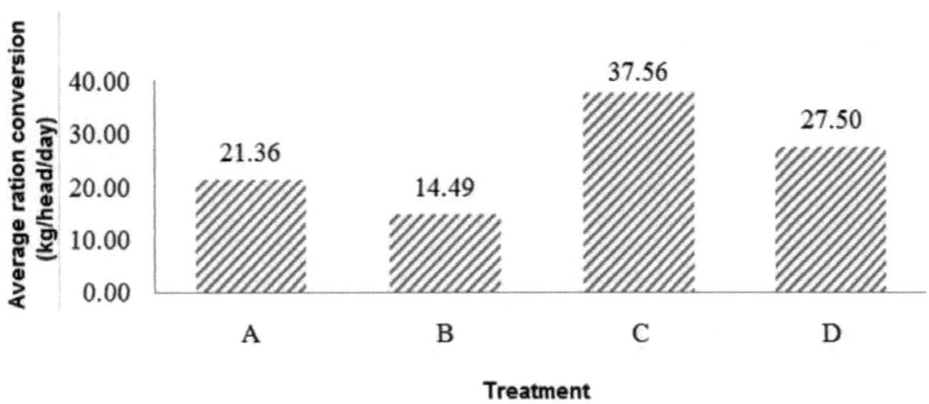

Figure 7. Average Food Conversion of Etawa Crossbred Goats Given Certain Amounts Mangosteen Peel Flour. A. 0%, B. 2.5%, C 5.0%, and D. 7.5%.

This is in agreement with [15] who suggested that higher nutritional values in the rations might reduce feed conversion, indicating better efficiency of feed use. The numbers of nutrients required and the ability to consume food of ruminant animals strongly depend on animal body weight

[16]. In other words, better food conversion represents low food consumption result in improved body weight.

Soewardi [17] also suggested that the ability of animals to covert feed closely relay on feed quality, in particular the contents of protein, energy and fibers. Among factors influence food conversion are genetic, environment (temperature, disease, feed, cage, and drink), feed nutritional values and feed energy content [18].

3.4. The Increase of Udder Volume

Udder is milk producing organ of dairy animals equipped by output ductus called nipple. The end of nipple excretes milk when milked. Average increase in udder volume of results of Etawa crossbred goats before milking ranged from 845-1045 cm^3. Statistical analysis results showed that the addition of mangosteen peel flour 2.5-7.5% insignificantly influenced (P>0.05) the increase of udder volume of Etawa crossbred goats. The lowest results was found in feed treatment A (1,045.00±126.79 cm^3), followed by treatment D (946.88 ± 118.95 cm^3), B (868.44±181.32 cm^3), and C (845.00±126.54 cm^3).

The increase in udder volume of Etawa crossbred goats after milking ranged from 570.00–774.06 cm^3 and were insignificantly different (P>0.05) among treatment groups. Pra- and post-milking udder volume, however, were significantly different. Average udder volume before milking was markedly bigger than those of after milking due to smaller, empty udders.

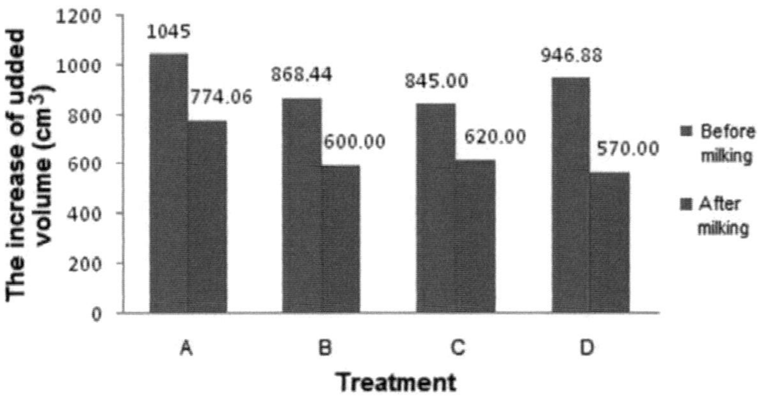

Figure 8. Average Udder Volume of Etawa Crossbred Goats Given Certain Amounts Mangosteen Peel Flour Before and After Milking. A. 0%, B. 2.5%, C 5.0%, and D. 7.5%.

Results obtained were in agreement with those reported by Judkinds [19] that big mammals do not always produce more milk, but body size and body weight do influence high milk production. According to Phalepi [20] animal milk production is influenced by genetic, age, udder dimension, live weight, lactation period, raising management (cage, food, and health), temperature, adaptability, and milking activity.

4. Conclusion

It can be concluded that the addition of mangosteen peel flour as food additive in the rations resulted in variations in the milk production, physiological condition (rectal temperature, heart rate and respiration frequency) and performances (daily weigh gain, food consumption, ration conversion and breast volume) of Etawa crossbreed goats, but significant effect was only observed in the respiration frequency. The addition of 2.6 % mangosteen peel flour in the ration caused the best, expected effects on milk production physiological condition and performance of Etawa crossbreed goats.

Acknowledgements

All praise and thanks to Allah the Almighty and Most Merciful, and may peace be upon to his prophet and messenger, Muhammad SAW and his companions. We would like to express our appreciation to the Ministry of Research, Technology and Higher Education of the Republic of Indonesia; the Rector of Syiah Kuala University; and the Head of Research Institute of Syiah Kuala University for funding this study under Higher Institution Research Grant scheme. We also would like to thank the Dean of Agricultural Faculty and the Head of Farm Animal Study Program of Syiah Kuala University, UD Aceh Livestock Farm and people who were directly and indirectly involve in the completion of this study.

References

[1] Abidin, Z and A. Sodiq. 2008. Increasing Etawa Cross Goat Milk Production. PT. Agro Media Pustaka. Jakarta.

[2] Qosim, W.A., 2007. Mangosteen Peel as Antioxidant. <http://anekaplanta.wordpress.com> Accessed on January 7, 2016.

[3] Steel, R.G.D. and J.H. Torrie. 1995. Statistical Principle and Procedure. PT Gramedia Pustaka Utama, Jakarta.

[4] Velmurugan, S. and T. Citarasu. 2010. Effect of herbal antibacterial extracts on the gut floral changes in Indian white shrimp Fenneropenaeus indicus. Rom.Biotech. Left.15: 5709-5717.

[5] Kellems, R.O. and D.C. Church. 2010. Livestock Feeds and Feeding. 6th Ed. Prentice Hall, Pearson, New Jersey.

[6] Frandson, R.D. 1996. Anatomy and Physiology of Animals. Gadjah Mada University, Yogyakarta.

[7] Smith, J.B., Mangkoewidjojo, S. 1988. Raising, Breeding and the Use of Experimental animals in Tropical Regions. Laboratory Rats (Rattus norvegicus): 37-57. Indonesian University Press, Jakarta.

[8] Anderson, 1970. Physiological Condition of Goats. Gajah Mada University Press, Yogyakarta.

[9] Dukes, H. 1957. The Physiology and Nutrition of Ruminants. 10th Ed. Cornell University Press. Ithaca, New York.

[10] Sulistyo, B. 2013. Animal Physiology. Universitas Padjajaran, Jatinangor.

[11] Samsudin, A. M. and Khoiruddin. 2009. Extraction, Membrane Filtration and Stability Test of Color Substance from Mangosteen (Garcinia mangostana). Research Report, Chemical Engineering Department, Faculty of Chemical Engineering, Faculty of Engineering, Diponegoro University, Semarang.

[12] Sumoprastowo, C. D. A. 1980. Successful Goat Farming. Printed I.Perberbit Bhatara Karya Aksara. Yogyakarta.

[13] Tillman, A. D. 1991. Livestock Feed Material Composition For Indonesia. Gadjah Mada University Press. Yogyakarta.

[14] Edjeng, S and Kartasudjana, R. 2006. Poultry Livestock Management. The Swadaya spreader. Jakarta.

[15] Maynard, L. A., J. K Loosly, H. F. Hintz and R. G. Werner. 1979. Animal Nutrition. 7th Ed, Tata McGraw Hill Publishing Company Limited, New Delhi.

[16] Siregar, S. 1994. Ruminant Livestock Rations. The Swadaya spreader. Jakarta.

[17] Soewardi, B. 1974. Ruminantia Nutrition. Volume I. Department of Animal Feed. Faculty of Animal Husbandry Bogor Agricultural University, Bogor.

[18] Nesheim, M. C., R. E. Austic and L. E. Card, 1979. Poultry Production. 12th ed. Lea and Febiger, Philadelphia.

[19] Judkins, H. F. and H. A. Keener. 1960. Milk Production and Processing White Palins, New York

[20] Phalepi, MA. 2004. Performance of Etwah Cross goat (A Case Study In Animal Husbandry Training Center and Village Swadaya Flavor) [thesis]. Bogor: Department of Animal Production, Faculty of Animal Husbandry, Bogor Agricultural University.

IOP Conf. Series: Earth and Environmental Science **119** (2018) 012041 doi:10.1088/1755-1315/119/1/012041

The Effect of the Gambir (*Uncaria gambir* (Hunt.) Roxb.) Leaves Waste and White Turmeric (*Curcuma zedoaria*) for The Productivity, Antioxidant Content and Mastitis Condition of The Fries Holland Dairy Cows

E. Nurdin[1*] and Fitrimawati[1]

[1]Faculty of Animal Science, Andalas University, Kampus Unand Limau Manis, Padang.

[*]Corresponding E-mail: ellyza_n@yahoo.com

Abstract

The aim of this research was to evaluate the effect of the gambir leaves waste and white turmeric on the productivity, antioxidant content and mastitis condition of dairy cows. The treatments were A. 0% gambir waste + 0.02% white turmeric; B. 20% of gambir waste + 0.02% white turmeric; C. 30% gambir waste + 0.02% white turmeric; D. 40% gambir waste + 0.02% white turmeric; and E. 50% gambir waste + 0.02% white turmeric. The results of *in vitro* experiment showed that the optimum ecological conditions rumen pH were ranging from 6.66 to 7.59, NH_3 concentrations were ranging from 23.42 to 31.54 nM and VFA concentrations were ranging from 66.15 to 119.82 mM. All treatments gave a real impact on the rumen ecology of dairy cows FH. *In-vivo* experiment for the best treatment from *in-vitro* experiment (30% gambir waste + 0.02% white turmeric) showed that similarly, the effect of administration of gambir leaves waste and white turmeric on dairy cattle FH, antioxidant activity affect the increase of milk production ($P<0.05$) from 9.75 liters to 11.52 liters (18.15%), increase the antioxidant content of milk ($P<0.05$) from 10.98 to 13.89 (26.50%) and a decline of the mastitis ($P<0.05$) from positive two to positive one. Gambir leaves decoction waste at the amount of 30% combined with white turmeric gave the optimal condition of rumen ecology. It was also increased milk production by 18.15%, the antioxidant content of milk by 26.50%, and decline mastitis.

1. Introduction

Maximum productivity is determined by the feed. Alternative way to balance the rumen ecology for optimal durability and animal productivity is using plants that contain antioxidants. Antioxidants work in a system to support each other. Therefore, the administration of combination of antioxidants will provide the better results. Antioxidants are commonly found in plants such as sunflower, curcuma, cumin [1,2,3,4] and gambir [5]. The numbers are very large and untapped, while white turmeric has an advantage in reducing cases of subclinical mastitis was 72.73 - 86.37% [3]

Gambir (*Uncaria gambir* (Hunt.) Roxb.) is a commodity of West Sumatra and contribute to gambir export in Indonesia by 80%. According to gambir crop area in West Sumatra is 21.412 hectares with a production of 14.220 tons / year, 6% of them is the yield of gambier exported and 94% or 168 087.67 tons left is gambir extraction waste [6]. It means that 168 087.67 tons of waste is not utilized and being wasted, whereas gambir waste such as leaves contains antioxidants and good nutrition if it is fed to livestock. Besides, the number of gambir processing units in West Sumatra is as much as 3571 units by 6908 workers. Therefore, the utilization of gambir substitution on livestock will improve the community economy.

Gambir has been used by the public as wound healing drug, anti-nematode, autonomic nerve stimulator, and an antioxidant. Gambir contains catechins which is a substance to be antioxidants [5]. Phytochemical content of flavonoids includes catechins, 50%, pyrocatechol 20% -30%, gambirin 1% - 3%, red kateku 3% -5%, quersetin 2% -4%, wax 1% -2% and alkaloids 2% -5%. The content of catechins in the leaves are 40% -50% [7]. Based on the results of qualitative analysis showed that gambir contains quinone, terpenoids, alkoloid, tannins, flavonoids and saponins [8] and antioxidant activity of extract gambir leaf are 172.62 ppm [21]. While the results of proximate analysis conducted by [9] on gambir extraction waste contains dry materials 56.43%, crude protein 10.66%, crude fat 4.90%, crude fiber 29.35%, BETN 35.73%, and tannin 5%.

The objective to be achieved was to obtain a formulation which is a combination of gambir leaf decoction waste and white turmeric in the feed of dairy cattle that can optimize rumen ecological conditions and will eventually be a parameter of dairy cattle productivity and produce antioxidant milk, so that it can utilize the available natural resources which are not utilized.

In the terms of the economic condition of the society, the utilization of gambir waste will increase farmers' income by utilizing the waste generated from gambir processing, become new business opportunities by producing a product to be given to livestock and will ultimately increase the productivity of dairy cattle. Besides, it can increase the income of dairy farmers because it can produce antioxidant milk which have a higher price than the normal one.

2. Materials and Method

2.1. *Phase I. is the phase of in-vitro studies.*

Materials: Materials was rumen fluid was taken from Fries Holland (FH) dairy cows of 4 years old, The waste decoction of the leaves of gambier (*Uncaria gambir* (Hunt.) Roxb.) come from home industry of gambir at Payakumbuh-West Sumatera and white turmeric (Curcuma zedoaria) come from herbal market at Bandung-West Java. This research was carried out at Balai Penelitian Ternak, Ciawi-Bogor, West Java, and Chemical Research Laboratory of the University of Padjadjaran in Bandung, West Java.

Research design: The design used was complete randomized design with 5 treatments and 8 replications. Treatments were several levels of combination of gambir waste and white turmeric by body weight. The treatments are A. 0% gambir waste + 0.02% white turmeric; B. 20% of gambir waste + 0.02% white turmeric; C. 30% gambir waste + 0.02% white turmeric; D. 40% gambir waste + 0.02% white turmeric; and E. 50% gambir waste + 0.02% white turmeric.

Sampling and analysis: The variables are rumen pH, rumen NH3 concentration, total VFA, and they were used as a parameter of the dairy cattle productivity before it was being given to cattle and the potential as an antioxidant. This research was carried out *in vitro* in the laboratory by *a pressure transducer* technique (Theodoreu and Brooks, 1990)

Statistical Analysis: The collected data was analyzed by Analysis of Variance. After a significant F test ($P<0.05$), Duncan's Multiple Range Test was used to inspect differences among treatments [10].

2.2. *Phase II is the phase of in-vivo studies.*

The experiment was done in smallholder dairy farm in Padang Panjang, West Sumatera; Faculty of Animal Science and laboratory of Agricultural Technology, Andalas University, Padang, West Sumatra for three months. The cow were kept in individual cage equipped with feeder and waterer.

Animals: Twenty Fries Holland lactating cows suspected with in mastitis subclinical (++) and positive contained bacterial pathogens (*Staphylococcus, Streptococcus, E.coli,* and *Corynecbacterium*) were used in the experiment. These cattle were in 2nd - 4rd lactation period, with body weight of 350 kg and milk production of 10 – 12 liters/head/day.

Feeding: They were fed 40 kg/head/day of King grass and 6 kg/head/day of concentrate feed produce. The feed contained 87.07% dry matter, 12.42% crude protein and 68.00% TDN. Before being given treatment, all animals in the given worm medicine and materials adapted to the treatment (gambir and white turmeric) for 1 week. The waste decoction of the leaves of gambier (*Uncaria gambir* (Hunt.) Roxb.) come from home industry of gambir at Payakumbuh-West Sumatera and white turmeric (Curcuma zedoaria) come from herbal market at Bandung-West Java.

Research design: used was complete randomized design with 2 treatments and 10 replications. The treatments using the best treatment of *in-vitro* studies (Phase I) are A. non antioxidant; and B. 30% of gambir waste + 0.02% white turmeric.

Sampling and analysis: The parameters measured were the production of milk (4% FCM), milk antioxidant content (HPLC) and the condition of mastitis (CMT reagen).

Statistical Analysis: The collected data was analyzed by Analysis of Variance. After a significant F test ($P<0.05$), Duncan's Multiple Range Test was used to inspect differences among treatments [10].

3. Results and Discussion

The effect of substitution treatment of gambir leaf decoction waste in the forage on ruminal pH, NH_3 concentration, total gas and Total VFA was shown in Table 1.

Table 1. The Effect of Treatments to Average Rumen pH, NH_3 Concentration and VFA

Treatments	Parameters		
	pH	NH_3 (mg/l)	VFA (mg/100ml)
A	6,66^{a}	23,42^{a}	113,38^{a}
B	6,94^{a}	28,12^{b}	112,31^{a}
C	6,81^{a}	31,54^{c}	119,82^{c}
D	7,08^{b}	27,85^{b}	87,62^{b}
E	7,59^{b}	21,35^{a}	66,15^{b}

Description: A. 0% gambir waste + white turmeric; B. 20% gambir waste + white turmeric; C.30% gambir waste + white turmeric; D. 40% gambir waste + white turmeric; and E. 50% gambir waste + white turmeric.

The results showed that the optimum ecological conditions are rumen pH ranging from 6.66 to 7.59, NH_3 ranging from 23.42 to 31.54 nM and VFA ranging from 66.15 to 119.82 mM. All treatments give a real impact on the rumen ecology of dairy cows FH. It means that the administration of gambir substitution with a concentration of 20%, 30%, 40% and 50% affect the ecology of the rumen ($P <0.05$). Gambir waste is an antioxidant ingredients that substitution of gambir waste can be used to 30% as an antioxidant. Because of its nature as an antioxidant, it will provide the antioxidant benefits at a certain level and in excessive doses will act as bactericidal agent [11,12]. As a result, rumen pH will increase at the dose above 30% gambir waste + white turmeric. VFA as a parameter of productivity also showed the highest production at a dose of 30% gambir waste + white turmeric and decreased at a dose above 30% gambir waste + white turmeric. Gambir leaves decoction waste also contains good nutritional value for dairy cattle. It does not contain ingredients that are malnourished and hazardous materials (Table 2 and Table 3). This is due to fast protein degradation of feed compared with microbial protein synthesis due to insufficient availability of carbon skeletons, so NH_3 will accumulate and exceed the optimum level.

Table 2. Proximat Analysis and Antioxidant of Gambir Leaf Decoction Waste

Sample	Content (%) [1)]					Antioxidant (ppm)
	Water	BK	PK	SK	BETN	
Fresh	13,97	86,03	10,62	18,28	67,06	172.62[2)]
Decocted	8,99	91,01	13,26	26.82	53,92	34.99[3)]

Source:
1) PUPT research result at Laboratory of Livestock Nutrition and Food, Faculty of Animal Husbandry, Padjadjaran University (2017).
2) Rahmawati at al. (2013) [21]
3) PUPT research result at Analysis result in Laboratory of Research and Services, Department of Chemistry, Faculty of Science, Padjadjaran University (2017)

Buffering fuction of gambir as antioxidant was able to maintain rumen pH remains in normal conditions to the level of 30%. pH range in the results showed that the treatment in the manufacturing process of gambir did not affect the active ingredients contained in the leaf decoction waste of gambir, so that pH conditions stayed in optimum state that have an impact on the condition of NH_3 and total VFA as a final product. Budinuryanto et al. (1999) stated that the saponins and tannins on gambir as the active ingredient have a function as a pH buffering [13]. Gambir processing will increase the active substance of antioxidants, tannins and polyphenols. Analysis result in Laboratory of Research and Services, Department of Chemistry, Faculty of Science, Padjadjaran University (2017) showed that the decoction waste of gambir leaves has a potential antioxidant activity by 17.22 - 34.99 ppm, while white turmeric itself has antioxidant content of 0.881 ppm [14,15].

It showed that the quantity of the active ingredient after processing is higher in gambir waste, where the decoction waste has a potential antioxidant activity by 6.56 ± 0.04 ppm accordance with the opinion of Nurdin et al. that conducting research on the effects of antioxidants and antioxidant ingredients processing, that the handling of antioxidant ingredients can increase or decrease the activity depending on the type of antioxidant. Some of them are heat resistant while the others are not [3,11,12,16].

In-vivo experiment for the best treatment from *in-vitro* experiment (30% gambir waste + 0.02% white turmeric) showed that similarly, the effect of administration of gambir leaves waste and white turmeric on dairy cattle FH, antioxidant activity affect the increase of milk production from 9.75 liters to 11.52 liters (18.15%), increase the antioxidant content of milk from 10.98 to 13.89 (26.50%) and a decline of the mastitis from positive two to positive one (Table 3).

Table 3. The Effect of Treatments to Milk Production, Antioxidant Content of Milk, and Mastitis Condition

Treatments	Parameters		
	Milk production	Milk antioxidant	Mastitis condition
A	9.75^{a}	10.98	Positive2
B	11.52^{b}	13.89	Positive 1

Description: A. 0% gambir + 0.02% white turmeric; B. 30% gambir + 0.02% white turmeric.

The optimum ecology of rumen would increased produced VFA (acetic acid and propionic acid). Acetic acid is function to milk fat precursor and propionic acid is function for glucosa syntesis [22]. The other hand, the condition of rumen ecology would be effect of quality and quantity dairy cows production [2, 23, 24]. Gambir leaves waste and white turmeric by virtue of its antioxidant compound and anti-inflammation potential can increase alveolar cell permeability and can assist in rumen ecological balance.

The greater the value of mastitis status showed good condition because the more healthy dairy cows (rating the status of mastitis: negatif = 4; + = 3; ++ = 2; +++ = 1; and ++++ = 0). Mastitis condition in this research showed a decrease in the level of mastitis compared with control (76.92%). Test of mastitis condition with CMT reagen's mastitis conducted at the beginning and end of research showed that treatment B were negative ($P<0.01$). This is possible because the cow's used in this research detected positif mastitis (++) and supported of microbiologik test found pathogenic bacteria in milk. But supplementation of herbs resulted in the decrease mastitis.

Provision of natural antioxidants of Curcuma sps. group, such as white turmeric and turmeric mango given to dairy cows, is significantly increase the productivity, endurance, and the levels of antioxidants of milk in different rate depending on dairy cows [3, 4, 12, 15, 17, 18]. White turmeric has advantages in reducing mastitis condition and gambir waste has higher antioxidant content, which can increase the antioxidant content of milk. Because feed contents consumed by livestock will be accumulated in the body and milk by 5-20%.

The administration of antioxidants in combination will give a good effect because each antioxidant will complement the lack of other antioxidant ingredients. The treatments increased antioxidant milk aboaut 26.50%. It is caused by the work mechanism of antioxidants that usually work within a system (antioxidant networking).

Supplementation of 30% of gambir waste + 0.02% white turmeric until 4 weeks showed improvement on mastitis status. The treatment of mastitis by using antibiotic, the new *Staph. aureus* infection occurred again and new intramammary infections occuring after treatment and bacteria become resistant [25]. The effect of antibiotics has improved and some of them are retained in the cow's body for along time and at higher concentration [26]. Antibiotic residues in livestock product was harmful to consumer health, In addition, this treatment improves milk yield and milk quality , decrease condition of mastitis in dairy cattle and cured mastitis.

The study discover that doses of combination natural antioxidant (Waste decoction of the leaves of gambier (*Uncaria gambir* (Hunt.) Roxb.) and white turmeric (Curcuma zedoaria)) that can be beneficial for increase the productivity and healthy of dairy cattle, produce antioxidant milk and increase the economic condition of the society. The utilization of gambir waste will increase farmers' income by utilizing the waste generated from gambir processing, become new business opportunities by producing a product to be given to livestock and will ultimately increase the productivity of dairy cattle. Besides, it can increase the income of dairy farmers because it can produce antioxidant milk which have a higher price than the normal one. This study will help the researcher to uncover the critical areas of productivity and health of dairy cows especially of mastitis problem by using the combination of natural antioxidant.

4. Conclusion

Gambir leaves decoction waste at the amount of 30% combined with white turmeric gave the optimal condition of rumen ecology. It was also increased milk production by 18.15%, the antioxidant content of milk by 26.50%, and decline mastitis.

Acknowledgements

Authors thank General Directorate of Higher Education, Ministry of Education and Culture in accordance with the agreement letter of Research Grant No. No.059/SP2H/LT/DRPM/IV/2017 in "Penelitian Unggulan Perguruan Tinggi" program.

References

[1] Nurdin E. 2007. The Effect of Supplementation Sunflowers (Helianthus annuus L.) and BIOPLUS to the Solid Non Fat, Total Solid and Milk Acidity of Sub-clinical Mastitis's Holstein Dairy Cows. J.Anim. Prod. Vol. **9** No.2: 79.

[2] Nurdin E, Arief. 2009. The Effectivity of Cumin as Natural Anti-oxidant to Improve Rumen Ecology of Mastitis Dairy Cow's. J. Anim. Prod. Vol **11**. No.3: 160-164.

[3] Nurdin E, Amelia T, Makin M. 2011. The Effect of Herbs on Milk Yield and Milk Quality of Mastitis Dairy Cows. Journal Indonesian Tropical Animal Agriculture. Vol **36** No. 2, Hal. 104.

[4] Susanty H. 2011. Pemanfaatan Temu Putih (*Curcuma Zedoaria* (Berg.) Roscoe) Dalam Meningkatkan Kuantitas Dan Kualitas Susu Sapi Perah Penderita Mastitis Subklinis. Thesis. Institut Pertanian Bogor. Bogor.

[5] Nurdin E dan F Rahim. 2015. Pemanfaatan limbah rebusan daun gambir (Uncaria gambir (Hunt.) RoXB.) sebagai substitusi hijauan terhadap Ekologi rumen dan sebagai bahan anthelmintik alami pada ternaK. sapi perah FH (in-vitro). Seminar Nasional Peternakan Berkelanjutan 7. Fakultas Peternakan Universitas Padjadjaran. Jatinangor.

[6] BPS Sumatera Barat. 2013. Sumatera Barat dalam angka tahun 2012. Badan PusatStatistik dan Badan Perencanaan Pembangunan Daerah Propinsi Sumatera Barat. Padang.

[7] Lucida H, A. Bakhtiar, Wina AP. 2007. Formulasi Sediaan Antiseptik Mulut dari Katekin Gambir. J. Sains Tek. Far, **12**(1): 1 .

[8] Nazir N. 2000. Gambir Budidaya, Pengolahan dan Prospek Diversifikasinya. Cetakan I. yayasan hutanku. Padang : 1-14.

[9] Kasim A. 2010. Reorientation of Research and Utilization of Gambier . Proceding International Seminar Food and Agricultural Sciences-ISAFAS 2010. Bukittinggi, Sumatera Barat.

[10] Gaspersz, V. 1995. Tekhnik Analisis dalam Penelitian Percobaan. Tarsito, Bandung. pp.62 – 111.

[11] Nurdin E, Putra DP, Amelia T. 2013. Analysis Of Heavy Metal Lead (Pb) Levels With Aas In Cow's Milk By Giving Cumin (*Cuminum Cyminum* L.*)*, White Turmeric (*Curcuma Zedoaria* Rosc.) And Mango Turmeric (*Curcuma Mangga* Val.). Pak. J. Biol. Sci. Vol **16/21** : 1373.

[12] Nurdin E, Susanty H. 2015. Effect of Curcuma zedoaria, Curcuma mangga and Cuminum cyminum on rumen ecology and Pb profile in the rumen of mastitis dairy cows (in-vitro). Pak. J. of Biol Sci. Vol. **18(3)**: 146.

[13] Budinuryanto, D. C., E. Komolowati dan Hermawan. 1999. *Potensi Antioksidan dan Antiflakmatirik alami untuk penanggUlangan Mastitis Subklinis*. Laporan Hasil Penelitian RUT. Jakarta.

[14] Nurdin E, Makin M, Tasripin D. 2013. Respon Pemberian Tanaman Obat terhadap Peningkatan Efisiensi Produksi Susu, Efisiensi Ransum dan Kadar Antioksidan Susu dalam Rangka Peningkatan Kualitas Pangan Nasional. Prosiding Seminar Nasional Akselarasi Pembangunan Pertanian Berkelanjutan Menuju kemandirian Pangan dan Energi. UNS. Solo.

[15] Nurdin E, Afriani T, Susanty H, Marbun F. 2015. Pemanfaatan Berbagai Metoda Pengolahan Kunyit Putih (*Curcuma Zedoaria)* Sebagai Sumber Antioksidan Terhadap Ekologi Rumen Ternak Kerbau (*In-Vitro).* Seminar Nasional Fakultas peternakan Universitas Jendral Soedirman. Purwokerto Jawa Tengah.

[16] Afriani A, Nurdin E, Susanty H. 2014. Pemberian Berbagai Bentuk Pengolahan Kunyit Mangga Pada Ternak Kerbau Dalam Rangka Penentuan Jenis Kelamin Gudel Dan Peningkatan Produktivitas(*In-Vitro).* Semnas Peternakan Berkelanjutan 6. Unpad, Jatinangor.

[17] Nurdin E, Susanti H. 2011. Determination of utilization level of Curcuma zedoaria Rosc. to Improve Rumen Ecology of Mastitis Dairy Cows (in-vitro). Proceding International Seminar, The 8th Biannual Meeting, The 3rd Congress and Workshop of AINI. Unpad Bandung.

[18] Nurdin E, Rahim F, Lumbantungkup J. 2014. Pengaruh Metoda Pengolahan Kunyit Kuning (Curcuma Domestica) Sebagai Suplementasi Pakan Pada Ekologi Rumen Kerbau (*In-Vitro).* Seminar Nasional Peternakan Berkelanjutan 6. Fakultas Peternakan UNPAD. Jatinangor-Sumedang. Jawa barat.

[19] Kasim A, Asben A, Mutiar S. 2015. Kajian Kualitas Gambir dan Hubungannya dengan Karakteristik Kulit Tersamak. Kulit, Karet dan Plastik Vol 31 No.1.

[20] Kresnawaty, Irma, Ahmad Z. 2009. Aktivitas antioksidandan anti bakteri dari Derivat metil Ekstrak Etanol Daun. (Uncaria gambir). J. Littri **15** (4): 145.

[21] Rahmawati,N., A.Fernando and Wachyuni. 2013. Kandungan Fenolik dan aktivitas Antioksidan Ekstrak Daun Gambir kering (Uncaria gambir (Hunter) Roxb.). J.Ind.Che.Acta Vol.**4** (1): 1.

[22] Kalscheur K.F., R.L. Baldwinvi, B.P. Glenn and R.A. Kohn. 2006. Milk Production of dairy cows differing concentration of rumen-degraded protein. J. Dairy Sci. **89**: 249.

[23] Kay, J.K., J.R. Roche, E.S. Kolver, N.A. Thomson and L.H. Baumgard. 2005. A Comparison Between Feeding System (pasture and TMR) and The Effect of Vitamin E Supplemntation on Plasma and Milk Fatty Acid Profiles in Dairy Cows. J. Dairy Res. **72**: 322.

[24] Wu, Z. and L.D. Satter. 2000. Milk Production during the complete lactation of dairy cows fed diets containing different amount of protein. J. Dairy Sci. **83**: 1042.

[25] Owens, W.E., S.C. Nickerson, R.L.Boddie, G.M.Tomita, and C.H.Ray. 2001. Prevalence of mastitis in dairy heifers and Effectiveness of Antibiotic Therapy. J. Dairy Sci. **84**: 814.

[26] Langford, F.M., D. M. Weary and L. Fisher. 2003. Antibiotic Resistance in Gut Bacteria from Dairy Calves: A Dose Response to the Level of Antibiotics Fed in Milk. J Dairy Sci **86**:3963.

Carcass Production of Cattle Slaughtered at Salatiga City Slaughter House, Salatiga, Central Java, Indonesia

E. Purbowati[1*], C. M. S. Lestari[1], M. J. Ma'ruf[1] and S. Sutaryo[1]

[1]Faculty of Animal and Agricultural Sciences, Diponegoro University, Semarang Indonesia

*Coresponding E-mail: purbowati@hotmail.com

Abstract

The objective of this study was to evaluate the breed, age, sex, slaughter weight, carcass weight, and carcass percentage of cattle which was slaughtered at Slaughter House in Salatiga, Central Java. The materials used in the study were 156 head of catlle. The sampling used was incidental sampling to identify the breed, age, sex, slaughter weight and carcass weight. The data gathered were analyzed descriptively. The result showed that the sex of all the cattle slaughtered were male. The breeds of the cattle were Frisian Holstein Grade (70.51%), Simmental (15.38+3.21), Simmental-Ongole Grade (5.13%), and Limousine-Ongole Grade (5.77%). The average age of the cattle were 2.34 year old, with an average of slaughter weight of 529.34 kg, while the averages of carcass weight were 277.61 kg. The average of carcass percentage was as high as 52.56%. The conclusion of the study was the highest number of breeds of the cattle slaughtered at Slaughter House in Salatiga were young Frisian Holstein, the body weights were included in large frame score, and the carcass percentage were moderate.

1. Introduction

The Indonesian government has many programs to increase the livestock production, especially meat, in reaching the dream in which Indonesia becoming a country with self-sufficiency of meat. The meat production indicator of cattle can be measured from the weight and carcass percentage, because the meat is a part of carcass components [1]. According to Soeparno (2005) [2], the definition of carcass is the slaughter production of cattle which the blood, skin, head, viscera, low parts of the legs, and tails are removed. Cattle can be classified as good cattle are if it the carcass yield is 59% of the slaughter weight, and from there, the amount edible meat is 46.5 % [3]. The carcass percentage is ratio of carcass weight to slaughter weight, multiplied 100 % [4].

Some factors that affect carcass percentage are feed, breed, age, sex, hormone, slaughter weight, and body conformation [5]. The kind of feeds, feed consumption, and feed composition has a great influence to the growth of cattle, where the higher protein and energy intake will accelerate the growth rate of the cattle [2]. The slaughter weight will increase as the cattle's age increase [3], so the carcass weight will increase too [2]. The carcass percentage will increase with the increase of the slaughter weight [6]. According to Saka *et al.* (2011) [7], Bali bull has higher carcass percentage those female Bali cattle. A big cattle breed will also produce big carcass [2]. The slaughter weight, carcass weight, and carcass percentage of Simmental-Bali Grade breed is higher than Bali-Bali Grade, Ongole-Bali, and Brahman-Bali breed [8]. The study objective is to determine breed, age, slaughter weight, and carcass percentage of cattle slaughtered at Slaughter House Salatiga, Salatiga, Central Java.

2. Material and Method

2.1. Time and Location

The study was done in Slaughter House Salatiga (SHS) Salatiga City, Central Java on January 20 to February 18, 2016.

2.2. Materials

The materials used in the study were 156 heads of cattle that were slaughtered at SHS Salatiga City. The equipments that were used to slaughter are knife, axe, digital scale, hanging hooks, scale, barrow, hydraulic clamp, hose, and working suit.

2.3. Methods

The study was done via observation and by following the whole activity of cattle slaughtering process in Slaughter House Salatiga. The sampling method that is used is incidental sampling, where the samples of the cattle are taken incidentally while conducting the study. The slaughtering process in Slaughter House Salatiga started from ante mortem inspection, cattle slaughtering and carcass formingprocess, until post mortem inspection. The data that are gathered were breed, age, sex, slaughter weight, carcass weight, and carcass percentage.

The age was determinedusing the observation to the teeth eruption included incisors, pre molar, molar, and permanent teeth [9]. Slaughtered and carcass weight were measured using digital scale. The definition of carcass is the slaughtered cattle which the skin, head, tail, low part of leg and viscera is removed. The dressing percentage is a proportion of carcass to body weight multiplied by 100% [10]. The data gathered was analyzed descriptively.

3. Result and Discussion

3.1. The Slaughter House Salatiga

The Slaughter House Salatiga (SHS) is located at Jl. Imam Bonjol No. 111 A. Salatiga. The SHS has been built in 1984-1985 on the area of 13,000 m^2, including the SHS building of 785 m^2. SHS was inaugurated officially by the Governor of Central Java in 1987. In 1993, the SHS was renovated to be type B Slaughter House by Director General of Livestock. According to Badan Standarisasi Nasional (1999) [11], the type B slaughter house must have special room to wash internal organs, a laboratory to detect antibiotic residue, physically and biologically waste processing process, chilling room, the inside wall of slaughter building should made of porcelain, hot water resources, vehicle with cooling box, and has a veterinarian.

The SHS has no laboratory to detect antibiotic residue, because the lab only has simple equipments. Hot water, chilling room, cooling room, and the cooling vehicle are not well utilized. The waste management in SHS was done by a third party. In addition, SHS was not distributing beef meat as per distribution guideline of type B Slaughter House, which is interprovincial.

The location of SHS is beside a traditional market and quite closed to housing area, so disruptions are likely to happen. Lestari (1994) [12] stated that a good slaughter house location is in city border where the human population is not high, near the river stream, or in the lowest part of city so that it will notproduce any disruption or contaminate the environment. According to Setiajatnika (2011) [13] the technical requirement of slaughter house location is 2-3 km from housing area. The goal of SHS is to serve people in getting a safe, healthy, complete, and *halal* meat. The average number of cattle which slaughtered in SHS was 8 heads/day. The number of cattle that was slaughtered in SHS was low. Based on the number of cattle slaughtered in SHS, the SHS could be grouped as category 1 slaughter house [14].

3.2. The Characteristics of Cattle and its Carcass at Slaughter House Salatiga

The result of the study (Table 1) showed that the average age of the cattle slaughtered at Slaughter House in Salatiga was 2.34 year old, male, and the average of body weight, carcass weight, carcass percentage were 529.94 kg, 277.61 kg, 52.56% respectively. Purbowati *et al.* (2015) [15] reported that the age of cattle that was slaughtered at Slaughter House Penggaron, Semarang is ranging from 2-3 year old (has 2 pairs of permanent incisors). The average body weight, carcass weight, and carcass percentage were 497.95 kg, 247.17 kg, and 49.59%, lower than those of this study.

Most of the cattle breeds slaughtered at SHS was Friesian Holstein Grade (FHG), as many as 70.51% followed by Simmental (15.38%), Limousine-Ongole (LimOG) Grade (5.77%), Simmental-Ongole (SimOG) Grade (5.13%), and Limousine (3.21%). The reason why most of cattle breed slaughtered at SHS was Friesian Holstein Grade because the Salatiga city was very closed with the Friesian Holstein Grade dairy cattle location, so that the bulls would be fattened as a beef cattle. The study of [15] showed that the cattle breeds slaughtered at Penggaron Slaughter House (PSH) were Simmental = 36%, Limousine = 22%, Limousine Grade = 18%, Simmental Grade = 16%, and other breeds (Ongole Grade and Brahman Angus Grade = 8%).

Table 1. The Characteristic of Cattle Slaughtered at Slaughter House Salatiga.

Cattle Breeds	n	Age	Slaughter Weight	Carcass Weight	Carcass Percentage
		(year)	----------------------- (kg)	------------------	(%)
FHG	110	2.40	477.91	254.23	53.99
Simmental	24	2.33	548.92	285.00	51.95
Limousin	5	2.50	622.20	328.80	52.90
Sim-OG	8	2.44	542.25	282.13	51.95
Lim-OG	9	2.06	458.44	237.89	52.00
Average		2.34	529.94	277.61	52.56

n = total sample

The age range of the cattle slaughtered at SHS was 1- 4.5 year old. It was showed that the cattle which were slaughtered were young. The young cattle have a good quality of meat, so that the consumer would like it. Arifin and Purbowati (2011) [14] noted that the prime carcass quality is received from the slaughtered cattle with an age of 9 months to 3.5 year old, and has a slightly abundant of marbling. Cattle can produce prime quality of beef by fattening it using high concentrate or grain. The cattle's age slaughtered at SHS is similar with the age of cattle slaughtered at PSH, which was 2-3 year old. Ali *et al.* [16] reported that the age of slaughter cattle slaughtered at Slaughter House in Makassar (South Sulawesi) was 4.76 year old in average, older than the age of the cattle from the study. The age percentage of cattle slaughtered at SHS for each cattle breed is presented at Illustration 1.

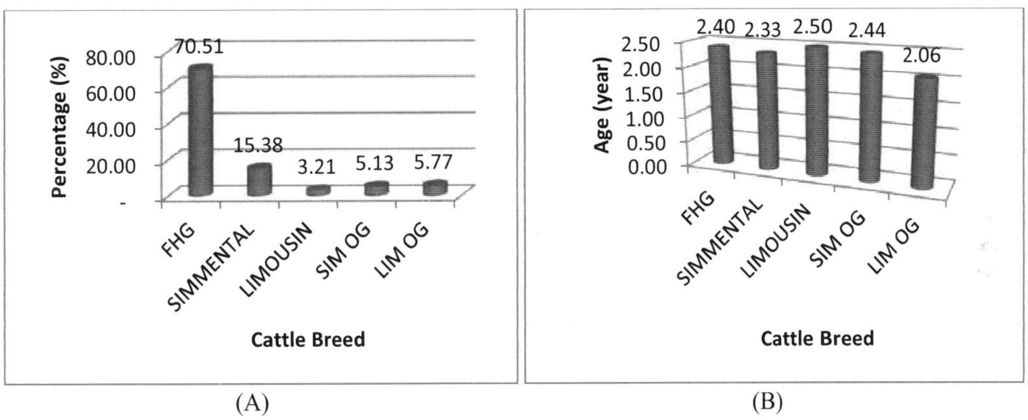

Illustration 1. Breed Percentage (A) and Age (B) of Cattle Slaughtered at SH Salatiga

The average slaughter weight of cattle slaughtered at SHS was as high as 529.94 kg. The highest slaughter weight was the Limousine (622.20 kg), and the lowest was Limousine-Ongole Grade (458.94 kg). According to Arifin and Purbowati [14], the slaughter weight of cattle slaughtered at SHS was included into medium score frame (405-495 kg) to large score frame (495-585 kg). Frame score of the suitable cattle for slaughter is used to describe how heavy (in kg) the cattle were when reaching maturity, when the back fat on 12th rib is 0.2 inch thick and the gastric was properly full. The study of Purbowati *et al.* [15] resulted that at the SHP Semarang, the average of the cattle slaughter weight was 497,95$\pm$44,97 kg (CV = 9,03%). The slaughtered weight was affected by feed, breeds, and age of cattle. The slaughter weight of each cattle breed at SHS could be seen at Illustration 2.

The highest and the lowest carcass weight were Limousine breed (328.80 kg), and Limousine-Ongole Grade (237.89), with an average carcass of 277.61 kg. The carcass weight is in line with the slaughter weight, as high slaughter weight will also be produce high carcass weight. This was shown by Purbowati *et al.* [15] that the highest carcass weight at SHP was Brahman-Angus Grade (295 kg), followed by Simmental (249.33 kg), Limousine (245.65 kg), Ongole Grade (236,25 kg), Limousine Grade (235,56 kg), and Simmental Grade (221,35 kg), in line with their slaughter weight.

The carcass weight of the study at SHS was higher compared to carcass weight in Slaughter House Penggaron (SHP). The average of dressing percentage (DP) of the study was as high as 52.66%. Frisian Holstein breed has the highest DP (53.99%), while Simmental and Simmental-Ongole Grade have DP of 51.95%. The average of DP in the study (52.56%) is included as medium DP. According to Zajulie *et al.* [3], cattle are considered as good if it has a DP of 59%. The DP of the study was better compared to DP from SHP as many as 49.59% (Purbowati *et al.*) [15]. The weight and DP of each cattle breed in SHS was presented at Illustration 3.

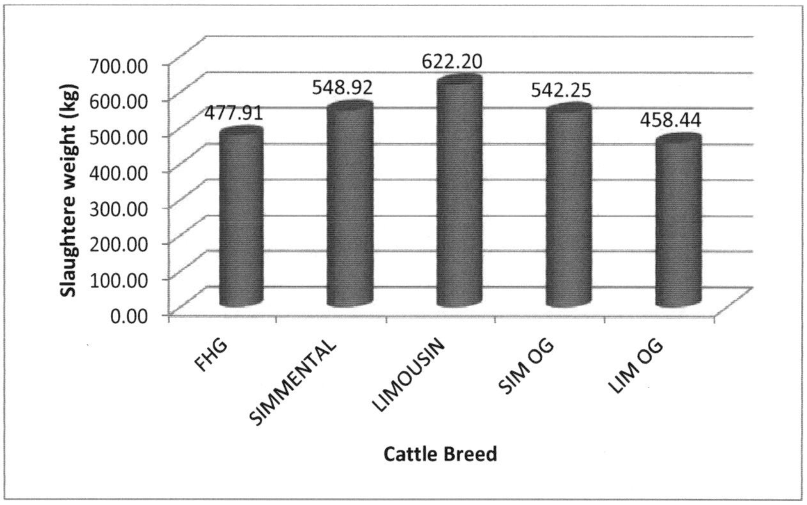

Illustration 2. The Average of Slaughter Weight of Cattle Breed at SHS

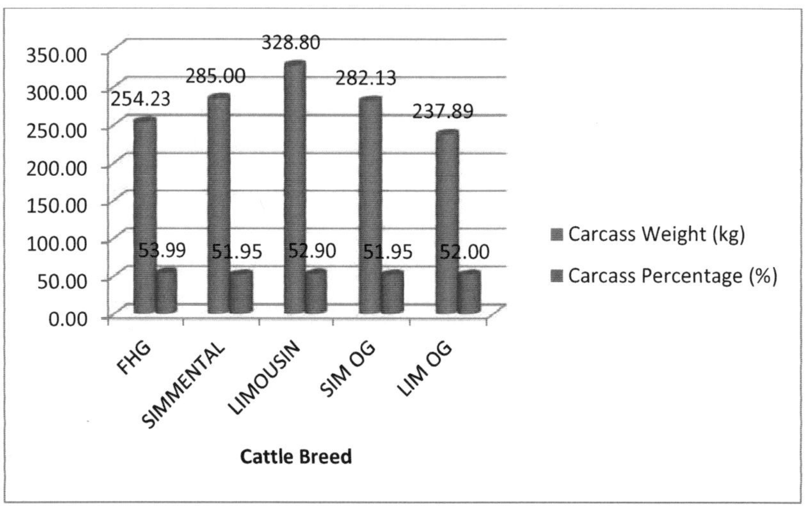

Illustration 3. The Carcass Weight and Carcass Percentage of Cattle at SHS

4. Conclusion

The conclusion of the study was the breeds of cattle slaughtered at Slaughter House in Salatiga were mostly young Frisian Holstein, considerably young age (2,34 years old), male, with the body weight of 529.94 kg, belongs to large frame score, and the DP (52.56%) were moderate.

Reference

[1] Hafid, H., Nuraini, and Herman. 2013. Karakteristik karkas dan bagian-bagian karkas sapi Peranakan Ongole jantan dan betina pada peternakan rakyat di Provinsi Sulawesi Tenggara. Prosiding Seminar Nasional Teknologi Peternakan dan Veteriner. Medan 3-5 September 2013. Medan (Indonesia). Puslitbangnak hlm: 116 – 121

[2] Soeparno, 2005.Ilmu dan Teknologi Daging.Gadjah Mada University Press.Yogyakarta (Indonesia).

[3] Zajulie, M.I., M. Nasich, T. Susilawati, dan Kuswati. 2015. Distribusi komponen karkas sapi *Brahman Cross* (BX) hasil penggemukan pada umur pemotongan yang berbeda. Jurnal Ilmu-ilmu Peternakan 25: 24-34

[4] Prado, I.N., J.A. Aricetti, P.P. Rotta, R.M. Prado, D. Perroto, J.V. Visentainer, and M. Matsushita, M. 2008. Carcass characteristics, chemical composition, and fatty acid profile of the Logissimus muscle of bulls (Bos taurus indicus vs Bos taurus taurus) finished in pasture systems. Asian-Aust J. Anim. Sci. 21 : 1449 – 1457.

[5] Preston, T.R. and M.B. Willis.1982. Intensif Beef Production. 2nd Ed. Pergamont Press, New York

[6] Forrest, J.C., E.D. Arbele, H.B. Hendrick, M.D. Judge and E.A. Merkel. 1975. Principles of meat science.W.M. Freeman and Company, San Fransisco (US)

[7] Saka, I.K., I.B. Mantra,I.N.T. Ariana, A.A. Oka,N.L.P. Sriyani,and S. Putra 2011. Karakteristik karkas sapi Bali betina dan jantan yang dipotong di rumah potong umum Pesanggaran, Denpasar. The Excellence Research Universitas Udayana, Denpasar (Indonesia). hlm: 39 – 47

[8] Munthalib, R.A. 2003. Karakteristik karkas dan daging turunan F1 empat bangsa pejantan dengan sapi Bali betina. J. Indon. Trop. Anim. Agric. 28 : 7 – 10.

[9] Frandson, R. D., 1996. Anatomi dan fisiologi ternak, Edisi ke-7.B. Srigandono dan K. Praseno, penterjemah.Gadjah Mada University Press, Yogyakarta (Indonesia)

[10] Irin. 2012. Survei karkas sapi potong dan kerbau tahun 2012. Newsletter Pusat Data dan Informasi (Pusdatin) Bulan Oktober Tahun 2012, 9: 1 – 4.

[11] Badan Standarisasi Nasional 1999. SNI. 01-6159-1999: Rumah Potong Hewan. Badan Standarisasi Nasional. Jakarta (Indonesia)

[12] Lestari, P.T.B.A., 1994. Rumah Potong Hewan Ruminansia Indonesia. Bina Aneka Lestari.Jakarta (Indonesia).

[13] Setiajatnika, E. 2011. Feasibility study pendirian rumah potong hewan (RPH) di Kabupaten Bandung Barat. J. Ilmiah Ekonomi, Koperasi dan Kewirausahaan. **2** (1) : 55-72.

[14] Arifin, M. and E. Purbowati. 2011. Abatoir dan Teknik Pemotongan Ternak. Buku Ajar.Semarang (Indonesia): Badan Penerbit Universitas Diponegoro.

[15] Purbowati, E., C.M.S. Lestari, N.A. Sarie, Y. Haryati, W.S. Saputra, M. Istiadi,M., Arifin, E. Rianto, and A. Purnomoadi. 2015. Karakteristik sapi yang dipotong di rumah potong hewan Kota Semarang. Proceeding Seminar Nasional Teknologi Peternakan dan Veteriner, Jakarta 8-9 Oktober 2015. Jakarta (Indonesia). Puslitbangnak.hlm: 42-47

[16] Ali, H.M., Nahariah, Zulkharnaim, andJ.A. Syamsu. 2015. Karakteristik ternak potong pada RPH Makasar. Dalam: L. Abdullah dkk. Prosiding Seminar Nasional Strategi Pemanfaatan Lahan Rawa dalam Mendukung Kedaulatan Pangan Nasional. Banjarmasin 17-18 Maret 2015. Banjarmasin (Indonesia).LPPM. Fakultas Pertanian, Universitas Islam Kalimantan Muhammad Arsyad Albanjary. hlm: 272 – 277.

Bali Cattle Carcass Characteristic of Different Butt Shape Condition

H. Hafid[1*], Nuraini[1], Inderawati[1], and W. Kurniawan[1]

[1]Faculty of Animal Science Universitas Halu Oleo Kendari, Southeast Sulawesi, Indonesia.

*Coresponding e-mail: harapinhafid14@gmail.com

Abstract

Carcass was main product on cattle slaughtering which contain beef for human consumption and it has high nutritional and economical value. Carcass production on cattle has been influenced by several factors, such as cattle breed, feed, and body conformation. Cattle *Butt Shape* was one of part cattle body conformation which allegedly has have positive correlation on produced carcass. This research was aimed to evaluate *Butt Shape* condition influenced on Bali cattle carcass characteristic. The research was using Bali cattle which slaughter in Kendari Slaughtering House (Rumah Potong Hewan/ RPH - Kendari). The observation includes weighing, and measuring parts of carcass was conducted on 60 heads of *Halal* process slaughtered Bali cattle which traditionally maintained. The research parameters were carcass productivity parameters *i.e:* slaughtering weight, carcass weight and length, leg length and circumstances. Obtained data were analyzed using Complete Randomized Design and post hoc analyzed using Least Significant Different if have any influence. The research result showed that cattle *Butt Shape* condition have significant (p<0, 05) influence on all Bali cattle carcass productivity parameters. Butt shape with B category was result the best carcass productivity compare the others, while C category was better than D. It can be concluded that body and carcass weight were linearly influenced by cattle *butt shape*.

1. Introduction

Cattle classified as good and productive if they can produce optimum carcass quantity and quality. Carcass gained from cattle slaughtering could be indicated the successful of cattle raising process and also showing its meat productivity and corresponding with livestock economical value. Beef cattle was one of protein source which is will always to be needed human in Indonesia and in the world. Based on [1] explained that Indonesia have about 15.419.720 heads beef cattle, which producing beef reach to 506.660 tons and it was the most favoured meat source after poultry.

Cattle productivity increasing can be obtain by combining good raising management on it. The management action program can be applied on initial or end of raising process by evaluating cattle Body Score Condition (BSC), estimating butt shape condition, and calculating meat index, based on that, we can estimating and get the real cattle productivity description.

Cattle grading and classification, especially on carcass, was uncommon and wasn't generally known in Indonesia. It is because majority beef consumer didn't consider about beef quality. Almost all of beef parts were cooked and consumed by the same traditionally methods, include carcass and non carcass (offal). These caused slow development of Indonesian beef industry which create only two market segmentation *i.e.*: (i) traditional local market for middle low economical segment, and (ii) special market for high end segment, restaurant, starred hotel, and franchise store [2]; [3]. Therefore, we need a rapid and easy doing methods which can estimate carcass composition. Butt Shape score evaluation was one of carcass composition determination method with easy to apply. The evaluation can be held by comparing between cattle butt shape and its standard. Steer which raising by feedlot in Indonesia generally have B to D conformation butt shape score [3]; [4].

Butt Shape was cattle butt plumpness degree, which have function to describe cattle fatness condition and its carcass gaining. Butt shape score for very fat until very lean cattle was A until E respectively. The supreme cattle carcass should be full of meat, with little bones proportion, optimum fat composition, and most of meat [2]. This research be expected can be clearly explain the correlation between butt shape and gained cattle carcass characteristic.

2. Method

This research was conducted on Slaughtering House (Rumah Potong Hewan /RPH) of Kendari City in October 2014 until December 2016. The observation was focused on 40 heads of traditional raising Bali cattle. The research tools which used in this research were knife, body weighing scale, meat weighing scale, and stationary. The research idea scheme was showed below:

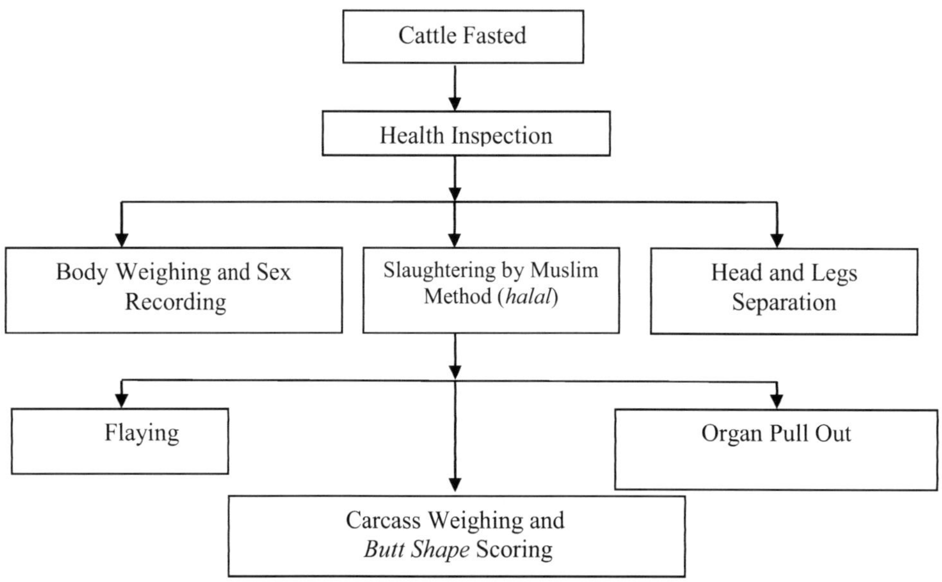

Figure 1. Research Main Idea

An ante mortem inspection applied to make sure those cattle was proper to be slaughtered, continued by cattle body weighing and sexes recording. Cattle slaughtering was held on Muslim method (halal) by cut their carotids' artery, jugulars venous, oesophagus, and trachea to obtain perfect bleeding. The research procedures was design based on [5] methods. Head and legs separation was performed on canon bones joint. Flaying process conducted on floor started by create long slice on chest and abdomen middle line, continuing slicing on medial (legs) section. Leather was separated start from ventral (bottom) to the back part. Organs pull out from abdomen and chest hollow were conducted, except the kidneys. Carcasses parts weighing were held after slaughtering, flaying and evisering process complete, and divide the parts to be has meat, flank, chest, front - rear legs, and totalize all parts to obtain warm carcass weight. The total weight of meat and bone in every part were obtained by separating them. In this research, carcass fat contain unable to determine due of it characteristic (merge on carcass meat) and there was no frozen room facility in slaughtering house.

2.1. *Butt Shape Determination*

Butt Shape of cattle could be defined by its upper leg conformation and shape. Butt Shape conformation defined as the conformity form of upper leg shape toward over all carcass conformation (skeleton structure, muscle and fat). Butt Shape determinations are useful to describing cattle fatness and its produced carcass. The cattle Butt Shape score presented as A for fatty and E for lean condition [6].

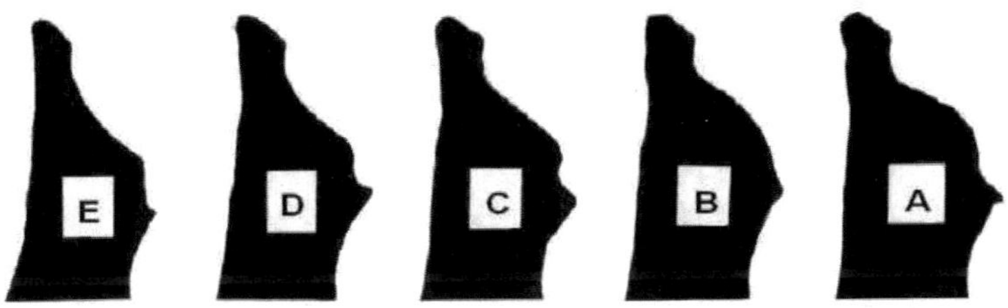

Figure 2. The Illustration of *Butt Shape* Conformation Score [4]

B. 1. B*utt Shape A category*, very convex
C. 2. B*utt Shape B category*, convex
D. 3. B*utt Shape C category*, flat/ average
E. 4. B*utt Shape D category*, deep-set
F. 5. B*utt Shape E category*, very deep-set

2.2. *The Observed Parameters*

1. Slaughtering Weight: the weight of cattle before slaughtered.
2. Carcass Weight: the weight of obtained carcass after slaughtering process.
3. Carcass Length: the measuring of carcass *os scapula* and *os pubis* range after slaughtering process.
4. Upper Leg Circumstance: the measuring of upper leg circumstance using ribbon scale
5. Leg Length: the measuring of carcass *os tibia* rear leg and *os pubis* range
6. *Butt Shape* Score: Scoring of butt shape condition based on Upper Leg Circumstance and Leg Length ratio.

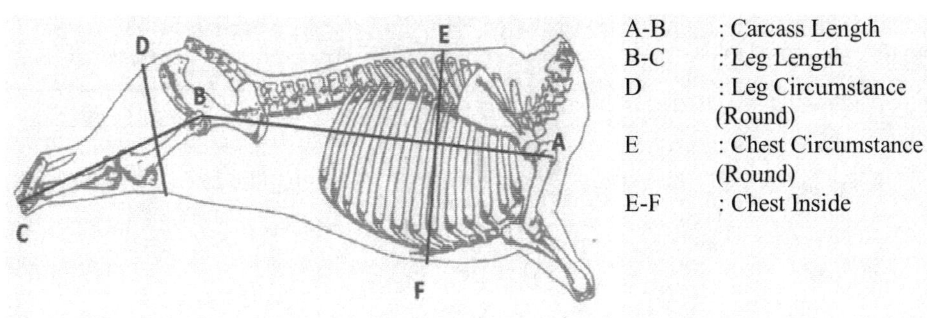

Figure 3. Carcass Measurement Illustration [7]

2.3. Data Analysis

The obtained data were analyzed using Analysis of Variance based on Complete Randomized Design through different replication numbers, and the significantly different were continued analyze using Least Significant Different (LSD) [8] as the models below:

$Y_{ij} = \mu + \tau_i + \varepsilon_{ij}$

Explanation:

$_i = 1, 2, 3,$

$_j = 1, 2, 3, \ldots\ldots, 18$

Y_{ij} = Observation value of *Butt Shape* and its Ith combination and Jth replication

μ = Observation value average

τ_i = Ith Treatment effect

ε_{ij} = Error

The post hoc model of LSD:

$LSD = q\alpha\ (p, fe)\ s\,\hat{Y}$

Explanation:

LSD = LSD value on significant α level.

qα = LSD table standard value

P = Treatments number

fe = Error freedom degree

$s\hat{Y}$ = Error mean square

3. Result and Discussion

3.1. *Slaughtering and Carcass Weight*

Slaughtering Weight was obtained while cattle ready to be slaughtered. The Butt Shape scoring was classified based on cattle's upper leg plumpness condition, which categorized on: A (very plump), B (plump), C (average), D (less plump), and E (poor). The research result showed that there was three Butt Shape categories on Kendari Slaughtering House, which was B, C, and D.

Table 1. The Average of Butt Shape Condition Effect on Carcass Characteristic.

Parameters	***Butt Shape* Condition**		
	B	**C**	**D**
Slaughtering Weight (kg)	205.0 ± 69.00^{a1}	180.0 ± 40.00^{b1}	150.0 ± 20.00^{b1}
Carcass Weight (kg)	105.3 ± 34.72^{a2}	88.6 ± 19.27^{b2}	64.7 ± 6.87^{c2}
Carcass Percentage (%)	51.10 ± 3.70^{a2}	49.03 ± 2.03^{b2}	42.1 ± 1.85^{c2}
Carcass Length (cm)	117.0 ± 4.32^{ns}	115.0 ± 1.75^{ns}	111.0 ± 1.80^{ns}
Upper Leg Circumstance (cm)	67.50 ± 1.60^{a1}	63.50 ± 4.70^{b1}	53.67 ± 1.21^{c1}
Leg Length (cm)	59.25 ± 3.50^{a1}	55.72 ± 3.20^{b1}	54.43 ± 3.10^{b1}

- Different superscript on same row showed significance different $^1(p<0,05)$ and $^2(p<0,01)$
- ns: not significance ($p>0,05$)

Based on research result (Table 1), the cattle Butt Shape categorized as B, C, D were have average slaughtering weight 205.0±69.00, 180.0±40.00, and 150.0±20.00 kg respectively. The variance analyses showed that Butt Shape condition have significant influence (p<0,05) on cattle slaughtering weight on Kendari Slaughtering House, which B categorized cattle Butt Shape significantly have higher on slaughtering weight compare with C and D category, meanwhile there was no significant difference between C and D category. There was Butt Shape significant influence (p<0, 01) on Bali Cattle carcass weight parameter which slaughtered on Kendari Slaughtering House. The result showed that B Butt Shape categorized cattle have higher carcass weight compare with C and D category, and C was significant higher than D Butt Shape category with the value 105.3±34.72; 88.6±19.27; and 64.7±6.87kg respectively. These indicated that slaughtering and carcass weight were increasing linearly based on the increase of cattle Butt Shape condition. Body conformation, such as Butt Shape, was related with body weight, which higher Butt Shape conformation would impact on the increasing of body weight and carcass weight [2]; [6]; [3];[9].

3.2. *Carcass Percentage and Carcass Length*

The research result showed that cattle slaughtered on Kendari Slaughtering House, which categorized as B, C, and D Butt Shape condition, have averages carcass weight percentage 51.10±3.70; 49.03±2.03; 42.1±1.85% respectively and Butt Shape condition have significant influence (p<0.01) on carcass weight percentage. In contrast with carcass percentage, Butt Shape condition didn't have significant influence (p>0.05) on carcass length. The different Butt Shape conditions have the same carcass length (B: 117.0±4.32; C: 115.0±1.75 and D: 111.0±1.80 cm)

Bali cattle with B Butt Shape condition have higher slaughtering weight compare to the others, and this condition also have higher carcass weight percentage. This result was in line [2];[3];[9];[10] which stated that the increasing of slaughtering weight would increase the carcass weight percentage. Higher slaughtering weight would have impact on the increasing of fresh carcass weight, and its percentage. In relation with cattle maturity, the increasing of cattle age should be accordance with the increasing of its body weight, and carcass weight.

There are various factors which have influence on carcass percentage; the main factors were cattle body conformation and fatness degree. Fat cattle should have high carcass weight percentage and generally have thick conformation (beams shaped). Lean/ slim shaped cattle, with long body and neck, have triangle shape (such as dairy cattle), generally have lower carcass weight percentage. The other factors which influenced carcass weight percentage were amount of feed and water on cattle digestion track, if there are so many of them, it could lower carcass weight percentage. Skin size and thickness also influenced carcass weight percentage [9].

3.3. Leg Circumstance and Lenght

The influence of Butt Shape condition B, C, and D on average Upper Leg Circumstance and Leg Lenght were (showed on Table 1.) 67.50±1.60 and 59.25±3.50; 63.50±4.70 and 55.72±3.20; 53.67±1.21 and 54.43±3.10 cm respectively. The variance analysis showed that Upper Leg Circumstance and Leg Lenght of Bali cattle with B Butt Shape category have significant (p<0.05) higher leg length, Flummness of Leg, and Fleshing Index compare to the other categories. These informed that Butt Shape condition have significant effect on leg circumstance of slaughtered Bali cattle in Kendari Slaughtering House, which B Butt Shape category was higher compare with C and D category, and C Butt Shape category was higher than D category. The leg lenght of Bali cattle categorized as B Butt Shape (59.25 ± 3.50 cm) was have significant higher compare with two other categories, while C (55.72 ± 3.20 cm) and D (54.43 ± 3.10cm) category was not significant different.

On living cattle, butt plumpness degree can be estimated using Butt Shape conformation or its leg shape. Butt Shape conformation defined as the conformity form of upper leg shape toward over all carcass conformation, involved skeleton, muscle and fatty [4]. There is some interesting case about the measurement method which nothing useful indication from using Butt Shape condition to estimated its produced and marketed beef although Butt Shape conditions generally choose as carcass marketing because of economical benefit. The A, B, and D shape score have higher price compare with D and E, with the same carcass weight price difference around US$ 40 [11].

4. Conclusion

Bali cattle butt plumpness degree (Butt Shape) have significant effect on all its carcass productivity parameters, which B category was better than C or D, while C category was better than D. Linearly, the higher cattle butt plumness would produced higher body and carcass weight, and have higher carcass weight percentage.

References

[1] BPS 2016 Statistik Indonesia. Badan Pusat Statistik Jakarta

[2] Hafid H dan R Priyanto 2006 Pengaruh konformasi Butt Shape terhadap karakterisitik karkas sapi Brahman Cross pada beberapa klasifikasi jenis kelamin. Media Peternakan 29(3)

[3] Hafid H, Nuraini, A M Tasse, Inderawati dan M Hasdar 2014 Karakteristik karkas sapi bali pada kondisi tubuh yang berbeda. Prosiding Seminar Nasional Peternakan Fakultas Peternakan dan Pertanian Universitas Diponegoro Semarang

[4] Hafid H 2005 Kajian pertumbuhan dan distribusi daging serta estimasi produktifitas karkas sapi hasil penggemukan. Disertasi Doktor Sekolah Pasca Sarjana IPB Bogor

[5] Hafid H dan N Rugayah 2009 Persentase karkas sapi bali pada berbagai berat badan dan lama pemuasaan sebelum pemotongan. Proseding Seminar Nasional Teknologi Peternakan dan Veteriner Pusat Penelitian Peternakan Deptan RI Bogor

[6] Hafid 2007 Bahan Ajar Teknologi Feedlot. Jurusan Peternakan Fakultas Pertanian Universitas Halu Oleo Kendari

[7] Hafid H 2011 Pengantar Evaluasi Karkas. Penerbit Unhalu Press Kendari

[8] Sudjana 1985 Desain dan analisis eksperimen. Penerbit Tarsito Bandung

[9] Soeparno 1992 Ilmu dan Teknologi Daging. Gadjah Mada University Press Yogyakarta

[10] Hafid H, R E Gurnadi, R Priyanto and A Saefuddin 2007 Identification of carcass characteristics for estimating composition of beef carcass. Jurnal Pengembangan Peternakan Tropis **35**(1) 22

[11] Thornton R F 1991 Report on Muscle Socers Trials. Australian Meat and Livestock Corporation Sydney

Effects of Feed Supplementation in Friesien Holstein Crossbreed Cows at the First Quarter on the Production and Quality of Milk

Suharyono[1*], S. N. W. Hardani[1], P. D. Sitoresmi[2] and Adiarto[2]

[1]Centre for the Application Isotopes and Radiation, BATAN.
[2]Faculty of Animal Science, Gadjah Mada University.

[*]Corresponding E-mail: suharyono@batan.go.id

Abstract

Nine heads of dairy cows were used in the study. The dairy cows were in 7-8 months pregnant condition, lactation II and were expected calving soon.The purpose of this study was to determine the effect of feed supplementation to increase the production and quality of milk. The feeding treatment was divided into 3 groups, G_I (control) given the usual feed given by livestock owners, G_{11} was given G_I+500 g UMMB/h/d and G_{111} was given GI + 500 g MFS/h/d. Variables observed were feed and nutritional consumption, average milk production, milk quality, cumulative milk production, average 4% fat corrected milk (FCM) production, peak milk production. The experimental design used a completely randomized design of direct pattern, continued by Duncan's new multiple range test (DMRT) if there were a significant difference of variable values between treatments. The results showed that the addition of dietary supplement significantly affected the mean consumption of crude protein between G_{III} and G_I and G_{11}, respectively 1.22 kg /d versus 0.99 and 0.33 kg/d. The average milk production was also influenced by the addition of dietary supplement that was between G_I; G_{11} and G_{111}, respectively 9.55; 10.69 and 11.85 l/d. Cumulative milk and 4% FCM production were also significantly different at P <0.05, with value of each G_I; G_{11} and G_{111} was 954.98; 1068.70 and 1184.57 l; 10.72; 12.06 and 13.70 kg /d respectively. The conclusion was MFS to be able to increase cumulative and 4% FCM milk production.

1. Introduction

One of the most important factors in livestock raising is feed. The main factor influencing the efficiency and profit in the dairy cattle business is the quantity and quality of feed given to dairy cattle [1] The success or failure of the livestock raising business is largely determined by the feed given as a source of energy and protein. The reality in the field shows that many farmers who feed their animals do not pay attention to the quality, quantity and nutritional adequacy required. As a result, livestock productivity that is maintained does not achieve as they expect.

The first trimester of lactation is a critical period for dairy cattle. In this period milk production increases but the cattle's ability to consume nutrients is not proportional to the need of cows to produce milk. As a result of this condition during this period the cows have a nutrient deficiency of feed, causing the mobilization of the body's reserve energy in the tissues that cause weight loss. This is supported [2,3] mentioned that dairy farmers are generally less concerned about the condition of the cows when the dry phase, consequently at the beginning of lactation there is a backup of body mobilization. This will cause a negative energy balance and simultaneously a decrease in body weight of the initial lactation cattle, resulting in a fall in body condition score. This means that dairy cows that do not get the feed as required, in the early period of lactation will not be able to show the milk production curve and weight changes as expected [4]. Weight loss is too high will cause productivity decreased. Furthermore, it has been suggested by previous researchers [5] who stated that in order to avoid NEB at the time of cow 2 weeks before and 4-12 weeks after calf was born, the feeding strategy in that phase must be done, because NEB is also a factor which causes the profit or loss of dairy cattle business for livestock farmers. Van Straten *et al.*, 2009 suggests that early dairy lactation of reproductive performance declines as a result of changes in BCS and body weight which are also associated with negative energy balance [6]. Early lactation dairy feeding strategies in order to reduce the energy of nagative balance and decrease in body weight and maintain milk production [7]. The strategy that has been done is by feeding individually on the cow's early lactation. Other researchers also stated that provision of higher energy sources in the ration at the beginning of lactation affects higher milk production and peak persistence of milk production is also longer [8]. One attempt to avoid NEB reduction should be added a high-quality feed that has nutritional content that is able to meet the nutritional needs required to increase the production of dairy cows in a late pregnant and the early lactation [1]. Lebih lanjut menyebutkan bahwa ransum yang diberikan pada sapi perah tersebut berupa konsentrat yang mengandung protein tinggi yang mana berbasis bahan baku lokal [1]. The improvement of feed on dairy cows that produce first-trimester milk with supplements based on local ingredients as well, the name of the feed supplement is urea molasses multi-nutrient block (UMMB) and multi nutrient feed supplement (MFS) [9,10].Other studies have stated that UMMB as a dietary supplement contains nutritional sources of minerals, proteins, and energy that had an important role to increase ruminant livestock [11]. Molasses urea blocks have utility for providing trace elements in ruminant diets [12].

Feed supplement of UMMB and MFS had been tested on ruminant livestock (sheep, goat, beef cattle and dairy cow) [10,13], the result shows a positive response to the increase in production of body weight and milk, and used as a business unit by livestock farmer group, especially UMMB [14,15] . The socio-economic assessment of this UMMB feed supplement has been conducted by [16]. The results reported that farmers as users of UMMB for beef cattle agree very well if this feed supplement was always given every day on the cow. Further, it was mentioned that UMMB was needed to be developed more widely because it had been able to increase the production of beef cattle. Technically, economically, socially and culturally the characteristics of this feed supplement technology have been accepted by user farmers.

Similarly for multi-nutrient feed supplements (MFS) which was able to increase milk production and maintain the persistence of milk production longer [9,10]. The high content of crude fiber caused the consumption of feed decreases, whereas when dairy cows got forage feed as much as 17% dry matter or 1.35% of body weight, then the quality and production of milk will increase [17]. The UMMB and MFS supplements contained crude protein, crude fiber, and energy, 24.5 each; 28.8% and 6.5; 17.2%, 2889 and 3997 K-cal [9,10], were expected to meet the nutritional requirements of dairy cattle used for research. The purpose of this study was to determine the effect of feed supplementation on increasing milk production and quality in the first lactation tri-semester period.

2. Materials and Method

Friesian holstein crossbreed cow were used in this study 9 heads with a mean body weight of 350 in lactation II. The average age of 2 to 3 years was pregnant 7-8 months and soon gave birth to a calf. Feeding was adapted to cow owner management. Forage was given in the form of cuttings with a size of 10-15 cm. Concentrate used was a commercial concentrate (Production Argasari Boyolali) with crude fiber (CF) content 22.61% and crude protein (CP) 14.71%. The feed treatment was divided into three groups, group I (G_I) : was given daily by the farmer, G_{II}: G_I + 500g UMMB and G_{III}: G_I + 500 g MFS. The content of CF and CP respectively for UMMB and MFS 19.06% and 19.58%;

14.96% and 16.90%. While the elephant grass content for CF and CP are 32.75 and 8.78%, respectively.The experimental design used a completely randomized design of direct pattern, which then when there was a difference between treatments followed by DMRT test [18] . Parameters observed included nutritional content [19], feed intake, nutritional balance, average milk production, milk quality (fat and protein), cumulative, average 4% FCM, and peak milk production, then losing body weight. For the composition of rations used in the study are presented in Table 1. Table 1 shows that the composition of rations in G_{11} and G_{111} were added 500 g of UMMB and MFS respectively.

Table 1. Composition of Rations for Dairy Cattle (Head/Day) in This Study

Feed ingredient	Feed treatment		
	Control G_I	G_I + 500 g UMMB (G_{II})	G_I + 500 g MFS (G_{III})
Elephant grass (kg)	14	14	14
Consentrate (kg)	8	8	8
Rice bran (%)	45	45	45
Coconut meal (%)	25	25	25
Oil palm cake (%)	10	10	10
Soy meal (%)	10	10	10
Coffee peels (%)	7	7	7
Vitamins and minerals (%)	3	3	3
Feed supplement (g)			
UMMB	0	500	0
MFS	0	0	500

3. Results and Discussion

3.1. Nutrient Content

The nutrient content of the feed ingredients used for the observations has been analyzed, the results obtained are presented in Table 2. The nutrient content of basal diet for CF content of 32.75% and CP 8.78%, and concentrate, respectively 22.61% and 14.71% (G_1). Therefore, with the addition of 500g UMMB or MFS provided information that G_{11} and G_{111}, the CF content was lower while the CP content was high. Because each got additional nutrients from CF and CP from G_1. Therefore, CF content would be lower, while the CP content was higher in each of G_{11} and G_{111}. The CF and CP content of G_{11} and G_{111} are presented in Table 2.

The nutrient content of this analysis shows that the added CP content of MFS 16.90% which is greater than the concentrate and concentrate + UMMB is 14.71 and 14.96%. This was due to the CP content of MFS and UMMB, respectively 28.8 and 24.5% [9].

3.2. Feed Consumption

The feed given was forage of *Pennisetum purpurium* and concentrate. The results of the observation of forage consumption and concentrates for the G_I, G_{II} and G_{III} groups are presented in Table 3.

Table 2. Nutrient Content of Feed Ingredients Used for Experiments (%)

Matterials	DM	Ash	CP	Extract ether	CF	TDN	OM
Elepant grass	25.41	11.39	8.78	1.71	32.75	54.93	88.61
Concentrate (C)	80.76	12.17	14.71	5.01	22.61	73.44	87.83
C + UMMB	81.27	11.43	14.96	5.95	19.06	76.54	88.57
C + MFS	81.14	10.86	16.90	4.01	19.58	74.37	89.14

TDN elephant grass was calculated using the formula of regression equation No.2 [20]
TDN concentrate was calculated by using formulas from [21]

Table 3. Average of Forage Consumption and Concentrate (Kg / day)

Feed Ingredient	Feed Consumption/DM		
	G_I : control	G_I + 500 g UMMB (G_{II})	G_I + 500 g MFS (G_{III})
Elephant grass (F)/kg^{ns}	2.93	2.86	2.84
Consentrate (C)/kg	4.96^{a}	4.50^{a}	5.74^{b}
Total^{ns}	7.89	7.37	8.58
Ratio F:C	37:63	39:61	33:67

a, b Different superscripts on the same line show the difference of P <0.05
ns was not significantly different P <0.05.

The treatment of feed supplementation did not significantly influence P <0.05 in forage consumption. This was because the forages given to the dairy cattle were not different both in quality and quantity from the three treatments. In contrast to consumption of concentrates, it appears that feeding supplementation had an effect on P <0.05, especially G_I and G_{II} differing from G_{III}. Each value was 4.96 and 4.50 kg / day compared to 5.74 kg/day. This was because the level of palatability MFS on G_{III} responded to appetite. There were several factors that influenced feed consumption such as feed form, nutrient content and palatability. Dairy heifers behavior to increase feed intake was done by adding oregano extract in concentrate feed, physiologically it did not change the parameters measured, including feed intake, body weight, and body performance [22]. This material has a distinctive flavor and aroma when used as an additive in the concentrate for the dairy cow [22]. Cows that consume concentrate, generally higher energy intake than forage, so if the proportion of concentrate in the feed is higher, the feed intake becomes lower [23]. Attempts to increase appetite and high DMI in dairy cows after calving need to be noted that some environmental conditions and body stay comfortable, BCS is not excessive, energy consumption in accordance with requirements that have been set at the time of the needs of the cow in the dry period, and forage intake adequate after calving [24].

The total consumption of the ration was also not significantly different, but G_{III} tended to be higher when compared with G_I and G_{II}, respectively 8.58 vs 7.89 and 7.37 kg/day. Forage given to the cow was limited to only 3.6 kg/day/DM, while the concentrate was 6.5 kg/day as fed. Based on the calculation of total DM rations given a percentage, feed treatment on G_I, G_{II} and G_{III} each value 1.95, 1.89 and 1.96% body weight. This means that it was not sufficient for what has been determined in [25] NRC (2001) that was for dairy cattle with body weight 400 kg, dry matter intake (DMI) 2 to 4%. This may be due to the limited supply of forages and the low palatability of the concentrate feed provided.

3.3. Consumption of Feed Nutrients

The result of observation of nutrient content of feeds presented are organic matter (OM), crude protein (CP), crude fiber (CF) and total digestible nutrient (TDN). The results are presented in Table 4.

Table 4. Consumption of OM, CP, CF and TDN (kg / DM / d)

	Feed treatments		
Parameters	G_I : control	G_{II} : G_I + 500 g UMMB	G_{III} : G_I + 500 g MFS
OM ^{ns}	6.95	6.53	7.63
CP	0.99^a	0.93^a	1.22^b
CF ^{ns}	2.08	1.79	2.05
TDN^{ns}	5.25	5.02	5.83

$^{a, b}$ Different superscripts on the same line show the difference of $P < 0.05$
^{ns} was not significantly different $P < 0.05$.

The feeding supplement of UMMB and MFS did not have a significant effect on nutrient consumption of OM, CF and TDN feed, only tended to be higher than G_I and G_{11} (Table 4), whereas CP consumption was significantly affected $P < 0.05$ as a result of the addition of MFS which differed significantly from G_I and G_{II} of 1.22 kg /DM /d compared with 0.99 and 0.93 kg / DM/d. This difference was likely due to increased consumption of CP and the ability of the cows to consume concentrates also increased. In addition, CP content of concentrate + MFS (G_{111}) was higher when compared with G_I and G_{II} of 16.9 vs. 14.71 and 14.96%. Further CP content of forage was the same from the three treatments (Table 2). The results of the study [26] reported that the soy bean meal present in the initial lactation dairy ration was an undegradable protein in the rumen containing amino acids by bypass to small intestine, a factor that might also affect CP consumption and concentrates.

3.4.Nutrient Balance

The balance of energy and protein could be known from the consumption and the needs of cows used in the study. The average consumption, need and balance of nutrient of cattle feed in the three treatment groups can be seen in Table 5.

Based on the table of lactation dairy requirement according to [25] NRC (2001), the calculation of DM, CP , and TDN in the treatment group G_1, G_{11} and G_{111} shows a negative balance on total DM, CP and TDN consumption of the three feed treatments, each of which was, -3.28, -2.73 and -4.24 kg DM/d, -0.34, -0.52 and -0.36 kg/d, -1.96, -3.5 and -2.21 kg/d respectively. It means that to obtain a negative balance as a result deficiency of DM, protein, and energy. Low consumption of DM, TDN and protein were likely due to limited feeding forage and consumption of less optimal concentrate, thus causing consumption of DM, TDN and CP were smaller than its needs.

Table 5. Balance of Consumption and Needs of DM, CP, and TDN (kg DM/d).

Parameters	Feed Treatments		
	G_I	G_{II}	G_{III}
Dry matter (DM) (kg/d)			
Consumption	7.89	7.37	8.58
Standard of minimum needed	11.17	10.15	12.82
Balance	-3.28	-2.78	-4.24
Organic matter (OM) (kg/d)			
Consumption	6.95	6.53	7.63
Crude protein (CP) (kg/d)			
Consumption	0.99	0.93	1.22
Standard of minimum needed	1.33	1.45	1.58
Balance	-0.34	-0.52	-0.36
Total Digestible Nutrient (TDN) (kg/d)			
Consumption	5.25	5.02	5.83
Standard of minimum needed	7.21	7.52	8.04
Balance	-1.96	-3.50	-2.21
Crude fiber (CF)			
Consumption	2.08	1.79	2.05
Standard of minimum needed	1.90	1.72	2.18
Balance	0.18	0.07	-0.13

3.5. The Balance of Dry Matter Ingredients (DM)

Based on the requirement of lactation dairy cows according to [25], the DM requirement corresponding to the standards for the cows in the treatment of G_I, G_{II}, and G_{III} was 11.17, 10.15, and 12.82 kg DM / d, respectively. While total DM consumption from research result to three treatments were 7.89, 7.37 and 8.58 kg DM / d, respectively. The result is a negative balance, of course, the impact of deficiency in each treatment is -3.28, -2.78 and -4.24 kg DM/d. DM deficiency was due to the limitations of forage feeding and low levels of palatability of the concentrate feed. This was what causes the low consumption of DM. Dairy cows need a number of nutrients to meet the needs of various body functions that is for basic needs of life and the need for production. The basic necessities of life were the need to fulfill the processes of life alone without the process of growth and milk production, while those for milk production were also used to produce milk also for growth and growth of the fetus when the cows were pregnant [27]. Feed consumption is an important parameter because feed consumption is related to the fulfillment of the need for basic living and producing. It was also important to determine the nutrient supply, function and response of livestock and use in feed [28]. According to [25], daily consumption of feed depended on the level of palatability of feed, weight, production, and livestock itself. Bath, *et.al.* (1985) [17] states that in high-yielding cows, the consumption per unit of body weight was higher than that of low-yielding cattle.

3.6. Consumption Balance Crude Protein (CP)

Based on the requirement of lactation dairy cows according to [25] about the required standard need, the requirement of each CP on treatment is 1.33 (G_1), 1.45 (G_{11}) and 1.58 kg DM / d. While total CP consumption in this study, each treatment was 0.99, 0.93; and 1.22 kg DM / d. Consequently each feed treatment occurs a negative balance. The values are -0.34 (G_1), -0.52 and -0.36 kg DM / d. This may be due to low DM consumption especially forage feeding and low concentration of feed concentrate, thus causing CP deficiency. However, the total CP consumption of G_{111} was higher at 1.22 kg DM / d and was significantly different (P <0.01) with G_1 and G_{11} of 0.99 and 0.93 kg DM / d, respectively. Meaning that MFS feed treatment can reduce CP deficiency. According to [29], low protein consumption in lactating cattle can decrease milk production, milk protein content and solid nonfat content (SNF). In some cases of deficiency, loss of body weight at the beginning of lactation will not be able to return to normal at the end of lactation.

3.7. The Balance of Total Digestible Nutrient Consumption (TDN)

Based on the requirement of lactation dairy cows according to [17] NRC (2001) has obtained the required TDN standard for each treatment G_1, G_{11}, and G_{111}, the value is 7.21, 7.52 and 8.04 kg DM/d. Total of TDN consumption G1: 5.25, G_{11}: 5.02 and G_{111}: 5.83 kg DM / d, resulting in a negative balance or deficiency in the treatment group, respectively -0.96, -2.50 and -2.21 kg DM / d. According to [29] deficiency in energy consumption caused low milk production, followed by a decline in body condition score, also caused a decrease in solid nonfat content (SNF) and milk protein content.

3.8. Crude Fiber Balance (CF)

The minimum requirement of CF 17% DM total, so for the G_1 group: 1.90, G_{11}: 1.72 and G_{111}: 2.18 kg DM/d, while total CF consumption of the three treatment groups, respectively 2.08, 1.79 and 2.05 kg DM / d, to obtain a negative balance or deficiency in group G_{111} of -0.13 kg DM / d, whereas G_1 and G_{11} have a surplus of 0.18 and 0.07 kg DM / d respectively. CF deficiency by group G_{111} was caused by the minimum requirement of CF namely 17% requirement of DM total. Group G_{111} higher than G_1 and G_{11} group whereas highest CF consumption occurs in group G_1, this was influenced by CF content contained in concentrate feed respectively - each group was G_1: 22.6%; G_{11}: 19.06 and G_{111}: 19.58%.

3.9. Milk Production

Results from observations on feeding treatment on G_1, G_{11}, and G_{111} on milk production and quality from early lactation dairy cow are presented in Table 6. Feed treatment for G_{III} was significantly different P <0.05 compared with G_I and G_{II}. The average milk production was 11.85 l/d compared with 9.55 and 10.69 l/d. Measurement of production in month II, total cumulative milk and 4% FCM production were significantly different at P <0.05 and P <0.01, respectively values for G_1, G_{11} and G_{111} were 9.62, 10.66 and 12.32 l / d; 954.98, 1038.70 and 1184.57 l /14 weeks and 10.72, 12.06 and 13.70 kg/h respectively. The milk production was influenced by the cow response to dietary supplements added in the treatment group concentrate especially the addition of MFS. This was probably because in the second month the G_{111} group experienced a period of peak production and this group has a better ability to maintain peak production. The addition of UMMB and MFS feed supplements in commercial concentrates was able to increase milk production and peak milk production achievement [10]. The response from the addition of MFS proves that the cow was more utilizing feed with high protein content to produce milk. The increase in production in G_{111} may also be affected by the soy bean meal contained in the G_{111} ration, as it is an undegradable feed material in the rumen which directly enters the small intestine [26] and supplementation with cotton seed meal can increase milk production [30]. Wanapat et al (2018) states that cotton seed meal instead of soy bean meal [1].

The response of dairy cows that got feed treatment with the addition of UMMB production was smaller (10.69 vs 11.85 l/d). This may be due to the difference in the crude protein content of which the MFS of its CP content was 16.90% compared to UMMB only 14.96% (Table 2).

Table 6. Mean of Milk Production and Quality

Parameters	Feed treatments		
	G_I	G_{II}	G_{III}
Avarage of milk production (liter/d)	9.55^a	10.69^{ab}	11.85^b
Milk production month I (I/d)^{ns}	9.98	10.13	11.78
Milk production month II (I/d)	9.62^c	10.66^c	12.32^d
Milk production month III (I/d)^{ns}	9.07	11.23	11.44
Cumulative Milk production (l/14 weeks)	954.98^a	1068.70^{ab}	1184.57^b
Average of milk production (kg/d)	9.77^a	10.93^{ab}	12.13^b
Average of milk production 4% FCM/kg/d	10.72^a	12.06^{ab}	13.70^b
Peak of milk production (l)^{ns}	11.33	12.67	13.00
Average of fat content in milk (%)^{ns}	4.65	4.69	4.87
Average of protein content in milk (%)^{ns}	2.77	2.79	2.86

^{a,b} a different superscript b on the same line shows a significant difference (P <0.05)
^{c,d} different superscript on the same line show very real difference (P <0.01)
^{ns} non significant (P> 0.05)

3.10. Dynamics Weight Loss

The change of dairy cattle weight in the control group (G_I), UMMB (G_{II}) and MFS as the average weighing result during the initial 14 weeks of lactation is presented in Figure 1. The change in body weight was related to the amount of feed consumption or the amount of nutrient and feed consumed, as well as the amount of milk production. The weight loss of cows in this study was calculated from the initial weight difference with the lowest body weight. The average weight of dairy cows from each treatment is shown in Table 7.

Table 7. The Lowest Average Weight Loss of Dairy Cows (kg)

Cows	G_I	Weeks to	G_{II}	Weeks to	G_{III}	Weeks
1	56	8	58	10	44	8
2	84	4	62	6	71	10
3	82	10	40	4	99	10
Average	74	7	53.33	7	71.33	9

Weight loss from treatment G1, G_{II} and G_{III} were 74, 53.33 and 71.33 kg at weeks 7, 7 and 9 respectively. The feed treatment did not show any significant difference, but the decrease in body weight obtained by MFS supplement tended to be longer. This was supported by the milk production in cows that got MFS also higher and also milk production peaked at 9 weeks. According to [25] NRC (2001), early weight loss lactation dairy cows were caused by the insufficient energy consumption of nutrients from the energy needed to be able to reach the peak production of milk. This condition can be seen from the TDN needs which had negative nutritional balance both G_I, G_{II} and G_{III}. The nutritional balance values for each energy were -1.96, -2.50 and -2.21 kg /d.

A description of the relationship between milk production dynamics and body weight reduction in the study is presented in Fig.1 and 2. The results of the control feeding studies, the production tended to be lower starting from 2 weeks after birth compared with the added treatment feed of UMMB and MFS. Addition of MFS at week 10 started to decrease, while for cow that get UMMB tends to fluctuate but production is still lower if compared to cow that got MFS (Fig 1). This was not followed by the dynamics of weight gain. In accordance with Fig 2, the weight gain in cows that received MFS, UMMB and control tended to decrease until week 12 for cows given MFS, while those given UMMB and control, each at 14 and 16 weeks tended to increase body weight (Fig.2). This means that the addition of MFS to the cow at the early of lactation tended to increase milk production and improved their body condition.

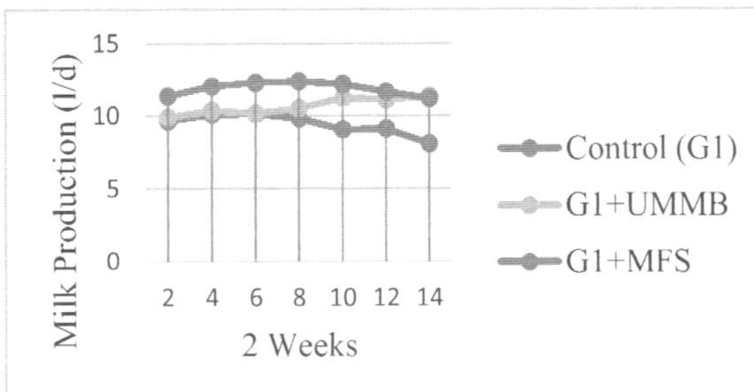

Figure 1. The Average Dynamics of Milk Production Every 2 Weeks

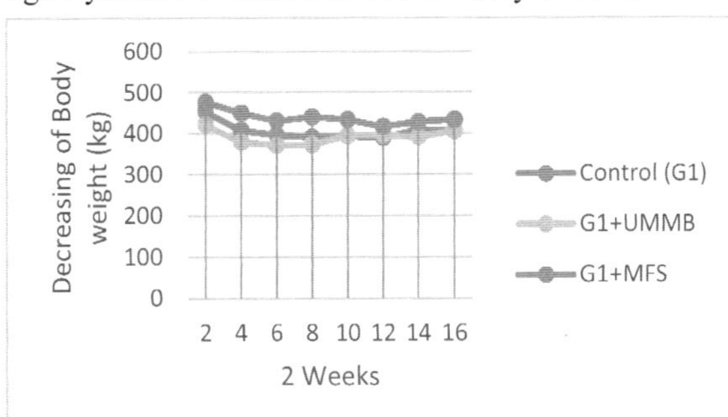

Figure 2. The Average Dynamics of Decreasing of Body Weight Every 2 Weeks

4. Conclusion

The addition of dietary supplements multi nutrient feed supplement (MFS) has a very real effect on CP consumption and was able to increase milk production and 4% FCM production. In addition, MFS tended to be able to improve the early body lactation weight loss at 10 weeks and this was faster when compared with cows fed with UMMB and control feed.

5. References

[1] Wanapat. M, Foiklang S., Sukjai, S.,, Tamkhonburi, P., Gunun, N., Gunun, P., Phesatcha, K., Norrapoke, T and Kang, S. 2018 Feeding tropical dairy cattle with local protein and energy sources for sustainable production. Journal of Applied Animal Research, 46:1, 232-236.

[2] Coffey, M.P., G. Simm and S. Brotherstone. 2002. Energy balance profiles for the first three lactations of dairy cows estimated using random regression. J. Dairy Sci. 85: 2669-2678.

[3]Pryce, J.E., M.P. Coffey and S. Brotherstone. 2000. The genetic relationship between calving interval, body condition score and linear type and management traits in registered Holsteins. J. Dairy Sci. 83: 2664-2671.

[4] Prihadi, S and Adiarto 2008 Dairy Farming Science Faculty of Animal Husbandry Gajah Mada University Press Yogyakarta

[5] Collard BL, Boettcher, P.J., Dekkers, J.C., Petitclerc, D., Schaeffer, L.R. Relationships between energy balance and health traits of dairy cattle in early lactation. J Dairy Sci. 2000, 83(11):2683-90.

[6] Van Straten M., Shpigel N.Y., Friger M. 2009 Associations among patterns in daily body weight, body condition scoring, and reproductive performance in high-producing dairy cows. Journal of Dairy Science, 92, 4375–4385

[7] Bossen d, Weisbjerg, M.R., Munksgaard, L., and Hojsgaard, S.2009. Allocation of feed based on individual dairy cowlive weight changes: Feed intake and live weight changes during lactation. Livestock Science 126,252-272.

[8] Gaillard, C., N. C. Friggens, M. Taghipoor, M. R. Weisbjerg, J. O. Lehmann,§ and J. Sehested 2016 Effects of an individual weight-adjusted feeding strategy in early lactation on milk production of Holstein cows during extended lactation. J. Dairy Sci. 99:2221–2236 http://dx.doi.org/10.3168/jds.2015-10359 .© 2016

[9] Suharyono 2010 Development of feed supplements for ruminants and introduction to farmers. Nuclear Sciences, scientific presentations of middle/major researchers.Center for the Dissemination of Nuclear Science and Technology. BATAN. **1**(1): 1

[10] Suharyono, L Farida, A Kurniawati and Adiarto 2008 Effect of feed supplement on peak milk production of dairy cattle at first lactation. Prospect of Dairy Industry toward Free Trade 2020 Puslitnak Jakarta

[11] Tekeba E, Wurzinger M and Zollitsch W 2012 Effects of urea-molasses multi-nutrient blocks as a dietary supplement for dairy cows in two milk production systems in north-western Ethiopia. Livestock Research for Rural Development Volume 24, Article #130. Retrieved September 30 2016 from http://www.lrrd.org/lrrd24/8/teke24130.htm

[12] Forsberg, N.E., R. Al-Maqbaly, A. Al-Halhali, A.Ritchie and Srikandakumar 2002 Assesment of molasses-urea blocks for goat and sheep production in the Sultanate of Oman: Intake and growth studies. Tropical Animal health and Production, 34 231-239

[13] Suharyono, N Litasova, A Kurniawati and Adiarto. 2014. Development of multi nutrient no molasses feed supplement for improving milk productivity on early lactation dairy goat. Proceedings The 2nd Asian-Australian Dairy Goat Conference April 25-27th Faculty of Animal Science Bogor Agricultural University. 209

[14] Suharyono, H. Sutanto, H. Sutanto., Purwanti, Martanti, A. Agus and R. Utomo. 2014. The effect of urea multi-nutrient and medicated block for beef cattle, beef and dairy cow. Atom Indonesia Journal Vol. 40. 2.49-56.

[15] Suharyono 2014 Development of feed supplement urea molasses multi nutrient block (UMMB) using protein from soy bean meal and Gliricidia sepium (Gs) for ruminant animal. A Scientific Journal for the Apllications of Isotopes and Radiation **10**(1): 11

[16] Lestari VS, Djoni PR, Tanrigiling R, Aslina A, Ikrar MS and Ilham R 2016 Beef cattle farmers perception toward urea mineral molasses block. World Academy of Science, Engineering and Technology International Journal of Biological, Biomolecular, Agricultural, Food and Biotechnological Engineering **10**(10)

[17] Bath DL, FN Dickinson, H A Tucker and RD Appleman 1985 *Dairy Cattle, Principles, Problem, Profit* Third Edition Lia and Febiger Philadelphia

[18] Astuti M 2007 Introduction to Statistics for Animal Husbandry and Animal Health. First Print Binasti Publisher Bogor

[19] AOAC 2005 Official Method of Analysis the Association of Official Analytical Chemists. 18th Ed. Maryland AOAC International, William Harwitz (ed) United States of America

[20] Hartadi H, S Reksohadiprojo and AD Tillman 2005 Table feed composition for Indonesia. Gajah Mada University Press. 5th Ed.

[21] Crampton E W and L E Harris 1969 Applied Animal Nutrition. W H Freeman and Company United States of America

[22] Giovani JK, Dejani MP, Alexandre MG, Marcelo T. S., Marcel B dos P, Eduardo A da C and Vivian F. 2016. Oregano Extract Added into the Diet of Dairy Heifers Changes Feeding Behavior and Concentrate Intake. The Scientific World Journal

[23] Chamberlain AT and Wilkinson JM 2009 Feeding the Dairy Cow. Published in Great Britain by Chalcombe Publications, Mounthwood House, Biddenfield Land Shedfield, Hampshire SO32 2HP

[24] Drackley JK and Cardoso FC 2014 Prepartum and postpartum nutritional management to optimize fertility in high-yielding dairy cows in confined TMR systems. The Animal Consortium. 85.

[25] NRC 2001 Nutrient requirement of Dairy Cattle, 7th Ed. National Academic Press Washington DC.

[26] Broderick, GA., Huhtanen, P., Ahvenjarvi, S., Reynal, S.M., Shingfield, K.J 2010 Quantifying ruminal nitrogen metabolism using the omasal sampling technique in cattle- a meta-analysis. J.Dairy Sci. 93:3216-3230.

[27] Siregar S 1992 Sapi perah Jenis, teknik pemeliharaan dan Analisis Usaha. PT Penebar Swadaya Jakarta

[28] Van Soest, P.J. 1994. Nutritional Ecology of Ruminant second edition. Cornell University Press Ithaca, New York.

[29] Miller WJ 1979 Dairy Cattle Feeding and Nutrition. Academic Press. NewYork San Fransisco

[30] Brito, A.F., and Broderick, G.A.2007. Effect of different protein supplement on milk production and nutrient utilization in lactating dairy cows. J.Dairy Sci. 90:1816-1827.

IOP Conf. Series: Earth and Environmental Science **119** (2018) 012045 doi:10.1088/1755-1315/119/1/012045

The Concentrations of Rumen Fluid Volatile Fatty Acids and Ammonia, and Rumen Microbial Protein Production in Sheep Given Feed During the Day and Night Time

D. A. K. W. Gumilar[1], E. Rianto[1*] and M. Arifin[1]

[1]Faculty of Animal and Agricultural Sciences, Diponegoro University, Semarang, Central Java, Indonesia.

*Correspending E-mail: erianto05@yahoo.com

Abstract

An experimental study was carried out to investigate the concentrations of volatile (VFA), ammonia and microbial protein production of rumen fluid in sheep given fedd during the day and at night. This study used 12 fat-tailed rams aged 12-18 months and weighed 24,12 ± 25 kg (CV= 10,51%). The rams were fed a complete feed containing 16.64% protein and 68,33% total digestible nutrients (TDN). The rams were allocated into a completely randomised design with 3 treatments and 4 replications. The treatments applied were: T1: day time feeding (6.00 hrs – 18.00 hrs); T2: night time feeding (18.00 hrs – 6.00 hrs); and T3: day and night time feedings (6.00 hrs – 6.00 hrs). The parameters observed were dry matter intake (DMI), rumen VFA concentration, rumen ammonia concentration, rumen rmicrobial protein production and the efficiency of rumen microbial protein production. The results showed that feeding time did not significantly affect (P>0.05) all the parameters observed. Dry matter intake, VFA concentration, ammonia concentration, the microbial protein production of rumen fluid and the efficiency of microbial protein production were 1,073g/d, 49.69 mmol; 4.77 mg N/100 ml, 12,111 g/d and 19.96 g per kg digestible organic matter intake (DOMI), respectively. It is concluded that feeding time did not affect DMI, condition of rumen fluid and rumen microbial protein production in sheep.

1. Introduction

The environmental temperature is one of the external factors that have a big influence on livestock productivity, especially related to feed intake and efficiency of feed utilization [1] [2] [3]. The timing strategy of feeding on livestock is one way to increase livestock productivity. A proper timing of feeding is considered necessary to be one of the strategies for the feed given to livestock can be utilised optimally.

The time difference between day and night feeding probably result in different thermoregulations in livestock. Livestock ruminants in the tropics face challenges with high external heat intakes, especially during the day due to higher ambient temperatures [2]. Animals fed during the day receive a high heat load derived from feed metabolism and environmental temperature. Sheep reduce the amount of feed intake when ambient the temperature increases. Animals under heat stress attempt to reduce body heat production by reducing feed intake [3], therefore a new strategy of timely feeding is needed to improve the efficiency of feed utilisation.

Feeding animal at night should be considered to avoid the animal from heat stress that is likely to happen during the day. The time difference between day and night feeding is considered to result in different thermoregulations in livestock. Ruminants in the tropics face challenges with high external heat exposure, especially during the day due to higher ambient temperatures [4]. The purpose of this study was to examine the comparison of feed digestion efficiency in terms of volatile fatty acids (VFA) concentration, rumen fluid ammonia and production of microbial protein in sheep treated with day and night feeding. The results of this study were expected to be used as a guide in managing feeding strategies.

2. Materials and Methods

This study used 12 fat tailed rams of 12-18 months old and 20.65 ± 1.88 kg body weight. The rams were fed a complete feed containing 16.64% protein and 66%total digestible nutrients (TDN).

The experimental design used was a complete randomized design (CRD), consisting of 3 treatments and 4 replications. The treatments applied were: T1: feeding at noon (06.00-18.00); T2: feeding at night (18.00-06.00); and T3: feeding on day and night at (06.00-06.00).

Parameters observed in this study were feed intake, live weight gain (LWG), rumen VFA concentration, rumen ammonia concentration and rumen microbial protein production. Calculation of feed intake was conducted by subtracting the feed offered with refusals. The LWG was calculation by subtracting the final live weight (kg) with the initial live weight (kg) divided by the length of observation (day). Measurements of VFA concentrations were measured by gas chromatographic techniques. Measurement of ammonia concentration was done by Connway method.

3. Results and Discussion

3.1. Dry Matter Intake

The data obtained showed the feeding time during day and night had no significant effect ($P>0.05$) on the dry matter (DM) intake (average 1,073 g/d). This is probably caused by the fact that there was only a small difference in ambient temperature between the day and night time, which were 27.7°C and 25.2°C, respectively. Such a difference was not big enough to alter DM intake. This was in accordance with the findings of [5] that sheep did not reduce feed intake when ambient temperature increased from 24° to 34°C. A study by [2] also showed that DM intake of the lamb decreased when the ambient temperature increased from 20 to 30°C.

The DM intake of the sheep in this study was comparable with the findings of [6], that thin tailed sheep with 18 kg body weight had 0.75 DMI. The difference can be attributed to the body weight of the animals; animals with higher body weight need and consume more feed that those with lower body weight [7].

3.2. Concentration of Rumen Fluid Volatile Fatty Acids

Day and night feeding did not significantly affect ($P>0.05$) VFA concentrations, except at 21.00 hrs ($P<0.05$; Table 2). It was attributed to the fact that DM intake in rams T1 and T2 was not significantly different. A study by [8] showed that rumen VFA concentration was influenced by rumen microbial population and DM intake.

The concentration of rumen VFA in this study (49.69 mM) was relatively lower than the normal limits to support the growth of rumen microbes. The range of normal rumen fluid VFA products that support microbial growth is 70–150 mM [7]. The VFA concentrations of this study were also lower when compared with studies conducted by [9] showing that the rumen VFA concentration in sheep that received grass and concentrate in the at 0, 4 and 8 hrs after feeding were 84.1, 101.1 and 97.2 mM, respectively. Such the differences can be attributed to the different composition of feed offered to the animals [10]. It is also suggested that feed digestion with non-structural carbohydrate content would result in higher VFA production [11].

Table 1. Concentration of Rumen Fluid Volatile Fatty Acids

Parameter	Treatment			Average
	T1	T2	T3	
Total VFA (mM)				
• 06.00 hrs	38.95	46.65	44.95	43.52
• 09.00 hrs	62.61	71.84	47.95	60.80
• 18.00 hrs	39.43	36.75	58.03	44.74
• 21.00 hrs	46.05[a]	51.15[a]	74.84[b]	-

Legend: different superscripts on the same row indicate significant difference (P<0.05)

3.3. Ammonia Concentrations of Rumen Fluid

The results of this study indicate that feeding time day and night have no significant effect (P>0.05) on rumen fluid ammonia concentration (Table 3). The condition was suspected to occur because the intake of dietary protein was not significantly different. One of the factors that influence the low concentration of ammonia in the ruminant is the amount of feed protein entering the rumen. High dietary protein content with high degradability will result in an increased rumen fluid ammonia concentration [7].

The mean rumen ammonia concentration in this study was 4.7 mg/100 ml. The condition is slightly lower than that required for normal rumen microbial growth. The minimum concentration of rumen fluid ammonia for rumen microbes to grow well is 5.0 mg per 100 ml rumen fluid, and the optimum concentration of rumen fluid ammonia ranges between 8.5 and 30.0 mg/100 ml [7]. Meanwhile, according to [12] the minimum ammonia concentration was 40 -240 mg/L.

Table 2. Rumen Fluid Ammonia Concentration

Parameter	Treatment			Average
	T1	T2	T3	
NH_3 (mg N/L)				
• 06.00 hrs	46.40	39.74	48.70	44.95
• 09.00 hrs	47.06	43.36	63.53	51.32
• 18.00 hrs	45.97	39.16	47.31	44.15
• 21.00 hrs	46.56	42.64	62.24	50.48

Notes: there is no significant difference (P>5%) among treatments.

The ammonia concentration in this study was lower than that of [13], who found N-ammonia concentration was 90.3 and 98.3 at ambient temperature of 20 and 40°C, respectively. It is suggested that the minimum rumen N-ammonia concentration is 50 mg/L [7]. The difference is due to the fact that the dietary protein content of in the study by [13]was higher, i.e. 17.89%, while the content of dietary protein in the current study was 16.64%. The rumen fluid ammonia concentration is influenced by the level of dietary protein intake, the degree of degradability, feed duration in the rumen and the acidity (pH) of rumen fluid [73].

3.4. Production of Microbial Protein

Feeding on day and night had no significant effect (P>0.05) on microbial protein production and its efficiency (P> 0.05; Table 4) with an average of 12.11 g/d and 19.96 g microbial protein/kg DOMI. This condition can be attributed to the fact that DM intake, dietary protein concentration and rumen fluid concentration in the animal were not significantly different (P>0.05) between the treatments. The synthesis of rumen microbial protein depends on protein and other nitrogen source intake, the speed of dietary protein degradation into ammonia, and the need of the microbes for amino acids [7] [10].

The rumen microbial protein production in this study (12.11 g/d) was lower than the finding of [13] which was 19.0 g/d. That was because the types of feed and animals used in this study were different. According to [7] and [10], the factors affecting microbial protein synthesis are fermentation types based on the type of feed and microbial requirements of amino acids.

Table 3. Microbial Protein Production

Parameter	Treatment			Average
	T1	T2	T3	
DOMI (kg/d)	0.735	0.559	0.689	0.661
Microbial protein production (g/d)	9.240	13.92	13.122	12.111
Efficiency of microbial protein production (g microbial protein/kg DOMI)	14.23	26.99	18.23	19.96

Legend: DOMI = digestible organic matter intake

The treatment of day and night in this study had no significant effect (P>0.05) on the efficiency of microbial protein production, it is suspected that the feeding of day and night had no significant effect on the production of microbial protein and the consumption of feed PK. The production of rumen microbial proteins is also inseparable from the use of dietary protein that is reorganized by the rumen microbes through degradation process. It is claimed by [14]) that such efficiency is affected by dietary protein content that is degraded into ammonia. The efficiency of rumen microbial protein production in this study was comparable with the finding of [13], i.e.19.0 g/kg DOMI. The differences in data from both studies are probably due to the use of different feed ingredients and different nutritional content of the feed.

4. Conclusion

Based on the results of this study. it can be concluded that the feeding time during the day or night does not affect the DM intake and the digestion process of feed in the rumen in sheep.

References

[1] Bhattacharya A N and F Hussain 1974 Intake and utilization of nutrients in sheep fed different levels of roughage under heat stress *J. Anim Sci* 38: 877

[2] Rianto E, M K Hill dan J V Nolan 2001 The Effect of Quality of Diet on Carcass and Meat Characteristics of Lambs at Ambient Temperatures of 20 and 30 °C *Bulletin Anim. Sci* **25**(3): 127

[3] Baumgard L H and R P Rhoads 2012 Ruminant Nutrition Symposium: ruminant production and metabolic responses to heat stress *J. Anim Sci*. 90:1855

[4] Al-Tamimi H J 2007 Thermoregulatory response of goat kids subjected to heat stress *Small Ruminant Res.* 71:280

[5] Susiloningsih 2009 Pemanfaatan Protein pada Domba Lokal Akibat Perbedaan Suhu Lingkungan. Program Pasca Sarjana Universitas Diponegoro Semarang (Tesis Magister Ilmu Ternak).

[6] Prakoso, F Mulia, F A Setyawatue, S Dartosukarno, S Mawati, E Rianto, R Adiwinarti and Soedarsono 2009 Prosiding Seminar Nasional Teknologi Peternakan dan Veteriner. Bogor 13-14 Agustus 2009 456

[7] McDonald P, R A Edwards, J F D Greenhalgh, C A Morgan, L A Sinclair and R G Wilkinson 2012 Animal Nutrition Seventh Edition. Prentice Hall, Harlow, England

[8] Yurleni, R Priyanto, E Gurnadi and K G Wiryawan 2013 Efektivitas minyak ikan lemuru terproteksi terhadap populasi mikrob rumen dan fermentasinya pada kerbau dan sapi *Jurnal Veteriner* 14(3) 285

[9] Giraldo L A, M. L Tejido, M J Ranilla, S Ramos and M D Carro 2008 Influence of direct-fed fibrolytic enzymes on diet digestibility and ruminal activity in sheep fed a grass hay-based diet *J. Anim. Sci.* 86 1617

[10] Bondi A A 1987 *Animal Nutrition*. John Wilay & Sons,Chichester

[11] Van Soest P J 1994 *Nutritional Ecology of the Ruminant.* 2nd Ed. Comstock Publishing Associates A division of Cornell University Press Ithaca

[12]Satter L D and L L Slyter 1974 Effect of ammonia concentration on rumen microbial protein production in vitro *British J. Nut.* 32 199

[13] Rianto E, C J Thwaites and J V Nolan 2005 Pengaruh Suhu Lingkungan dan Ketersediaan Air Minum terhadap Kecernaan Pakan, Protein Nitrogen Mikroba Rumen dan Kondisi Rumen pada Domba *Bulletin Anim. Sci.* **29**(1)10

[14] Yulianti A 2006 Kinetika Volatile Fatty Acid Cairan Rumen dan Estimasi Sintesis Protein Mikrobia pada Sapi Perah Dara Peranakan Friesian Holstein yang Diberi Pakan Basal Rumput Raja, Jerami Jagung, dan Jerami Padi yang Disuplementasi Konsentrat Protein Tinggi *Jurnal Teknologi Pertanian* **6**(1)25

The Effect of Antioxidant Activity of Liquid Smoke in Feed Supplement Block on Meat Functional of Muscle *Longissimus dorsi*

E. Abustam[1*], M.I. Said[1] and M Yusuf[1]

[1]Departement of Animal Production Faculty of Animal Science Hasanuddin University, Makassar.

Corresponding E-mail: effendiabu@hotmail.com

Abstract

This study aims to look at the role of liquid smoke as an antioxidant added in feed supplement block and administered to cattle for 45 days on the functional properties of meat. The level of liquid smoke in the feed and the time of maturation in Muscle *Longissimus dorsi* after slaughtering cattle were the two treatment factors observed for the functional properties of meat. The study used a complete randomized design in which factor 1 was a 10% smoke level in the feed (0, 1, 2%) and factor 2 was maturation time (0, 2, 4, 6, 8 days). The parameters observed were water holding capacity (WHC), raw meat shear force (RMSF), fat oxidation rate (thiobarbituric acid reactive substance) and antioxidant activity (DPPH). The results showed that liquid smoke levels lowered the WHC, RMSF more or less the same, increased fat oxidation rate, and antioxidant activity more or less the same. While maturation tends to increase WHC, increase RMSF, fat oxidation rate, and antioxidant activity. It can be concluded that liquid smoke as an antioxidant in the diet of block supplements can maintain the functional properties of Muscle *Longissimus dorsi* of Bali cattle during maturation.

1. Introduction

The functional properties of meat decreased postharvest, especially water holding capacity and shear force value of meat. One of the causes of the decline in the functional properties of meat is stress due to the uncomfortable treatment of livestock before slaughter resulting in oxidation of the functional properties. Oxidation in the functional properties of meat can occur in fats and proteins; fat oxidation results in increased rancidity, while oxidation of proteins results in decreased functional properties of meat [1].

Efforts to provide concentrate feed during cattle fattening before slaughter and post-slaughter handling are expected to reduce or inhibit stress due to transport and uncomfortable treatment of livestock. Cattle that experience stress during transport and do not get adequate rest before slaughter resulted in decreased meat quality marked by dark red meat, firm and dry texture [2]. WHC decline and tenderness of the meat and increased cooking loss is an implication of livestock who experience stress before slaughter [2,3].

Giving antioxidants to the feed or the meat and its processed products is an attempt made to reduce or inhibit oxidation in the functional properties of meat. The researchers provided antioxidants to the meat to inhibit the oxidation in which free radicals would be bound by antioxidants so that the functional properties of the muscles persisted during postharvest handling. One such antioxidant is liquid smoke from pyrolysis at 400^{0}C, producing 4.13% phenol, 11.3% carbonyl and 10.2% acid [4].

The phenol compound is one of the chemical content of liquid smoke that acts as an antioxidant and as a preservative.

Some researchers use liquid smoke as an antioxidant in meat and its processed products; [5] suggest that liquid smoke as an antioxidant can increase WHC and decrease cooking loss in meatballs products. Similarly, [6] reported that the functional properties of Bali beef increase with the administration of liquid smoke in *Longissimus dorsi* muscle during maturation.

The use of liquid smoke in mice was conducted by [7] in which in 14 days of observation, the body weight of mice increased at all levels of liquid smoke exposures. Meanwhile, the liquid smoke as an antioxidant in catfish was studied by [8]. [9] concluded that the protein of tilapia (*Oreochromis niloticus*) increased to 79% with the use of liquid smoke concentration of 5% as a preservative.

Another way of giving liquid smoke as an antioxidant is to add it to the block supplement feed. Previous studies have shown that the higher concentrations of liquid smoke in feed supplement block decreased shear force value to 3.23 kg /cm^2 at a concentration of 20%. Similarly, WHC increases with increasing maturation time [10]

This study aims to improve or maintain the functional properties of post-aging flesh through the provision of liquid smoke as an antioxidant into the block supplement feed; UCSMB.

2. Materials and Methods

This study utilized the post-rigor *Longissimus dorsi* (LD) muscle, from 9 males of 2-year-old cattle fed with Urea Coconut Smoke Multi-nutrient Blocks (UCSMB) and elephant grass (*Pennisetum purpureum)* as a source of roughages for 45 days. *Longissimus dorsi* muscle-derived from Bali cattle raised in Maiwa Breeding Center Hasanuddin University, 198 km north of Makassar capital of South Sulawesi. Muscle extraction and retrieval were performed at the slaughterhouse Tamangapa Makassar. The liquid smoke concentration of 10% added to the UCSMB feed at 1 to 2% (w/w) functioned as an antioxidant to preserve meat, especially during the maturation time. *Longissimus dorsi* muscles (LD) were dissected from all of the nine male cattle after the slaughtering and aged for eight days with two days interval the measurement.

Table 1. Composition of feed materials in UCSMB[1)]

Feed materials	Composition (g/kg) at Levels of liquid smoke concentration of 10%		
	0%	1%	2%
1. Coconut water	30	29	28
2. Urea	5	5	5
3. Rice Bran	30	30	5
4. Cornmeal	10	10	10
5. Copra meal	10	10	10
6. Cement	10	10	10
7. Cow Mineral	2	2	2
8. Table salt	3	3	3
9. Liquid smoke	0	1	2

[1)] Abustam *et al*. (2015b)

Urea, coconut water, liquid smoke multi-nutrient block (UCSMB) is a modification of UMMB (urea-molasses block multi-nutrient) where molasses replaced with coconut water, and liquid smoke was added as an antioxidant [11]. There were three kinds of UCSMB used with liquid smoke in concentration 10% treatment adapted to the levels of 0% (control), 1%, and two % in the formulation (w/w). The composition of the feed material in UCSMB shows in Table 1.

This research used an entirely randomized design of factorial pattern of 3 x 5. The first factor was the levels of liquid smoke in the feed block (0, 1%, and 2%) and the second factor was the aging time (0, 2, 4, 6, 8 days) at a temperature of 2 - 5^0C replicated for three times. Thus, the variables measured included raw meat shear force value (RMSF), water holding capacity (WHC), fat oxidation by the thiobarbituric acid reactive substance (TBARS) and antioxidant activity (DPPH). Every animal received feeds in the form of 500 g of UCSMB, and 20 kg of *Pennisetum purpureum* per day for 45 days.

The shear force measured by using CD Shear Force, in which the meat samples in cylinder form with 1 cm length and a diameter of 0.5 inches placed in the hole on CD shear force. Thus, the samples have torn by using CD shear force knife with 1 mm thickness. The bigger the weight to tear the meat samples, the harder the meat tenderness. SF value was in kg/cm^2 [2].

Measurements on WHC of raw meat was carried out by using Filter-Papers Press Method as proposed by Hamm 1986 [2]. TBARS test was performed to measure the level of rancidity caused by fat oxidation during maturation [12].

Measurement of antioxidant activity by methods DPPH (1, 1-diphenyl-2-picrylhydrazyl) [13] was slightly modified. An aliquot of 3.9 mL of 0.1 mM DPPH radical in methanol was mixed with 0.1 mL of methanol extract samples. The mixture of DPPH reagent was shaken vigorously and allowed to stand at ambient temperature in the dark for three h. Measurement of the absorbance of the sample used with a UV-VIS spectrophotometer (Shimadzu brand) at 515 nm.

Processing of the data by utilizing analysis of variance (ANOVA) and testing between average uses LSD based on [14] with SPSS (SPSS 16.0, SPSS Ltd., West Street Woking, Surrey, UK).

3. Results and discussion

Variance analysis showed that liquid smoke level did not significantly affect RMSF even though there was a tendency of RMSF to increase with increasing liquid smoke level (Table 2). It indicates that liquid smoke in UCSMB feeds up to 2% has not been able to trigger cathepsin enzymes to digest muscle protein so that the RMSF value even increases. In theory, maturation will increase the tenderness of meat due to the work of cathepsin enzymes. The provision of liquid smoke in the block feed needs to be increased by the amount given by the weight of the cow body, which is likely to have a noticeable effect. As the RMSF increases with increasing levels of liquid smoke suppressed by protein oxidation, as suggested by [1] stated that if the protein undergoes oxidation, it can decrease the tenderness of the flesh. [15] suggest that the protein oxidation will change the WHC and the tenderness of the meat.

The maturation time had a very significant effect ($P < 0.001$) on the RMSF where there was a marked difference between day maturation to zero by day four, day six and day eight (Table 2). The more maturation time the RMSF increases even though there is no real difference between the fourth day with the sixth day and the eighth day. This condition indicates that the level of liquid smoke in the feed shows two groups of RMSF namely the maturation group 0 and two days and the maturation group 4, 6, and eight days. An increase in RMSF in the second group reached an average of 0.48 points from the first group. It concludes that the RMSF increases with increasing maturation time until the fourth day.

In this case liquid smoke, up to 2% in the feed has not been able to improve the tenderness during maturation. Oxidation of proteins may be the cause. [1] stated that if the protein undergoes oxidation, it can decrease the tenderness of the flesh. [15] suggest that the protein oxidation will change the WHC and the tenderness of the meat.

Increasing the level of liquid smoke in UCSMB feed, the WHC values decrease even though there is no real difference between liquid smoke level 0 and 1%. At the 2% level, it lowers the WHC to 6.11 point of the no-smoke liquid. This increase indicates that the provision of liquid smoke in the feed to 2% has not been able to maintain a high WHC when the new cattle finish slaughtering [2,3]. The amount of liquid smoke provided is relatively low compared with the weight of the cow body, and the high activity of rumen microbes may be the cause.

The maturation time has no significant effect on WHC values , but there is a tendency for WHC to increase. The results of this study do not support previous studies where the WHC increased significantly with increasing maturation time [10]. Differences in treatment wherein the previous study using the treatment of differences in smoke concentration of 0, 10, 20% as much as 2% in the diet of block supplements can explain this.

The higher the liquid smoke level increases the TBARS value reaching 65.62% at 2% level (Table 2), indicating that the delivery of liquid smoke in UMSCB feed has not been able to inhibit the increase of TBARS value, although the value of TBARS is still very low. This value is really low when compared with the critical value of 3 mg MDA/kg in which the level of rancidity is detected [16]. [17] point out that the low intensity on fat oxidation during storage occurs if TBARS values remain low in the threshold to detect rancidity but remain high in meat maturation sampling.

Increased maturation time resulted in increased TBARS value reaching 0.93 points on the eighth day compared to day zero. The increase in TBARS values is quite high but is still within the limits of low-fat oxidation rates. [17] point out that the low intensity on fat oxidation during storage occurs if TBARS values remain low in the threshold to detect rancidity but remain high in meat maturation sampling. An addition of vitamin E to the pasture as an antioxidant in cattle increases the vitamin E in muscle and decreases the impact of fat oxidation on beef [18].

Analysis of variance showed a significant interaction ($P < 0.05$) between liquid smoke level and maturation time on TBARS value (Figure 1). Until the fourth day of maturation, the TBARS values are more or less the same at all three levels of liquid smoke. The three levels of liquid smoke though not significantly different, increase the value of TBARS from the fourth day to the eighth day of maturation. At 0% level, the increase of TBARS value until the eighth day of maturation is lowest among the three levels of liquid smoke. At the end of maturation, the value of TBARS at 1% and 2% levels increased significantly with the 0% level.

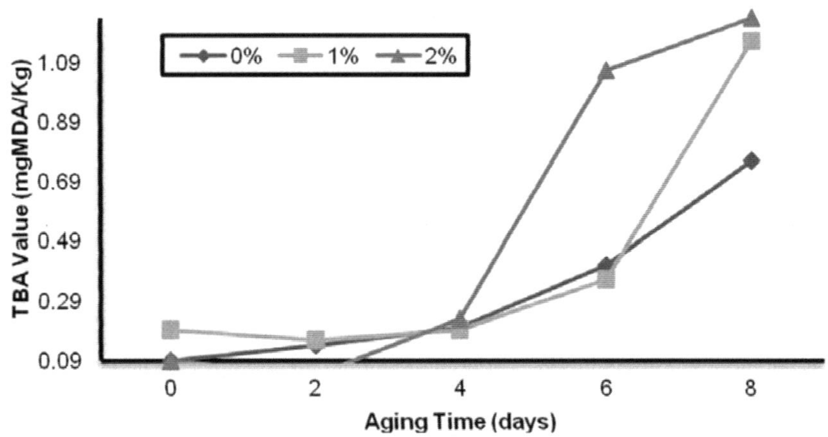

Figure 1. Interaction between levels liquid smoke and time of aging on TBARS value

Analysis of variance showed that liquid smoke level did not significantly influence, while maturation time had a significant effect (P <0.001) on antioxidant activity. There is a tendency for antioxidant activity to decrease with increasing levels of liquid smoke in UCSMB feed. Differences between liquid smoke treatments that did not differ in antioxidant activity may cause the WHC to decrease, the value of TBARS increases and RMSF is approximately equal to the increase of liquid smoke level (Table 2).

Antioxidant activity increased with increasing maturation time reaching 57.20% on the eighth day compared to days zero (Table 2). The high antioxidant activity reached 88.44% on the eighth day of maturation indicating that beef protein is a good source of antioxidant. [19] argued that food of animal origin that contains high protein is a good source of antioxidants. This increase has implications for the RMSF decrease until the second day of maturation, the WHC trend more or less the same during maturation and the increasing TBARS in the sixth day of maturation.

Table 2. Effects of liquid smoke levels in UCSMB and aging time of *Longissimus dorsi* on the functional properties of meat quality of Bali cattle (means and SE)

Treatments	RMSF (kg/cm^2)	WHC (%)	TBARS (mg MDA/kg)	DPPH (%)
LS Levels:	Sig:NS	Sig:0.001	Sig:0.05	Sig:NS
0%	1.20±0.38	30.28±3.80 ^{a}	0.32±0.25 ^{a}	60.42±18.47
1%	1.24±0.32	30.12±4.04 ^{a}	0.41±0.41 ^{ab}	60.72±20.98
2%	1.26±0.33	24.17±3.61 ^{b}	0.53±0.26 ^{b}	57.74±21.36
Aging:	Sig:0.001	Sig:NS	Sig:0.001	Sig:0.001
0 days	1.03±0.24 ^{a}	26.06±2.94	0.12±0.09 ^{a}	31.24±7.39 ^{a}
2 days	0.86±0.18 ^{a}	28.70±5.38	0.11±0.08 ^{a}	58.93±7.17 ^{b}
4 days	1.40±0.20 ^{bc}	29.71±5.88	0.21±0.02 ^{a}	52.44±6.55 ^{c}
6 days	1.45±0.22 ^{bc}	28.37±5.10	0.61±0.50 ^{b}	67.07±7.70 ^{d}
8 days	1.41±0.35 ^{bd}	28.12±4.00	1.05±0.25 ^{c}	88.44±2.56 ^{e}

Description: Numbers with different superscripts in the same column stated a significant difference (P <0.05) and a highly significant difference (P <0.001).
LS Levels: Liquid smoke levels; RMSF, raw meat shear force; WHC, water holding capacity; TBA, thiobarbituric acid reactive substance; DPPH, 1,1-diphenyl-2-picrylhydrazyl.

4. Conclusion

This study revealed that liquid smoke levels lowered the WHC, RMSF more or less the same, increased fat oxidation rate, and antioxidant activity more or less the same. While maturation tends to increase WHC, increase RMSF, fat oxidation rate, and antioxidant activity. It can be concluded that liquid smoke as an antioxidant in the diet of block supplements can maintain the functional properties of Muscle *Longissimus dorsi* of Bali cattle during maturation.

Acknowledgements

The authors would like to thank DRPM Kemenristek Dikti who has funded this research through Competency-Based Research Program FY 2017 with contract No. 005 / SP2H / PPM / DPRM / IV / 2017.

References

[1] Bhattacharya D, Kandeepan G and Vishnuraj M R 2016 Protein oxidation in meat and meat products- A Review. *J. Meat Sci. Technol.* April-June, 2016.**Vol 4** Issue 2 Pages 44

[2] Abustam E 2012 *Meat Science: Aspects of Production, Chemistry, Biochemistry, and Quality.* 1st Ed. Masagena Press. Makassar (Indonesian)

[3] Soeparno 2005 *Meat Science and Technology*. Gajah Mada University Press, Yogyakarta (Indonesian).

[4] Setiadji B A H 2000 Liquid coconut shell. Liquid smoke as the natural preservative safe for humans *(www.asapcair.com)*, PPKT, Jogjakarta (Indonesian)

[5] Abustam E, Yusuf M, Ali HM and Yuliati F N 2015a Effect of muscle types of Bali beef pre and post-rigor on the quality of meatballs during storage. *Pak. J. Nutrition* **14**(3)170

[6] Abustam E, Yusuf M, Ali HM and Yuliati FN 2012 Increased durability and quality of meat and meat products processed through the addition of liquid smoke as an eco-friendly natural preservative. *National strategic research report*. Hasanuddin University (Indonesian)

[7] Budijanto S, Hasbullah R, Prabawati S, Setyadjit S, Sukarno and Zuraida I 2008 Identification and safety test on liquid smoke made from coconut shell for a food product. *J. Pascapanen* **5** (1) 32

[8] Ernawati, Purnomo H and Estiasih T 2012 Antioxidant effect of liquid smoke on the stability of oxidation sausage catfish (*Clarias gariepinus*) dumbo during storage. *J. Tech. Agriculture* **13** (2)119

[9] Budaraga I K 2017 Effect of combination treatment of concentrated liquid smoke, immersion duration, packaging and old type storage different levels of protein Nila fish fillet (*Oreochromis niloticus*). *International J. ChemTech Res.* **10**(3)1

[10] Abustam E, Said MI, Yusuf M, Nahariah 2016 Effect of liquid smoke concentration in feed supplement and maturation time M. *Longissimus dorsi* against the quality of Bali beef. Prosiding Nasional conference on *Optimalisasi Teknologi dan Agribisnis Peternakan Dalam Rangka Pemenuhan Protein Hewan Asal Ternak* ISBN 978-602-1004-42-5 642

[11] Abustam E, Yusuf M, Said MI and Mide MZ 2015b IbM Showroom Bali cattle in district Tanete Riaja of Barru Regency. *Reports of PPM DP2M Dikti*, Faculty of Animal Science Hasanuddin University (Indonesian)

[12] Apriyantono A, Fardiaz D, Puspitasari N.I, Sedarnawati, Budiyanto S 1989 *Food Analysis*. IPB Press, Bogor (Indonesian)

[13] Pajak P, Socha R, Galkowska D, Roznowski J and Fortuna T 2013 Phenolic profile and antioxidant activity in selected seeds and sprouts. *Food Chem.*, **143** 300

[14] Steel RGD and Torrie JH 1991.*Principles and Procedures of Statistics.* McGraw-Hill, Book Co. Inc, New York

[15] Rowe L, Maddock J, Lonergan KRSM, Huff-Lonergan E 2004 Influence of early post-mortem protein oxidation on beef quality. *J. Anim. Sci.*, **82**(3) 785

[16] Wong JW, Hashimoto K, Shibamoto T 1995 Antioxidant activities of rosemary and sage extracts and vitamin E in a model meat system. *J. Agric. Food Chem.*, **43**(10) 2707

[17] Popova T, Marinova P, Vasileva V, Gorinov Y, Krasimira L 2009 Oxidative changes in lipids and proteins in beef during storage. *Archiva Zootechnica* 12:3, 30

[18] Gatellier P, Mercier J, Juin H, Renerre M 2005 Effect of finishing mode (pasture -or mixed diet) on lipid composition, color stability and lipid oxidation in meat from Charolais cattle. *Meat Sci.*, **69**(1) 175

[19] Hernandez L, Zermeno B, Arino, Blasco 2003 Antioxidant, lipolytic and proteolytic enzyme activities in pork meat from different genotypes. *Meat Sci.*, **66**: 525.

The Productivity of Male Thin-Tailed Lambs and Sheep Fed Complete Feed

M.S. Aluns[1] and N. Luthfi[1].

[1]Faculty of Animal and Agricultural Sciences, Diponegoro University, Semarang, Indonesia.

*Corresponding E-mail: agung194@yahoo.com

Abstract

The aim of this study is to examine the productivity of thin tailed lambs and sheep fed complete feed. The material used in this study were 6 thin tailed lambs aged ± 5 months with an average body weight of 15.41 ± 2.11 kg (CV 37.04%) and 6 thin tailed rams aged ± 1 year with an average body weight of 23, 01 ± 1.91 kg (CV 28.83%). The animals were raised intensively with complete feed as much as 3.5% of body weight. The feed contained 12% crude protein (CP) and 55% Total Digestible Nutrients (TDNs). The results showed that the average feed intake was 567.10 g DMI/d in lambs and 726.24 g DMI per day in sheep. An average of body gain of lambs was 0.02 kg/d and 0.01 kg/d. The day matter digestibility of lambs achieves 50.23% and 50.74% in sheep. Based on results, it can be inferred that lambs has same digestibility with sheep and has more efficient to be fattened than sheep.

1. Introduction

Thin tailed sheep were local sheep that adaptable to tropical climates, more resistant to some diseases and be acceptable to the consumer even though growth rates are low [1]. Generally, consumer choose sheep aged 1 year old [2] but it requires long time to raise and need a lot of cost. According to [3], early fattening system in sheep can improve growth rate and feed efficiency and giving complete feed is expected to fullfil the needs of sheep in the process of fattening. According to [4], complete feed is feed containing enough nutrients for animals and given as animal feed to find requirements and production without additional substance other than water [5] stated that the complete feed consists of fibrous (forage) feed and ingredients which are concentrated and additional feed that has been calculated its nutrient.

The purpose of this study was to examine the productivity of lambs and sheep given complete feed. The benefit of this study is to provide information the nutrient needed of the lambs and sheep to achieve good productivity and efficiency.

2. Materials and Methods

The materials used in this study were 6 thin tailed lambs aged ± 5 months with an average body weight of 15.41 ± 2.11 kg (CV 37.04%) and 6 thin tailed sheep aged ± 1 year with an average body weight of 23, 01 ± 1.91 kg (CV 28.83%). They were raised intensively and fed complete feed that contained of 12% CP and 55% TDN. The complete feeed consisted of 50% elephant grass, 8% soybean meal, 9% rice bran, 2% mineral mix, 8% dried cassava, 18% wheat bran and 5% molasses. This study used group randomized design with age as treatment. The parameters observed were dry matter intake, dry matter digestibility, feed conversion and body gain.

The study was conducted in four stages: preparation, adaptation, preliminary and treatment stage. Feed was given as much as 3.5% of body weight and water was given ad libitum. Sheep were weighed weekly to obtain the body weight that was used to calculate feed needed. The rest of the feed and water were weighed every morning before feed and water were given. The data obtained will be analyzed by using T test significance level 5%.

3. Results and Discussion

3.1. Feed Intake and Feed Digestibility

The average dry matter intake and dry matter digestibility in lambs and sheep are shown in Table 1. Dry matter intake and dry matter digestibility in lambs and sheep are not significant (P> 0.05). It illustrates that lambs aged 3-4 months already have ability to consume and digest the dry matter intake as much as those of sheep. High level of dry matter digestibility was do to the high percentage of digestible nutrients described by total digestible nutrients (TDN) [6]. The same consumption value indicates that lambs have been able as well as the sheep. Feed consumption is used to determine the requirement for animal production. [7] claimed that feeding was conducted to know the optimum of prouction and the amount of nutrients needed of sheep.

3.2 Body Weight Gain and Feed Conversion

The average of body weight gain of sheep is shown in Table 1. Body weight gain of lambs and sheep shows no significant differences (P> 0.05). It indicates that the growth of lambs has been able to same with sheep. [8] stated that on the growth rate charts, lambs are optimum for bone and meat growth, along with increasing age then will grow also into fat like a subcutaneous fat, pelvic fat and intramuscular fat. Feed conversion of lambs lower than that of sheep (P <0.05). It indicates that lambs are more efficient than that of sheep. This is in accordance with the opinion of [9] that the feed conversion is the amount of feed needed to increase one unit of weight gain, the higher the feed conversion, the lower feed efficiency.

Table 1. Average of dry matter intake, dry matter digestibility, feed conversion and body weight gain of lambs and rams.

Parameters	Lambs	Rams
Dry matter intake (%)	3,50	3,50
Dry matter intake (g)	567,10^{a}	726,24^{b}
Dry Matter Digestibility (%)	50,23	50,74
Body Gain (kg/day)	0,02	0,01
Feed Conversion	30,10^{a}	51,90^{b}

^{a,b} Different letters on the same line show significant differences (P <0.05)

4. Conclusion

Based on the results, it can be inferred that lambs have same ability on production with sheep. Lambs was more efficient that sheep on production.

References

[1] Hersandika P, A Prima, C M S Lestari dan E Purbowati 2016 Pertumbuhan domba ekor tipis jantan lepas sapih dengan protein pakan yang berbeda. Prosiding Seminar Nasional Peternakan Berkelanjutan 8 Bogor 475

[2] Purbowati E dan T P M T Farm 2009 Usaha Penggemukan Domba Penebar Swadaya Jakarta

[3] Schandbacher B D dan J D Course 1980 Growth and performance of growing finishing lambs exposed to long or short photoperiods. J. of Animal Science **50** (4) 943

[4] Hartadi H 2005 Tabel Komposisi Pakan Untuk Indonesia Edisi kelima. Gajahmada University Press Yogyakarta

[5] Karimizadeh E, M Chaji, dan T Mohammadabadi 2017 Effects of physical form of diet on nutrient digestibility, rumen fermentation, rumination, growth performance and protozoa population of finishing lambs. J. Animal Production. **30** 1

[6] Prima A 2017 Konsumsi Pakan, Kecernaan Pakan dan Tingkah Laku Makan Pada Domba Muda dan Dewasa yang Diberi Pakan dengan Kualitas Berbeda. Fakultas Peternakan dan Pertanian Universitas Diponegoro: Semarang (Thesis Magister Ilmu Ternak)

[7] Davis M S, T L Mader, S M Holt and A M Parkhurst 2003 Strategies to reduce feedlot cattle heat stress: Effects on tynamic temperature. J. Anim. Sci. **81** 649

[8] Owens F N, P Dubeski, and C F Hanson 1993 Factors that alter the growth and development of ruminant. J. Anim. Sci. **71** 3138

[9] Purbowati E, Sutrisno C I, Baliarti E, Budhi S P S, dan Lestariana W. 2007. Pengaruh pakan komplit dengan kadar protein dan energi yang berbeda pada penggemukan domba lokal jantan secara *feedlot* terhadap konversi pakan. Pada Prosiding Seminar Nasional Teknologi Peternakan dan Veteriner. Pusat Penelitian dan Pengembangan Peternakan, Badan Penelitian dan Pengembangan Pertanian, Departemen Pertanian. Bogor 394

Body Composition of Lambs and Rams Fed Complete Feed

F. Rahmawati[1*] and E. Rianto[1]

[1]Faculty of Animal and Agricultural Sciences, Diponegoro University, Semarang, Indonesia

*Corresponding e-mail : firdarahmawati012@gmail.com

Abstract

This study was aimed to investigate the proportion of body water, protein and fat of lambs and rams. The material used in this study were 6 lambs (weighed of 15.92 kg), and 6 rams (weighed 23.31 kg). The sheep were fed a complete feed. In weeks 1, 5 and 9, the body composition (water, protein and body fat) was measured using the "*urea space*" method. The results showed that the proportion of water, protein and fat in the body of lambs and sheep differed significantly (P <0.01). The proportion of lambs' body water in weeks 1, 5 and 9 were 58.71, 58.43 and 58.43%, respectively; while thoset of the rams were 58.40, 58.14% and 58.14%, respectively. The proportion of body protein of lambs in weeks 1, 5 and 9 were 11.84, 11.94 and 11.99%, respectively; while those of rams were 13.01; 13.01 and 13.03%, respectively. The proportion of body fat of lambs in weeks 1, 5 and 9 were 20.51, 20.87 and 20.87%, respectively; while those of rams were 20.9, 21.26 and 21.25%, respectively. Body water of lambs and rams week decreased from week 1 to 9. On the other hand, the proportion of body protein and fat increased. It is concluded that the proportion of body water was higher while the proportions of body protein and fat were lower in lambs as compared with those in rams.

1. Introductions

Consumers generally want meat that contains low fat, but marketed meat tends to be rich in fat. This happens because the animals are fattened after having been fully grown. Lambs have a growth rate that tends to be faster than rams and it is estimated that lambs are still in a period of increased muscle in the body to produce meat. The growth rate is related to the development of body composition in the form of water, protein and body fat.

Nutritional composition of the diet given to the animal will greatly affect growth rate and body composition. When forage feed is fed to more livestock, livestock will be bulky, causing difficulty to digest in the body. In contrast to concentrates that tend to be more easily digested so much absorbed by the body to be able to synthesize or the formation of body tissues with a maximum of meat.

Livestock has a good growth if the nutrients provided in accordance with the needs. The amount of livestock feed should be adjusted to the body weight and age of the livestock. If the livestock is short of feed then the need of animal feed is also not fulfilled. The low weight of livestock body is related to the body composition, ie the higher the body weight percentage of water decreases, the percentage of fat increases, while the percentage of body protein is relatively constant [1].

Therefore, a study is needed to be done to find out the difference of body composition between lambs and rams as reference in the selection of meat in accordance with consumer demand.

This study was aimed to compare the body composition of the sheep of different age and body weight. The benefits of this study was to provide information about the body composition of lambs and rams, so it could be used as a reference in determining the age and weight of sheep to be slaughtered to fulfill the demand of consumers.

2. Materials and methods

This study used 12 thin tailed sheep consisting of 6 lambs with an average body weight of 15.92 ± 1.59 kg and 6 rams with an average body weight of 23.31 ± 0.77 kg. The sheep were kept in individual cages equipped with a feed and drinking water bunks. The sheep were fed complete feed with 12.07% crude protein (CP), 55.03% Total Digestible Nutrient (TDN) and 19.61% crude fiber (CP).

This study used an independent sample comparison design, namely comparing lambs and rams in proportion and changes of body water, protein and fat content. Every day the sheep were fed a complete feed at amount of 3.5% body weight, which was offered in the morning and evening. The refusals was weighed on the following morning.

In weeks 1, 5 and 9, the measurement of body water, protein and fat content was performed using urea space method. Blood plasma obtained wass analysed by Berthelot method based on the principle of calorimetric enzymatic test.

The parameters observed in this study were dry matter intake (DMI), live weight gain (LWG) and body composition (covering the proportion of body water, body protein and body fat). Dry matter intake was calculated from the difference between the weight of the offer and that of the refusals. Live weight gain was measured from the difference between the final body weight and the initial body weight divided by length of maintenance. Water content, protein and body fat were calculated using [2] equation:

Body water content = 16,519 + 1,047 US (%)
Body protein content = 6,714 + 0,2082 US (%)
Body fat content = 23,837 – 0,2965 US (%)

3. Results and Discussion

There a significant difference (P<0.05) DMI between lambs and rams. This happened because of differences in body weight between lambs and rams. Having heavier body weight, the rams ate more than the lambs (Table 1). However, in term of the percentage of body weight, there was no significant difference (P> 0.05) between the lamb and the ram (Table 1). This suggested that the capacity of the lamb's digestive tract was similar to that of the rams. The DMI of the rams this study was higher than the findings of [3] that DMI of sheep with 18 – 24.5 kg body weight was 817 g/day. This difference may be attributed to the quality of the diet.

Table 1. Dry Matter Intake and Live Weight Gain of Lambs and Rams During Week 1 - 9

Parameter	Lambs	Rams	Significations
Dry Matter Intake (g/day)	585.78	822.59	**
Dry Matter Intake (% BB)	2.91	2.61	Ns
Live Weight Gain (g/day)	18.00	1.00	*

Notes: *) significant (P> 0.05); **) very significant (P> 0.01); ns: not significant

There was significant difference (P<0.05) in LWG between the lambs and the rams. The LWG of lambs were higher than that of the rams (Table 1). The rams had a very low LWG because the majority of dietary nutrients were used for maintenance, so that only a little part could be used for production. It is claimed by [4] that nutrients consumed by the animal are primarily used for maintenance, while the rest is used for production.

3.1.Proportion of Water Content, Protein and Body Fat Content in Lamb and Rams

The results showed that there was a very significant difference (P<0.01) between lambs and rams in the proportion of body water, protein and fat of lambs and rams (P>0.01). The lambs were higher in body water content but lower in body protein and fat content as compared with the rams (Table 2). This was in agreement with opinion [1] that the high proportion of body water is affected by age, growth rate and feed consumption; the older the animal the higher the body weight, followed by a decrease in the proportion of body water.

The proportion of body water in this study was lower than the claim of [5], that body water content of an animal was 59%, and higher than the findings of [6] being 52.8%. These differences may be caused by body weight of the animals. This is in agreement with the opinion of [5] that the body's water content is affected by age, body weight and feed intake.

Table 2. Proportion and Changes of Body Water, Body Protein and Body Fat Content on Lambs and Rams

Parameter	Lambs	Rams	Signification
Proportion of Body Water Content (%)			
Week 1	58.71	58.40	SN
Week 5	58.43	58.14	SN
Week 9	58.43	58.14	SN
Significations	SN	SN	
Proportion of Body Protein Content (%)			
Week 1	11.84	13.01	SN
Week 5	11.94	13.01	SN
Week 9	11.99	13.03	SN
Significations	SN	SN	
Proportion of Body Fat Content (%)			
Week 1	20.51	20.91	SN
Week 5	20.87	21.26	SN
Week 9	20.87	21.25	SN
Significations	SN	SN	

Notes : SN) very significant (P>0.01)

There was highly significant difference (P<0.01) in proportion of body protein between the lambs and the rams. The body protein content of the lambs was lower than that of the rams (Table 2). This suggested that the lambs used in this study is still on the accelerated growth process [7]. In that phase the sheep still mendeposisikan protein on body tissues quickly.

The proportion of body protein in this study is lower than the results of research [8] who obtained body protein levels of 16.50%. This is due to the level of fat in the body has a negative correlation with the body's water content, if the body water decreases then the fat increases but the protein content tends to be constant. This is in accordance with the opinion [1] the percentage of body protein remains even though sheep grow older.

The proportion of body fat in the lambs was lower very significantly (Table 2) than that in the rams. This indicated that the increasing age and increase in body weight result in fat accumulation in the body. It is suggested by [1] that the increased percentage of fat in the body is associated with increased body weight.

The proportion of body fat in this study was slightly higher than that of [8] who obtained the proportion of body fat by 20.82%. One of factors that affect the fat content is the age of the animal. It stated by [7] that age, body weight, growth rate are factors that have a close relationship with body composition.

3.2.Changes in the proportion of body water, protein and body fat content of lambs and rams

In week 1, 5, and 9 consecutively, the proportion of body water of lambs and rams decreased (Table 2), while the proportion of body protein and body fat increased. These phenomena were in agreement with the opinion of [5] that water and fat have a negative relationship, along with the age and weight of the body, the proportion of body fat tends to increase, while the proportion of body water decreases.

The proportion of body protein of lambs and rams in week 1.5 and 9 were still increasing (Table 2). This indicated that the animal were still growing. The increase of body protein in the lambs was higher than that in the rams, indicating that the lambs were in the accelerated growth phase, while the rams had reached the decelerated growth phase. It is stated by [9] that the livestock at the accelerated phase growth rate has the ability to synthesize proteins larger than those with slower phase growth rates.

The proportion of body fat in weeks 1, 5 and 9 increased continuously, both in lambs and in rams (Table 2). This indicated until the sheep weighed about 23 kg, the fattening process still continued. The proportion of body fat in this study was similar to that of [8], who obtained body fat content about 21%. This was so since the body weight of the sheep in the current study and that in the study of [8] were almost the same.

4. Conclusion

Based on the study it can be concluded that the proportion and changes of body water, protein and fat in lambs and rams differ. The proportion of body water of lambs was higher than that of rams, while the proportion of body protein and fat in lambs was lower than that of the rams. During the study period, the proportion of body water decreased, while the proportion of protein and body fat increased, both in the lambs and in the rams.

References

[1] Pond W G, D C Church and K R Pond 1995 Basic Animal Nutrition and Feeding 4th Edition. John Wiley & Sons New York

[2] Astuti D A, dan D Sastradipradja 1999 Evaluasi komposisi tubuh dengan menggunakan teknik ruang urea dan pemotongan pada domba Priangan. J. Media Vet. **6** (3) 5

[3] Utomo R 2004 Pengaruh penggunaan jerami padi terfermentasi sebagai bahan dasar pembuatan pakan komplit pada kinerja domba. Buletin Peternakan **28** (4) 162

[4] McDonald P, R A Edwards, J F D Greenhalgh, C A Morgen, L A Sinclair and R G Wilkinson 2012 *Animal Nutrition*, Seventh Edition. Prentice Hall Harlow England

[5] Tillman A D, H Hartadi, S Prawirokusumo, S Reksohadiprodjo dan S Lebdosoekojo 1991 Ilmu Makanan Ternak Dasar Cetakan Ke-4. Gadjah Mada University Press Yogyakarta

[6] Widnyana I G, N N Suryani dan I K M Budiasa 2014 Pengaruh komposisi hijauan dengan level konsentrat yang berbeda pada ransum terhadap komposisi tubuh dan retensi nutrien kambing Peranakan Etawah (PE). J. Peternakan Tropika **2** (3) 375

[7] Soeparno 2005 Ilmu dan Teknologi Daging. Gadjah Mada University Press Yogyakarta

[8] Haryati Y, D Heriyadi dan E Hernawan 2015 Komposisi tubuh domba Priangan yang diberi larutan elektrolit berbasis air kelapa (*Cocos nucifera*) dan ekstrak rosela (*Hibiscus sabdariffa*) sebelum transportasi **4** (3) 1

[9] Bulle F C P C, P V Paulino, A C Sanches and R D Sainz 2007 Growth, carcass quality, protein and energy metabolism in beef cattle with different growth potentials and residual feed intakes. J. Anim. Sci. **85** 928

Profile of Rumen Fermentation and Blood Urea Nitrogen Concentration of Kacang Goat Fed Total Mixed Ration Vs. Roughage

R. Adiwinarti[1]*, Kustantinah[2], I.G.S. Budisatria[3], Rusman[4], E. Indarto[2]

[1]Faculty of Animal and Agricultural Sciences, Diponegoro University, Semarang, Indonesia,
[2]Department of Animal Nutrition and Feeds Science, Faculty of Animal Science, Gadjah Mada University, Yogyakarta, Indonesia
[3]Department of Animal Production, Faculty of Animal Science, Gadjah Mada University, Yogyakarta, Indonesia
[4]Department of Animal Product Technology, Faculty of Animal Science, Gadjah Mada University, Yogyakarta, Indonesia

*Corresponding E-mail: retno_adiwinarti@undip.ac.id

Abstract

Kacang goat is usually reared traditionally fed natural grass and having inferior performance. Many researches had been done to improve the performance. Total mixed ration (TMR) containing soybean meal (SBM) and fish meal (FM) could increase the performance of Kacang goat, but the profile of rumen fermentation has not been published. Therefore, this study investigated the profile of rumen fermentation and blood urea Nitrogen concentration (BUN) in Kacang goat fed natural grass (roughage) and TMR. Twelve yearling Kacang buck, 15.2-19.6 kg were arranged in completely randomized design. The treatments were NG (natural grass), FM (TMR containing FM), and SBM (TMR containing SBM). The TMR were formulated containing 15% crude protein and 56-58% TDN. Data were analyzed by one way ANOVA. Rumen pH 6 hours after feeding of NG (7.4) was higher ($P<0.01$) than that of FM (6.2) and SBM (6.4). This lowering pH of TMR was caused by increasing volatile fatty acids (VFA). The VFA total of FM (129.7 mmol/l) and SBM (153.1 mmol/l) were higher than that of NG (86.4 mmol/l). At 3 and 6 hours after feeding, ammonia in the rumen of SBM was higher than that of NG and FM, indicating higher degraded protein. The BUN at 3 hours after feeding of SBM was higher than that of NG. It can be concluded that protein in SBM was degraded higher than others and the lower pH in rumen of TMR goats was caused by higher VFA produced by TMR goats compared to NG goats.

1. Introduction

Kacang goat is usually reared traditionally fed natural grass and having inferior performance. Many researches had been done to improve the performance. Total mixed ration (TMR) containing soybean meal (SBM) and fish meal (FM) could increase the performance of Kacang goat [1,2], but the profile of rumen fermentation has not been published yet. Therefore, this study investigated the profile of rumen fermentation and blood urea Nitrogen (BUN) concentration in Kacang goat fed natural grass (roughage) and TMR (TMR containing FM and SBM).

2. Materials and Methods

The procedures of animal experiment were based on the guidelines of the Faculty's Livestock Ethics Committee. Twelve yearling Kacang buck, 15.2-19.6 kg were arranged in completely randomized design. The goats were penned individually provided by water *ad libitum* and different feeding treatments. The treatments were NG (natural grass), FM (TMR containing FM), and SBM (TMR containing SBM). The TMR were formulated containing 15% crude protein and 56-58% TDN consisted of *Pennisetum purpureum*, gliricidia leaves, cassava waste product, wheat bran, SBM or FM, and mineral mix (Table 1). Volatile fatty acids (acetic, propionic, and butyric acids) were assayed from rumen fluid collected at 0, 3, and 6 h after feeding and determined by gas chromatography (Shimadzu GC-8). Rumen ammonia was analyzed by Conway procedure [3]. Blood urea Nitrogen concentration was determined at 0, 3, and 6 h after feeding using Spectro MicroLab. Data were analyzed by one way ANOVA using SPSS statistics software version 19.

Table 1. Feed composition and nutrients contents (CP and TDN) of the ration [1,2].

Parameters	NG	FM	SBM
	----------	%	----------
Feed composition:			
Natural grass	100	-	-
Pennisetum purpureum	-	30.00	30.00
Gliricidia leaves	-	30.00	30
Cassava waste product	-	20.10	19.20
Wheat bran	-	13.75	13.80
Fish meal	-	6.15	-
Soybean meal	-	-	7.00
Ration nutrient content:			
Crude fiber (CF)	34.62	29.68	29.18
Crude protein (CP)	10.92	15.26	15.59
Total digestible nutrients (TDN)	63.23	56.21	57.95

3. Results and Discussion

Rumen pH, volatile fatty acids, rumen ammonia and blood urea nitrogen (BUN) concentration was affected by the ration (Table 2 and 3). Rumen pH 6 hours after feeding of NG (7.4) was higher ($P<0.01$) than that of FM (6.2) and SBM (6.4). This lowering pH in TMR goats was caused by increasing volatile fatty acids (VFA) in the rumen (Table 2). The VFA total at 6 h after feeding in FM goats (129.7 mmol/l) and SBM goats (153.1 mmol/l) were higher than that in NG goats (86.4 mmol/l). Higher ruminal VFA concentration in TMR groups (FM and SBM) because the ration contained less fiber (Table 1.) and digested easier than in roughage group (NG). Concentration of VFA tended to increase from 3 to 6 h after feeding as Widiawati *et al.* [4] reported. Widiawati *et al.* [4] stated that using *in vitro* method, the VFA concentration peaked at 6 h of incubation for gliricidia.

Ajayi and Ogunleke [5] reported rumen pH of West African Dwarf goats fed concentrate containing corncob 0-45% around 5.8-6.1 that is lower than this research. Abubakr *et al.* [6] stated rumen pH in goats fed dietary palm oil by-products about 6.1-6.4, while Darlis *et al.* [7] reported that rumen pH of goats fed rice straw supplemented with SBM was 6.5.

Total of VFA in this study was higher than those in Ariyani *et al.* [8] and Jelantik *et al.* [9], while similar to Darlis *et al.* [7]. Other researchers reported that VFA total was about 58.4-69.1 mM in Jawarandu goats fed diet containing sugarcane bagasse [8], 44.1-61.1 mM in Kacang goats fed ration containing fish meal [9], and 117.5-119.1 mM in goats fed rice straw supplemented with SBM or SBM and sago meal [7].

Table 2. Rumen pH, volatile fatty acids total, and rumen ammonia of Kacang goats fed roughage (NG) and TMR (FM and SBM).

Parameters	NG	FM	SBM
Rumen pH:			
0 hour	7.3	6.5	6.4
3 hours after feeding	7.2	6.3	6.3
6 hours after feeding	7.4^{A}	6.2^{B}	6.4^{B}
Volatile fatty acids (VFA) total (mM/L):			
0 hour	85.5	114.0	135.4
3 hours after feeding	79.2	118.6	141.6
6 hours after feeding	86.4^{Aa}	129.7^{b}	153.1^{Bb}
Rumen ammonia (mg/100mL):			
0 hour	20.8^{a}	24.6^{ab}	39.1^{b}
3 hours after feeding	22.0^{a}	23.8^{a}	38.3^{b}
6 hours after feeding	23.3^{a}	24.2^{a}	35.1^{b}

$^{A, B}$ Means with different superscripts within a row are highly significantly different ($P<0.01$)
$^{a, b}$ Means with different superscripts within a row are significantly different ($P<0.05$).

At 3 and 6 hours after feeding, ammonia in the rumen of SBM was higher than that of NG and FM, indicating higher degraded protein. This condition was supported by the BUN concentration at 3 hours after feeding of SBM that was higher than that of NG. In this study, high ammonia concentration in the rumen and BUN concentration of SBM goat was caused by high dry matter intake (DMI) and crude protein intake. The DMI of SBM goats (707.5 g) was higher than that of NG goats (502.2 g) and FM goat (597.86) [1], while crude protein intake of SBM, FM, and NG goats was 115.8 g, 94.7 g, and 52.6 g, respectively [2]. High nitrogen intake had increased rumen degradable nitrogen, therefore increasing ruminal ammonia concentration and BUN concentration. Abubakr *et al.* [6] reported that ammonia concentration in the goats' rumen was influenced by the diet. Ariyani *et al.*[8] stated that ruminal ammonia concentration was affected by dry matter intake and total degraded nitrogen in the rumen. Akhsan *et al.* [10] reported that the increase of protein in the ration from 8.69% to 17.00% have increased protein intake and digested protein in the rumen.

Table 3. Blood urea Nitrogen concentration of Kacang goats fed roughage (NG) and TMR (FM and SBM).

Parameters	NG	FM	SBM
Blood urea Nitrogen (BUN) (mg/dL):			
0 hour	34.4	40.4	48.1
3 hours after feeding	41.4[a]	45.0[ab]	56.6[b]
6 hours after feeding	40.7	43.3	54.1

[a, b] Means with different superscripts within a row are significantly different ($P<0.05$).

There is a tendency that BUN concentration increased at 3 h after feeding, then decrease at 6 h after feeding (Table 3). Usually, the concentration of urea tended to increase 2 h after feeding because of dietary protein degradation in the rumen [11]. Borja *et al.* [11] reported BUN concentration of 3/4 Boer goats fed 0-45% licury cake (*Syagrus coronate* (Martius) Beccari) about 20.9-25.7 mg/dL that was lower than BUN concentration in this study (34.4-56.6 mg/dL).

4. Conclusion

It can be concluded that protein in SBM was degraded higher than others. It was indicated by high ammonia concentration in the rumen and BUN concentration. The lower pH in rumen of TMR goats (FM and SBM) was caused by higher VFA produced by TMR goats compared to NG (roughage) goats.

Acknowledgments

The authors gratefully appreciate the Directorate Research and Community Service, Directorate General of Higher Education, the Ministry of Research, Technology, and Higher Education, Republic of Indonesia for financially supported by the Research Grant "Penelitian Unggulan Perguruan Tinggi" (PUPT) Universitas Gadjah Mada No. 001/SP2H/LT/DRPM/IV/ 2017, "Kacang goat" team and Meat Sci. Lab. assistance that had assisted for data collection.

References

[1] Adiwinarti R, Kustantinah, Budisatria I G S, Rusman and Indarto E 2016 Improving the performance of local Kacang goats using ruminally undegradable protein feeds Asian. J. Anim. Sci. **10(4)**, 262.

[2] Kustantinah, Adiwinarti R, Budisatria I G S, Rusman and Indarto E 2017 Improved productivity of kacang goats reared by farmers using balanced rations with different sources of protein Pakistan. J. Nut. **16** (**9**), 672.

[3] Conway E and O'Malley E 1942 Microdiffusion method. ammonia and urea using buffered absorbents. Biochem. **36**.

[4] Widiawati Y, Winugroho M, Teleni E and Thalib A 2007 Fermentation kinetics (*in vitro*) of *Leucaena leucocephala, Gliricidia sepium,* and *Calliandra callothyrsus* leaves (3) the pattern of gas production, organic matter degradation, pH, NH_3 and VFA

concentration; estimated CH_4 and microbial biomass production Indonesian J. Anim. Vet. Sci. (IJAVS) **12** (**3**), 202.

[5] Ajayi F T and Ogunleke F O 2015 Performance, nutrient utilization and rumen fermentation characteristics of West African Dwarf goats fed concentrate diet with corncob in varying inclusion levels The Pacific J. Sci. Technol. **16** (**1**), 274.

[6] Abubakr A R, Alimon A R, Yaakub H, Abdullah N and Ivan M 2013 Digestibility, rumen protozoa, and ruminal fermentation in goats receiving dietary palm oil by-products J. Saudi Society of Agric. Sci. **12**, 147.

[7] Darlis, Abdullah N, Halim R A, Jalaludin S and Ho Y W 2000 Rumen parameters and urea kinetics in goats and sheep Asian-Aus. J. Anim. Sci. **13** (**7**), 922.

[8] Ariyani S A, Nuswantara L K, Pangestu E, Wahyono F and Achmadi J 2014 Parameters of protein metabolism in goats fed diets with different portion of sugarcane bagasse J. Indonesian Trop. Anim. Agric. **39** (**2**), 111.

[9] Jelantik I G N, Weibsjerg M R and Madsen J 2012 Intake, rumen degradation and utilisation of urea-ammoniated grass hay by Kacang-goats as affected by supplementation of sun-dried fish or fishmeal Anim. Prod. **14** (**2**), 77.

[10] Akhsan F, Nuswantara L K and Achmadi J 2015 Combination of soybean meal and *Hibiscus tiliaceus* leaf in the goat diet: effect on some parameters of protein metabolism J. Indonesian Trop. Anim. Agric. **40** (**2**), 100.

[11] Borja M S, Oliveira R L, Bagaldo A R, Pereira M L A, Portela R W, Barbosa A M, Ribeiro C V D M and Carvalho G G P 2014 Microbial protein and blood parameters of goats fed with licury cake Ciências Agrárias. **35** (**1**), 519.

Fermentation of Anaerobic Cow Waste as Bio-Slurry Organic Fertilizer and Nitrogen Chemical Fertilizer on Soybean

Yafizham[1*] and Sutarno[1]

[1]Faculty of Animal and Agricultural Sciences, Diponegoro University, Semarang, Central Java, Indonesia

*Corresponding E-mail: yafiz60@yahoo.com

Abstract

The study aimed was to evaluate the effect of bio-slurry organic fertilizer and urea chemical fertilizer combination on fresh material weight, phosphorus and potassium soybean straw, and seed weight per soybean plant plot. The experiment was conducted with a randomized block design with a single treatment repeated 5 times consisting of P0: control (without fertilizer), P1: bio-slurry 10 t/ha + 25 kg of N/ha, P2: bio-slurry 10 t/ha + 50 kg of N/ha, P3: bio-slurry 10 t/ha + 75 kg of N/ha, P4: bio-slurry 10 t/ha + 100 kg of N/ha and P5: bio-slurry 10 t/ha. The results showed that bio-slurry treatment of 10 t/ha + 25 kg of N/ha resulted in the highest fresh weight and dry weight of soybean plants, respectively of 240.7 g and 22.33 g, but not significantly different from the bio-slurry treatment of 10 t/ha + 50 kg of N/ha which yielded fresh weight of 197.7 g and a dry weight of 19.08 g. P production of 10.23 g per plant was significantly higher than other treatments but didn't differ significantly between P2 and P4 treatments of 8.05 and 7.17 g per plant. The bio-slurry treatment of 10 t/ha + 25 kg of N/ha also yielded K of 6.46 g per plant butn't unlike the bio-slurry treatment of 10 t /ha + 50 kg of N/ha. While the number of pods per plant and weight of 100 grains of the highest soybean seeds were also produced from bio-slurry treatment of 10 t /ha + 25 kg of N /ha.

1. Introduction

Soybean is an important component of the national food supply in Indonesia. Soybean is one of the main food commodities after rice and corn. Soybean is the main food source of vegetable protein for the community. Soybean plants are also known as seasonal crops that can grow on various soil types. Soil nutrient content and soil organic matter also greatly affect the increase of soybean plant productivity. In soils containing soil nutrients from very low to low and soil organic materials from low to slightly low require additional chemical fertilizers and more organic fertilizers than soil-containing soils and higher organic matter.

Low productivity and perceptions of farmers who consider soy not profitable to be cultivated as staple crops cause they do not want to spend the cost to complete the production facilities in the form of fertilizer whose needs are quite large and the price is not affordable [1]. The cultivation technique that needs to be done in soybean farming is fertilization. Fertilization is the addition of nutrients for plants one of them by providing organic or inorganic fertilizers. Organic and inorganic fertilizers that can be given are bio-slurry and nitrogen.

In many developing countries, animal manure is anaerobically digested, sometimes in combination with human excreta, to produce biogas, which is used for cooking and/or basic lighting. Besides biogas, anaerobic digestion produces a residue, the so-called di gestate or bio-slurry, which contains most of the nutrients of the original manure, partially in a converted form. This bio-slurry can potentially be used as an organic fertilizer [2].

Bio-slurry is a product of biogas processing made from a mixture of manure and water through a process without oxygen (anaerobic) in a confined space. Bio-slurry fertilizer is a final product of cow dung waste processing in the form of solid and liquid which is very useful as a source of nutrients for plants [3]. Bio-slurry is very useful as a source of nutrients for plants. Utilization of bio-slurry in addition to helping to improve the physical, chemical and biological properties of the soil, will also reduce the negative effects of environmental pollution and the spread of disease sources. Solid bio-slurry has a macro nutrient content of N: 2.05%, P_2O_5: 2.70%, K_2O: 0.58%. The benefits of bio-slurry include adding humus as much as 10-12%, supporting the development activity of worms and microbes that are beneficial to plants. Fermented slurry, sometimes called bio-slurry, as a product of anerobic fermentation of animal excrements in the biogas digester, is an excellent organic fertilizer which can make an important contribution to better crop yields and lasting soil fertility [3]. Provision of bio-slurry fertilizer alone in the short term has not been able to meet the nutrient needs for soybean crops, so it is necessary to add inorganic fertilizers such as nitrogen fertilizer.

Islam [4] reported that N contents in composted bio-slurry of both cow dung and poultry manure in Bangladesh were higher than in non-composted manure, whereas concentrations of most other nutrients were not higher. It might be that this additional N origins from the biomass which has been added to the bio-slurry for composting. Unfortunately, they did not report which biomass was used for composting. In the same report they also showed that similar results were found in a study in Nepal. Here, they concluded that the increase in N was due to the higher N concentrations in added organic material, which was evidenced by the given compositions of commonly added materials [2].

The combination of bio-slurry organic fertilizer with nitrogen fertilizer is a good alternative in soybean cultivation. The combination has the potential to increase production if applied at the right time. Thus, a research to determine the effect of dosage of bio-slurry organic fertilizer and chemical fertilizer nitrogen on growth and yield of soybean crop.

2. Materials and Methods

The research material used is bio-slurry organic fertilizer that is derived from biogas reactor Faculty of Animal Husbandry and Agriculture University of Diponegoro and Urea chemical fertilizer as source of nitrogen. and soybean seed varieties Grobogan as plant tested. The experiment was conducted at Experimental Field of Faculty of Animal Husbandry and Agriculture, Diponegoro University, Tembalang Sub-district, Semarang City, Central Java Province, from July to November 2016.

The study used a randomized monophonic group design. The treatment provided is a combination of bio-slurry organic fertilizer with nitrogen ie P0: Control (without fertilizer), P1: bio-slurry 10 t / ha + 25 kg of N / ha, P2: bio-slurry 10 t / ha + 50 kg of N / ha, P3: bio-slurry 10 t / ha + 75 kg of N / ha, P4: bio-slurry 10 t / ha + 100 kg of Nbha. P5: Bio-slurry 10 t / ha. All treatment combinations were repeated 5 times, so the number of experimental units was 25 experiments.

The first research was done by cultivating the soil and making experimental plot. Application of bio-slurry fertilizer was done simultaneously with soil processing, while the application of nitrogen fertilizer was done on 7 days after planting and 35 days after planting with dose of each half. For the maintenance of plants is done watering, thinning, embroidery, weeding and controlling plant-disturbing organisms using chemical pesticides.

Plant parameters observed were fresh weight and dry weight of soybean plants, P and K absorption, number of pods per plant and weight of 100 grains of soybean seed. Data were analyzed by Duncan's Multiple Range Test (DMRT) at the level of $\alpha = 5\%$.

3. Results and Discussion

Application of bio-slurry of 10 t/ha with each nitrogen 25 and 50 kg of N/ha able to produce fresh weight and dry weight significantly higher than others. This indicates that the application of bio-slurry fertilizer and nitrogen fertilizer (25 and 50) kg of N/ha can increase fresh weight and dry weight of soybean crop, while the increase of nitrogen fertilizer does not show the addition of fresh weight and dry weight of plant. This is in line with opinion [5] that excessive fertilization can have a negative effect on the plant. Fertilization is added continuously so that the amount exceeds the needs of the plant will cause the growth and the unfavorable plant crops.

The dry weight of the plant is the resultant of three processes: the accumulation of asymilat through photosynthesis, the decrease of assimilate due to respiration and accumulation to the food reserves. According to [6] the dry weight of plants depends on the rate of photosynthesis and respiration, because the respiration process uses energy derived from photosynthesis. In line with the opinion of [7] that the dry weight of a plant is influenced by the optimum of photosynthesis because the dry weight of a plant depends on the amount of carbohydrate accumulation in the plant body.

The N element is indispensable in the early stadia of plant growth, where N is the compiler of important organic compounds such as amino acids, proteins and plays a role in the formation of chlorophyll which affects the photosynthesis process so as to increase the assimilates utilized for the growth of plant vegetative organs which will support the dry weight of plants [8].

Table 1. Average Fresh Weight and Dry Weight of Soybean Plants on Combination Fertilizing Bio- slurry with Nitrogen

Treatment	Fresh Weight Plant (g)	Dry Weight Plant (g)
Control (without fertilizer)	$78,4^{e}$	$7,75^{e}$
Bio-slurry 10 ton/ha + 25 kg of N/ha	$240,7^{a}$	$22,33^{a}$
Bio-slurry 10 ton/ha + 50 kg of N/ha	$197,7^{ab}$	$19,08^{ab}$
Bio-slurry 10 ton/ha + 75 kg of N/ha	$155,7^{bc}$	$15,15^{bc}$
Bio-slurry 10 ton/ha + 100 kg of N/ha	$137,1^{cd}$	$13,05^{cd}$
Bio-slurry 10 ton/ha	$92,3^{d}$	$9,35^{d}$

Description: Different superscripts on the same column show significant differences (P> 0.05)

Table 2 shows that bio-slurry fertilization of 10 t/ha and nitrogen 25 kg of N/ha in soybean plants yielded the highest P and K absorption of 10.23 g per plant and 41.59 g per plant, respectively the uptake of P and K is no different from that of bio-slurry crop 10 t/ha and 50 kg of N/ha. This is presumably because the nutrient content of P contained in bio-slurry fertilizer can be absorbed more optimally in plants. While the nutrients P contained in chemical fertilizers are not expected to be absorbed by plants optimally, due to the relatively low availability of nutrients P in the soil. The right dosage combinations can help the absorption of P nutrients by maximum by soybean crops. Phosphorus increases the rate of photosynthesis, enzymatic activity, energy transfer, root development, uptake and other nutrient transfer, According to [9] that a plants require the optimum type, dosage and concentration to trigger maximum productivity and growth. If the dose and concentration is given more then the growth rate will decrease.

The effect of bio-slurry fertilizer and its combination with nitrogen fertilizer on K uptake showed that bio-slurry treatment yielded the highest K nutrient uptake of 41.59 g per plant. This is presumably because of the K nutrient contained in bio-slurry fertilizer is the main source of K nutrients absorbed by plants. According to [10] potassium regulates several plant processes including: water and nutrient transport across cell walls, water vapor regulation and carbon dioxide (CO2) through stomates, uptake and nutrient transfer, helps maintain plant defenses against damaging effects of nematodes in minimizing yield loss. In short, P and K help improve farmers' profits mainly through increased soybean yield.

In general, N application increases the P uptake. Application of 20 kg of N/ha increases the uptake of N, P, Ca and Mg and. subsequent grain yield and 60 kg of N/ha increases the N uptake indicates the usefulness of N application [11]. Although soy is a leguminous plant, it responds to N fertilizers in addition to P and K requirements. In general, N applications increase uptake P. Phosphorus fertilization into soybean increases the absorption of N, P and K. The application of N, P and K together significantly increases the attributes of growth and yield of soybeans.

Table 2. Average Absorption of P and K Uptake on Combination of Bio-slurry Fertilization with Nitrogen

Treatment	Absorption P (g/plant)	Absorption K (g/plant)
Control (without fertilizer)	$2{,}69^{d}$	15.31^{e}
Bio-slurry 10 ton/ha + 25 kg of N/ha	$10{,}23^{a}$	41.59^{a}
Bio-slurry 10 ton/ha + 50 kg of N/ha	$8{,}05^{ab}$	39.71^{ab}
Bio-slurry 10 ton/ha + 75 kg of N/ha	$6{,}15^{bc}$	23.26^{cd}
Bio-slurry 10 ton/ha + 100 kg of N/ha	$7{,}17^{abc}$	28.63^{bc}
Bio-slurry 10 ton/ha	$4{,}16^{cd}$	16.65^{d}

Description: Different superscripts on the same column show significant differences (P> 0.05)

The bio-slurry treatment of 10 t/ha with each nitrogen of 25 and 50 kg of N/ha was able to produce the number of pods and 100 grains weight significantly higher than the others. The data in Table 3 shows that the application of bio-slurry 10 t/ha and nitrogen 25 kg of N/ha gave the highest number of pods per plant (25.11 fruits) significantly without fertilizer application. Meanwhile, bio-slurry 10 t/ha fertilizer and nitrogen 25 kg of N/ha resulted in the highest 100 grain weight but not significantly different from bio-slurry 10 t/ha and nitrogen 50 kg of N/ha. This indicates an increase in the number of plant pods more determined by the application of nitrogen fertilizer. The bio-slurry fertilizer provided does not meet the nutrient requirements required for pod formation. This is due to the slow organic material available for plants so it needs to be balanced with nitrogen fertilizer.

According to [12] the nitrogen element enhances the formation of proteins, enzymes, and as a chlorophyll-forming element, in addition the availability of N can increase the uptake of P. The administration of N on fertilized soil will further dissolve P so that P is more available and can be utilized by plants for metabolism including photosynthesis, especially in CO2 fixation so that carbohydrates are formed and transocated for pod formation. Potassium plays a role in the process of plant metabolism, which acts as an enzyme activator in plant metabolism reactions, regulating osmotic pressure of cells, where the osmotic pressure-preserved cells increase protein synthesis for metabolism, increasing the number of pods.

The weight of 100 seeds is one of the observation parameters closely related to the production achieved. If the weight of 100 seeds is high then the more the results will be obtained. Yet all of that is partly still influenced by the cultivation system, the genotype and the varieties of the plant itself. One of the factors that determine the quality of plant materials such as seeds is the amount of substrates such as carbohydrates available for metabolism that support early plant growth. This makes seed size or weight often used as a benchmark to get a uniform planting material.

Table 3. Average Number of Grains of Fill and Weight of 100 Seeds on Combination of Bio-slurry Fertilization with Nitrogen

Treatment	Number of pods/plant	Weight of 100 Seeds (g)
Control (without fertilizer)	7.70[e]	12[e]
Bio-slurry 10 ton/ha + 25 kg of N/ha	25.11[a]	20[a]
Bio-slurry 10 ton/ha + 50 kg of N/ha	21.06[b]	19[ab]
Bio-slurry 10 ton/ha + 75 kg of N/ha	15.68[c]	18,2[bc]
Bio-slurry 10 ton/ha + 100 kg of N/ha	14.77[cd]	17,2[cd]
Bio-slurry 10 ton/ha	11.65[d]	16,6[d]

Description: Different superscripts on the same column show significant difference (P> 0.05)

According to [13] macro nutrients present in bio-slurry fertilizer is very important for the plant. Besides producing macro nutrients, bio-slurry fertilizers also produce a number of micro nutrients, such as Fe, Zn, Bo, Mn, Cu, and Mo. So it can be said that this bio-slurry fertilizer can be considered as an alternative fertilizer to maintain maximum crop production while maintaining the safety and sustainability of organic products without changing the genetic engineering so familiar to the environment. Next according to [14], the availability of nutrients complete with the amount of each nutrient in accordance with the needs of plants will be able to stimulate the growth and production of plants.

4. Conclusion

The effects of bioslurry applications are proportional to the effects of the application of chemical fertilizers. Thus, bioslurry can be a serious alternative to chemical fertilizers. Bio-slurry organic fertilizer and chemical fertilizer nitrogen can increase the growth and production of soybeans. The combination of 10 t / ha bio-slurry fertilizer and 25 kg of N / ha nitrogen is the best combination of fertilizer that can increase fresh weight and dry weight, P and K absorption, and number of pods and weight of 100 grains of soybean seed.

Acknowledgments

Special thanks to Agroekoteknology Studi Program of Agriculture Department of Faculty Animal and Agricultural Sciences, Diponegoro University. We thank, Widyati Slamet, Adriani Darmawati and Bagus Herwibawa for their encouragement.

References

[1] Nurlisan, Aslim Rasyad and Sri Yoseva 2013 Pengaruh Pemberian Pupuk Organik Terhadap Pertumbuhan dan Hasil Tanaman Kedelai (*Glycine max (L.) Merril*)

[2] Bonten L T C, K B Zwart, R P J J Rietra, R Postma and M J G de Haas 2014 Bio-slurry as fertilizer Is bio-slurry from household digesters a better fertilizer than manure a literature review Wageningen Alterra Wageningen UR (University & Research centre). Alterra report 2519

[3] Choke Mikled, Suchart Jiraporncharoen and Nirandorn Potikanond 2002 Utilization of fermented slurry as bio-fertilizer. Department of Animal Science and Department of Soil

Science and Conservation Faculty of Agriculture Chiang Mai University Thailand Proceedings Biodigester Workshop March 2002

[4] Islam M F, 2009 Proceedings of International Bio-slurry Workshop and Study Tour Netherlands Development Organization (SNV) Dhaka Bangladesh

[5] Binford G D, Hearn B K, Isaacs M A, H Ansen D J, Tayor R W 2004 Foliar fertilization of roundup ready soybeans. Plant Management Network

[6] Pedersen P, Lauer J G 2004 Response of soybean yield components to management system and planting date. Agron J. **96** 1372

[7] Gai Z J, Zhang W Jiang F F, Cai L J 2016 Response of soybean root nodule and yield to biological-sugar nitrogen fertilizer and urea. J. Nuclear Agric. Sci. **30** (4) 822

[8] Schmitt M A, Lamb J A, Randall G W, Orf J H, Rehm G W 2001 In-season fertilizer nitrogen applications for soybean in Minnesota. Agronomy Journal **93** 983

[9] Fu C F, Sun C, Dong Y M 2011 Effects of nutrient management on NPK uptake and yield of soybean. Heilongjiang Agr. Sci. **10** 33

[10] Snyder C S 2000 Raise Soybean Yields and Profit Potential with Phosphorus and Potassium Fertilization. News & Views. Potash & Phosphate Institute

[11] Andrew G Ristvey, John D Lea-Cox and David S Ross 2007 Nitrogen and Phosphorus Uptake Efficiency and Partitioning of Container-grown Azalea During Spring Growth. Journal of the American Society for Horticultural Science

[12] Lakitan B 2007 Dasar-Dasar Fisiologi Tumbuhan. Raja Grafindo Persada Jakarta

[13] Gibson G Taroreh 2010 Pemanfaatan Limbah Biogas Sebagai Substitusi Pupuk Pada Tanaman Kedelai di Kabupaten Bolaang Mongondow Seminar Regional Inovasi Teknologi Pertanian mendukung Program Pembangunan Pertanian Propinsi Sulawesi Utara

[14] Supadma A A Nyoman, Pudja I Nyoman dan Mega I Made. 2014. Peningkatan Hasil Kedelai Melalui Pemberian Organik Cair dan Dosis Pupuk Fosfat Fakultas Pertanian Universitas Udayana Bali

Milk Production of Ettawah Grade Goat Fed Diet Containing Different Protein and Energy Contents Supplemented with Organic Mineral and Grapes Seed Oil

A. Muktiani[1*], E. Kusumanti[1] and D. W. Harjanti[1]

[1]Faculty of Animal and Agricultural Sciences, Diponegoro University, Semarang, Central Java, Indonesia.

*Corresponding Email: anismuktiani@undip.ac.id

Abstract

This study aimed to investigate the effect of different protein and energy contents and supplementation ofZinc (Zn), Chromium (Cr) and grapes seed oil on milk production in dairy goat.Randomized block design(RBD) was used in this study. Twelve lactating Ettawah crossbreed goats divided into three groupsbased on milk production. The treatment ration were:T1 = ration containing 16% CPand 66% TDN ;T2 = ration containing 14% CP and 63% TDN supplemented with Zn 20 ppm + Cr 2 ppm; and T3 = T2 + 22 ml grapes seed oil/head/day. The ration wasa dry complete feed in the form of pellet. The feed ingredients used were rice bran, cassava, wheat pollard, soybean meal, coconut meal, molasses, coffee husk and corn straw.Experiment was conducted over 30 days. Results revealed that the goat fed the ration supplemented with Zn and Cr(T2) producehigher milk yield (1012.29 g/day) and better in milk fat production ($P<0.05$)compared to those fedT1 and T3. Feed intake was decreased in treatment supplemented with grapes seed oil in the T3 ($P<0.05$), but no significant in milk fat production compared with T1. In conclusion,drycomplete feed containingof CP 14%, TDN 63% supplemented with Zn 20 ppm + Cr 2ppm is recommended for lactating dairy goat.

1. Introduction

Goats are foliage eaters (browsers), but in the lowlands especially in the dry season, this feedstuff is very limited. Therefore, alternative feeding is required, which is available throughout the year.Complete feed is one of the right solutions. Complete feed is a nutritionally adequate feed for animal by specific formula, to be fed as the sole ration and is capable of maintaining life and promoting production without any additional substance being consumed except water [1]. The complete feed is a blend of concentrate feed and fiber source [2].

In Indonesia the most common type of dairy goat is the EttawahCrossbreed goat, which is a descendant of Ettawah goat from India mated to local goat. The EttawahCrossbreedgoat has the ability to produce milk which is not different from the Ettawah goat, which is 1.5 - 3.7 kg / day [3] but in the lowlands the production of goat milk of PE is less than 1 kg / day, ie between 90-572 grams / day [4] or 440-590 ml / day [5]. This is due to the failure of goat maid adaptation to the environment, both to the environment temperature and feed according to the nutrient requirement. Complete nutrient feed composition for childcare, fattening and breeding purposes is different, especially in crude protein content (PK) and total digestible nutrients (TDN) [7].Therefore it is necessary to find level energy and protein content for lactating goats.

Zinc supplementation (Zn) needs to be done, because the mineral content of Zinc in feed ingredients in Indonesia is generally less, while the requirement of goats is 10-50 ppm depending on the physiological conditions [7]. Chromium (Cr) minerals in physiological form as glucose tolerance factor (GTF) play a role in increasing insulin activity and insulin like growth factor-I (IGF-I), it cause uptake of amino acid and glucose into cells mammary gland increased [8].Glucose is the main precursor of milk lactose synthesis,therefore the supplementation of Cr will increase the synthesis of milk lactose and milk production.The addition of oil to lactating dairy goat ration was one method to increase energy. Grapes seed oil containinghigherpolyunsaturated fatty acids compared with sunflower seed oil. Polyunsaturated fatty acidhave no negative effect to rumen fermentation.Morsy et al.[9] obtained that giving sunflower seed oil to goats as much as 20 ml per head per day could increase the production of acetate and NH_3 rumen. The purpose of this study was to find the proper energy and protein content and to examine the effect supplementation of mineral organic and grapes seed oil on dairy goat to support milk production.

2. Materials and Methods

The research was conducted in experimental enclosure of Faculty of Animal Husbandry and Agriculture Diponegoro University of Semarang. Twelve lactating Ettawah crossbreed goats with an average live body weight 42.14$\pm$3.25 kg and in first lactating seasons were used in this experiment. Goats were divided into three groups based on milk prodoctionto be fed treatment ration.

2.1.Experimental design.

The study was prepared based on randomized block design with 3 treatment and 4 replications. The experimental dietswere :T1 = ration containing 16% CP and 66% TDN ; T2 = ration containing 14% CP and 63% TDN supplemented with Zn 20 ppm + Cr 2 ppm; and T3 = T2 + 22 ml grapes seed oil/head/day.

2.2.*Ration*

The shape of ration is dry complete feed in the form of pellet. The feed ingredients used were rice bran, cassava, wheat pollard, soybean meal, coconut meal, molasses, coffee husk and corn straw. Samples of treatment ration were analyzed with proximate analysis method [10].The nutrient composition of the rations as listed in Table 1.

Zn and Cr minerals used were Zn-proteinate and Cr-proteinate produced by Laboratory of Nutrition and Feed Science , Faculty of Animal Science and Agriculture, Diponegoro University. The Znproteinate containing Zn at 1501.36 ppm, while Crproteinatecontaining at 166,67 ppm.The feedingexperiment was conducted for 7 days of preliminary and 30daysof treatment period.

2.3. Parameter and data analysis

The parameters observed were daily dry matter intake and daily milk yield. Fat,protein and lactosa contentsof milk composition were analyzed with Lactoscan. All the data obtained were statistically analyzedusing a variance analysis and if there were differences between treatments followed by Duncan's Multiple Range Test [11].

Table 1. Ration nutrient compositions with different protein and energy level.

Nutrient	Treatments	
	T1	T2
Crude protein (%)	16.67	13.92
Total Digestible Nutriens ,TDN (%)	66.58	63.22
Crude fiber (%)	22.06	24.75
Ether extract (%)	3.5	2.87
Ash (%)	8.39	8.76
Nitrogen free extract (%)	49.38	49.7
Calcium (%)	0.54	0.61
Phosphor (%)	0.47	0.52
Zinc (ppm)	22.57	46.11
Chromium (ppm)	1.87	3.29
Neutral Detergent Fiber (NDF)	42.34	40.11
Acid DetergentFiber (ADF)	26.02	23.96

3. Results and Discussion

Feed intake, milk yield and feedconversion are listed in Table 2. It can be seen in the table that ration treatment has a significant effect on dry matter intake, milk production and feed conversion. The dry matter intake was decreased in the addition of grapes seed oil (T3). The decrease of dry matter intake was caused by the increase of the ration energy, due to the addition of grapes seed oil. Addition of 2% oil can cause TDN increase between 4-4,5%. The fact occurred was agree to that described by [12], that the higher level of energy in the diet gives the lower feed intake. On the other hand dry matter consumption between T1 and T2 are not significantly different. This means that Zn and Cr supplementation in low quality rations (CP 14%, TDN 63%) can increase consumption equivalent to high quality rations (CP 16%, TDN 66%). Dry matter intake in this study was lower than [9] and [4], which found that goats weighing 44-51 kg could consume dry matter above 1000 g/day. The low intake in this study was possibly due to the form of ration, which is dry complete feed without fresh forage.

Milk yield was increased with supplementation of Zn and Cr also with grapes seed oil ($P<0.05$). The increased of milk yield obtained with Zn and Cr may be due to the increased digestibility of feed. [13] explains that Zn minerals are trace minerals that are the compilers of 300 kinds of digestive and metabolismenzymes. Furthermore,the addition of Cr will increase the availability of glucose for lactose synthesis. This is in agreement with the statement of [8] that Cr in the form of glucose tolerance factor can increase the potential of insulin activity that causes uptake of glucose and amino acids into udder gland cells increases.Supplementation of grapes seed oil increased the energy but decreased dry matter intake, this is due to increase milk yield higher than T1 but lower than T2. In relation to both parameters, T2 and T3 treatments resulted in better feed convention than T1 treatment. Although the T2 and T3 feed conversions were not significantly different, T2 treatment was recommended for lactating goats because they could enhance milk production.

Table 2. Dry matter intake and milk yield due to treatment effect.

Parameter	Ration treatment		
	T1	T2	T3
Dry matter intake (g/d)	939.92^{a}	963.40^{a}	859.62^{b}
Milk yield (ml/d)	801.96^{c}	984.17^{a}	891.85^{b}
Milk yield (g/d)	826.74^{c}	1013.40^{a}	919.51^{b}
Feed conversion	1.14^{a}	0.95^{b}	0.93^{b}

Mean with different superscript indicates significantly different (p<0,05).T1 = ration containing 16% CP and 66% TDN; T2 = ration containing 14% CP and 63% TDN supplemented with Zn 20 ppm + Cr 2 ppm; and T3 = T2 + 22 ml grapes seed oil/head/day

Based on the contents of milk fat, protein and lactose, Table 3 listed production of milk fat, milk protein and milk lactose. Milk fat production was higher in ration supplemented with Zn and Cr (T2) compared withthose inT1 and T3 (p<0.05).

Table 3. Milk fat, milk protein and milk lactose yield due to treatment effect

Parameter	Treatment		
	T1	T2	T3
Milk fat yield (g/day)	44.00^{b}	59.08^{a}	49.22^{b}
Milk protein yield (g/day)	26.47	31.03	28.64
Milk lactosa yield (g/day)	48.56	52.07	47.80

Mean with different superscript indicates significantly different (p<0,05).T1 = ration containing 16% CP and 66% TDN ; T2 = ration containing 14% CP and 63% TDN supplemented with Zn 20 ppm + Cr 2 ppm; and T3 = T2 + 22 ml grapes seed oil/head/day

Milk fat contains long-chain fatty acids and short chains. Short chain fatty acids are synthesized from acetate and beta hydroxybutyrate. Both fatty acids are the result of digestion of fiber feed in the rumen. [14] found that Zn supplementation of 20 mg per kg increased the digestibility of acid detergent fiber (ADF). While Cr supplementation in ration can contribute uptake of glucose, which is the raw material of alpha glycerol phosphate in milk fat synthesis. The addition of grapes seed oil can produced milk fat equivalent to T1 treatment, this is beneficially for farmers because the price of ration is cheaper.

4. Conclusion

It could be concluded that feeding dry complete feed containing CP 14% and TDN 63% suplemented with Zn 20 ppm and Cr2 ppm recommended for lactating dairy goat.

Acknowledgments

Author would like to thank to Diponegoro University for funding this project through professorship research program.

References

[1] AAFCO (Association of American Feed Control Officials)2000 Official Publication, Association of American Feed Control Officials Inc. West Lafayette. http://www.aafco.org

[2] Riwantoro. 2007 Technology Complete Feed Solution for Feeding Problems Sheep and Goat Livestock. http://www.disnak.jabar.go.id.

[3] Budiarsana I G M and I K Sutama2001Efficiency of Goat Milk Production Ettawah*Proc.* National Seminar on Animal Husbandry and Veterinary. Bogor September 17-18: 417.

[4] Sulistyowati E,A.Sudarman, K G Wiryawan and T Toharmad2013 Quality of milk fatty acid during late lactation in dairy goat fed on pufa-diet supplemented with yeast and *Curcuma xanthorrhiza*roxb, J.Indonesian Trop. Anim. Agric. **38**(4): 247.

[5] Yuniarti, E; D. Evvyernie, and D.A. Astuti 2016 Production and Energy Partition of Lactating Dairy Goats Fed Rations Containing Date Fruit Waste. Media Peternakan,**39**(1):27

[6] Adriani, A. Sudono, T. Sutardi, IK. Sutama, and W. Manalu2004 Influence Superovulation and Zinc Mineral Supplementation In Rations In Parent Goat Against Growth Of His Child. J. Pengembangan Peternakan Tropis, **29**: 177.

[7] M I Lengarite, P N Mbugua, C K Gachuiri and L W Kabuage 2013 Mineral intake of sheep and goats grazing in the arid rangelands of northern Kenya. Livestock Res. Rural Dev. **25** (10).

[8] Mordenti A, A Piva and G Piva1997 The European perspective on organic chromium in animal nutrition. Proc. Alltech's 13th annual Symp.page : 227.

[9] TA Morsy, S M Kholif, A E Kholif, O H Matloup, A Z M Saleem and Abu Elella2015 Influence of grapes whole seeds or oil on ruminal fermentation, milk production, composition, and fatty acis profile in lactating goats. Asian Australas. J. Anim. Sci. **28**:1116

[10] AOAC 1990 Official Methods of Analysis. Association of Official Analytical Chemist Inc, Virginia

[11] Steel, R.G.D. and J.H. Torrie. 1981 Principles and Procedures of Statistic. Mc Grow Hill Book Co. Inc. N.Y.

[12] McDonald P, R A Edward and J F D Greenhalgh 1996 Animal Nutrition 5th ed Longman Singapore LTD, Singapore.

[13] Fu-yu Xin, H Ming-hai, L Wen-li, L Yan-qin, W Ling-ling, S Jie and Z Ji-feng 2007 Effect of different levels of zinc on blood physiological and biochemical parameters in stud holstein bulls Chinese J. Anim. Nutr., **5**:19.

[14] Jia, Xiaoping Zhu, Wei Zhang, Jianbo Cheng1, CuihuaGuo and ZhihaiJia 2009 Effects of Source of Supplemental Zinc on Performance, Nutrient Digestibility and Plasma Mineral Profile in Cashmere Goats Asian-Aust. J. Anim. Sci.Vol. **22**, No. 12 : 164.

IOP Conf. Series: Earth and Environmental Science **119** (2018) 012052 doi:10.1088/1755-1315/119/1/012052

The Effect of Concentrate Supplementation on Creatinine Excretion in Thin-Tailed Lambs and Sheep

N. A. Purnami[1*] and A. Prima[1]

[1]Faculty of Animal and Agricultural Sciences, Diponegoro University, Indonesia.

Coresponding E-mail: nuningarii@gmail.com

Abstract

An experimental study was carried out to examine the effect of concentrates supplementation on creatinine excretion in lambs and sheep. The study used 8 male thin tailed lambs (aged ±3-4 months, weighed ±15.20 kg) and 8 sheep (aged ±1 year and weighed ±22.71kg). The animals were fed the diet contained 100% napier grass (100G) and 50% napier grass 50% concentrate as much as 3.5% of body weight (50G50C). This study used a complete randomized nested design. The parameters observed were dry matter intake (DMI) and creatinine excretion. The results showed that the diet significantly affect (P<0.05) DMI. The consumption of 100G was lower than that of 50G50C both lambs (0.32 vs 0.62 kg/d) and sheep (0.47 vs 0.88 kg/d). On the other hand, the diet did not affect the creatinine excretion (P 0.05) either G100 or G50C50 in lambs or sheep. However, the creatinine excretion in sheep (185.66 mg/d) was higher than that of lambs (299.1 md/ d) (P<0.05). It can be concluded that concentrate supplementation of grass diet increased DMI but did not affect creatinine excretion. The creatinine excretion of sheep was higher than that of lambs .

1. Introduction

The productivity of animals is affected by the age and quality of feed. Nutrients consumed by an animal are used for metabolism in the muscle tissue. Muscular metabolism requires energy fulfilled by ATP, which formed from reaction of ADP and creatine phosphate. The side product of those reactions is creatinine, that will be excreted through the urine [1]. According to [2], creatinine excretion has positive and high correlation with body weight of sheep. [3] [4] and [5] found that the high correlation value is influenced by the proportion of muscle tissue in animals that are accelerating growth. The high body weight will be followed by the amount of protein in the body. At the time of acceleration of growth, the older animals the more body weight but that facts occur until they have reached the peak of growth, so the increase in body weight tends to be stable. Therefore, it is necessary to examine the effect of feed and age on creatinine excretion of lambs and mature sheep.

2. Materials and Methods

This study used 8 male thin tailed lambs (aged ±3-5 months, weighed 15.20±1.59 kg), and 8 mature sheep (aged ±1 year, weighed 22.71±0.77 kg). The animals were raised in individual cages equipped with feeding and drinking equipment. The experimental design used in this study was complete randomized nested design. The treatments applied were 2 types of diets: 100% napier grass (100G) and 50% napier grass 50% concentrate as much as 3.5% of body weight (50G50C). The nutrient contents of the diets are shown in Tables 1. The diets were offered to the animals at amount of 3.5%

body weight at dry matter basis. In week 0 and 2 the urine excreted by the animals was collected for creatinine concentration analysis.

The parameters measured in this study were dry matter intake (DMI) and creatinine excretion by the urine. The DMI was measured by the feed intake times DM content. The creatinine concentration in the urine was analyzed using Juffe method.

Table 1. Nutritional Content of Feedstuffs and Diet (100% Dry Matter Basis)

	Nutritional content					
	Protein	Fat	CF	Ash	NFE	TDN
Feedstuff						
Napier grass	8.77	1.18	34.43	24.01	31.601	42.36
Concentrate	16.50	3.46	8.57	7.67	59.80	69.31
Diet						
100G	8.77	1.18	34.43	24.01	31.601	42.36
50G50C	12.64	2.32	21.50	15.84	45.70	55.84

Legends: CF: Crude Fibre, NFE: Nitrogen Free Extract, TDN: Total Digestible Nutrients, 100G: 100% grass, and 50G50C: 50% grass and 50% concentrate

3. Results and Discussion

There was a significant difference ($P<0.05$) in DMI between 100G and 50G50C both lambs and sheep. It is due to 50G50C contained concentrate. Therefore, the digestibility was higher than that of 100G. Dry matter intake is influenced by feed digestibility. The higher digestibility, the higher DMI. Feed of higher digestibility is digested and absorbed by the digestive tracts more rapidly than that of lower digestibility, causing the digestive tract to be empty more quickly and stimulate animal to be hungry [6]. The DMI of lambs was lower ($P<0.05$) than that of sheep, because the lambs had smaller body size and lower body weight that of sheep. However, in term of percentage of body weight, the lambs and sheep had no significant different DMI ($P>0.05$). It indicates that the condition of digestive tracts of the lambs had grown to be similar with sheep [6] [7].

The results showed that there was no significant effect of diet treatment (P 0.05) in creatinine excretion between diets, both in lambs and sheep. This indicates that the animals fed 100G and 50G50C had similar physiological condition. The physiological condition of the lambs and sheep could also being similar. One of factors affecting creatinine excretion is the physiological condition of the animal.

Nutrients of the feed consumed by the animal are metabolized in the muscle tissue. The energy source used for such metabolism process in the muscle was ATP, which is formed from the reaction of ADP and creatine phosphate and releases creatinine as the waste product that was excreted through the urine. The correlation of DMI and creatinine levels as found by [8] can be seen in Figure 1.

Physiological responses in lambs and adult sheep were not significantly different (P 0.05). It indicates that lambs and sheep in this study have same activity and ATP needs. The response of physiology of animals can be influenced by several factors including factors of the body, growth and development, activity, and environmental factors. The lambs and sheep were in the same environment, so the frequency of breath, pulse rate and rectal the sheep was not different. So it can be said the lambs and sheep adapted well to the environment. Animals show a physiological response when they receive stimulation from outside or when there is environmental alteration [9]

Table 2. Dry Matter Intake and Creatinine Excretion.

Parameters	Lambs		Sheep		Effect of Diet	Effect of Age
	100G	50G50C	100G	50G50C		
DMI (kg)	0.32	0.62	0.47	0.88	s	s
Percent of DMI to BW (%)	1.87	3.41	2.06	3.38	s	ns
Creatinine Excretion (mg/d)	181.90	269.28	189.42	328,92	ns	s

Legends:
s : significant
ns : not significant
DMI : dry matter intake
BW : body weight

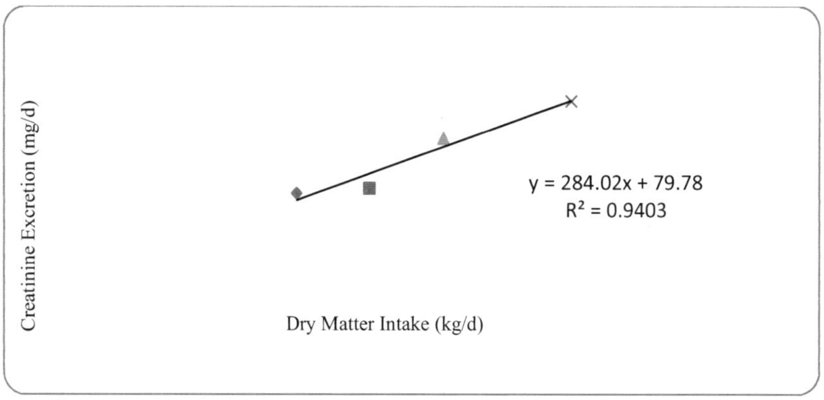

Figure 1. Correlation between Dry Matter Intake and Creatinine Excretion

Table 3. Phyisiological Response upon Diet in Lambs and Adult Sheep

Parameters	Lambs		Adult Sheep		Effect of Diet	Effect of Age
	T0	T1	T0	T1		
Respiration rate (times/min.)	22.35	36.13	29.4	35.15	ns	ns
Pulse rate (times/min.)	55.27	83.28	58.55	69.3	ns	ns
Rectal temperature (°C)	37.31	38.62	37.66	38.47	ns	ns

4. Conclusion

Based on the results, it can be concluded that concentrate supplementation of grass diet increased DMI but did not affect creatinine excretion. However, the creatinine excretion in the sheep was higher than that of lambs.

References

[1] Cetinkaya N, S Yaman and N H O Baber 2006 The use of purine derivatives/creatinine ratio in spot urine samples as an index of microbial protein supply in Yerli Kara crossbred cattle. Livest. Sci. **100** 91

[2] Luthfi N, W L Salido, A Prima, C M S Lestari, E Purbowati, E Rianto, M Arifin, and A Purnomoadi 2016 Correlation of creatinine excretion and metabolic body weight and body protein in goats and sheep (in the 9th Vietnamese – Hungarian International conference : Research for Developing Sustainable Agriculture Tra Vinh Viet Nam September 22nd 2016

[3] Chizzotti, M L, S C V Filho, R F D Valadares, F H M Chizzotti, and L O Tedeschi 2008 Determination of creatinine excretion and evaluation of spot urine sampling in Holstein cattle. Livest. Sci. **113** 218

[4] Ma T, Kai-dong Deng,1 Yan Tu, Cheng-gang Jiang, Nai-feng Zhang, Yan-ling Li, Bing-wen Si, Can Lou, and Qi-yu Diao 2014 Effect of Dietary Concentrate:forage Ratios and Undegraded Dietary Protein on Nitrogen Balance and Urinary Excretion of Purine Derivatives in Dorper × thin-tailed Han Crossbred Lambs. Asian-Australas. J. Anim. Sci. **27** (2) 161

[5] Luthfi N, N Mukminah, E Rianto, E Purbowati, C M S Lestari and A Purnomoadi 2015. Hubungan bobot badan dengan kreatinin kambing Kacang muda dan dewasa dengan level pemberian pakan berbeda. Seminar Nasional Universitas Padjajaran Bandung 10-11 November 2015

[6] McDonald P, R A Edwards, J F D Greenhalgh, C A Morgan, L A Sinclair and R G Wilkinson 2012 Animal Nutrition Seventh Edition. Prentice Hall Harlow England

[7] Church D C 1988 The ruminant animal: digestive physiology and nutrition. No. SF 768.2. R8. R85 1988

[8] Rahmawati K S, E Rianto, S Mawati dan A Purnomoadi 2009 Keluaran kreatinin lewat urin dan hubungannya dengan protein tubuh pada domba pada berbagai imbangan protein-energi. In Prosiding Seminar Nasional Teknologi Peternakan dan Veteriner. Pusat Penelitian Peternakan Balai Penelitian dan Pengembangan Pertanian Bogor

[9] Nurmi A 2016 Respon fisiologi domba lokal dengan perbedaan waktu pemberian pakan dan panjang pemotongan bulu. J. Eksakta. **1** (1) 58

Fermentation Kinetic of Maize Straw-Gliricidia Feed Mixture Supplemented by Fermentable Carbohydrate Measured by In Vitro Gas Production

D. Yulistiani[1*] and Nurhayati[1]

[1]Balai Penelitian Ternak, Jl Veteran III.Ciawi-Bogor 16720. P.O. Box 221.

*Corresponding E-mail: dwiyulistiani@yahoo.com

Abstract

Utilization of crop by-products such as maize straw mixed with legume is expected to be able to overcome the limitation of forage availability during dry season and have similar nutritional value with grass. Addition of fermentable carbohydrate in this diet can be improved fermentability and reduced methane production. The objective of this study was to evaluate supplementation of ground corn grain or rice bran as fermentable carbohydrate in maize straw-gliricidiamixture. Treatment diets evaluated were: Maize straw + gliricidialeaf meal (Control/RO); Control + 10% ground maize grain (ROC); Control + 10% rice bran (RORB). Maize straw was chopped and ground then mixed with gliricidia leaf meal at ratio 60:40% DM. Maize straw-gliricidia mixture then supplemented either with ground corn grain or rice bran at 10% of DM basal diet (control). Sample was incubated for 48 hours, gas production was recorded at 4, 8,12, 16, 24, 36 and 48 hours. Study was conducted in randomized complete design. Results of the study showed that supplementation of fermentable carbohydrate from corn grain or rice bran was able to increased ($P<0.05$) rate of gas production by 24 and 18% respectively. However only in ROC potential gas production was increased ($P<0.05$) by 32% and percentage of methane production was decreased. From this study it can be concluded that supplementation of ground corn grain at 10% in maize straw-gliricidia mixture was able to improve diet fermentation and reduced methane production.

1. Introduction

Maize straw is a second widest crop by products used as ruminants feed in Indonesia. However the utilization of straw as animal feed is limited due to low nutritional quality. As reported by [1] maize straw had high fibercontent (74-86%), low protein content (2.76%) consequently low digestibility (40-53%).This low protein content will caused low feed intake and utilization.Feeding ruminants with low quality diet contained high fiber caused higher production of methane which is one of greenhouse gasses. Many methods to reduce methane production by supplementation of legume forage containing secondary compounds[2]. Beside that supplementing legumes forage to straw diet provide protein source thereby increased diet digestibility.Gliricidia foliage is suitable to be used as supplement in low quality feed such as corn stover. Gliricidia has high protein content (24.59%), but it had high protein degradability in the rumen (76,55%) [3] the rapid fermentation of N caused rapid production of ammonia. The high rate of ammonia production requires a matching of rumen fermentable energy supply to improve microbial protein production [4]. Nitrogen and energy should be simultaneously available for utilization to enhance microbial growth rate and nutrient utilization efficiency [5]. Beside that supplementation energy in the form of starch also could reduce methane production [6]. Rice bran is abundantly available as by-products from rice production and widely used as source of carbohydrate for livestock feed in Indonesia. Rice bran contained high DM water soluble in the rumen (60%) therefore can be used as fermentable energy source [7]. Beside rice bran, corn is a second grain produced in Indonesia for food and animal feed. Corn grain is most common used in concentrate for non ruminants and ruminants feed because of its high energy and starch content [8]. The objective of this study was to evaluate the effect of supplementation of rice bran or corn supplementation to maize straw-gliricidia mixed feed on in vitro gas production kinetics and methane production.

2. Materials and Methods

2.1. Diet treatments and experimental design

A control diet consisted of maize straw mixed with gliricidia foliage at ratio 60:40% DM (RO). Treatment diets were formulated by supplementation of control diet with ground corn grain (ROC) or rice bran (RORB) at 10% to control diet. The study was conducted in randomized complete design.

2.2. In vitro Study

Sample of 500 mg was incubated for 48 hours, gas production was recorded at 4, 8,12, 16, 24, 36 and 48 hours. The fermentation was conducted for 48 hours under anaerobic condition at 39°C according to procedure ofTheodoru and Brooks [9], medium for incubation was prepared according Menke and Steingass[10]. Fermentation kinetics for gas production was analyzed using the non-linear model [11]. The equation was $P = A + B\ (1 - e^{ct})$ where p is the amount of gas produced (ml) at time t, *A*is the intercept of the gas production curve at time zero and represent as gas produced from the fermentation of soluble fraction (%), B is the gas produced from fermentation of rumen-insoluble, but slowly degradable fraction (%), *c* is the rate constant gas production of the *b* fraction (%/h) and t is the incubation time (h). Data collected were analyzed using general linier model procedure in SAS package version 9.1. Means were compared using Duncan's Multiple Range Test

3. Results and Discussion

Fermentation kinetics of the diet is shown in Table 1. It shows that supplementation of corn or rice bran reduces gas production from soluble fraction (A). However in B values (gas produced from the fermentation of insoluble fraction) there was significantly incrased by 31% in corn supplementation. On the other hand, rice bran supplementation did not significantly increased B values. The increased of gas production from insoluble fraction in corn supplementation might due to synchronization of the availability of fermentable energy in corn and rapidly fermentable nitrogen from gliricidia[12] this stimulate the fermentation of the diet. This resulted in the increased of gas production from insoluble fraction (B fraction) in corn supplemented diet.

Table 1. In vitro Fermentation kinetics of gas production of diet treatments incubated for 48 hours

Diets	A	B	c	A+B
RO	-13.3a	106.9b	0.054b	93.61b
ROC	-17.4b	140.5a	0.067a	123.5a
RORB	- 15.4ab	109.9b	0.064a	94.5b

Different letter in the same column indicates significantly different (P<0.05); RO=maize stover + gliricidia; ROC = RO + 10% ground corn grain; ROBC = RO + 10% rice bran; A = considered as gas produced from the fermentation of immediately soluble materials; B = gas produced from the fermentation insoluble fraction, but fermentable; C = the rate of gas production of B fraction; A+B = extent gas production

Rate of gas production (c) was increased either by corn (ROC) or rice bran (RORB) supplementation, this indicates that carbohydrate supplementation in corn or rice bran enhanced cellulolytic activity of microbes [13]. However, rice bran was only able to stimulate fermentation at the early fermentation as shown in figure 1 the increase of gas production in RORB from 4 hours to 8 hours incubation was higher than RO (12.16 vs 16.86 ml), while ROC was consistently higher than RO and ROC. Though rate of gas production was increased by corn or rice bran supplementation, however the extent of gas production was only significantly higher in corn supplementation (123.5, 94.5 and 93.6 ml; respectively for ROC, RORB and RO).

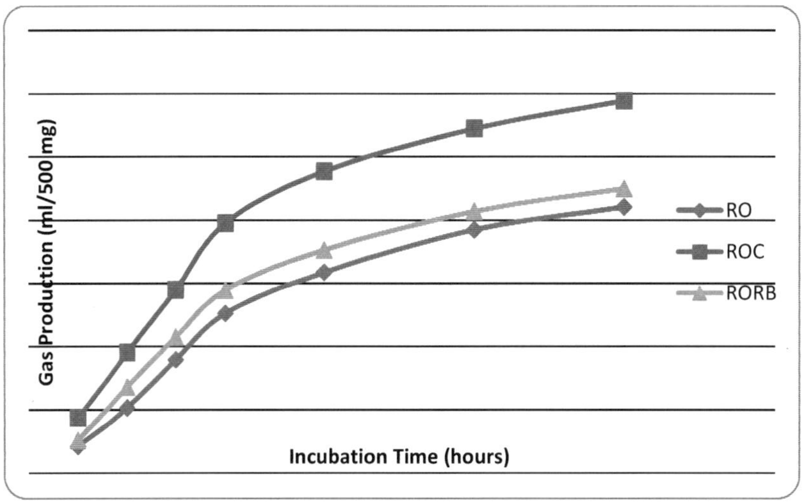

Figure 1. Cumulative gas production of maize straw+gliricidia mixed feed (RO), R0+corn grain (ROC)and R0+ rice bran (R0RB) incubated for 48 hours.change of interleukin in animals.

Methane production was significantly highest in corn supplementation and the lowest was in control feed (Table 2), however when it was calculated into percentage of total gas production corn supplementation resulted in the lowest proportion of methane compare to control (RO) and rice bran supplementation (RORB). Methane production can be reduced indirectly by defaunation. This was due to association between protozoa and methanogens. Hydrogen produced by protozoa during carbohydrate fermentation, was used was used by methanogens to produce methane. Therefore, when protozoa population can be reduced it is expected methane production can be decreased consequently [14]. Supplementation of gliricidiasepium did not affect protozoa diversity and dencity in rumen of sheep fed on corn stover silage and concentrate diet [15]. Therefore the lower proportion of methane in corn supplementation could be due to the higher carbohydrate content of corn than rice bran [8]. The higher content of carbohydrate in corn might be resulted in high production of propionic acid, propionic production is considered as competitive pathways of hydrogen utilization by methanogens in the rumen, propionic production involved hydrogen utilization [6], therefore proportion of methane production in ROC diet was the lowest. Rice bran on the other hand contain higher NDF (21.5 %; 16) compare to corn (11.77-12.7;17) cell wall carbohydrate (NDF) degradation produced hydrogen which used by methanogen to produce methane [6].

Table 2. Total gas and methane production of treatment diets incubated for 48 hours

Diets	Total gas (ml/500 mg)	Total CH_4 (ml/500 mg)	% CH_4/total gas
RO	84.2^c	33.7^c	40.0^b
ROC	117.8^a	44.9^a	38.2^c
RORB	90.0^b	37.0^b	41.7^a

^{abc}Different letter in the same column indicates significantly different (P<0.05); RO=maize stover + gliricidia; ROC = RO + 10% ground corn grain; ROBC = RO + 10% rice bran

4. Conclusion

Supplementation of corn grain at 10% to corn straw gliricidia basal diet was able to increase diet fermentation as indicated by the increase of rate of gas production and potential gas production. This supplementation was also able to decrease proportion of methane production.

References

[1] GeletiD, ATolera, A Mengistu andMHailemariam 2011 Effect of variety of maize on yield of grain, residue fractions and the nutritive value of the whole stover. Ethiop J. Appl. Sci. Technol. **2**(2) 91

[2] Goel G,andMakkar HPS 2012 Methane mitigation from ruminants using tannins and saponins. Trop.Anim. Health Prod. **44** 729

[3] KhamsekieewB, JB Liang, CC Wong, and ZAJelan 2001Ruminal and intestinal digestibility of some tropical legume forages. Asian-Austr J. Anim. Sci. **14** 321

[4] Corbett JL 1987 Energy and protein utilization by grazing animalsIn: Temperates Pastures: Their Production, Use and Management Eds Wheeler JL, Pearson CJ, and Robarts GE Aust. Wool Corporation/CSIRO Australia. 341 DELjungdahl, LG Wilson, JR Harris, PJ 325 New York Elsevier Science Publishing Co, Inc

[5] Sniffen CJ and PH Robinson 1987 Symposium: Protein and fiber digestion, passage, and utilization in lactating cows. J. Dairy Sci. **70** 425

[6] Moss A R, JJPierre, JNewbold 2000 Methane production by ruminants: Its contribution to global warming. Ann. Zootech. **49** 231-253

[7] Yulistiani D,ZAJelan, and JB Liang 2008 Degradability of Mulberry (Morus alba) and Rice Bran in the Rumen of Sheep Fed Different Diets. JITV **13** (4) 264

[8] Arcari MA, CMMR Martins, T Tomazi, JLGoncalves, MV Santo 2016 Effect of substituting dry corn with rehydrated ensiled corn on dairy cow milk yield and nutrient digestibility. Anim. Feed.Scie. Technol. **221** 167

[9] Theodoru, MK and AE Brooks 1990 Evaluation of a new procedure for estimating the fermentation kinetics of tropical feeds. The Natural Resources Institute Chatham UK

[10] Menke KH and H Steingass 1988 Estimation of the energetic feed value obtained from chemical analysis and in vitro gas production using rumen fluid. Anim. Res. Dev. **28** 7

[11] Orskov ER, and I MC Donald 1979 The estimation of protein degradability in the rumen from incubation measurements weighted according rate of passage. J. Agric. Sci. **92** 499

[12] Jonathan N A, Nieto, J L VCerdán,F CPineda,SACampos,and E VBPérez 2013 Digestibility of Buffel grass (Cenchrusciliaris)-based diets supplemented with four levels of Gliricidiasepium hay in hair sheep lambs. Trop. Anim. Health Prod. **45** 1357

[13] Getachew G, EJ Depeters, PH Robinson and JG Fadel 2005 Use of an in vitro rumen gas production technique to evaluate microbial fermentation of ruminant feeds and its impact on fermentation products. Anim. Feed Sci. Technol. **123** 547

[14] PatraAKandSaxena J 2009 Effect and mode of action of saponins on the microbial populations and fermentation in the rumen and ruminant production. Nut. Res. Rev. **22** 204

[15] Martinele Isabel, Lucas Feitosa Silva, Marta D'Agosto, EvandroNeves Muniz, José Luiz de Sá, Gladston Rafael de Arruda Santos. 2014. Abundance and diversity of rumen protozoa in lambs fed *Gliricidiasepium* Silage. R. Braz. Zootec **43**(8) 436

[16] Warren BE and Farrel DJ 1990The nutritive value of full fat and defatted Australian rice bran: chemical composition. Anim. Feed Sci. Technol. **27** 219

Texture Profile and Color Determination on Local and Imported Meat Available in Semarang City, Indonesia

Mukh Arifin[1], Ahmad Ni'matullah Al-Baarri[2*], Bhakti Etza Setiani[2], Risa Fazriyati Siregar[3]

[1]Animal Sciences Department, Faculty of Animal and Agricultural Sciences, Diponegoro University, Indonesia
[2]Food Technology Department, Faculty of Animal and Agricultural Sciences, Diponegoro University, Indonesia
[3]Food Technology Laboratory, Integrated Laboratory at Diponegoro University, Indonesia

*Corresponding Email: albari@undip.ac.id

Abstract

This study was done for analysing the texture profile and colour performance in local and imported meat in Semarang, Indonesia. Two types of available meat were compared in the hardness, cohesiveness, springiness, adhesiveness and the colour L*a*b* performance. Five fresh beef cut of round meats from local and imported meat were used in this experiments. Data were analysed statistically using T-test. The results showed that local beef exhibit higher in the springiness than imported beef resulting in the remarkable differences. The colour analysis showed that imported beef provided remarkable higher in L* value than local beef. Resulting significant differences among two types of beef. As conclusion, these value might provide the notable of differences among local and imported meat and may give preferences status to the user for further application in meat processing.

1. Introduction

Tenderness is important factor as well as colour of meat to determine visual and sensory acceptability resulting the consideration for meat industry to use in their commercial application [1, 2]. The tenderness is influenced by genetics, meat handling, and the process of slaughtering [3]. The tenderization phase is known to be occurred during post-mortem storage [2], therefore the method for storage should be considered.

Meat colour has been widely known not only able to attract consumers preferences but also could be used for determining the changes in the meat composition [4]. Meat colour may be affected by pH and the type of fibre, and also chilling method that may have been treated on the meat [4].

Indonesia has imported beef dominantly from Australia [5] therefore imported beef meat could be easily founded in local market and the price was almost equal to the local beef resulting the reasonable price for all consumers and widely available in market. Since this meat is easily obtained, small scale meat industry was also utilized this meat as sources. However the user of small scale industries pay its preference to use local beef since the differences in appearance was easily recognized.

The objectives of this study were to evaluate the texture and colour on imported and local meat available in local market at Semarang City, Indonesia. This evaluation may provide the beneficial information to the further meat application.

2. Materials and methods

Local beef from Ongole was obtained from slaughter house in Semarang and imported beef from Australia was obtained from local modern market. Both of places were located about 10 km from laboratory. Obtained meat was then prepared for analysis without any preservations. For imported meat, the chilled meat with less than one month of preservation was used.

2.1. Determination of texture profile and colour parameter

Determination of texture was done using texture analyser Brookfield CT-3 to measure the hardness, cohesiveness, springiness, and adhesiveness. To measure the colour of L*a*b*, Digital Color Meter software by Macintosh was used. The luminous of lightness was set to 100 lumen. The method for determining texture profile and colour parameter was adopted from other researcher [6].

2.2. Meat sampling

The sampling of meat was done in four weeks and started from August to September 2017. Meat from slaughtering house was obtained from morning slaughtering process (at 03.00 AM) during five days of sampling. Imported meat was also obtained during five days of collection and was immediately thawed for 8 hour prior to the determination of texture and colour.

Data were obtained from three time replication using five time of meat sampling and were analysed using T-test. The calculation was done using open source of R software for Macintosh.

3. Result and Discussion

3.1. Texture profile in local and imported meat

Hardness, cohesiveness, springiness, and adhesiveness from local and imported meat were measured using texture analyser to obtain the data of texture profile. The value of hardness was 0.36 and 0.21 for local and imported meat resulting no remarkable differences (Figure 1). The less notable differences was also found in cohesiveness value; 0.73 and 0.67 for local and imported meat, respectively. It was found that the chilling provided less change in hardness resulting the less change in hardness [7]. This evidence may explain the less differences in both meat sources. The final cohesiveness of common meat product such as sausage was 0.41 [8], so the utilization of both meat are less worries to provide the cohesiveness differences.

The remarkable significant differences was found in springiness that was found a value of 1.97 and 1.08 for local and imported meat, respectively. It seems the storage was contributed to the remarkable value of this springiness however, it was found that the value of springiness close connection to differences in meat composition [7]. This is the evidence that user of small scale meat industry provided its preference to the local meat. The adhesiveness was also found to had remarkable differences among two meat source since the thawing process was conducted in imported meat leading to an elevation in the outflow of protein content and moisture in meat [9].

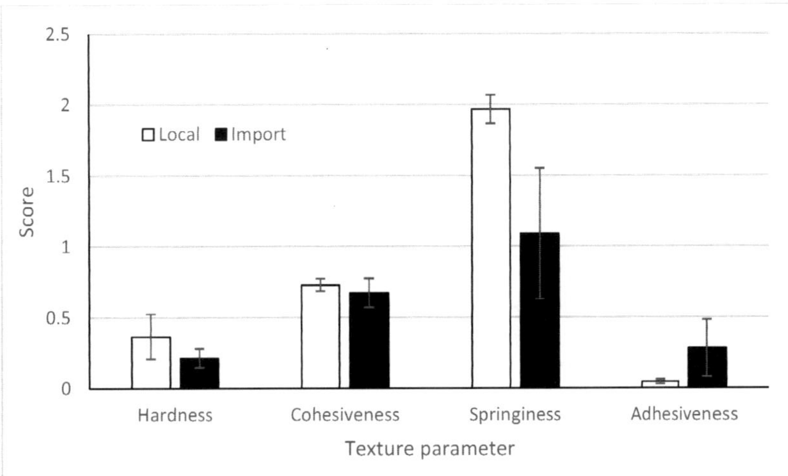

Figure 1. Hardness (N/cm^2), cohesiveness, springiness (cm), adhesiveness (Ns) of local and imported meat available in local market. Data were obtained from the average of 3 times replication and 5 times of sampling. Error bars represents the standard deviation.

3.2. Colour performance of local and imported meat

Colour measurement in sample was expressed as L*, a*, and b* that represented as lightness, redness/greenness, and yellowness/blueness, respectively. The lightness of local meat appeared to be remarkable lower that of imported meat with the value of 29.20 and 40.29 representing the simple to recognize the source of meat (Figure 2). As 0 value represents as black and 100 as white, the local meat was recognized as darker than imported meat. The lightness tended to low as handling and the manufactured beef might have a value of lightness about 23 [6].

The value of a* in local and imported meat was 2.08 and 2.17, respectively and the value of b* was 1.09 and 1.90, respectively. There was no remarkable differences in the value of a* and b* between local and imported meat. Since the value of a* and b* represents as redness and blueness, respectively, the local and imported meat could not be notified in its redness and blueness. The value of a* and b* was reasonable determined since other researcher reported its value as 4.4 and 2.9, respectively [6]

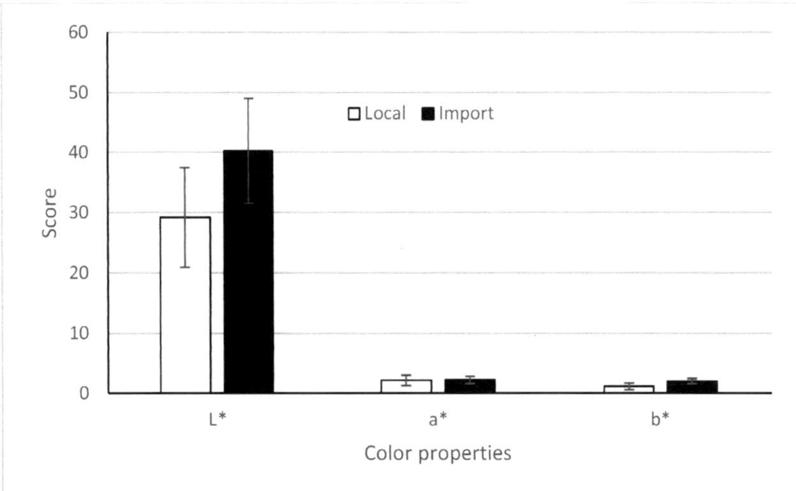

Figure 2. The colour properties of L*a*b* of local and imported meat available in local market. Data were obtained from the average of 3 times replication and 5 times of sampling. Error bars represents the standard deviation.

4. Conclusion

Based on the texture profile, the springiness of local meat showed remarkable higher than imported meat. The local meat showed remarkable darker than imported meat. These value might provide the notable of differences among local and imported meat and may give preferences status to the user for further application in meat processing.

Reference

[1] J.M. Hughes, S.K. Oiseth, P.P. Purslow, R.D. Warner, 2014 A structural approach to understanding the interactions between colour, water-holding capacity and tenderness, Meat Science **98** 520.

[2] E. Veiseth-Kent, M.E. Pedersen, S.B. Rønning, R. Rødbotten, 2018 Can postmortem proteolysis explain tenderness differences in various bovine muscles?, Meat Science **137** 114.

[3] R.D. Warner, P.L. Greenwood, D.W. Pethick, D.M. 2010 Ferguson, Genetic and environmental effects on meat quality, Meat Science **86**(1) 171.

[4] J.K. Apple, J.B. Machete, R.J. Stackhouse, T.M. Johnson, C.A. Keys, J.W.S. Yancey, 2014 Color stability and tenderness variations within the gluteus medius from beef top sirloin butts, Meat Science **96** 56.

[5] Data-Center-and-Information-in-Agriculture, 2016 Outlook beef meat as agricultural comodity in sub animal husbandry sector, Secretary General – Agriculture Ministry of Indonesia Republic (In Bahasa Indonesia), Jakarta,.

[6] K.S. Ojha, D. Granato, G. Rajuria, F.J. Barba, J.P. Kerry, B.K. Tiwari, 2018 Application of chemometrics to assess the influence of ultrasound frequency, Lactobacillus sakei culture and

drying on beef jerky manufacture: Impact on amino acid profile, organic acids, texture and colour, Food Chemistry **239** 544.

[7] J. Weiss, M. Gibis, V. Schuh, H. Salminen, 2010 Advances in ingredient and processing systems for meat and meat products, Meat Science **86** 196.

[8] M. Cierach, M. Modzelewska-Kapituła, K. Szaciło, 2009 The influence of carrageenan on the properties of low-fat frankfurters, Meat Science **82** 295.

[9] S. Savadkoohi, H. Hoogenkamp, K. Shamsi, A. Farahnaky, 2014 Color, sensory and textural attributes of beef frankfurter, beef ham and meat-free sausage containing tomato pomace, Meat Science **97** 410.

IOP Conf. Series: Earth and Environmental Science **119** (2018) 012055 doi:10.1088/1755-1315/119/1/012055

Methane Respiration Emission, Feces Production and Feces Quality of Sheep Fed with Different Level of Energy and Protein Content

S. Sutaryo[1*], A. Prima[1] and C.B. Listianto[1]

[1]Faculty of Animal and Agricultural Sciences, Diponegoro University, Semarang, Central Java, Indonesia.

*Corresponding E-mail: soeta@undip.ac.id

Abstract

Study investigating methane respiration emission, feces production and feces quality of sheep that fed with combination different level of energy and protein content was done. Combination of two level total digestible nutrient (TDN) (±60 and 70%) and three level protein (±14, 16 and 18%) were investigated. The result showed that feed with higher TDN content gave effect on methane production in term of L/h/d but no significant effect in term of L/dry matter intake. Feces nutrients from treatment with higher TDN content were higher than that from lower TDN content. In term of sustainable agriculture feed with lower TDN and protein content that investigated in this study is more recommended than that with higher TDN and protein content.

1. Introduction

Livestock production particularly ruminant animal has the highest contribution to the greenhouse gas (GHG) emission from agricultural sector and small ruminants share about 12.25% of the total GHG emission from livestock's enteric fermentation and manure CH_4 and manure N_2O [1].

A strategy to reduce GHG emission from animal is by increase animal productivity, therefore it can decrease GHG emission per kg animal product. In the case of sheep production, that part of small ruminant, this animal usually started to fattened in age of 6 – 8 months. Early sheep fattening (3 months) probably can be alternative to reduce GHG emission from this animal, since this strategy can produce earlier readily slaughtered sheep. In addition, by early sheep fattening probably can produce lamb with low fat content.

Methane, an important GHG, is produced during feed fermentation in animal digestive tract. Changes in the efficiency of feed energy utilization can influence CH_4 emissions of animals and the feed energy utilization depends on the type of animal, the type or quality and quantity of feed, environmental condition, etc. [2]. Therefore, the aim of this paper is to determine the effect of different level of energy and protein feed content on methane respiration emission, feces production and feces quality.

2. Materials and Methods

2.1. Animals and diets

Twenty four thin tailed bucks (14.19 ± 1.17 kg) were randomly divided into six combination treatment. Two level total digestible nutrient (A) and 3 level protein (B). Diet composition and the nutrient content for each treatment can be seen in Table 1. The sheep were individually kept in metabolic cages. The animals were fed with pellet ration two times a day at 8 a.m. and 4 p.m. Both feed and water were accessible at all times. Sheep were adapted for four weeks followed by 14 weeks of data collection. Feces and urine were collected in the week 9th for 7 days and feces were kept in the freezer until for its chemical composition analysis.

Table 1. The Ration Composition and Nutrient Content (Nutrient Composition are Expressed as Percentage of Dry Matter (DM).

Feedstuffs / Nutrients	Treatments					
(%)	A1B1	A1B2	A1B3	A2B1	A2B2	A2B3
Molasses	6.00	6.00	6.00	8.00	8.00	8.00
Cassava	11.50	9.50	7.00	38.50	36.40	34.30
Sugar cane leave	30.20	29.00	28.50	10.35	8.90	7.00
Rice bran	18.00	16.00	14.00	19.65	18.00	17.30
Cassava peel	15.00	15.00	15.00	3.00	3.00	3.00
Soybean meal	13.50	17.50	21.50	14.50	18.50	22.20
Fish meal	3.80	5.00	6.00	4.00	5.20	6.20
Mineral mix	2.00	2.00	2.00	2.00	2.00	2.00
Nutrients content in the rations:						
Crude protein	13.37	15.58	17.69	13.45	15.67	17.72
Total Digestible Nutrients (TDN)	60.39	61.15	61.67	68.93	69.63	70.17

2.2. Respiratory Methane Emission

Evaluation of respiratory methane emissions of sheep were evaluated using facemask method according to [3]. The methane concentration was analyzed using methane analyser (VIA-510, HoribLtd., Japan), while air volume was measured by air flow meter (STEC SEF-6470, Horiba Ltd., Japan). The data were recorded continuously using IBM PC/AT compatible computer by Test Point TM program (Test Point TM Technique & Reference, 1999). Data were collected for 10 minutes at 3-hour intervals over two days [4].

2.3. Analytical procedure

Crude protein, fiber, fat, total solid and ash content were analysed using proximate analysis. Data were analysed using ANOVA with 95% confidence level. Duncan multiple range tests (DMRT) were used in post ANOVA analysis when differences were found to be significant [5].

3. Results and Discussion

3.1. Methane respiration emission

There was no interaction effect of combination treatment on methane respiration emission ($p>0.05$). The significant effect ($p<0.05$) was only found in the treatment level of TDN on methane respiration emission in term of L/h/d (Table 2). Methane respiration emission in this study is in accordance with [6]Shibata et al. (1992) who reported that methane emission in sheep and goat is 34.3 and 25.2 L/d respectively while in term of L/kg DMI is 25.9 and 27.1 respectively.

The higher methane production in treatment A1 than that in treatment A2 in term L/h/d can be attributed by the fact that ration in treatment A2 has higher TDN content than TDN content in treatment A1. As reported by [7]Shibata and Terada (2010) that ruminant methane production is affected by feed quality, forage species, forage processing, the proportion of forage in the diet and the source of grain in the ration. Feed with high quality will promote the microbial growth and propionate production which lead to reduced CH_4 production.

Table 2. Respiration Methane Emissions

Respiration Methane Emissions		Treatments combination B1	B2	B3	Average
L/h/d	A1	45.08	38.62	34.07	39.26[a]
	A2	30.68	29.80	25.42	28.63[b]
	Average	37.88[a]	25.40[a]	29.75[a]	
L/DMI	A1	28.71	30.57	30.06	29.78[a]
	A2	35.39	38.21	26.67	33.42[a]
	Average	32.05[a]	34.39[a]	28.36[a]	

[a,b]: Values in each raw and column in the same parameter followed by the same letters are not significantly different (p>0.05)

3.2. Feces production and its quality

Feces production and its quality of sheep fed with different protein and energy content are presented in Table 3. There was no interaction of combination treatment between level TDN and protein content of sheep ration on feces production and its quality. There was no significant effect (p>0.05) of protein content on sheep feces production. However, there was significant effect (p<0.05) of level TDN on feces production. Feces production from treatment A1 was 613.36 g DM/h/d while it's from treatment A2 was 251.33 g DM/h/d. The higher feces production in treatment A1 than that from treatment A2 probably was attributed by higher feed consumption ($p<0.05$) (data not shown) in treatment A1 than feed consumption in treatment A2. In addition, higher TDN content in ration A2 probably can be limiting factor of feed consumption therefore feed consumption in this treatment was significantly lower ($p<0.05$) than that in treatment A1.

On average, feces nutrients of sheep fed with higher TDN content (treatment A2) were significantly higher ($p<0.05$) than that from treatment A1. Feed with higher TDN content means that this ration contain more part of nutrient that can be digest by animal, however since the absorption of nutrient capacity of sheep digestible tract is limited therefore part of this nutrient will undergo to the feces or urine. In this study there was significant effect of combination treatment on sheep productivity (data not shown) and since higher nutrient content in sheep feces treatment with higher TDN content probably can produce higher CH_4 and N_2O therefore in respect of agriculture sustainability a lower protein and TDN content in this study is more recommended than the higher one.

Table 3. Feces Production and Its Quality

Parameters (%)		Treatments combination			Average
		B1	B2	B3	
Crude protein	A1	8.77	9.12	9.22	9.04[a]
	A2	12.32	12.36	13.59	12.76[b]
	Average	10.55[a]	10.74[a]	11.41[a]	
Crude fat	A1	1.89	1.17	1.97	1.68[a]
	A2	2.54	2.92	3.25	2.90[b]
	Average	3.16[a]	2.05[a]	2.61[a]	
Fiber content	A1	26.44	26.86	25.82	26.37[a]
	A2	26.32	23.93	22.82	24.36[b]
	Average	26.38[a]	25.40[b]	24.32[c]	
Feces Production (g DM/h/d)	A1	613.15	659.11	567.83	613.36[a]
	A2	257.86	239.34	256.78	251.33[b]
	Average	435.86[a]	449.23[a]	412.31[a]	

[a,b]: Values in each raw and column in the same parameter followed by the same letters are not significantly different (p>0.05)

4. Conclusion

Feed with higher quality produced lower methane production in term L/h/d but no effect in term L/DMI. This feed also produced higher nutrient content on the feces that probably can produce more manure CH_4 and N_2O emissions. Since no effect on the sheep productivity therefore in term of livestock environmentally friendly, ration with lower TDN and protein content in this study is more recommended than that with higher TDN and protein content.

Acknowledgments

This study was funded by the Ministry of Research, Technology and Higher Education - Republic of Indonesia (grant number 315-02/UN7.5.1/PP/2017) under the PMDSU Program Batch II.

References

[1] Zervas G, Tsiplakou E 2012 An Assessment of GHG emissions from small ruminants in compararison with GHG emissions from large ruminants and monogastric livestock. Atm. Environ. **49** 13

[2] ShibataM, Terada F 2010 Factors affecting methane production and mitigation in ruminant. J. Anim. Sci. **82** 2

[3] Kawashima T, Sumawal W, Terada F, Shibata M 2001 Respiration trial system using ventilated flow-through method using face mask. J. Jircas **9** 53

[4] Purnomoadi A 2013 Current and future inventory data activities on emission of methane from liveztock in Indonesia; do we need a spesificcalcultaion. IAARD Press Jakarta

[5] Gomez KA, Gomez AA 2007Prosedur statistic untukpenelitianpertanianTranslated bySjamsuddin, E and Baharsjah JS. UI Press Jakarta

[6] Shibata M, Terada F, Iwasaki K, Kurihara M, Nishida T 1992 Methane production in heifers, sheep and goats consuming diets of various hay – concentrate ratios. Anim. Sci. Technol. **63** 1221

[7] Shibata M, Terada F 2010 Factors affecting methane production and mitigation in ruminant. J. Anim. Sci. **82** 2

IOP Conf. Series: Earth and Environmental Science **119** (2018) 012056 doi:10.1088/1755-1315/119/1/012056

The Optimation of Crude Fiber Content of Diet for Fattening Madura Beef Cattle to Achieve Good A:P Ratio and Low Methane Production

N. Luthfi[1*], V. Restitrisnani[1] and M. Umar[1]

[1]Faculty of Animal and Agricultural Sciences, Diponegoro University, Semarang, Indonesia

*Corresponding E-mail: agung194@yahoo.com

Abtract

Methane (CH_4) is one of the major greenhouse gases being reducted. This study was carried out to determine the optimum of crude fiber needed for fattening Madura beef cattle to achieve low methane production. Twenty-four male madura beef cattles with an average body weight of 206.89 ± 7.82 kg were used in this study. Cattle were fed contained 47.65 – 70.23 % Total Digestible Nutrients (TDN), 9.22 – 13.20% Crude Protein (CP), and 10.25 – 28.53% Crude Fiber (CF). Correlation analysis was used to describe the correlations among crude fiber,Methane, and Acetate:Propionate acid ratio. The result showed that crude fiber (%) had medium possitive correlation with methane and the equation was y = 0.0936x + 5.4853 with determination as much as 33.98%. The Acetate: propionate acid had positive correlation withmethane and the equation was y = 1.1121x + 3.581 with determination as much as 61.64%. Crude fiber (%) had low possitive correlation with A/P ratio and the equation was y = 0.039x + 2.4437 with determination as much as 9.56%. According to the results, it can be concluded that the optimum ofcrude fiberof diet for Madura beef cattle should be 15.38% to obtain A:P ratio of 3 or less and methane production of 6.91 MJ/d.

1. Introduction

The high demand of meat production increases the beef cattle industry. Due to its high increases, methane emission plays a role in the environment impact as a contributor of GHG's emissions [1].Methane is always produced in the rumen as lost energy around 8 – 12% from the digestible energy ingested [2]. Methane is the feed energy utilization description of animals as well. Changes in the efficiency of feed energy utilization can influence CH4 emissions [3].

The factors that can impact on the efficiency of feed energy utilization are the type or quality and quantity of feed. One of the most useful predictors of CH_4 Production is the intake of crude fiber on the diet or its apparently digested. The fiber fraction in the ruminant diet, has the greatest impact on CH4 production [4].The amount of the diet fiber content affects fermentation acid production in the rumen [5].Volatile fatty acid (VFA) and methane, are end-products of fermentation that resulted in the rumenbut both are differently related in feed utilization and production [6]. Therefore, the improvements of feed energy efficiency and lowering the environmental impact become important and should be achieved through a higher production per animal.

It is such a great consideration that there is correlation among crude fiber and gross energy lost as CH4 and acetate:proionate ratios. Thus, it is essential to find out anoptimum ofcrude fiber that can achieve lowmethane production of cattle.

2. Materials and Methods

This study used 24 heads of madura beef cattle, weighed 203.29±4.82 kg body weight. Cattle were fed contained 47.65 – 70.23 % Total Digestible Nutrients (TDN), 9.22 – 13.20% Crude Protein (CP), and 10.25 – 28.53% Crude Fiber (CF). Feed given in this study consisted of 70% of concentrate and 30% Napiergrass.

Methane emission was measured using facemask method for two days (2x24 h) using a mask which was connected to a methaneanalyzer (Horiba Ltd., Japan) and airflow meter. Methane emissions were recorded automatically by IBM PC for 10 minute at 3 hours intervals. The results obtained were converted into energy units (1 liter CH_4 = 9.45 kcal) [7].Feed was analyzed using proximate analysis [8]. Taking of rumen liquor and analysis of Volatile fatty acids (VFA) was analyzed using Gas Chromatography [9].

Variables data measured were methane from the rumen, rumen acetate to propionate ratio and crude fiber. The data were analysed by correlating [10] the value of methane to the crude fiber and crude fiber to A/P ratio. The strength of relationship among parameters was determined based on the correlation coefficient which was classified medium (0.40 to 0.599); strong (0.60 to 0.799); and very strong (0.80 to 1).

3. Result and Discussion

3.1. Crude fiber toward A:P ratios

The results are presented in figure 1 and Table2. The figure 1 showed that crude fiber had low and positive correlation with A:P ratio ($r = 0.35$, $P = 0.09$). The higher the crude fiber, the higher the A:P ratio. Acids that are produced in the rumen are derived from feedstuffs as end products of OM fermentation in the rumen (primarily acetic acid) [11]. The formation of acetic acid is always accompanied by H_2 and CO_2 production, whereas propionic acid production requires H2 to be formed as propionic acid [12]. Thus, the fermentation of cell wall fiber increase acetic : propionic acid ratios [13]ghly influencedby interactions among the diet, the animal,and ruminal microbes, affecting residence time in therumen and microbial activity[14].

The prediction of A:P ratio from dietary characteristics might be more useful for diet formulation to meet the fiber requirements of beef cattle. This paper addresses the fiber requirements of beef cattle by considering the balance between production that can be indicated using A:P ratio of rumen fluid and crude fiber. This study showed that every increase 1% of crude fiber increases A:P ratio as much as 0.039. The optimum A:P ratio for fattening beef cattle is 3 or less. Based on the equation, at least it requires 15.38% crude fiber to get the optimum of A:P ratio on Madure beef cattle.

3.2. A: P Ratios toward methane production

The figure 2 showed that A:P ratio had positive and strong correlation with methane production ($r = 0.78$, $P > 0.0001$). The higher the A:P ratio, the higher the methane production. A shift in VFA pattern from propionic towards acetic will increase the amount of excess hydrogen available for conversion to CH4 [15]. Methanogens utilize H_2 and CO_2 produced in the rumen to generate methane gas and hence to reduce hydrogen accumulation, which otherwise can negatively affect fermentation in the rumen [16]. Therefore, feed efficiency in cattle production can be seen by asetate: propionate ratio of rumen liquid. Propionic acid has a positive relationship with dietary energy utilization efficiency. The higher propionic acid indicates that the efficiency of energy use in cattle is better [16].

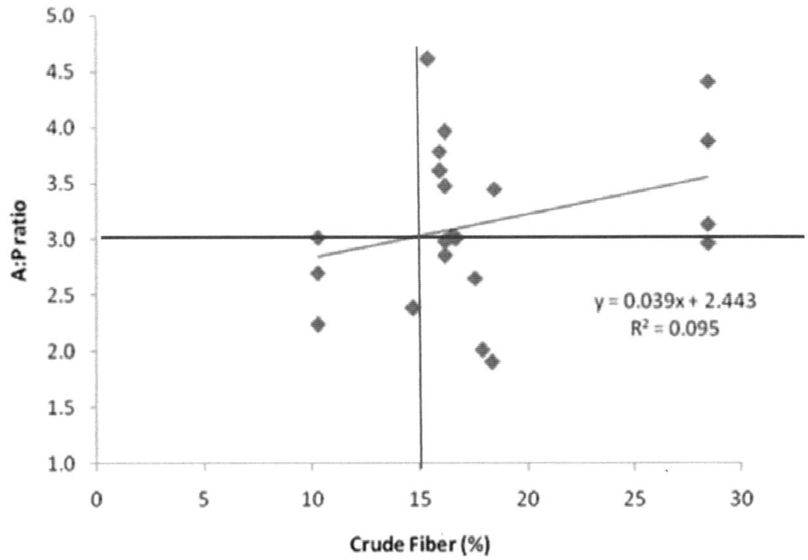

Figure 1. Correlation between crude fiber (%) and A:P ratio

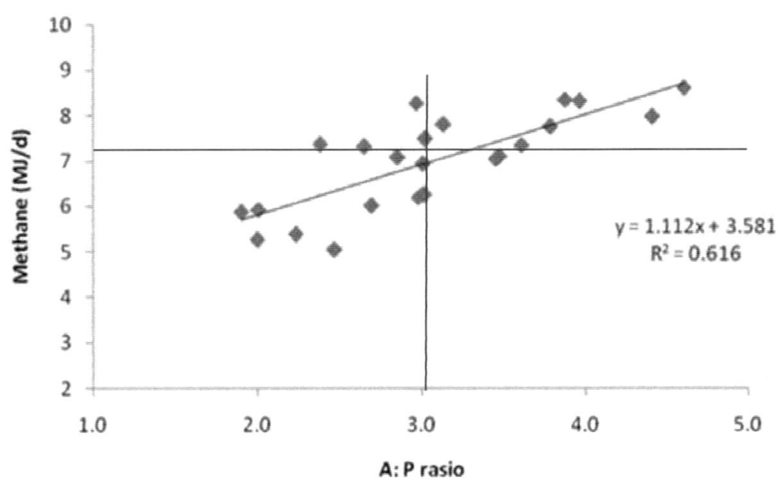

Figure 2. Correlation between A: P ratio and methane production

It is important to predict methane production through those correlation . Every increase of A:P ratio increases methane in the rumen as much as 1.11 MJ/d on the Madura beef catlle. The optimum A:P ratio for fattening cattle is 3 or less. Based on the equation, 3 of A:P ratio results methane production as much as 6.91 MJ/d on madura beef cattle.

4. Conclusion

Based on this study, it can be conluded that crude fiber impact on the A: P ratio and its A: P ratio of rumen fluid had positive and strong correlation with methane production of Madura beef cattle. At least, it requires 15.38% of crude fiber to get 3 of A: P ratio of rumen fluid and 6.9 of methane production.

References

[1] Grainger C and K A Beauchemin 2011Can enteric methane emissions from ruminants be lowered withoutlowering their production. Anim. Fd. Sci. Technol.**167**308

[2] Tamminga S 1992 Nutrition management of dairy cows as a contribution to pollution control. J Dairy Sci.**75**345

[3] Shibata Mand F Terada 2010Factors affecting methane production and mitigationin ruminantsAnim. Sci. J.**81** 2

[4] Jentsch W, M Schweigel, F Weissbach, H Scholze, W Pitroff and M Derno2007Methane production in cattlecalculated by the nutrient compositionof the diet. J. Anim. Nut.**61**(1)10

[5] Allen M S 1997Relationship Between Fermentation Acid Productionin the Rumen and the Requirementfor Physically Effective Fiber. J. Dairy Sci.**80**1447

[6] Khiaosa-ardR and Q Zebeli 2014Cattle's variation in rumen ecology and metabolism and its contributions to feed efficiency.Lives. Sci.**162** 66

[7] Kawashima T, Sumamal W, Terada F, Shibata M 2001 Respiration trial system using ventilated flow-through method with facemask. JIRCAS. **9** 53

[8] AOAC 2005 Official Methods of Analysis: 18th edition. Association of Official AnalyticalChemists International Maryland: USA.

[9] Luthfi N, CMS Lestari and A Purnomoadi 2014Ruminal fermentation and blood glucose at low and high level intake growing and mature Kacang goat. J. Indonesian Trop. Anim. Agric.**39**(3) 152

[10] Steel, R G D and J H Torrie 1993 Prinsip dan Prosedur Statistika Suatu Pendekatan Biometrik Edisi Kedua. PT Gramedia: Jakarta

[11] Allen M S 1997Relationship Between Fermentation Acid Productionin the Rumen and the Requirementfor Physically Effective Fiber.J. Dairy Sci.**80**1447

[12] Miller, T L 1995 Ecology of methane production and hydrogen sinks in the rumen In: Engelhardt W, Leonhard-Marek S, Breves G, Giesecke D, editors. Ruminant physiology: Digestion, metabolism, growth andreproduction. Proc 8th Intern Symp Ruminant Physiology. Stuttgart: Ferdinand Enke Verlag 317

[13] Johnson KA, and DE Johnson 1995 Methane emissions from cattle. J. Anim. Sci.**73** 2483

[14] Wilkerson VA, DP Casper, and DR Mertens 1995 The prediction of methane production of Holstein cows by severalequations.J. Dairy Sci.**78**2402

[15] Khiaosa-ardR and Q Zebeli 2014Cattle's variation in rumen ecology and metabolism and its contributions to feed efficiency. Lives. Sci.**162** 66

[16] Cherdthong A, M Wanapat, W Wongwungchun, S Yeekeng, T Niltho, D Rakwongrit, W Khota, S Khantharin, G Tangmutthapattharakun, K Phesatcha, S Foiklang, S Kang, N Anantasook and P Gunun. Anim. Feed Sci. Technol. **198** 151

IOP Conf. Series: Earth and Environmental Science **119** (2018) 012057 doi:10.1088/1755-1315/119/1/012057

The Effect of Crude Protein Content on Meat and Fat Production in Sheep

Mawati S[1*], V. Restitrisnani[1] and Soedarsono[1]

[1]Faculty of Animal and Agricultural Sciences, Diponegoro University, Semarang, Indonesia

*Corresponding Email: agung194@yahoo.com

Abstract

This study was undertaken to evaluate the effect of crude protein (CP) content on meat protein and fat production in sheep. Twenty four male thin tail sheep aged 6-7 months with average body weight of 13±1.56 kg were used in this study. The sheep were fed 10-14% CP. Sheep with the average body weight amount 16.75 kg were slaughter after 4 months rising. Parameters observed in this study were carcass weight, meat weight and fat weight of thin tail sheep. The data were analyzed using correlation analysis. The result of this study showed that CP content on diet had weak and negative correlation with meat production (r= -0.06) (y= -0.148x + 62.54) but had weak and possitive correlation with fat production (r= 0.3) ($y= 0.807x^2$ -18.40x + 119.1). Based on the result, it can be concluded that the optimum CP content for sheep is 12.5% CP.

1. Introduction

Increasing economic growth will stimulate increasing the demand of livestock product such as meat. One of meat which people like is sheep meat. Nowadays, people prefer to consume lean meat. This is due to public interest in nutrition and healthy eating [1,2]. Since consumers need to be assured the quality of meat, understanding how the factors affect carcass and meat quality is very important for the lamb meat industry.

One of the factors that affect on meat and fat production is feed quality.Commonly, the quality of feed is determined from its protein content.One of the functions of protein is to support the growth of livestock. When metabolizable protein supply of the basal diet can fulfill the protein requirements [3]therefore, the growth performance will increase as well [4,5].Increase of the growth efficiency will affect meat and fat production[6, 7].Several researches were done to found the optimum feed protein level to get high meat production and low fat meat. As reported by [8]Anggora goats fed 14.7% CP give 13.5% body protein, 26.5% body fat and 52.8% body water. Other results presented by [9] in Tunisia local goats fed different level protein diet (10, 13 and 16%) give 74.3% to76% body water 22.4% DM to 23.4% body protein and 2.0% to 4.5% body fat. However, more sheep industry has had the difficulty of balancing the heavy slaughter weights of lambs with a desirable amount of fat thickness. In the other hand, optimizing nutrient efficiency is essential to increase sustainability, since it may reduce nutrient excretion, environmental pollution and production costs.

Based on the explanation above, it's necessary to known the optimum protein content on feed to get high meat production and low fat meat.In this study we investigated the effect of dietary protein on sheep meat quantity and quality.

2. Materials and Methods

Twenty four male thin tail sheep aged 6-7 months with average body weight of 13±1.56 kg were used in this study. The sheep penned individually and fed 10-14% crude protein (CP). Sheep were slaughtered after 4 months risingand they reached body weight amount 16.75 kg.Lambs were fasted for 6 hours before slaughtered.The slaughter was done followingIslamic methods.

Immediately after slaughter, hot carcasses were weighed. Carcasses were weighed before and after chilling at 18°C for 10 hours. Then carcass was separated into meat, fat and bone. The *Longisimusdorsi* (LD) and *Biceps femoris* (BF) of the right part of carcass were excised to measure protein and fat (%) content.

Parameters observed in this study were carcass weight, meat weight and fat weight of thin tail sheep. The data were analyzed using correlation analysis.

3. Result and Discussion

The correlation of dietary crude protein with meat productivity and fat content shows on Figure 1.Crude protein dietary content had weak and negative correlation with meat production (r= -0.06) (y= -0.148x + 62.54) but had weak and possitive correlation with fat production (r= 0.3) (y= 0.807x^2 - 18.40x + 119.1).Figure 1.showed that percent of meat was stable, but percent of fat was increase by increasing dietary crude protein. It is due to muscle cell have the maximum size capacity to growth. Therefore, the nutrient intake did not used for meat growth. Protein intake in this study was used to fat growth. It is caused by sheep's age. In this study, sheep is in the maximum muscle growth phase, so the nutrient intake was allocated to the growth fat [10]. From the equation above, it is known that the optimum crude protein content in diet is 12.5%.

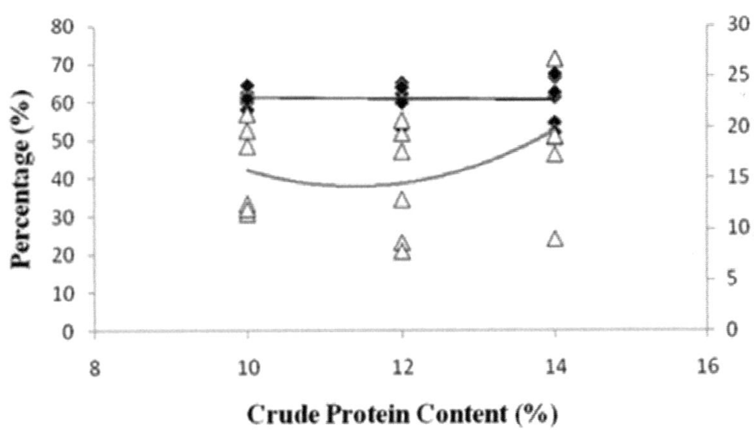

Figure 1. Correlation of Dietary Crude Protein with Meat and Fat Production on Sheep

4. Conclusion

Based on the result, it can be concluded that the optimum CP content for sheep is 12.5% CP.

IOP Conf. Series: Earth and Environmental Science **119** (2018) 012057 doi:10.1088/1755-1315/119/1/012057

Reference

[1] Verbeke W, L Van Wezemael, MD De Barcellos, JO Kugler, JFHocquette, O Ueland, and KG Grunert 2010European beef consumers' interest in a beef eating-quality guarantee.Insights from a qualitative study in four EU countries. Appetite **54**289

[2] Hocquette JF, R Botreau, B Picard, A Jacquet, DW Pethick, and ND Scollan 2012Opportunities for predicting and manipulating beef quality. Meat Sci. **92**197

[3] NRC 2007 Nutrient Requirement of Small Ruminant Sheep, Goats, Cervids, and New World Camelids**.** National Academy Press: Washington DC

[4] R Barajas, B Ortiz, A Camacho, NE Villalba, RL Flores, JJLomeli, and JARomo 2011Influence of additional tannins-extract level on feedlot-performance of finishing lambs. J. Anim. Sci.**89**497

[5] Ortiz B, A Camacho, NE Villalba, LR Flores, MA Mariezcurrena, MD Mariezcurrena, and R Barajas 2013 Influence of tannins extract supplementation at low level on feedlot performance of Katahdin × Pelibuey hair-lambs**.**J. Anim. Sci.**91** 497

[6] Owens F N, P Dubeski, and C F Hanson 1993 Factors that alter the growth and development of ruminants. J. Anim. Sci. **71** 3138

[7] Manso T, A R Manteco´n, F JGiraldez, P Lav´ın, T Castro. 1998. Animal performance and chemical body composition of lambs fed diets with different protein supplements. Small.Rum. Res. **29** 185

[8] Sahlu T, S P Hart, A L Goetsch 1999 Effects of level of feed intake on body weight, components, and mohair growth in Angora goats during realimentation. Small Rum. Res. **32** 251

[9] Atti N, H Rouissi, M Mahouachi 2004 The effect of dietary crude protein level on growth, carcass and meat composition of male goat kids in Tunisia.Small Rum. Res.**54** 89

[10] Ponnampalam E N, B J Hosking,and A R Egan 2003 Rate of carcass components gain, carcass characteristics and muscle longissimus tenderness in lambs fed dietary protein scources with a low quality roughage diet. Meat Sci. **63** 143

IOP Conf. Series: Earth and Environmental Science **119** (2018) 012058 doi:10.1088/1755-1315/119/1/012058

Study of Local Herb Potency as Rumen Modifier: The Effect of Red Ginger (*Zingiber officinale Var.Rubrum*) on Parameters of Ruminal Fermentation In Vitro

A. Kurniawati[1*], Widodo[2], W. T. Artama[3] and L. M. Yusiati[1]

[1)]Department of Animal Nutrition and Feed Science, Faculty of Animal Science, Universitas Gadjah Mada. Jl Fauna 3, Kampus UGM, Bulaksumur Yogyakarta, 55281 Indonesia
[2)]Department of Animal Products Technology, Faculty of Animal Science, Universitas Gadjah Mada. Jl Fauna 3, Kampus UGM, Bulaksumur Yogyakarta, 55281 Indonesia
[3)]Department of Biochemistry, Faculty of Veterinary, Universitas Gadjah Mada. Jl Fauna 2, Kampus UGM, Bulaksumur Yogyakarta, 55281 Indonesia

*Corresponding E-mail : asihkurniawati@ugm.ac.id

Abstract

Essential oil is one of rumen modifier alternatives due to its antimicrobial property. Red ginger is one of local herbs with high essential oil content. The effect of red ginger on rumen fermentation parameters was studied in this research using in vitro gas production method. Five level of red ginger meal was added to the diet to meet final essential oil concentration in fermentation medium of 0, 25, 50, 75 and 100 mg/L. Substrate of fermentation as microbial feed composed of *Penisetum hybride*, rice bran, and wheat pollard in ratio 60:20:20 DM basis. Fermentation was carried out for 24 h at 39°C. Total gas production was measured at the end of incubation and sample for methane analysis was taken. Medium sample was taken for analysis of pH, ammonium and VFA concentration, microbial protein and protozoa number. Data showed that addition of red ginger in the diet did not affect the pH, ammonia and VFA concentration, microbial protein and also protozoa number. However, red ginger addition significantly decrease ammonia concentration in all treatment. It could be concluded that addition of red ginger in the diet reduced degradation protein in the rumen as illustrated in lower ammonia concentration.

1. Introduction

In ruminant, microbes inhabit in the rumen is important regarding their activity. Rumen microbe degrade and ferment almost all of the nutrient which enter to the rumen. Modification of fermentation process, addressed to increase efficiency of feed utilization, can be done by administration of feed additive which altered rumen microbe activity. Antibiotic is one of proven feed additive in modification of rumen fermentation [1]. Monensin, one of ionophore antibiotic, increase efficiency of feed utilization by decrease 30% of ruminal methanogenesis, reduce ammonia concentration in the rumen by interfere proteolytic bacteria and mostly deamination bacteria. Reduction of ammonia in the rumen lead to decreasing of protein loss in urine. In growing cattle monensin increasing rumen VFA concentration, digestibility and protein retention, thus improving food use and weight gain [2]. Ionophore in feedlot cattle increase of daily gain 1.6% and feed efficiency 7.5% respectively [3]. But utilization of antibiotic recently banned in several country regarding of antibiotic residue in animal product and emerging of resistance bacteria. Hence safely feed additive as antibiotic alternative are needed. Essential oils have antimicrobial activities and are currently considered safe for human and animal consumption, and are categorized as Generally Recognized as Safe [4].

Essential oil is plant secondary metabolite, a natural product which exhibit activity against a wide range of microorganisms, including bacteria, protozoa and, fungi (Chao et al. 2012; Cosentino et al. 1999; Deans and Ritchie 1987; Sivropoulou et al. 1996). Naturally, essential oil is part of plant defending mechanism again predator as antibacterial, antifungals, insecticide, antivirals [9]. Therefore based on the activity, essential oil has chance to be used as antibiotic alternative to modified rumen microbes and ruminal fermentation.

Several researches in essential oil utilization as feed additive perform positive effect i.e. increasing of productivity and decreasing of methane production, in vitro and or in vivo [10,11]. Mix oil of cinnamaldehyde, eugenol, and capsicum in feedlot cattle have the same effect as monensin [10] increase growth and health performance optimized of feed fermentation in the rumen and increased immune system [12] hence essential oils could be used as antibiotic alternatives in feed additive [13].

Essential oil from rosemary, oregano, ceylon cinnamon, dill seeds, cinnamon leaves, cinnamon bark, and eucalyptus, at level 1,125 ml/L of fermentation media, reduce methane production and ammonia concentration in rumen medium with no detrimental effect in NDF degradability except eucalyptus. Oregano individually and combinations of several those essential oil reduce abundance of archaea (Cobellis et al. 2016).

Essential oil of clove, eucalyptus, garlic, oregano and peppermint at concentration of 0.25, 0.5, and 1 g/L of medium reduced methane production in line with the increasing of essential oil level, reduced population of protozoa and archaea, but NDF digestibility also decreased except in garlic oil due to the decreasing of cellulolytic bacteria. Those essential oils have no effect on volatile fatty acid production except in addition of clove and oregano oil [15,16]

Biological activity of essential oil in rumen fermentation vary. Effects of essential oils vary depend on their chemical composition. Same essential oil which obtained from different plant in same genus may have opposite effect, stimulatory or inhibitor (Ferme et al. 2004; Patra A. K. 2011). Purity and dose also influenced activity of essential oil [20].

Among other two kinds of ginger in Indonesia red ginger or *sunti* ginger (*Z. officinale* var. Rubrum) has the highest content of essential oil. Two other ginger are elephant the large white ginger or *gajah* ginger (*Z. officinale* var. Roscoe), small white or yellow ginger (called *emprit*) ginger (*Z. officinale* var. Amarum)[21]. Essential oil of ginger could be found either in leaf or in rhizomes. Dominant essential oil component in rhizome are monoterpenoid, with camphene (14.5%), geranial (14.3%), and geranyl acetate (13.7%), which active against the Gram-positive bacteria *Bacillus licheniformis*, *Bacillus spizizenii*, and the Gram-negative bacteria *Escherichia coli*, *Klebsiella pneumoniae* and *Pseudomonas stutzeri* [22].

2. Materials and Methods

2.1. *Feed, Treatments and In Vitro Fermentation*

The effect of red ginger addition in the diet toward ruminal fermentation and methane production were studied using batch culture of in vitro gas production technique. Red ginger meal was added in to in vitro feed substrate which consist of *Pennisetum purpureum* (cut before flowering stage), rice bran and wheat pollard (obtained from feed shop). Composition of those three feed stuffs were in ratio 60:20:29 based on dry matter. Red ginger meal was prepared by drying in dryer incubator at 55°C and ground to pass through a 1 mm pore size sieves. Final concentration of essential oil in fermentation media i.e. 0, 25, 50, 75, and 100 mg/l were used as basis in calculation of red ginger requirement to be added in feed substrate.

2..2. In vitro gas production

Mix of feed sample and red ginger meal according to the treatment were incubated in syringe glass containing buffered rumen fluid according to Menke [23] in triplicate. Rumen fluid as rumen microbes source was obtained from two ruminally fistulated Ongole grade cattle fed twice daily with a diet consisting of *Pennisetum purpureum* and beef cattle concentrate 60:40 DM bases.

Contents Rumen microbes as inoculum for the in vitro gas production Rumen fluid was collected before morning feeding, and squeezed through polyester cloth into a vacuum flask thermos, and immediately sent to the laboratory.

Artificial rumen fluid as fermentation medium consisted of (added in order) 500 ml H2O, 0.1 ml micro mineral solution (consisted of 13.2 g CuCl22H2O, 10.0 g MnCl24H2O, 1.0 g CoCl26H2O, 8.0 g FeCl26H2O and made up to 100 ml with water), 200 ml buffer solution (consisted of 35 g NaHCO3 and 4 g NH4HCO3 added up to 1,000 ml with water), 200 ml macro mineral solution (Solution C consisted of 5.7 g Na2HPO4, 6.2 g KH2PO4, 0.6 g MgSO47H2O added up to 1,000 ml with water), 1 ml resazurin (0.1% w/v), and 40 ml reduction solution (consisted of 95 ml H2O, 4 ml 1 N-NaOH and 625 mg Na2S9H2O). This mixture was then kept under CO2 in a water bath at 39°C and stirred using a magnetic stirrer.

Calibrated glass syringes of 100 ml (Fortuna®, Häberle Labortechnik, Germany) contained approximately 0,300 g dry weight of sample were prewarmed at 39°C before the injection of 30 ml of buffered rumen liquor a mixture of artificial rumen fluid and rumen liquid in ratio of 20:10 followed by incubation in a water bath at 39°C. Volume of gas production were recorded before incubation (0) and after 2, 4, 6, 8, 12, 24 h of incubation. Total gas values were corrected for blank incubation.

At the end of incubation gas were collected using calibrated syringe and 5 ml of gas were transferred into 5 ml plain vacuum tube (Becton Dickinson Vacutainer System) for methane analysis. Sample for protozoa calculation were prepared by pipetting 1 ml of syringe content and be added to 0.8ml of formaldehyde saline (1ml of 37%formaldehyde + 9 ml 0.9% NaCl). One microliter sample then transferred to haemocytometer for direct calculation under microscope according to method explained by[24].

For ammonia measurement 1 ml of syringe content were preserve with 1 ml NaCl 20% and be frozen until later analysis of ammonia base on phenol hypochlorite reaction as explained by Chaney and Marbach (1962). Media, as much as 1 ml for volatile fatty acid (VFA) analysis were added into tube containing 1 ml of 20% metha-phosphoric acid and stored in freezer for further analysis using gas chromatography. Prior to sampling for ammonia, VFA, microbial protein and protozoa, pH media were measured. Rumen microbial protein was determined by Lowry method (Alexander and Griffiths, 1993). Microbial cell were separated from residual feed by centrifugation 1.5 ml of syringe content at 500g. Cell were precipitated from supernatant by spin down at 15.000 g. Pellets were re-suspension in physiology solution and re centrifuge. Re-suspension was repeated for twice. The last suspension was subjected for protein determination.

2.3.Statistical analysis

Parameters of ruminal fermentation in this studied were total VFA, acetate, propionate and butyrate concentration as mmol/100ml, rumen microbial protein, ammonia concentration as mg/100l, methane production as ml/g DM digested, and protozoa number. Data were subjected to one-way analysis of variance with level of red ginger as treatment factor. Comparisons between means were analysis using *t*-test of Duncan Multiple Range test.

3. Result and Discussion

pH of medium in vitro did not affected by addition of red ginger, It was range from 6.75 to 6.80 (Table 1.). According to Puniya et al [25] pH is one of variable factor in rumen environment and physiologically range from 5.5 to 6.9 to support normal feed fermentation in rumen. Those range is in the physiology pH for activity rumen microbes that support for feed metabolism in rumen.

Total VFA production and its composition as well as ratio of acetate/propionate did not affected by addition of red ginger (Table 1.) even though crude fiber digestibility increased (data not shown). Volatile fatty acids in the rumen are come from digestion and fermentation of carbohydrate by rumen microbe [26]. Effects of essential oils on VFA production in the rumen VFA are inconsistent. Blend essential oil consist of oregano, cinnamon, thyme, orange peel were reported decreases in VFA concentrations [27]. Decreasing of VFA also reported as effect of addition single essential oils i.e. dill seeds; cinnamon leaves; cinnamon bark; Ceylon cinnamon bark; eucalyptus leaves; oregano leaves; rosemary leaves, essential oil of except eucalyptus [14]. Addition of several individual essential oil in several doses showed addition at 0 to 30 mg/l did not changed VFA production, at level 300 mg/l some essential oil reduced or increased slightly and addition at 3000 mg/l almost all essential oil changed VFA production and VFA composition depend on essential oil source [28]. Addition spices as source of essential oil at low of cinnamon have no effect on VFA, clove and coriander lower VFA, whereas cumin and turmeric increased VFA and all treatment reduced acetate: propionate ratio except cumin and turmeric [29].

Rumen microbial protein synthesis did not affected by treatment meanwhile protozoa number were reduced (Table 1.). Essential oils effects on rumen microbe population are varies. Some studies report different effect of herb and extract and essential oil on protozoal numbers. Anise extract reduced protozoa number while capsicum and blend of cinnamon and eugenol did not affect [30]. Effect of essential oil on protozoa number depend on essential oil source and the doses [31]. Eucalyptus and clove oil have stimulatory effect at level 250 mg/l but at 500 and 1000 mg/l inhibit protozoa. Garlic, oregano and peppermint inhibit protozoa at level 250 up to 1000 mg/l. Individual essential oil have varies effect while combination of those essential oil reduce protozoa number [14]. Similar tendency as in protozoa number effect of essential oil on microbial synthesis is inconsistent. Some research shows stimulation [32], some have no effect [14,33] and some decreased microbial protein synthesis [34]. Even essential oils were reported have ability to inhibit rumen microbes [5–8] but the activities are depend on doses and functional group of compounds. Kalemba dan Kunicka [35] explain antimicrobial activity of functional group from the strongest are phenol>cinamic aldehide> alcohol> aldehid=ketone>eter>hydroarbon. Phenol is component of essential which frequently found in essential oil and broadest spectrum activity. In general, the mechanism of action is considered to be the disturbance of the cytoplasmic membrane, disrupting the proton motive force, electron flow, active transport, and coagulation of cell contents [36].

As data shown in Table 1. ammonia in all treatment were lower than control ($p<0.05$). Mostly in several research, addition of herb and essential oil reduced ammonia concentration [14,29,32,37]. Reduction of ammonia concentration might be caused of essential oil effect in lowering peptidolytic activity of ruminal bacteria [28], and inhibition of hyper ammonia producing bacteria growth. Both of microbial group are involve in ammonia production [38,39].

Methane production were not affected by the treatment although protozoa number in red ginger treatment group were lower than control ($P<0.05$). About 9% to 25% of the rumen methanogens are associated with protozoa and around 37% of methane in the rumen are produced by this methanogens group. Methanogens archaea in rumen are associated with protozoa by endosymbiosis [40].

Table 1. Effect of red ginger on parameters of rumen fermentation

Parameters	Level of essential oil				
	0	25	50	75	100
pH	6.80	6.80	6.78	6.77	6.75
Gas production (ml/g DM digested)	386.57	442.49	434.30	395.42	352.40
Total VFA (mmol/100ml)	19.77	17.35	15.54	18.09	16.92
Acetate	14.78	13.63	11.47	13.03	11.29
Butyrate	1.59	1.51	1.28	1.64	1.43
Propionate	3.40	2.21	2.80	3.42	3.20
Acetate/propionate	6.38	4.73	4.09	3.81	3.63
Rumen microbe protein (mg/l)	324.41	233.75	239.07	262.99	303.94
Protozoa (cel x 10^4)*	11.53^b	8.36^a	8.09^a	10.71^{ab}	9.18^{ab}
Ammonia (mg/10ml)*	28.93^b	25.08^a	24.29^a	24.26^a	25.02^a
Methane production/DM digested (ml/g)	49.70	59,28	59.93	64.77	44.69

^{a,b} different superscript in the same row indicated significant different (P<0.05)

4. Conclussion

Utilization of red ginger in diet equal to EOs up to level 100 mg/l have do detrimental effect on ruminal fermentation and tend to improve feed efficiency particularly N efficiency and reduced protozoa number. Furthermore red ginger have potency to be developed as ruminal fermentation modifier agent.

Acknowledgments

Authors gratefully acknowledge Directorate General of Higher Education (DGHE), Ministry of Research Technology and Higher Education for awarding Doctoral scholarship. This research is part of doctoral research. Authors also acknowledge Nutritional Laboratory, Faculty of Animal Science, Universitas Gadjah Mada for the supports and provide the facilities for this research

References

[1] Russell J B and Strobel H J 1989 Effect of ionophores on ruminal fermentation.Appl. Environ. Microbiol. **55** 1

[2] Salles M S V, Zanetti M A, Titto E A L and Conti R M C 2008 Effect of monensin on performance in growing ruminants reared under different environmental temperatures Anim. Feed Sci. Technol. **147** 279–91

[3] Jouany J-P and Morgavi D P 2007Use of 'natural'products as alternatives to antibiotic feed additives in ruminant production Animal **1** 1443

[4] U.S. Food & Drug Administration 2017 Food Drug Adm. Dep. Heal. Hum. Serv. Subchapter B Food Hum. Consum. revised April 1, 2017

[5] Cosentino S, Tuberoso C, Pisano B, Satta M, Mascia V, Arzedi E and Palmas F 1999 In-vitro antimicrobial activity and chemical composition of Sardinian thymus essential oils Lett. Appl. Microbiol. **29** 130

[6] Deans, S.G. and Ritchie G 1987 Antibacterial properties of plant essential oils Int. J. Food Microbiol. **5** 165

[7] Sivropoulou A, Papanikolaou E, Nikolaou C, Kokkini S, Lanaras T and Arsenakis M 1996 Antimicrobial and Cytotoxic Activities of Origanum Essential OilsJ. Agric. Food Chem. **44** 1202

[8] Chao S C, Young D G and Oberg C J 2000 Screening for inhibitory activity of essential oils on selected bacteria, fungi and viruses J. Essent. Oil Res. **12** 37

[9] Bakkali F, Averbeck S, Averbeck D and Idaomar M 2008 Biological effects of essential oils–a review Food Chem. Toxicol. **46** 446

[10] Geraci J I, Garciarena A D, Gagliostro G A, Beauchemin K A and Colombatto D 2012 Plant extracts containing cinnamaldehyde, eugenol and capsicum oleoresin added to feedlot cattle diets: Ruminal environment, short term intake pattern and animal performance Anim. Feed Sci. Technol. **176** 123

[11] Bodas R, Prieto N, García-González R, Andrés S, Giráldez F J and López S 2012 Manipulation of rumen fermentation and methane production with plant secondary metabolitesAnim. Feed Sci. Technol. **176** 78

[12] Compiani R, Sgoifo Rossi C A, Pizzi A and Dell'Orto V 2013 Trends in Veterinary Sciences: Current Aspects in Veterinary Morphophysiology, Biochemistry, Animal Production, Food Hygiene and Clinical Sciences pp 177

[13] Khorrami B, Vakili A R, Mesgaran M D and Klevenhusen F 2015 Thyme and cinnamon essential oils: Potential alternatives for monensin as a rumen modifier in beef production systems Anim. Feed Sci. Technol. **200** 8

[14] Cobellis G, Massimo T, Marcotullioc M C and Yu Z 2016 Evaluation of different essential oils in modulating methane and ammonia production, rumen fermentation, and rumen bacteria in vitro Anim. Feed Sci. Technol. **215** 25

[15] Calsamiglia S, Busquet M, Cardozo P W, Castillejos L and Ferret A 2007 The use of essential oils in ruminants as modifiers of rumen microbial fermentation J. Dairy Sci. **90** 2580

[16] Benchaar C and Greathead H 2011 Essential oils and opportunities to mitigate enteric methane emissions from ruminants Anim. Feed Sci. Technol. **166–167** 338

[17] Ferme D, Banjac M, Calsamaglia S, Busquet M, Kamel C and Augustin G 2004 The effects of plant extracts on microbial community structure in a rumen-simulating continuous-culture system as revealed by molecular profiling Folia Microbiol. **49** 151

[18] Patra A. K. 2011 Effects of essential oils on rumen fermentation, microbial ecology and ruminant production Asian J. Anim. Vet. Adv. **6** 416

[19] Mahmud. L. E. 1994 Antifungal action and antiaflatoxigenic properties of some essential oil constituents Lett. Appl. Microbiol. **19** 110

[20] Macheboeuf D, Morgavi D P, Papon Y, Mousset J L and Arturo-Schaan M 2008 Dose–response effects of essential oils on in vitro fermentation activity of the rumen microbial population Anim. Feed Sci. Technol. **145** 335

[21] Wiedhayati D 2016 Export News Indonesia ed T Prahastuti, Sugiarti and R A Marlena

(Directorate General Of National Export Development) pp 1–12

[22] Sivasothy Y, Chong W K, Hamid A, Eldeen I M, Sulaiman S F and Awang K 2011 Essential oils of Zingiber officinale var. rubrum Theilade and their antibacterial activities Food Chem. **124** 514

[23] Menke K H, Raab L, Salewski A, Steingass H, Fritz D and Schneider W 1979 The estimation of the digestibility and metabolizable energy content of ruminant feedingstuffs from the gas production when they are incubated with rumen liquor in vitro J. Agric. Sci. **93** 217

[24] Abreu A, Carulla J E, Lascano C E, Díaz T E, Kreuzer M and Hess H D 2004 Effects of fruits on ruminal fermentation and duodenal nitrogen flow of sheep fed a tropical grass diet with and without legume J. Anim. Sci. **82** 1392

[25] Puniya A K, Singh R and Kamra D N 2015 Rumen Microbiology: From Evolution to Revolution ed P K Choudhury, A Z M Salem, R Jena, S Kumar, R Singh, A K Puniya and 1 Abstract (Springer India) pp 3

[26] Moran J 2005 How the rumen works in :Tropical Dairy Farming: Feeding Management for Small Holder Dairy Farmers in the Humid Tropics. Landlinks Press pp 41-49 pp 41–9

[27] Spanghero M, Zanfi C, Fabbro E, Scicutella N and Camellini C 2008 Effects of a blend of essential oils on some end products of in vitro rumen fermentation Anim. Feed Sci. Technol. **145** 364–74

[28] Busquet M, Calsamiglia S, Ferret A and Kamel C 2006 Plant extracts affect in vitro rumen microbial fermentation J. Dairy Sci. **89** 761

[29] Chaudhry A S, Mehedi M and Khan H 2012 Impacts of different spices on in vitro rumen dry matter disappearance, fermentation and methane of wheat or ryegrass hay based substrates Livest. Sci. **146** 84

[30] Cardozo P W, Calsamiglia S, Ferret A and Kamel C 2006 Effects of alfalfa extract, anise, capsicum, and a mixture of cinnamaldehyde and eugenol on ruminal fermentation and protein degradation in beef heifers fed a high-concentrate diet J. Anim. Sci. **84** 2801

[31] Patra A K and Yu Z 2012 Effects of quillaja and yucca saponins on communities and select populations of rumen bacteria and archaea, and fermentation in vitro Appl. Environ. Microbiol. **78** 4271

[32] Fraser G R, Chaves A V, Wang Y, Mcallister T A, Beauchemin K A and Benchaar C 2007 Effects of essential oils and their components on in vitro rumen microbial fermentation J. Dairy Sci. **90** 2315

[33] Castillejos L, Calsamiglia S, Ferret A and Losa R 2005 Effects of a specific blend of essential oil compounds and the type of diet on rumen microbial fermentation and nutrient flow from a continuous culture system Anim. Feed Sci. Technol. **119** 29

[34] Wanapat M, Kongmun P, Poungchompu O, Cherdthong A, Khejornsart P, Pilajun R and Kaenpakdee S 2012 Effects of plants containing secondary compounds and plant oils on rumen fermentation and ecology Trop. Anim. Health Prod. **44** 399

[35] Kalemba D and Kunicka A 2003 Antibacterial and antifungal properties of essential oils Curr. Med. Chem. **10** 813

[36] Kotzekidou P, Giannakidis P and Boulamatsis A 2008 Antimicrobial activity of some plant extracts and essential oils against foodborne pathogens in vitro and on the fate of inoculated pathogens in chocolate Food Sci. Technol. **LWT 41** 119

[37] Wanapat M, Kang S, Khejornsart P and Wanapat S 2013 Effects of plant herb combination supplementation on rumen fermentation and nutrient digestibility in beef cattle Asian-Australasian J. Anim. Sci. **26** 1127

[38] Mcintosh F M, Williams P, Losa R, Wallace R J, Newbold C J and Beever D A 2003 Effects of essential oils on ruminal microorganisms and their protein metabolism Appl. Environ. Microbiol. **69** 5011

[39] Newbold C J, Mcintosh F M, Williams P, Losa R and Wallace R J 2004 Effects of a specific blend of essential oil compounds on rumen fermentation Anim. Feed Sci. Technol. **114** 105

[40] Belanche A, de la Fuente G and Newbold C J 2014 Study of methanogen communities associated with different rumen protozoal populations FEMS Microbiol. Ecol. **90** 663

IOP Conf. Series: Earth and Environmental Science **119** (2018) 012059 doi:10.1088/1755-1315/119/1/012059

The Function of Credit Scheme to Improve Family Income among Beef Cattle Farmers in Central Java Province

E. Prasetyo[1], T. Ekowati[1], W. Roessali[1], and S. Gayatri[1]

[1]Fakultas Peternakan dan Pertanian, Universitas Diponegoro, Kampus Tembalang, Semarang 50275 - Indonesia.

Corresponding E-mail: edyprsty@yahoo.com

Abstract

The aims of study were: (i) identify of beef cattle fattening credit scheme, (ii) calculating and analyze of beef cattle farmers' income, (iii) analyze of factors influencing beef cattle credit scheme towards farmer's income. The research was held in five regencies in Central Java Province. Beef cattle fattening farm was standardized as an elementary unit. Survey method was used, while Two Stage Cluster Purposive Sampling was used for determining of sample. Data were analyzed using statistical method of quantitative descriptive and inferential statistics in term of income analysis and multiple linear regression models. The result showed that farmers used their own capital to run the farm. The average amount was IDR 10,769,871. *Kredit Ketahanan Pangan dan Energi* was credit scheme which was dominantly access by farmers. The average credit was IDR 23,312,200/farmer with rate of credit equal to 6.46%, the time of credit returning equal to 24.60 monthand the prediction of average collateral equal to IDR 35,800,00. The average of farmers' income was IDR 4,361,611.60/2.96 head of beef cattle/fattening period. If the labour cost did not calculate as a cost production, hence the farmer' income was IDR 7,608,630.41 or in other word the farmer' income increase 74.44%. Factors of credit scheme which partially significant influence to the farmers' income were number of own capital usage and value of credit collateral. Meanwhile, name of credit scheme, financing institution as a creditor, amount of credit, rate of credit scheme and time of returning credit were not significantly influence towards farmers' income.

1. Introduction

Beef cattle have been played as one of important income for villagers in Indonesia as well as family nutrient sources. Meat consumption from beef product have been increased, however national meat production have not been fulfil national consumption. Hence, there was gab between supply and demand of beef product [1; 2].

Beef cattle farming system have been run by the farmers and their family in Central Java, and it occupied both lowland and highland. There are two type of beef cattle farming systems in Central Java Province, namely *pola penggemukan* (dry lot fattening)and *pola perbibitan* (cow-calf operation). [3] told that the average of cattle ownership for dry lot fattening and cow-calf operation farming system are equal to 2,96 head/cattle and 2,40 head/cattle, respectivelly. Beef cattle farming system in

Central Java is based on smallholder farming system and most of the farmers did not think about being commercial farming system [4].

Farmers faced several problems, such as low productivity, poor management practices, and limited access to bank loan. Several efforts have been done by the government to improve productivity of beef cattle farming system, for example: (i) Artificial Insemination Programme; (ii) feed subsidies; (iii) improve farmers' knowledge through agriculture extension program; (iv) Some of the services provided by the government, such as training and credit scheme. Credit scheme was lauched by Indonesian government to farmers to increase farming' productivity with the loans coming collateral-free and subsidised interest rates. Several credits for farmers were called *Kredit Usaha Pembibitan Sapi* (KUPS), *Kredit Ketahanan Pangan dan Energi* (KKPE). The aims of the credit scheme were to improve farmers' access to the bank loan as well as to create opportunities by providing loans which will help the farmers to utilise their resources and skills. However, how does the actual function of credit scheme to improve farmers' income among beef cattle farmers have not been identified. Based on consideration above mention, the aims of study were: (i) identify of beef cattle fattening credit scheme, (ii) calculating and analyze of beef cattle farmers' income, (iii) analyze of factors influencing beef cattle credit scheme towards farmer's income.

2. Materials and Methods

The study was focus on beef cattle fattening farming system. Data were gathered from primary data through interview. In addition, secondary data were gathered to improve during data analysis through government offices or farmers' group. It followed by data entry process and data analysis. Survey method was used in this research based on interview with the beef cattle farmers who got benefit from credit scheme. Two stage's clustered purposive quota sampling was used for sampling methods. There were five regencies in Central Java Province as primary unit (Grobogan Regency, Blora Regency, Rembang Regency, Wonogiri Regency, dan Boyolali Regency). Five regencies were chosen based on five biggest beef cattle population in Central java Province. Meanwhile secondary unit was the beef cattle farmers who got benefit from credit scheme. In total, there were 50 respondents in this research with 10 farmers/each regency. Data were analyzed through editing, coding, and tabulating. Moreover, data were analyzed using financial analysis and multiple regression analysis.

TC : TVC + TFC
TR : $\Sigma (Q_i . Hq_i)$
TC : Total cost (rupiah)
TVC : Total variable cost (rupiah)
TFC : Total fixed cost (rupiah)
TR : Total revenue (rupiah)
Qi : Another income from farming activities, such as selling manure
Hq_i : Price/product (Rp/kg)

Income analysis:

π : TR – TC

Multiple regression analysis:

$$Y = \alpha + \beta_1 X_1 + \beta_2 X_2 + \beta_3 X_3 + \beta_4 X_4 + \beta_5 X_5 + \beta_6 X_6 + \beta_7 X_7 + \mu$$

Note :

Y	:	Income (rupiah)-dependent variable.
α	:	intercept
β_1 s/d β_6	:	Coefficient regression
X_1	:	Total asset (rupiah)
X_2	:	Name of credit scheme (score)
X_3	:	Name of bank institution (score)
X_4	:	Total loan (rupiah)

X_5	:	Credit interest rate (persen)
X_6	:	Repayment period (bulan)
X_7	:	Collateral value (rupiah)
μ	:	Disturbance variables

3. Result and Discussion

There were three types of cattle breeds to rise in Central Java. Simmental – Ongole Crossbreed or *simmental-peranakan ongole* (SPO) was the biggest cattle bread to raise (58%), it followed by Ongole Crossbreed (*peranakan ongole,* PO) (32%,) and limousine-Ongole Crossbreed or *persilangan limousine dengan peranakan ongole* (10%). Most of the farmers had 2.96 head/cattle and it was raised for 8.18 months and average daily gain equal to 0,64 kg/cattle/day. Average daily gain was lower than a research by [5]. The average daily gain was 0.72kg/cattle/day with forage and feed concentrat of 4kg/cattle/day as the main source of feeding resources [5]. In his research, [6] concluded that average daily gain was amounted to1,18 kg/cattle/day for LPO and 0,90 kg/cattle/day of SPO. The low productivity of fattening farming system in Central Java can be explained by the low feed quality resources, limited access to high-quality genetics, cattel feed efficiency, and the age of cattle [7]

Increase capital of the farmers is one of efforts to improve their saving and investment. The average asset was IDR10.769.871,- (31,60% of capital). It would motivate the farmer to improve their productivity, the lower asset is the better. The credit schemes had been offered by the government were 80% of KKPE (*Kredit Ketahanan Pangan dan Enerji*), followed by 14% of CSR *(Corporate Social Resposibility*), 4% credit scheme offered by private institution, and 2% of KUR (*Kredit Usaha Rakyat*). 74% of the farmers got the loan from BRI (*Bank Rakyat Indonesia*), followed by *Bank Negara Indonesia* (BNI) 8,00%, 2,0% from Bank Jateng, and 2,00% from private institution. BRI was one of government financial institution which was always commit to help farmers through credit scheme with low credit interest rate (6,46%/year). The government wished would improve farmers' livelihood through low interest rate. Based on the result, the average repayment period was 24,60 month. Most of the farmers (94%) used their land as collateral value. It had higher financial value than the credit value. The average collateral value was IDR 35.800.000,- and the average credit value was IDR 23.312.000,-. 56% of the farmer had loan lower than IDR 25.000.000,- 42% of the farmers had loan amounted to IDR 25.000.000,- until IDR 50.000.000,- and only 2% had loan above IDR 50.000.000,-. Meanwhile, the average income of the farmers was IDR 4.361.611,60 (equal to IDR 533.204,35/month). Total Cost, total revenue and income shows in Table 1.

Based on the result, farmers' income without labour cost being calculated was IDR 7.608.630,41 (equal to IDR 930.150,42/month). This research tried not to include labour cost into income analysis because the farmers used family labour in their farming activities and the farmers did not pay for labour cost. Moreover, it had low opportunity cost. The farmers' income was higer than a research among PO cattle breed' farmers in Eromoko District Wonogiri Regency by [8]. The research in 2005 told that (i) The cows had 100% ad libitum of forage and mixed with three times feed concentrate per day would gained 0,785 kg/day with famers' income amounted to IDR 637.230,95/head/3months; (ii) The cows had 100% ad libitum of forage and mixed with twice feed concentrate per day day would gained 0,629 kg/day with famers' income amounted to IDR 613.153,25/head/3 bulan; (iii) The cows had twice feed resources per day day would gained 0,547 kg/day with famers' income amounted to IDR 412.739,97/head/3 bulan.

Regression analysis shows that total asset (X_1), type of credit scheme (X_2), and collateral value (X_7), had positive relationship towards farmers' income (Y). Meanwhile, name of bank institution (X_3), total loan (X_4), credit interest rate (X_5), dan collateral value (X_6) had negative relationship towards farmers' income (Y).

$$Y = 0{,}324\ X_1 + 0{,}026\ X_2 - 0{,}025\ X_3 - 0{,}998\ X_4 - 0{,}259\ X_5 - 0{,}107\ X_6 + 1{,}316\ X_7$$

Table 1. Total Cost, total revenue and income of beef cattle fattening farming system in Central Java.

No.	Detail		Rupiah
1.	Variable costs		34.946.064,42
	▪ Feeder cattle price	23.758.221,60	
	▪ Forage costs	3.066.597,30	
	▪ Feed concentrat cost	4.874.226,72	
	▪ Labour cost	3.247.018,80	
2.	Fixed costs		402.530,02
3.	Revenue		39.710.206,04
	▪ Main product (the cows)	39.312.275,00	
	▪ Other product (manure)	397.931,04	
4.	Farmers' income		4.361.611,60
5.	Farmers' income without labour cost		7.608.630,41

The result of overall F test for the null hypothesis shows that all independent variables had significant influence towards farmers' income in beef cattle fattening system. Result of t test, total asset (X_1) and collateral value (X_7) had significant influence ($P < 0{,}05$), meanwhile type of credit scheme (X_2), name of bank institution (X_3), total loan (X_4), credit interest rate (X_5), repayment period (X_6), had no statistically significant influence ($P > 0{,}05$) towards farmers' income in beef cattle fattening system (Y).

Farmers used their own asset and the loan for beef cattle farming activities. Based on empirical data, average farmers' asset had been allocated for farming activities were IDR 10.769.871,- (31,60% of total capital or amounted to IDR 34.081.870,-). T test shows that total asset had significant influence ($P < 0{,}05$) towards farmers' income in beef cattel farming system. Hence, if the farmers want to increase their asset while others factors are constant, it would increase farmers' income. Collateral value variable (X_7) is the market value of anything used as collateral by the farmer to support a loan. Collateral value had influence toward farmers' income. Most of the farmers had their land as collateral value or equal to IDR 35.800.000,-. Based on t test value, Collateral value variable (X_7) had statistically significant influence ($P < 0{,}05$) towards farmers' income. It can be said, if the farmers wants to increase their loan while others factors are constant, it would increase farmers' income.

4. Conclusion

4.1. Conclusion:

1. Farmers owned 2,96 head/cattle with lenght for fattening period of 8,18 month/period and average daily gain (*Pertambahan Bobot Badan*/PBB) of 0,64 kg/day.
2. Data analysis showed that farmers had average total asset amount to IDR 10.769.871, meanwhile KKPE (*Kredit Ketahanan Pangan dan Enerji*) was the dominant credit scheme among beef cattle farmers and BRI (*Bank Rakyat Indonesia*) was dominant financial institution which gave credit to farmers. In addition, the average total loan among beef cattle farmers was IDR 23.312.000,- with

credit interest rate of 6,46%, repayment period of 24,60 month, and collateral value of IDR 35.800.000,-.

3. Average income of beef cattle farmers was IDR 4.361.611,60/2,96 head/fattening period. Meanwhile if the labour cost had not been calculated in production cost analysis, hence income would be IDR 7.608.630,41 (the calculation would be increased amount to 74,44%).
4. Based on multiple regression analysis, the variables of average total loanandcollateral value had significantly influence towards farmers' income. Moreover, type of credit scheme, financial institution, total loan, credit interest rate, repayment period were not significant influence towards farmers' income.

4.2. Recommendations:

1. It needs more efforts to improve farmers' productivity and their income through beef cattle fattening system, sush as: increase total asset, and increase collateral value.
2. The credit scheme program is still relevant to improve farmers' livelihood and need sustainable effort to improve the policy.

References

[1] Mersyah, R. 2005. Desain Sistem Budidaya Sapi Potong Berkelanjutan untuk Mendukung Pelaksanaan Otonomi Daerah di Kabupaten Bengkulu Selatan. Disertasi. Institut Pertanian Bogor, Bogor.

[2] Setiyono, B.W.H.E. Putranto, Suryahadi, T. Torahmat, dan R. Syarief. 2007. Strategi suplementasi protein ransum sapi potong berbasis jerami dan dedak padi. Jurnal Ilmu Pengetahuan dan Teknologi Peternakan. **30**(3): 207.

[3] Prasetyo, E., Sunarso, P.B. Santosa, E. Rianto, T. Ekowati, D.M. Yuwana, and B. Mulyatno. 2012[b]. The influence of 5-C factors on rate of credit return in beef cattle farming in Central Java. J. Indonesian Trop. Anim. Agric. **37**(3): 213.

[4] Prasetyo, E., Sunarso, P.B. Santosa, and E. Rianto. 2012[a]. The influence of agribusiness subsystem on beef cattle fattening farm's profit in Central Java. J. Indonesian Trop. Anim. Agric. **37** (2): 121.

[5] Daryanti, S., M. Arifin, dan Sunarso. 2002. Respon produksi Sapi PO terhadap aras pemberian konsentrat dan pakan basal jerami padi fermentasi. Prosiding Seminar Nasional Inovasi Teknologi dalam Mendukung Agribisnis. BPTP Yogyakarta – UMY Yogyakarta. Hal. 263 268.

[6] Subiharta, U. Nuschati, B. Utomo, D. Pramono, S. Prawirodigdo, T. Prasetyo, A. Musofie, Ernawati, J.Purmiyanto, dan Suharno. 2000. Sistem Usahatani Pertanian Sapi Potong di Daerah Lahan Kering. Laporan Hasil Penelitian. BPTP Provinsi Jawa Tengah, Ungaran.

[7] Soeparno dan H.L. Davies. 1987. Studies on the growth and carcass composition in Daldale Wether Lambs I: The effect of dietary energy concentration and pasture species. Aust. J. Agric. Res. **38**: 403.

[8] Prasetyo, T., S. Prawirodigdo, dan U. Nuschati. 2005. Pengaruh pola dan preparasi pakan pada penggemukan di Kecamatan Eromoko, Wonogiri. Prosiding Seminar Nasional Memacu Pembangunan Peternakan di Era Pasar Global. BPTP Jawa Tengah, Ungaran. Hal. 662 – 669.

Dairy Cows Productivity and Socio-Economic Profile of Dairy Smallholder's Communities in Yogyakarta, Indonesia

B. P. Widyobroto[1*], Rochijan[1], C. T. Noviandi[2] and A. Astuti[2]

[1]Department of Animal Production, Faculty of Animal Sciences, Universitas Gadjah Mada, Jl. Fauna No. 3, Bulaksumur, Yogyakarta 55281 – Indonesia.
[2]Department of Animal Nutrition and Feed Science, Faculty of Animal Sciences, Universitas Gadjah Mada, Jl. Fauna No. 3, Bulaksumur, Yogyakarta 55281 – Indonesia.

*Corresponding E-mail: budi_widyobroto@ugm.ac.id

Abstract

The objective of this field questionnaire survey was to describe the dairy cow productivity and socio-economic profile of dairy cattle farmers in Daerah Istimewa Yogyakarta smallholder farming communities which have been targeted dairy development policy. The study was conducted on 190 Friesian Holstein (FH) cows maintained under smallholder's management system in Daerah Istimewa Yogyakarta, Indonesia. A total of 83 farmers were randomly selected and interviewed with structured questionnaire to assess the socio-economic dairy farmer and productivity performance of dairy cows. The number of dairy productivity performance within the normal. Shortages as well as high cost of feed, occurrence of disease, scarce information about feeding and high medicament cost were the main constraints which might have contributed considerably to delayed age at first service, late age at first calving, long calving interval, short lactation length and low milk production. Therefore, strategies designed to solve the existing problem should be important by involving all stakeholders in the formulation and implementation of improvement strategiesor dairy development policy was being implemented and necessary respect to environmental factors affecting agricultural activities such as a constraint on land use and access to water resources.

1. Introduction

The Indonesia government has developed a policy aimed at the expansion of dairy production (dairy cattle, dairy goat and dairy buffalo) in order to satisfy the demand for dairy products and fresh milk in urban areas such as major cities and industrial areas in Indonesia, and alleviate poverty amongst dairy smallholder's communities. As a result, the number of dairy cattle has increased in recent years, example in 2015 the dairy population increased by 3.2% (0.52 million heads) [1], but the likelihood of further increases would depend on whether it can be demonstrated that farming smallholder's do benefit economically from this development.

Daerah Istimewa Yogyakarta, one of the provinces in Indonesia that is used to development dairy cow population. The population of dairy cattle in DIY is the fourth largest after the provinces of Central Java, West Java and East Java with DIY population in 2016 is 4,066 head [1]. Dairy farmers in DIY are mostly categorized as dairy farmer group and independent. The condition of the farm is in the highlands and low, most of which are incorporated in the dairy cooperative. The carrying capacity of the government is also enough to change and assist the farmers in managing their livestock, unfavorable environmental conditions, unstable livestock productivity and socio-economic conditions of unequal breeders, became a separate object for research.

The objective of the current study was to profile smallholder farming communities in which on the Daerah Istimewa Yogyakarta Province (DIY) dairy development policy was being implemented, and know the condition of dairy cattle productivity that exist in farmer. Particular emphasis was placed on condition of dairy cattle productivity and socio-economic characteristics dairy farmer.

2. Materials and Methods

1. Respondent and location of study area

A total of 83 farmers were interviewed randomly with scheduled questionnaire which was mainly based on the socio-economic dairy farmer and productivity performance of dairy cows. Atotal of 190 Friesian Holstein (FH) cows were included in this study were maintained under smallholder'smanagementsystem,locatedinlocated in Sleman Regency, Bantul Regency and Yogyakarta City, Daerah Istimewa Yogyakarta, Indonesia.

2. Questionnaire and interviews

The questionnaire was developed in accordance with the objectives of the study and designed in a simple manner to get accurate information from the dairy owners. Each respondent was given a brief description about the nature and purpose of the study and the responses were recorded directly on the survey schedule. Major sections of this questionnaire study were on characteristics respondent and characteristics of the dairy cattle population, basic management practices, knowledge and family support received in dairy farming production. Data on reproductive performance such as calving intervals, abortions and calf survival were obtained from farmers through careful questioning about past reproductive history and occurrence of calf losses

2.1. Data analysis

AllcollecteddatawereenteredintoMicrosoftExcel2013anddescriptive statistics such as mean, frequency and percentage were used to analyze the data using SPSS softwareforstatisticalanalysis(version16.0).

3. Results and Discussion

3.1. Socio-economic characteristic of respondent

The general characteristics associated with household respondents were distributed by sex, age, marital status, experience in dairy keeping and educational status was presented in Table 1. From the total interviewed respondents (N = 83), the majority of the (69.90%) respondents were male while the remaining (30.10%) were female. The majority age of the respondents in the study area ranges between 36–55 years (60.53%). This result showed that people in the most productive age are actively engaged in dairy activities. Of the total households interviewed, 98.79% are married. Concerning to level of education, the highest percentage (34.72%) of the respondents had diploma Elementary school, 30.56% had diploma Junior high school, 25% had diploma Senior high school, 4.17% had diploma university/college and above 5.56% of those respondents had not attended any formal or informal education. A 42.47% of the respondents had 11–20 years experiences in dairy keeping.

Table 1. Socio-economiccharacteristicsofrespondent

Parameters	∑ Person	Percentage (%)
Age (years)		
≤ 15	0	0
16 – 25	2	2.63
26 – 35	6	7.89
36 – 45	19	25.00
46 – 55	27	35.53
56 – 65	13	17.11
≥ 65	9	11.84
Formal education		
No schooling	4	5.56
Elementary school	25	34.72
Junior high school	22	30.56
Senior high school	18	25.00
College/University	3	4.17
Experience raising dairy cattle (year)		
≤ 10	26	35.62
11 – 20	31	42.47
≥ 21	16	21.92
Average	15.92±10.14	
Farming purpose / priority in order		
Priority 1	Milk production	
Priority 2	Save	
Priority 3	Children's education	
Priority 4	Companion animal	
Priority 5	Fertilizer	
Number of family member (person)		
≤ 2	18	24.32
3 – 4	40	54.05
5 – 6	14	18.92
≥ 7	2	2.70
Average	3.57±1.28	

AI = Artificial insemination

IOP Conf. Series: Earth and Environmental Science **119** (2018) 012060 doi:10.1088/1755-1315/119/1/012060

The purpose of raising dairy cattle by farmers (Table 1) prioritized on milk production (priority 1), savings (priority 2) and cost of children education (priority 3). The purpose other than maintenance is for companion animals or for pleasure (hobby) priority 4, as well as manure as fertilizer priority 5.

3.2. Description of study population or data ownership of cattle

On average, the household heads were 46 years old (range 15–65 years old) and had to look after five household members (range 3–4 people), including themselves. Not all of the households owned pasture land and most of the forage fed to dairy cattle came from pasture land is irrigated, native pasture cut along roadsides and river banks. Average cattle ownership in DIY is 4.03±1.97 head equal to 2.83±1.32 AU (animal unit). Conditions of livestock ownership is low partly caused by lack of capital and difficulty finding forage due to limited land for fodder crops to supply the quantity and quality of feed is also limited, which is also influenced by the availability of water in the dry season. This situation led to dairy farming has not been efficient [2]. The percent ownership of dairy cattle was presented in Table 2.

Table 2. Data ownership of dairy cattle

Animal Status	∑ Animal Head per Respondent
Cows	
lactation	1.72±0.83
non-lactation (dry)	1.32±0.48
Heifer	1.49±0.74
Bull	1.20±0.63
Calf	1.52±0.69
Average cattle ownership	4.03±1.97
Average cattle ownership (AU)	2.83±1.32

AU = Animal unit.

3.3. Productivity performance of dairy cows

The mean milk yield in the present study was found to be 11.77±3.47 L per cow days, respectively (Table 2). The mean milk yield in the present study is in accordance with the findings of Kebede [3] and who reported that the average daily milk production of crossbred cows was 8 L per cow days. In this study, the main reasons for low daily milk production as indicated by the respondents were shortage of feed and the interaction of poor health and management. Variation in milk production is mainly due to genetic and various non-genetic causes [4].

Success of reproduction performance is mainly determined by days open, the longer days open means the longer calving interval which affects the lowering amount of milk production throughout the cow life [5]. Reproductive performance of dairy cattle at the smallholder vary greatly, it is possible for a lot of factors that don't support the maintenance management of dairy cattle, especially the fulfillment of feed and environmental needs. The results overall of reproduction performance was presented in Table 3.

Table 3. Productivity performance of dairy cows in DIY

Parameters	Value
Milk yield (L/head/day)	11.77±3.47
Breeding system	AI
Postpartum mating (day)	72.18±31.51
Service per conception	2.88±1.68
Days open (day)	114.76±45.13
Calving interval (month)	15.61±6.38

AI = Artificial insemination

The mean postpartum mating in this research of 72.18±31.51 days. The result are within the optimum (idea) value range of 60-90 days [6] this could be attributed to different factors. Most researchers suggest that the reason for the delay in interval to first service is greater negative energy balance in modern dairy cows. Negative energy balance delays the continuation of ovarian activity. Then the value of open days, service per conception and calving interval is still within normal range.

4. Conclusion

Overall, productivity performance dairy cow was below the standard. However, the number of service per conception is within the normal that expected from commercial dairy herd. Shortages as well as high cost of feed, occurrence of disease, scarce information about feeding and high medicament cost were the main constraints which might have contributed considerably to delayed age at first service, late age at first calving, long calving interval, short lactation length and low milk production. Therefore, strategies designed to solve the existing problem should be important by involving all stakeholders in the formulation and implementation of improvement strategies or dairy development policy was being implemented and necessary respect to environmental factors affecting agricultural activities such as a constraint on land use and access to water resources.

Acknowledgments

This study funded by the Directorate of Research and Community Development, The Ministry of Research, Technology and Higher Education, Republic of Indonesia (through PUPT Grant 2017) and Gadjah Mada University.

References

[1] Direktorat Jenderal Peternakan dan Kesehatan Hewan 2016 Statistika Peternakan dan Kesehatan Hewan (Livestock and Animal Health Statistic) 2016. Kementerian Pertanian RI pp 71–76

[2] Priyanti A,Nurtini S and Firman A 2009 Analisis Ekonomi dan Aspek Sosial Usaha Sapi Perah.Profil Usaha Peternakan di Indonesia. Pusat Penelitian Pengembangan Peternakan. Badan Penelitian dan Pengembangan Pertanian. LIPI Press. Jakarta

[3] Kebede A B 2009 Characterization of milk production systems,marketing and on-farm evaluation of the effect of feed supplementation on milk yield and milk composition of cows at Buredistrict, Ethiopia.M.Sc.Thesis,Bahir Dar University

[4] Kumar N, Yemane A,Berihu GandDesalew T.T2014Production performance of dairy cows underfarmer's management in and around Mekelle, Ethiopia.Global Vet.**12**:207.

[5] Rochijan, Widyobroto B.P and Ismaya 2016 The impact of high rumen undegraded protein (HRUP) supplementation to blood urea nitrogen and reproduction performance in early lactation dairy cows. Int. J. Dairy Sci. **11**: 28.

[6] Berry D.P, Buckley F, Dillon P, Evans R.D, Rath M andVeerkamp R.F 2003 Genetic relationship among body condition score, body weight, milk yield and fertility in dairy cows. J.Dairy Sci. **86**: 2193.

Determinant of Household Business Scale of Moa Buffaloes at Moa Island Southwest Maluku Regency

J. Lainsamputty[1], W. Roessali[2], S. I. Santosa[2] and B. T. Eddy[2]

[1]Faculty of Agricultural Sciences, Pattimura University, Ambon, Maluku Province, Indonesia
[2]Faculty of Animal and Agricultural Sciences, Diponegoro University, Semarang Central Java, Indonesia

Corresponding E-mail : wroessali@gmail.com

Abstract

The objective of this research was to analyze factors that affect the business scale of Moa buffaloes at Moa Island, Regency of Southwest Maluku. The research used a survey method with multistage random sampling. The location chosen was the District of Moa Lakor at Moa Island based on its largest buffalo population. Respondents were randomly drawn in a total of 120 respondents. The variables measured were years of experience in rearing animals, costs of production, farmer's participation in group activities, animal housing systems, farmer's income and farmer's age. The statistical test used was the multiple linear regressions. The results showed that the mean of business scale in the area of high density of buffaloes population was 12.6 AU, in the moderate was 12.4 AU and in the low was 11.0 AU. The average of production cost was IDR 1.893.536.00/year, the average of revenue was IDR14.083.333.00/year, while the average of income was IDR 12.189.797.00/year. The independent variables simultaneously influence the business scale ($P<0.01$). Partially, experience, costs of production, participation and housing systems had highly significant influences on the business scale ($P<0.01$).

1. Introduction

Economic growth and urbanization has led to the changes in consumption patterns where the consumption of livestock products in the form of meat, eggs and milk increased. Hypothetically, the increase in social welfare will be followed by the increase in consumption of livestock products, which in turn support the economic progress in animal husbandry. Consumption of livestock products (especially meat) in Indonesia in 2008-2012 increased with growth of 1.06 per cent [1]. There are 3 uses of buffaloes, namely dairy, meat producers and animal labor [2]. Buffalo rearing has become an integral part of farms systems in Asia over 5,000 years in producing dairy, meat and skin [3].

Southwest Maluku as one of the district in Maluku Province has some livestock commodities that can be categorized as local high yield animal, one of them is buffaloes, mostly located on the island of Moa. Moa buffalo has been established as one of the original buffalo Indonesia based on the Ministry of Agriculture statutes No.2911 of 2011. The population of Moa buffalo in Southwest Maluku District is 14,009 [4]. This is a potential local resources owned by the region, and is a force in the development of the local economy.

Moa buffalo has a uniformity in physical form and genetic composition, that different with other local buffaloes. Moa buffalo also has a good adaptability to the extreme environment. This type is a wealth of local animal genetic resources of Indonesia which need to be protected and developed [1]. The body color of Moa buffaloes varies greatly from black (43.67%), gray (8.74%), white

(15.52%) and a mixture of those three colors 12.07% [5].Buffalo is a component in rural farming because it is the part of rural incomes by exploiting natural resources available in the surroundings [6].

The Moa buffalo small scale enterprises development can be approached by developing the existing businesses with regard with the potential of natural resources included climatic conditions and geographic region, land use and availability of forage. The business scale in the form of the number of buffaloes that the farmers raised is an important element in small scale business. The factors that assumed to have influence the farmers in deciding the number of buffalo they reared, are the experience in raising animals, costs of production, farmer's participation in group activities, animal housing systems, farmer's income and farmer's age. The buffalo business development should be conducted for the benefit of the regional development in order to meet the needs of the demand for meat, the increase in income and welfare of the farmers.

2. Materials And Methods

2.1. *Location and Sampling*

The research was using a survey method with multistage random sampling. The location chosen purposively was the District of Moa Lakor at Moa Island based on its largest buffalo population. The second phase was to determine the villages using stratified random sampling, by dividing Moa District into three strata of farmer population. The selected villages were Klis (as high population), Pati (as moderate population) and Werwaru (as low population). Respondents were randomly drawn in a total of 120 respondents. The determination of the number of respondents in each village using Slovin formula as cited by [7], as follows:

$$n = \frac{N}{1+N\,(e)^2} \quad \text{...(1)}$$

where:
n = sample size
N = Population size
e^2 = critical value)
The used of the formula in each village resulted:

Klis (high population): $n = \frac{85}{1+85\,(0{,}05)^2} = 70$ respondents

Pati (moderate): $n = \frac{39}{1+39\,(0{,}05)^2} = 36$ respondents

Werwaru (low population): $n = \frac{14}{1+14\,(0{,}05)^2} = 14$ respondents

The criteria of the samples were:1). actively raising buffalo at the time of the study, 2). have experience in rearing buffalo more than 2 years, 3). have conducted livestock marketing activities.

2.2. ***Data Collection***

The data collected in this study consisted primary and secondary data. Primary data obtained from thefield survey and questionnaire-based interviews. Secondary data were obtained from relevantagencies such as the office of the Central Bureau of Statistics, officers of Animal Husbandry and Health headquarter, reports and results of previous studies.

IOP Conf. Series: Earth and Environmental Science **119** (2018) 012061 doi:10.1088/1755-1315/119/1/012061

2.3. Analysis

Buffalo farm income was calculated based on the difference between the gross farm income and farm expenses (cost). The mathematical formula refer to Soekartawi [2002] were:

TI = TR – TC.. (2)

Where:

TI =Total Income

TR = Total Revenue

TC = Total Cost

The relationship between factors that affect the business scale of Moa buffalo were analyzed by multiple linear regressions using SPSS version 15 [8;9].As the dependent variable was the scale of business (AU), while the indepent variables were raising experiences (years), cost of production (IDR), Total Income (IDR), farmer's age (years), farmer's participation in group activities (dummy) and animal housing system (dummy).

$Y = \beta 0+\beta_1X_1+ \beta_2X_2+ \beta_3X_3+ \beta_4X_4+ \beta_5D_1+ \beta_6D_2+ e$... (3)

Where:

Y = business scale (Animal Unit, AU)

$\beta 0$= Parameter intercept

$\beta_1,\beta_2,\beta_3,...\beta_6$ = Regresion coefficient for factors studied

X_1 = Experience in raising buffalo (years)

X_2 = Cost of production (IDR)

X_3 = Income (IDR)

X_4 = Farmer's age (years)

D_1 = Farmer's participation in group activities,

Where 0 = not participated

1 = participated

D_2 = Animal housing system,

Where 0 = communal housing

1 = individual housing

3. Results and Discussion

3.1. Respondent's characteristics

A total of 73 farmers (60.83%) were in the productive age, ranged between 21-50 years. At these ages, farmers are likely to be able to expand its business in a better manner. Most farmers have a low level in formal education, 57.50 percent were only passed the elementary school. This condition can be an obstacle in the process of adoption of innovation in order to improve farming techniques. The efforts to improve the farmers' knowledge and skills can be approached through training and other non-formal education.

There were 21 respondents (17 percent) who attended extension education organyzed by the farmer' group. This condition is regrettable, given the extension can increase knowledge and skills of the farmers. As [10] found in his study in Enrekang Regency, farmers who appreciate the extension program on improvement of human resources has the advantage of their farming more than others. Farmers who attended the counseling had higher incomes than those who are inactive. The earned income was IDR 515,909.00 per head per month to the farmers who maintain 1-2 buffaloes, IDR. 977,385.00 per head for the farmers who raise 3-4 buffaloes. A total of 70.83% respondents had the main livelihood as crops farmers, 22.50% as animal raisers, 3.33% as as civil servants, 2.5% as entrepreneurs and 0.83% as construction laborers. A total of 80.83% farmers have experience morethan 10 years in raising animal, long enough to be able to learn to improve their buffaloesmanagement.

Table1. Frequency of Farmers Based on some Characteristics

Characeristics	Frequency (people)	Percent (%)
Ages		
21-35	24	20.00
36-50	49	40.83
51-65	17	14.17
Education		
Not passed the elementary	6	5.00
Elementary school	69	57.50
Junior high school	36	30.00
High school	8	6.67
Bachelor degree	1	0.83
Main livelihood		
Crops farmers	85	70.83
Animal raisers	27	22.50
Civil servants	4	3.33
Small business	3	2.50
Construction workers	1	0.85
Experiences in raising animal (years)		
5-9	23	19.17
10-14	20	16.67
15-19	17	14.47
20-24	13	10.80
25-29	14	11.67
>29	33	27.50

Source : Primary data (2015)

3.2. Business Scale

The average number of buffaloes that raised by breeders in Moa Island was 12.4 Animal Unit (AU, equivalent to ± 12 adult buffaloes), which when it was broken down per region will be

12.6 AU on highly population, 12.4 AU in moderate population and 11.0 AU on the sparse population of the farmers.

Table2. The rate of number of buffalo kept per family at Moa Island

Category	Male	Female	Total
	---------- (Animal Unit/AU) ----------		
Calves	0.28	0.36	0.64
Young	1.08	1.49	2.57
Adult	2.68	6.51	9.19
Total	4.04	8.36	12.4

The buffaloes that they kept generally comes from the legacy of parents (84.2%) and bought their own from local market (15.8%). Polhaupessy [11] found that 93.98% of farmers at Moa Island had their buffaloes from legacy, 4.9% from their own buying and 1.09% from exchange among them. A similar situation occurred in Mamasa Island, where more than 50% of the spotted buffaloes had their buffaloes from heritage [12].

3.3. Analysis of Factors Affecting Business Scale of Buffalo Rearing

Multiple linear regressions were used to analyze the factors affecting the business scale of buffalo rearing. The result of F-test showed that the independent variables (experience in raising buffalo, cost of production, income, farmer's age, farmers' participation in group activities and animal housing system) simultaneously had very significant influence ($P \leq 0,01$) to the scale of business. Tabel 4showed that the value of R = 0731, meaning that 73.1% of the business scale of buffalo raising were influenced by variables in the model, while the rest of 26.9% were explained by other factors not included in the regression model.

3.4. Years of experience in rearing buffalo

The farmers' experience in raising buffalo had a highly significant influence on the business scale. Regression coefficient of 0.288 indicated that every increase of 1 year of experience in raising buffalo will increase the scale of business 2.88%. As stated by [13], that the longer the experience of raising livestock, the farmers will be more skilled in dealing with issues of reproduction, production and disease prevention, so farmers can increase the number of animals that they raised. Hernanto [14] stated that the duration of doing business will increase the knowledge the farmers gained from seeing and socializing with other group members.

3.5. Production Cost

Production cost consisted of fixed cost and variable cost. Fixed cost paid by farmers included depreciation of equipment and animal housing. Variable costs covered the cost of breeds and medicines. The average cost of production of farmers was IDR. 1.893.536,00/year (from fix cost of IDR 665.201,00 and variable cost IDR 1.228.335,00 (Table 3). The rate of production cost per region was IDR 1.917.250,00-/year at Klis Village, IDR 1.860.432,00 at Pati Village and IDR 1.902.925,00 at Werwaru Village.

Table 3. The Rate of Revenue, Production Cost and Income of Buffalo Farmers at Moa Island

Variables	Region						The Rate/ Farmer
	Klis		Pati		Werwaru		
	(IDR)	(%)	(IDR)	(%)	(IDR)	(%)	(IDR)
A. Revenue:							
Sales of calves	28.571	0.20	0	0.00	0	0.00	16.667
Sales of young bufalloes	1.628.571	11.21	1.500.000	11.25	1.500.000	10.88	1.575.000
Sales of adult bufalloes	12.871.429	88.59	11.833.333	88.75	12.285.714	89.12	12.491.667
Total revenue	14.528.571	100.00	13.333.333	100.00	13.785.714	100.00	14.083.333
Production Cost							
B. Fixed Cost:							
Housing depreciation	741.989	38.70	568.548	30.56	571.482	30.03	627.340
Equipment depreciation	37.877	1.98	38.550	2.07	37.157	1.95	37.861
Total fixed cost	779.866		607.098		608.639		665.201
C. Variable cost:							
Purchasing of breeds	326.812	17.05	433.333	23.29	514.286	27.03	358.144
Purchasing of medicines	519.143	27.08	536.667	28.85	508.571	26.73	588.127
Purchasing of Naker	291.429	15.20	283.333	15.23	271.429	14.26	282.064
Total variable cost	1.137.384		1.253.333		1.294.286		1.228.335
D. Total production cost B+C)	1.917.250	100.00	1.860.431	100.00	1.902.925	100.00	1.893.536
Income (A-D)	12.611.321		11.472.901		11.882.789		12.189.797

The cost of production has a very significant influence on the business scale of buffaloes (P≤0,01). Regression coefficient of 0.138 showed that any increase in cost of production by 10% will increase business scale by 1.38%. The farmers in Moa Island just spent the cost to build animal housing, equipment and medications, otherwise farmers did not spend the cost of animal feed, because cattle released to feed themselves in the forest. This is not in line with those found by [15] which stated that the biggest component of production cost was feed.

The biggest cost came from depreciation of animal housing (33,25%). This can occur because the cost of building animal housing in Moa Island is relatively expensive due to the social cost that must be paid in building animal housing in accordance with the prevailing tradition in the region. The cost of production in the village of Klis was greater than in other regions, because the average size of the animal housing in this village was bigger. However, different test independent samples (Table 5) showed that the cost of production has no significant influence on the business scale.

Table 4. Factors Influencing the Business Scale of Moa Bufalloes at Moa Island

Variable	Regression Coeficient	Significance Value
Constants		0.000
Experience (X_1)	0.228	0.008**
Production cost (X_2)	0.138	0.010**
Income (X_3)	0.075	0.153
Farmers' age (X_4)	0.012	0.876
Farmers' participation (D_1)	0.519	0.000**
Housing system (D_2)	-0.278	0.000**
F counted 51.087	0.000	
R Square (R^2) = 0.731		

*($P \leq 0.01$)

Table 5. T-test Amongst The Strata of Location

Grup	Value of Significance				
	Business scale	Income	Cost	Participation	Housing
Klis – Pati	0.229	0.820	0.796	0.123	0.093
Klis - Werwaru	0.515	0.195	0.105	0.701	0.375
Pati - Werwaru	0.198	0.341	0.097	0.409	0.785

3.6. Income

The average revenue of the bufallo farmers was IDR 14.083.333,00/year (Table 3). The revenue was varied, influenced by the age and buffalo body shape and also the region. Each region has different revenue, at Klis Village the average revenue was IDR 14.528.571,00/year, while at Pati and Werwaru Villages respectivelly was IDR 13.333.333,00/year and IDR 13.785.714,00/year.

Income is the difference between revenue and expenditures for the maintenance of buffaloes for 1 year. Profit occurs when the amount of revenue is greater than the total cost. The average income earned by farmers in Moa Island Southwest Maluku district is IDR 12.189.797.00/year or IDR 1.015.816,4/month. Broken down by the region, then the farmers at village of Klis (high density of farmers population) had an average income of IDR 12.611.321.00/year village Pati (moderate population) of IDR 11.472.901,00/year and villages Werwaru (region less) of IDR11.882.789,00/year. In contrast to the results conducted by [16] in the district of Gunung Sindur Bogor, which found that the average net income for farmers who raise buffalo 3.27 heads was IDR 8.19 million/year, the income of farmers in Moa Island were far bigger. Based on the significance testing of T-test (Table 5), in which the significance value of Klis – Pati, Klis – Werwaru and Pati – Werwaru villages respectively was 0.820, 0.195 and 0.341, the value of income among strata had no significant differences.

3.7. Farmer's participation in group activities

Partially, the participation of farmers in group activities had significant influence on the scale of business (P≤0.01). Business scale of farmers who were not involved was51.9%higher than the scale of business of farmers who were participated in the group activities. Most of farmers who had participation in the group were the farmers who received facilities in some forms of governmentscheme. The form of participation was shown by their presence in the group meetings. There were no significant differences in participation among villages. The research should be developed by linking participation to the number of family members, because [17], found that the number of family members have significant influence on the number of animals reared.

3.8. Animal Housing System

Housing system has a highly significant influence on the business scale of buffalo farming (P≤0,01). Signs of expectation were negative, means that farmers with individual housing tend to have smaller scale businesses from farmers of communal housing system. According to Muis (2015 [18], there were several advantages of using the communal housing, because communal housing is more environmentally friendly, there is always communication between farmers, make it easier for extension workers to coordinate their program. Hermawan *et al.* (2011) [19] also stated that the communal housingencourage farmers to increase the scale of their business. T-test (Table 5) stated that there were no significant differences amongst the strata of location.

4. Conclusion

The average number of buffaloes that raised by farmers in Moa Island was 12.4 Animal Unit. The average of revenue in buffalo farming was IDR 14.083.333.00/year; cost of production was IDR 1.893.536.00/year so the average of income was IDR 12.189.797.00/year or IDR 1.015.816.00/month. Variables of years of experience in rearing buffalo, cost of production, farmers' participation in group activities and animal housing system had a significant effect on business scale.

References

[1] Directorate General of Animal Husbandry and Health. 2013. Kerbau Moa. Rumpun Kerbau Indonesia dengan Sebaran Asli. Geografis di Kabupaten Maluku Barat Daya Provinsi Maluku. Keputusan Mentri No. 2911/KPTS/OT. 140/6/2011. Kementrian Pertanian Republik Indonesia_ ditjennak@pertaniango.id . Jakarta.

[2] Pasha, T. N. and Z. Hayat. 2012. Present situation and future perspective of buffalo production in Asia. J. Anim. Plant Sci. **22**: 250.

[3] Nanda, A. S., and T. Nakao. 2003. Role of buffalo in the socioeconomic development of rural Asia: Curren status and future prospectus. J. Anim. Sci. **74**: 443.

[4] CBS Southwest Maluku District. 2012. Kabupaten Maluku Barat Daya Dalam Angka. Central Bureau of Statistics of Southwest Maluku Regency

[5] Salamena J. F. and B. J. Papilaja. 2010. Characterization and genetic relationships analysis of buffalo population in Moa Island of South-West Maluku Regency of Maluku Province. J.Indonesian Trop.Anim.Agric. **35** (2): 75.

[6] Kusnadi, U. 2004. Kontribusi Ternak dalam Meningkatkan Pendapatan Petani di Lahan Marginal Kabupaten Tangerang, Propinsi Banten.J. Pembangunan Peternakan Tropis SpecialEdition Oktober 2004. Seminar Nasional Ruminansia buku 3.

[7] Sangadji, M.S. and Sopiah, 2010. Metodologi Penelitian. Penerbit Andi. Yogyakarta.

[8] Ghozali, I. 2001. Aplikasi Analisis Multivariate dengan Program SPSS. Edisi II. Badan Penerbit Universitas Diponegoro. Semarang

[9] Sugiono. 2002. Metode Penelitian Bisnis. CV. Alfabeta. Bandung

[10] Baba, S. 2012. Penerapan participatory technology development (PTD) untuk meningkatkan kapasitas petani dan adopsi teknologi terpadu jagung dan ternak sapi di Sulawesi Selatan.

[11] Polhaupessy U. 2004. Pengembangan Plasma Nutfa Kerbau Moa. Kerja sama Dinas Peternakan dan Jurusan Peternakan Fakultas Pertanian Universitas Pattimura. Ambon.

[12] Rompis J.E.G., J.F. Paat, M. M. Kawatu and Demmalona. 2013. Tatalaksana Pemeliharaan Ternak Kerbau Belang Di Kecamatan Mamasa Kabupaten Mamasa Provinsi Sulawesi Barat. Jurnal Zootek. **33** (1): 68.

[13] Rahayu R. S., W. Roessali, A. Setiadi and Mukson. 2014. Kontribusi usaha sapi perah terhadap pendapatankeluarga peternak di Kecamatan GetasanKabupaten Semarang. Agriekonomika. **3** (1): 45-54

[14] Hernanto, F. 1996. Ilmu Usaha Tani. Penebar Swadaya. Jakarta

[15] Saputro, K. C., T. Ekowati and E. Rianto. 2002. Analisis Break Event Point Usaha Ternak Kerbau pada Anggota Kelompok Tani Ternak di Kabupaten Pemalang. J. Trop. Anim. Sci. **27**(4): 23-29

[16] Rusdiana S., I.G.A.P. Mahendri and C. Talib. 2011.Pendapatan Usaha Ternak Kerbau di Kecamatan Gunung Sindur Kabupaten Bogor. Seminar dan Lokakarya Nasional Kerbau. 21-22 June. Samarinda

[17] Roessali W., Masyhuri, S. Nurtini, D.H. Darwanto. 2011. Factors influencing farmers' decision to increase beef cattlebusiness scale in Central Java Province. J.Indonesian Trop.Anim.Agric. **36**(1):27.

[18] Muis, J.M. 2015. Performance and prospects for cattle farm business friendly environment in West Sumatra. Widyariset, **18** (1): 59.

[19 Hermawan A., Subiharta and B. Utomo. 2011. Masalah ketidakberlanjutan kandang komunal dalam pengembangan ternak sapi di Jawa. National Seminar on Animal Husbandry and Veterinary Technology, 7-8 June. Bogor

Study of Goat Herdering System in Sawohan Village, Buduran District, Sidoarjo

M. Nasich[1*], Suyadi, G. Ciptadi[1] dan A. Budiarto[1]

[1]Faculty of Animal Husbandry, Brawijaya University, Malang, East Java, Indonesia

*Corresponding E-mail: nasich@ub.ac.id; m.nasich@yahoo.com

Abstract

The aim of this research was to evaluate and analyze goat herdering system in group of goat "Oro-oro" in Sawohan Village, Buduran District, Sidoarjo. This research was conducted by case study method, using 5 members of goat farmers "Oro-oro" with 298 goats. The results of this study indicate that the average number of goat ownership was more than 50 head/farmer. The goat breed that raising by the farmers was Kacang goats. Average daily gains was 53.32 + 36.01 gram, litter size 1.34 + 0.51 head, mean of kidding interval 7,41 + 0,99 month and mortality of kid prenatal period was 20,4%. From this research can be concluded that productivity of goat in Sawohan village was good with herdering system, so it can be used as one of alternative in developing goat farming system in other area.

1. Introduction

Goat is a popular livestock for small farmers, because it is easy to maintain, small capital required and can be used as a savings [1]. The population of goat in East Java is the second largets after Central Java (3.178.197 head in 2015). There was increasing population in 2014 by 1.5% [2]. Of the total population, it is based on smallholder farming system.

Kacang goats are one of the native Indonesian goat breeds. According to [3], it has several advantages, for example produced of meat and skin, it is prolific or high litter size and has the ability to adapt on poor feed conditions.

The "Oro-oro" goat breeding group is one of a farmers group which speciality in goats farming system, including improving feeding management. The goat are grazed on the edge of the pond more than 1700 tails. Patterns of maintenance like this does not require a lot of cost for the purchase of feed, or it can be said to be zero. Every day the breeders only need to open the door of the cage and close it in the afternoon.

Looking at this kind of fact, it is very interesting to observe on how to maintenancemanagement and productivity.This study aims to examine and evaluate goat farming business from "Oro-oro" farming group in Sawohan Village, Buduran District, Sidoarjo Regency. The results of this study are expected to be used as information in the development of goats in other areas

2. Materials and Methods

The research was conducted in Sawohan Village, Buduran District, Sidoarjo Regency from early December 2016 until the end of January 2017. The material used in this study was 146 does and 196 kids, possession to 5 farmers

Case Study method was used in this research. The variables observed were litter size, mortalityat pre-weaning period, kidding interval and body weight gain at pre-weaning period. The data obtained were analyzed descriptively based on the formula [4]:

$$\chi = \frac{\Sigma Xi}{n} \qquad S = \sqrt{\frac{\Sigma(Xi-X)2}{n-1}}$$

Description: S = Standard Deviation
ΣXi = Total data

Daily Weight Gain (PBBH) is calculated using the formula[5] as follows:

$$PBBH = \frac{W2 \quad W1}{T2 \quad T1}$$

Description: PBBH = Daily Weight Gain
W2 = End body weight
W1 = Initial body weight
T2 = end time weighing
T1 = Start time weighing

Chi-square analysis (X2) as directed [6], was used to evaluate the incidence of male and female births.

3. Results and Discussion

From the results of this study, the "Oro-oro" located in Sawohan Village, Buduran District, Sidoarjo Regency, East Java. Sawohan Village data. The location had altitude 4 m above sea level (dpl), with average annual rainfall of 2000 mm and average temperature and humidity were 30^0C and 79%. This village located in the southern city of Sidoarjo, 15 km away from the city center. In particular the "Oro-oro" goat breeding group is located in aquaculture area, where the goats are kept released and feeding on the edge of the pond.The Kacang goat profile kept by goat group "Oro-oro" can be seen in Figure 1.

Figure1.Kacang goat performance belongs to one member of the goat group "Oro-oro"

From Figure 1, It's can be seen that the Kacang goat in Sawohan village are some semilar from the Kacang goat which is conveyed by Devendra and Burn. In Sawohan the ear rather drop (not erect).The average body weight of adults (PI1 and PI2) are 22.14 ± 4.51 kg for female and 26.00 ± 3.24 kg for males, the average litter size was lower at 1.34 ± 0 , 51 heads[8]

The average number and structure of population and number of goats per group member can be seen in Table 1.

Table 1. Population structure and number of livestock ownership of 5 breeders of "Oro-oro" breeders group

Farmer	Pre-weaning Kids		Post-weaning		Adult		Total (heads)
	Male	Female	Male	Female	Male	Female	
Amang	1	4	27	7	3	5	47
An'am	11	17	7	8	5	21	69
Majid	13	22	14	17	3	27	96
Mat Jaini	7	8	5	2	2	10	34
Kurniawan	5	10	8	11	4	14	52
Total	37	61	61	45	17	77	298
			Average				59.6

From Table 1shows that the average number of livestock ownership of each group member is above 50 head/goats. The average amount of ownership is considered very high when compared with the results of research by [9]. It stated that the amount of goat ownership between 3-16 heads/goat with an average of 9.2 head/goat. The ownership of goats in Sawohan, because the farmers do not need to find the feed This way of maintenance is certainly very easy for farmers and does not require a long time to find the feed resources.

The average daily weight gain of Goat Breeders of the pre-weaning period from the results of this study is presented in Table 2.

Table 2. Average daily body weight gain of the male and female kids at pre-weaning period

Sex	N (heads)	PBBH (g/head/day)
Male	54	64.87 ± 40.93
Female	31	41.77 ±29.10
Average	85	53.32 ± 36.01

From Table 2 above, it is seen that the average of daily body weight gain was 53.32 + 36.01 g/head/goat.This average PBBH was lower when it is compared with the results of the research from [10].Itwas equal to 63,1 g /head. It was caused by the additional of feed concentrate, in addition to forage. This is in accordance with the opinion of [11].It states that the factors influencing the increase of livestock body weight are the parent age, breeding season, the type of birth and the availability or adequacy of feed. The average of PBBH results of this study was still higher, when compared with the results of research by [12] which is only 26.28 g / head.

While daily body weights gain of males wee higher than females. This seems to be due to differences in birth weight and the resulting of hormonal. This situation is similar to the previous research results of [12] and [13], which showed that post-weaned goats' PBBH showed that males were higher than females.

4. Conclusion

From the results of this study, it can be concluded:

1. Shepherding system can increase the ability of breeder to keep goat to more than 50 goats/farmers.
2. Kacang goats was the goat breed that was raised in the breeding group of "Oro-oro".
3. The average daily gain is also quite good that is 53.32 $\pm$ 36.01 g/head/day

Acknowledgments

Goat breeding systems such as Sawohan Village can be used as a goat breeding model.

References

[1] Setiawan T dan A Tanisius 2003 Beternak Kambing Perah Peranakan Etawah Edisi 1 Penebar Swadaya, Jakarta

[2] Anonimous 2016 Capra hircus Domestic Goat. http://animaldiversity.org. Diakses 15 Nopember 2016

[3] Batubara A, RR Noor, A Farajallah, B Tiesnamurti dan M Doloksaribu 2011 Morphometric and Phylogenic Analysis of Six Population Indonesia Local Goats. Media Peternakan **34** (3) : 165.

[4] Gulo,W. 2000. Metodologi Penelitian. Grasindo. Jakarta.

[5] Edey TN 1983 Tropical Sheep and Goat Production. Australia University International, Development Program. Camberra.

[6] Hadi, S. 1995. Metodologi Research: untuk penulisan paper, Skripsi, Thesis dan Disertasi. Universitas Gajah Mada. Yogyakarta.

[7] Devendra C. and M. Burn. 1994. Produksi Kambing Di Daerah Tropis.Terjemahan IDK Haryaputra. Penerbit ITB Bandung.

[8] Dzikri, M.H. 2017. Performan Reproduksi Induk Kambing Kacang di Desa Sawohan, Kecamatan Buduran, Sidoarjo. Skripsi Fakultas Peternakan, Universitas Brawijaya, Malang.

[9] Makatita, J. 2013. Hubungan antara Karakteristik Peternak dengan Skala Usaha Peternakan Kambing di Kecamatan Leihitu, Kabupaten Maluku Tengah.Agrinimal, **3** (2): 78-83.

[10] Martawidjaja, M., B. Setiadi dan S.S. Sitorus. 1999. Pengaruh Tingkat Protein Energi Ransum terhadap Kinerja Produksi Kambing Kacang Muda. Jurnal Ilmu Ternak dan Veteriner. **4** (3): 167-173.

[11] Setiadi, B. 1996. Pertumbuhan, Perkembangan dan Komposisi Karkas Kambing. Balai Penelitian Ternak. Wartazoa, **5** (1): 12.

[12] Mahmilia, F., F.A. Pamungkas dan M. Doloksaribu. 2007. Laju Pertumbuhan Pra-sapih dan Sapih kambing Boer, Kacang dan Boerka-1.Seminar Nasional Teknologi Peternakan dan Veteriner. Loka Penelitian Kambing Potong.

[13] Adriani, 2014. Bobot Lahir dan Pertumbuhan Anak Kambing Peranakan Etawah sampai lepas sapih, berdasarkan litter size dan jenis kelamin. Jurnal Penelitian Universitas Jambi Seri Sains, **16** (2): 51.

IOP Conf. Series: Earth and Environmental Science **119** (2018) 012063 doi:10.1088/1755-1315/119/1/012063

Determinants Affecting on Smallholder Madura Cattle Farming at Pamekasan Regency, East Java, Indonesia

H. D. Utami[1*], A. Yakin[1] and A. P. Seruni[1]

[1]Animal Husbandry Faculty, Brawijaya University, Malang, East Java, Indonesia

[*]Corresponding e mail : hmamiek@yahoo.co.uk

Abstract

Study was case study at Pamekasan Regency of Madura Island, East Java, Indonesia. The research aimed to examine income of smallholder beef cattle farming and its influencing factors. The research used 30 members of the *Pancong Jaya* group farmer that obtained by purposive sampling method. Research regarded descriptive analysis using economic formulation and multiple regression technique. Results found that R^2 Adjusted obtained 73.70% and F-calculation (12.625) indicated significant (P< 0.000). Partial analysis confirmed that five out of nine variables have effected on the beef cattle farming income. Farmer's experience in raising cattle and their education revealed a positive and high significant (P< 0.000 and P<0.005, respectively), associated with the farm earning. Farmer's age, the number of household members, and purchasing breeding stock appeared a negative and significant (P <0.05) and (P<0.1, accordingly) influence on farmer's income. This study concluded that producing per AU cattle required capital approximately IDR 16,861,718 (structured by 59.99% of fixed and 40.01 % of working capital) and IDR 6,745,512 of production costs (composed by 57.47% of feeder cattle procurement, 17.78% of feed concentrate and 14.48 % of forage expenses) to obtain IDR 8,834 of revenue and IDR 2,089,512 of income. The efficient beef cattle farming also existed in corresponding with 1.31 of R/C Ratio. Farmers' education and experience in handling cattle strongly influenced on increasing income. However, the contrast view come from farmers' age, the number of family members, and purchasing feeder cattle in which these seemed to reduce farmers' income.

1. Introduction

Pamekasan regency of Madura Island was notoriously with two culture festivals namely "Karapan Sapi" (Bull race) and "Sapi Sonok" (female cattle contest), however feedlot cattle still emerged as the favourable venture among smallholder farmers. Besides natural mating, applying technology of Artificial Insemination might become useful in reducing feeder calves price [1] since both the feed and breeding stock required approximately 92% expenses [2]. The "Karapan Sapi" (Bull race) and "Sapi Sonok" (female cattle contest) preferred male and female best grade calves, respectively, while the rest calves with the multi quality went to fattening cattle farming. Even though, beef farming has diverse magnitudes, the profit motive as the ultimate purpose in order to contribute in fulfilling household needs [3]. However, [4] discovered that cattle traders have obtained more share margin approximately 0.61 in comparison with 0.39 those of farmers. The beef cattle income has remained low because of the less bargaining power in livestock marketing [5].

However, [4] discovered that cattle traders have obtained more share margin approximately 0.61 in comparison with 0.39 those of farmers. It means that farmers received only 39% of price share, while intermediary trader had the large proportion about 61% of those. Therefore, beef cattle farmers' income has remained low because of the less bargaining power in livestock marketing [5].

The governance in beef cattle marketing as well as empowering beef cattle farmers will be essential factors to encounter these obstacles including limited access to market-related information and lack of capital. [6] recognised the market value chain system to access to market information involving production systems, prices, value chains, competitors, and consumer preferences might require to enhance farmer's participation in beef marketing system In addition, the existence of the decentralization of cattle marketing information centre and the involvement of communal farmers in the dissemination of information plays a critical role in improving framer's access to formal market [7]. [1] suggested the importance of risk management toward forward pricing strategy to reduce grain feed and fed cattle price and results in cheaper feeder cattle price and increase the beef cattle price. [8] emphasised on the importance of production quality to access formal beef cattle market. Moreover, coordination and marketing information indicated a dominant factor that supports the strategy in beef cattle marketing [9] [10]. Hence, it becomes urgent to analyse the smallholder beef cattle farming in terms of income and its influence factors. [1] suggested the importance of risk management toward procurement pricing strategy to reduce grain feed and fed cattle price and results in cheaper feeder cattle price and increase the beef cattle price.

2. Materials and Methods

Case Study of small scale beef cattle faming located at East Waru Village, Waru Sub-District, Pamekasan Regency of East Java Province which selected by purposive method. This research area employed three motives cattle farming namely "Karapan Sapi" (Bull race), "Sapi Sonok" (Cow cattle contest), and feedlot cattle. Natural mating and Artificial Insemination (AI) methods also appeared more prevalent in producing calves to supply breeding stock for developing beef cattle farming.

Multistage sampling method aimed to reach 30 farmers who had at least five years' experience in rearing beef cattle and they also participated in *Pancong Jaya* group farmer. Survey method employed a structured questionnaire to obtain primary data, for instance production cost, revenue, etc. Secondary data were available in *Pancong Jaya* group farmer and the related institutions. Descriptive analysis was technique to delineate farmer's income, while exploring the determinants influencing on farmer's profit used regression analysis. The income in beef cattle farming is function of the internal and external variables of the model:

$$Y = \beta 0 + \beta_1 X_1 + \beta_2 X_2 + \beta_3 X_3 + \beta_4 X_4 + \beta 5 X_5 + \beta_6 X_6 + \beta_7 X_7 + \beta_8 X_8 + \beta_9 X_9 + \varepsilon_i$$

Where:

Y : income in smallholder beef cattle farming per farm (IDR /year),
X_1 : farmer's age (year),
X_2 : farmer's education (year),
X_3 : Number of household members (person),
X_4 : farmer's experience in raising beef cattle farming (year),
X_5 : Purchasing feeder calf (IDR / year),
X_6 : Forage feed cost (IDR//year),
X_7 : Rice bran cost (IDR/year),
X_8 : the number of beef cattle (AU/year),
X_9 : farming motives (0= Breeding farm; 1= mixing between breeding stock and fattening)
β_0 : constant
$\beta_1 \ldots \beta_9$: estimation parameter
ε : error

3. Results and Discussion

3.1. Beef Cattle farmers' characteristics

Majority farmers (57%) aged between 51-60 years old, while only 43% of them were aged less than 51 year in which structured with 33% and 10% of youth farmers aged 41-50 years and 31-40 years, respectively. Majority (57%) older farmers (aged between 51-60 years old) might prefer to apply simple technology as those usually presence among smallholder farmers. This circumstance therefore, effected on the quite low beef cattle productivity and hence decreased the farming profit.

Mostly older farmers (aged between 51-60 years old) might prefer to apply simple technology in raising their livestock which usually emerged among smallholder farming. This management practice therefore, effected on the quite low beef cattle productivity, and hence decreased the farming profit .

Mostly (40%) farmers have already obtained high education (High school). It means that farmers have the thinking pattern and ability to address better management practice in rearing beef cattle farming. It is evidence that 7% of farmers has graduated from university in that they have capability in access knowledge and adopting technology in developing beef cattle farming. Similarly, education has positively influenced on the farmers way thinking and measures in increasing productivity [11] [12]. Similarly, farmer's education attainment has a positive impact in achieving optimal yields and enhancing their beef cattle farms.

Majority farmers (43%) have possessed longer experience ranging 7-9 years in raising beef cattle farming. They might possess a better ability than those with low experience in encountering problem and making appropriate solution in carrying out the beef cattle farming. Likewise, the experience will effect on the behaviour including knowledge, attitude, and skill [12]. So, the longer experience in engaging beef cattle farming might associate with the better knowledge and ability in operating their farm.to acquire the inclining the farm production and obtaining optimal earning.

The high proportion (60%) of farmers comes from the large family member with 3-4 persons than those from the small household members (40%) with 1-2 persons. The allocation time of family labour can reduce the hired worker expenses and hence, reducing production costs. [13] stated that the more household members might be useful in helping daily farming operation in one side, whereas they become burden in fulfilling the family needs in other side. Hence, as the number of household members increase, a greater amount of time may be devoted in beef cattle farming to achieve high production and more farm income.

3.2. The structure of beef cattle faming

The composition of beef cattle plays an important role in determining the farm performance. It is not only gauged the production ability, but also established the farm income.
The structure of total beef cattle that owned by 30 smallholder farmers appear in table 1.

Table 1. Beef cattle structure for 30 farmers in year 2014

Livestock	Number (head)	Animal Unit (AU)	Percentage (%)
Calf	12	3	7.2
Young cattle	27	13,5	32.5
Mature cattle	25	25	60.3
Total	64	41,5	100
Average		1,38	

Beef cattle farming consisted more with mature cattle (60.3%) than those of smallest number of calf (7.2%). It was interesting that contribution of young cattle was about one third (32.5%) out of total beef cattle. Farmers were running their farm with 1.38 AU, and therefore they constituted as smallholder farming. This farming has characteristic with rearing small number of cattle, little capital, lack of information and using simple management. It is evidence that the number cattle indicated relatively constant by the year because of economic reasons.

3.3. Factors influencing on smallholder beef cattle farming income

The results of regression analysis showed that coefficient determination (R^2 Adjusted) were about 73.70% (Table 2). F-calculation (12.625) was significant ($P<0.000$). It implied that the nine determinants namely farmer's age (X_1), farmer's education (X_2), the number of household members (X_3), and farmer's experience in raising beef cattle (X_4), purchasing feeder calf (X_5). forage feed cost (X_6), and rice bran cost (X_7)), the number of beef cattle (X_8), and farming motives (0= breeding stock; 1= mixing between breeding stock and fattening) (X_9).

Partial analysis however clarified that five out of nine variables have effected on the beef cattle farming income. Farmer's experience in raising beef cattle farming (X_4) exhibited a positive and high significant ($P< 0.000$) association with the return of beef cattle farming. Similarly, the smallholder beef cattle profit was positively and significantly ($P<0.005$) related to farmer's education (X_2). The three predictor variables including farmer's age (X_1), the number of household members (X_3), and purchasing breeding stock (X_5) appeared a negative and significant ($P <0.05$) and ($P<0.1$, respectively) influence on the income of smallholder beef cattle farming. The relationships between profit in smallholder beef cattle farming and the predictor's variable can be formulated as:

$Y = 1,229,877 - 0.25\ X_1 + 0.204\ X_2 - 0.143\ X_3 + 0.644\ X_4 - 0.278\ X_5$

where:

Y : income in smallholder beef cattle farming per farm (IDR /year),
X_1 : farmer's age (year),
X_2 : farmer's education (year),
X_3 : Number of household members (person),
X_4 : farmer's experience in raising beef cattle farming (year),
X_5 : Purchasing breeding stock (IDR / year).

3.4. Farmer's education

This variable significantly ($P<0.05$)) and positively associated with smallholder beef cattle farming income. The high education might link with increase in the management livestock farming operation. It means that a 10% increase of the farmer's education would result on the 2.04% improvement in the smallholder beef cattle farming income. The finding has accepted hypothesis 1: " It is predicted that farmer's education has positive affected on small scale beef cattle farming profit" as when controlling all other variables. The strong relationship between those variables indicated that the higher education attainment, the great productivity, and result in the beef cattle farming income improvement.

The discovery reported that farmer's education play an important role in determining livestock income. This finding was similar to [11] [12] in that education has positively influenced on the farmers way thinking and measures in increasing productivity. Likewise, the study demonstrated that the accruing farmer's education influenced the great income in beef cattle farming. Mostly (40%) farmers have already obtained high education (High school) and it influenced on their thinking pattern and ability to address better management practice in rearing beef cattle farming. It is evidence that 7% of farmers has graduated from university in which they might capable in access knowledge and adopting technology to pursue beef cattle farming development.

3.5. Farmer's experience in raising beef cattle

Farmer's experience in rearing beef cattle revealed highly significant (P<0.000) and positive influenced in the profit of the smallholder beef cattle farming. This variable has a dominant contribution in the livestock farming practices. The study delineated that a 10% increase of the farmer's education would effect on the 6.44% enhancement in the livestock farming income. The sturdy relationship between those variables implied that the longer experience increased in their capability to encounter problem and the skill in making appropriate solution in operating the beef cattle farming. Hence, it had a positive impact in the great productivity, and result in the inclining farm profit. Likewise, the experience will effect on the behaviour including knowledge, attitude, and skill [12]. [13] clarified that experience in rearing livestock become foundation in developing farming which has higher influence in shaping successful farmer. Farmers 'experiences in raising beef cattle provide a good understanding for their attempt in supporting beef cattle management, and therefore influenced on their behaviour in engaging their farming. The finding suggest that the longer experience in engaging beef cattle farming have impact on the better knowledge and ability in operating their farm.to acquire the inclining the farm production. Mostly (43%) farmers had 7- 9 years' experience in rearing beef cattle which allow them to possess a better ability in engaging the farming. They become capable to encounter problem and competent in making appropriate solution in operating the beef cattle farming. It might be represented in the beef cattle farming productivity, and result in the income improvement in beef cattle farming.

Table 2. Determinants influencing in smallholder beef cattle farming income

Explanation	Regression Coefficients
Constant	1,229,877
X1 :Farmer's Age	-0.25***)
X2: Farmer's Education	0.204**)
X3:Household's members	-0.143**)
X4: Experience in raising beef cattle Farming	0.644****)
X5: Feeder calf expenses	-0.278*)
X6: Forage Feed cost	0.64
X7: Rice bran cost	0.162
X8: Number of beef cattle	0.143
X9: The purpose in raising beef cattle	0.04
R2 Adjusted = 73.70%	
R Square =80.10%	
n = 30	
Fcalculation = 12.625	

*) P . 0.1; **) P < 0.05; ***) P < 0.005; ****) P < 0.000

3.6. Feeder calf expenses

Purchasing cattle stock was significant (P< 0.1) and negative (-0.278) associated with the income of smallholder beef farming. This invention accepted hypothesis 2 in that: "It is predicted that feeder calf expenses have reduced the income of small scale beef cattle farming. It means that a 10% improvement of the breeding stock expenses would have a corresponding 2.78% reduction in the profit of smallholder beef cattle farming, if other factors are assumed constant. Breeding stock procurement has the reverse trend with the accruing profit of the small-scale beef cattle farming. It is fact that variable costs composed the highest proportion (90.69 %) with the greatest being purchase feeder calf (54.47%), followed by concentrate feed (17.78%) and forage expenses (14,48%). The high expenses in the procurement of feeder calf have an impact on high production cost and reducing the farm profit. Study of [2] added that 92% both the feed and breeding stock expenses indicated quite sensitive in which a bit decrease (6%) in the beef cattle price has effected on the zero of farming income. It is important to use technology in beef cattle breeding in order to reduce feeder calves price [1], and improving the availability of feeder cattle [14].

3.7. Farmer's age

Farmer's age reported moderately significant (P<0.05) and had a negative impact on the profit of the small-scale beef cattle farming. The 10% increase in the farmer's age would reduce about 2.78% decrease in the profit of smallholder beef cattle farming, if other factors were kept constant. Study suggested that predominantly (57%) farmers aged between 51-60 years old were unlikely to adopt technology in farming practice that impact on low productivity and hence decreasing farm profit. Alternatively, youth farmers (aged less than 50 years) were more likely to access the technology in operating the livestock farming and may lead to high farm productivy and the optimal profit. The young people reveals enthusiastic about high curiosity, and hence they easy to accept the change and adopt technology [15]. Similarly, young farmers have tendency to try something new as the effort to increase knowledge and skill in operating livestock farming, and impacting on better beef cattle productivity as well as improving the farming profit.

3.8. Number of household members

This variable appeared sufficient significant (P<0.05) and negative associated in the income of the smallholder farming. The number of family members has on average contribution in the livestock farming practice. It means that a 10% improvement of the number of family members would have a corresponding 1.43% decline in the cattle farming earning, if other factors are held constant. The large number of household members might relate to more labour allocation in operating beef cattle farming in which reduced hired labour expenses and hence, save production costs. However, allocating a greater amount time in operating beef cattle farming tasks might effect on the less time devoted in others earning activities in fulfilling the family needs [13]. This study confirms that small household (1-2 people) were more likely to obtain more income from beef cattle farming because they considered beef cattle farming as secondary livelihood. The discovering offers insight into the improvement of beef cattle productivity and a better income through more allocating the household members in running this farm.

4. Conclusion

Study on smallholder beef cattle farming at provided the following conclusions:

1. Beef cattle farming required capital of IDR 16,861,718 per Animal Unit which structured by 59.99% of fixed capital and 40.01% of working capital.
2. This farming constituted a profitable venture with IDR 2,089,512 of income, IDR 6,745,512 of production costs (composed by 57.47% of feeder cattle procurement, 17.78% of feed concentrate and 14.48 % of forage expenses) and IDR 8,834 of revenue in producing one AU in yearly basis.
3. The efficient beef cattle farming were also more likely to be reached in corresponding with 1.31 of R/C Ratio.
4. Farmers' education and experience in handling cattle strongly influenced on their income to be increase. However, the adverse view came from farmers' age, the number of family members, and purchasing cost of feeder cattle in which these seemed to reduce farmers' income.

References

[1] Marsh J M 2011. U.S. Feeder Cattle Prices: Effect on Finance and Risk, Cow Calf and Feedlot Technologies, and Mexican Feeder Imports. Journal of Agricultural and Resources Economics. Vol. 26 (2): 463-477.

[2] Malope P, Tsopito C M, Aganga A A and Madibela O R 2007. Profitability of dry season beef feed-lotting in grain deficit countries: the case of Botswana. Livestock Research for Rural Development. Vol. 19 (6).

[3] Setiawan H M 2014. Contribution of Beef Cattle Profit into Household Income (Case Study at Sukolilo Village, Jabung Sub-District, Malang Regency). Undergraduat Thesis. Faculty of Animal Husbandry, Brawijaya University.

[4] Utami H D and Seruni A P 2013. Prospect of Small-Holder Beef Cattle Farming at East Java, Indonesia. Proceeding. In The 14 International Conference on Sustainable Agriculture for Developing Countries (SAADC2013). 27-31 July 2013. Lanzhou, China.

[5] Utami H D, Nugroho B A, Hartono B and Fanani Z 2010. Analysis of Beef Cattle Farming At Poncokusumo Sub-District, Malang Regency. Research Report. Faculty of Animal Husbandry, Brawijaya University.

[6] Alemayehu K 2011. Value chain assessment of beef cattle production and marketing in Ethiopia: Challenges and opportunities of linking smallholder farmers to the markets. Livestock Research for Rural Development. Vol.23 (12).

[7] Musemwa I, Mushunje A, Chiminyo M, Fraser G, Mapiye C and Muchenje V 2008. Nguni Cattle Marketing Constrains and Opportunities in the Communal Areas of South Africa : Review. African Journal of Agricultural Research. Vol. 3 (4). Pages:.239-245.

[8] Teweldemedhin M Y and Kafidi L 2009. Risk Management Strategies of Cattle Farmers in Nambia : Case Study from Omaheke and Otjozondjupa Regions. Journal of Agricultural Extension and Rural Development. Vol. 1 (2), 00.063-070.

[9] Utami H D, Nugroho B A, Hartono B and Fanani Z, 2012. Feasibility Study of Beef Cattle Farming (Case Study at Pakis Village, Tambak Rejo Sub-District, and Bojonegoro Province). Research Report. Faculty of Animal Husbandry, Brawijaya University.

[10] Musemwa I, Chagwiza C, Sikuka W, Fraser G, Chimonyo M and Mzileni N 2007. Analysis of cattle marketing channels used by small scale farmers in the Eastern Cape Province, South Africa. Livestock Research for Rural Development . Vol.19 (9).

[11] Banna M M H 2014. Analysis of Beef Cattle Income At Mojojejer Village, Mojowarno Sub-District, Jombang Regency, Indonesia. Undergraduate Thesis. Faculty of Animal Husbandry, Brawijaya University.

[12] Yuhanria T 2013. The Role of Female Labour in Smallholder Dairy Farming At Kuningan District, Indonesia. Jurnal Ilmiah Peternakan 1(2): 536 – 545.

[13] Alam A 2014. Farmer Motivation Toward Beef Cattle Farming At Buru District Maluku Province, Indonesia. Agromedia. Vol 32 (2).

[14] Mlote S N, Mdoe N SY, Isinika A and Mtenga L A 2012. Value addition of beef cattle fattening in the Lake Zone in Tanzania: Challenges and opportunities. Livestock Research for Rural Development. Vol. 24 (6).

[15] Ngadiyono N 2010. Labour Allocation between Resettled and Local Household at KonaweDistrict, South East Sulawesi, Indonesia. Buletin Pet

IOP Conf. Series: Earth and Environmental Science **119** (2018) 012064 doi:10.1088/1755-1315/119/1/012064

Impact of Small Holder Dairy Farm on the Air Quality in Gunungpati District, Semarang Municipality

E. Widiastuti[1], Kustono[2], Adiarto[2], Nurliyani[2] and S. Sugiharto[1]

[1]Faculty of Animal and Agricultural Sciences, Diponegoro University, Kampus Tembalang, Semarang 50275
[2]Faculty of Animal Science, Gadjah Mada University, Jl. Fauna Bulaksumur, Yogyakarta 55281

Abstract

The study aimed to investigate the impact of small holder dairy farm on the air quality in the farm and surrounding area. The study was conducted in three farmer groups in the District of Gunungpati, including the farmer groups in the villages of Nangkasawit, Plalangan and Sumurejo. Samplings of air quality were conducted in four points (locations), i.e., inside the barn, 100 m, 200 m and 300 m from the area of farm. Parameters observed were emission of NH_3 CO_2, H_2S and CH_4. Results showed that the levels of NH_3 in the barn, 100 m, 200 m and 300 m from the farm area in Nangkasawit village were 0.211, 0.107, 0.104 and 0,035 ppm, respectively. The levels of NH_3 in Plalangan village were 0.289, 0.231, 0.13 and 0.108 ppm, respectively, and in Sumurejo village were 0.109, 0.110, 0.082 and 0.046 ppm, respectively. The levels of H_2S in Nangkasawit, Plalangan and Sumurejo villages at the entire points of observations were <0,002 ppm, with the acceptable standard level of H_2S was 0.02 ppm. In four locations of observations, the levels of CH_4 in Nangkasawit village were 809, 603, 599 and 521 ppb, in Plalangan village 999, 720, 645 and 582 ppb and in Sumurejo village 932, 824, 526 and 521 ppb. The levels of CO_2 in Nangkasawit village were 26.55, 28.35, 28.44 and 30.05 ppm, in Plalangan village 24.65, 25.10, 23.44, 21.05 ppm and in Sumurejo village 28.50, 27.35, 30.68, 31.50 ppm, with the standard level of CO_2 should be 30.000 ppm. In conclusion, the air quality was better (lower contamination) with the farther distance of locations from the barn.

1. Introduction

It has commonly been known that livestock sector produce waste that may pollute the environment. Waste derived from livestock activities can be solid waste, wastewater and gas. Ironically, most farmers in Indonesia do not care about the waste generated from their activities. Most of traditional farmers do not manage their waste properly, in that almost all the waste generated from the livestock husbandry is just directly disposed to the nearest river. This may consequently pollute the water especially around the farm area.

The concept of sustainable development requires realization of integration of economic interests, ecological interests and social interests. On the other hand, the principle and development pattern should be able to guarantee the implementation of justice and equity principles, increasing the quality of biodiversity, the application of integrative approach and must have a long-term perspective. The orientation of technological and institutional changes are conducted in such a way to ensure sustainable fulfillment and satisfaction of human needs for present and future generations. Hence, the public farming system with the environmental insights has a sense and purpose in the framework of food security and environmental sustainability (Budinuryanto, 2010).

Waste as a negative factor in the livestock industry is a fact that cannot be eliminated easily. Apart from profits (in terms of business), livestock sector has a negative impact on the environment and public health. Waste directly disposed to the environment (without any proper treatment) will contaminate air, water and soil causing environmental pollution. Other than causing pollution to water, livestock husbandry also produces CH_4, NH_3 and H_2S that may cause unpleasant odor. Livestock houses, stacking and waste dumps are the source of these unpleasant odors. These above mentioned gases are actually the waste of metabolic process in the body of animals and the breakdown of organic materials by microorganisms in an anaerobic conditions (Widyastuti et al, 2013). Several factors have been identified to affect the rate of odor emission from the livestock activities, including livestock species, environmental conditions and waste management (Sarwanto and Tuswati, 2011). The present study aimed to investigate the impact of small holder dairy farm on the air quality in the farm and surrounding area.

2. Materials and methods

The study was conducted in three farmer groups in the District of Gunungpati, including the farmer groups in the villages of Nangkasawit, Plalangan and Sumurejo. Samplings of air quality were conducted in four points (locations), i.e., inside the barn (ST0), 100 m (ST1), 200 m (ST2) and 300 m (ST3) from the area of farm. Parameters observed were emission of NH_3 CO_2 H_2S and CH_4.

The collections of air for analysis of NH_3, CO_2, H_2S and CH_4 were carried out by air capture in the field with the aid of chemical reagents. The air capture methods for NH_3, SO_2 and CO_2 were based on impinger method (Ruslinda, 2005). The principle of this method is to absorb the contaminated air into the catcher in the impinger. Syringe was used for catching of air samples for CH_4 analysis. Analyses of CH_4, NH_3, H_2S and CO_2 were carried out in the Laboratory of Central Pollution Prevention Technology Industry Semarang.

3. Results and discussion

Based on the Minister of Environment Decree nr. 50 / MenLH / 11/1996 about odor, air quality parameters that need to be monitored in a farm are as follows:

3.1 Ammonia (NH_3)

The results of observation and measurement of NH_3 parameter in Nangkasawit village were 0.211, 0.107, 0.104 and 0.035 ppm, in the village of Plalangan 0.289, 0.231, 0.13 and 0.108 ppm and in Sumurejo village 0.109, 0.110, 0.082 and 0.046 ppm, with a permissible standard of 2 ppm. Based on these data, the ammonia content at the study sites was still below the permissible quality standard, which means uncontaminated condition. The results of air quality analysis around dairy farms in Nangkasawit village are presented in Table 1.

In the Plalangan village, NH_3 levels at each sampling points was the highest, followed by Nangkasawit and Sumurejo villages. The fact that fecal waste was only moved to the edge of the barn made the NH_3 level in Plalangan village higher than other village. In Sumurejo village the waste was sprayed using water flow at the time of washing the livestock before milking. This was in accordance with Sarwanto and Tuswati (2011) suggesting that the levels of NH_3 is lower in the farm with good husbandry and management practices. The description of air quality in three villages based on NH_3 content is presented in Figure 1.

Table 1. The Air Quality in Nangkasawit Village

Nr.	Parameters	unit	Gas levels				Permissible standard
			ST0	ST1	ST2	ST3	
1.	NH_3	ppm	0.211	0.107	0.104	0.035	2
2.	H_2S	ppm	<0.002	<0.002	<0.002	<0.002	0.02
3.	CH_4	ppb	809	603	599	521	
4.	CO_2	µg/m^3	26.55	28.35	28.44	30.05	30,000

ST0: sample collection was conducted inside the barn, ST1: sample collection 100 m from the area of farm, ST2: 200 m, ST3: 300 m

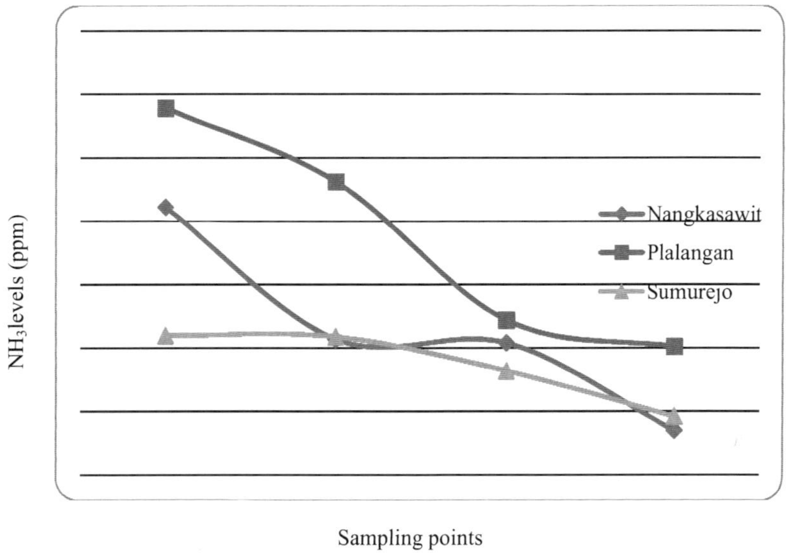

Figure 1. The levels of NH_3 in three villages at different sampling points

Based on our observations and measurements, it could be inferred that NH_3 levels tended to decrease with the farther of sampling point from the barn. This was in accordance with Latief et al. (2011) mentioning that NH_3 concentrations obtained from various places increase following the amount of feces present in the barn.

3.2 Hydrogen sulfide (H_2S)

Results from the observation and measurement of H_2S concentrations in Nangkasawit, Plalangan and Sumurejo villages at all observation points were below 0.002 ppm; with a permissible quality standard of 0.02 ppm. These results indicated that H_2S levels are still below the permissible standard, which means the environment is not polluted. The results of air quality analysis around dairy farms in Plalangan village are presented in Table 2.

IOP Conf. Series: Earth and Environmental Science **119** (2018) 012064 doi:10.1088/1755-1315/119/1/012064

Table 2. The air quality in Plalangan village

Nr.	Parameters	unit	Gas levels				Permissible standard
			ST0	**ST1**	ST2	ST3	
1.	NH_3	ppm	0.289	0.231	0.130	0.108	2
2.	H_2S	ppm	<0.002	<0.002	<0.002	<0.002	0.02
3.	CH_4	ppb	999	720	645	582	
4.	CO_2	µg/m^3	24.65	27.10	28.44	30.05	30,000

ST0: sample collection was conducted inside the barn, ST1: sample collection 100 m from the area of farm, ST2: 200 m, ST3: 300 m

Based on the above results, it can be inferred that the emission of H_2S in the study site has not caused unpleasant odor. Setyawan (1996) suggested that H_2S at the levels of 0.47 ppm in the air is the limit of concentrations that can produce unpleasant smell, while at the concentration of 1.0 ppm dimethyl sulfide begin to generate bad smell.

3.3 Carbon dioxide (CO_2)

The results of observation and measurement of CO_2 content in Nangkasawit village were 26.55, 28.35, 28.44 and 30.05 ppm, in Plalangan village 24.65, 25.10, 23.44, 21.05 ppm and in Sumurejo village 28.50, 29.35, 30.68 and 31.50 ppm for ST0, ST1, ST2 and ST3, respectively. Note that the permissible quality standard of CO_2 is 30,000 ppm. Based on the above data, the level of CO_2 in the research location was therefore still below the permissible quality standard (not polluted). The data of air quality around dairy farms in Sumurejo village are presented in Table 3.

Table 3. The Air Quality in Sumurejo Village

Nr.	Parameters	unit	Gas levels				Permissible standard
			ST0	**ST1**	**ST2**	**ST3**	
1.	NH_3	ppm	0.109	0.110	0.046	0.082	2
2.	H_2S	ppm	<0.002	<0.002	<0.002	<0.002	0.02
3.	CH_4	Ppb	932	824	526	521	
4.	CO_2	µg/m^3	28.50	29.35	30.68	31.50	30,000

ST0: sample collection was conducted inside the barn, ST1: sample collection 100 m from the area of farm, ST2: 200 m, ST3: 300 m

Our observations showed that Sumurejo village has a higher CO_2 levels compared to Plalangan and Nangkasawit villages. This was most likely because the sample collections in Sumurejo village followed the village road, so that there was a possibility of exposure of CO_2 emissions from the vehicle.

IOP Conf. Series: Earth and Environmental Science **119** (2018) 012064 doi:10.1088/1755-1315/119/1/012064

3.4 Methane (CH_4)

Our measurement of CH_4 content in Nangkasawit village in four observation points were 809, 603, 599 and 521 ppb, in Plalangan village 999, 720, 645 and 582 ppb and in Sumurejo village 932, 824, 526 and 521 ppb. Based on these data, the CH_4 content at the study sites was still below the permissible quality standard, which means not polluted.

In ruminant animals (beef, buffalo, sheep and goats), organic compounds in feed materials are fermented by rumen microbes to produce volatile fatty acids, CO_2, hydrogen (H_2) and microbial mass. Through methanogenic process (by methanogenic bacteria), CO_2 is reduced by H_2 to form CH_4, which emits through eructation (about 83%), respiration (about 16%) and anus (about 1%) (Haryanto and Talib, 2009).

Sumurejo village has the lowest methane (CH_4) emission throughout the sampling points, followed by Nangkasawit and Plalangan villages. The low methane emissions in Sumurejo village were closely related to feed management applied to dairy cows. In Sumurejo village, forage given to the cows was elephant grass, while field grass was provided to cows in Nangkasawit and Plalangan villages. Herawati et al. (2012) revealed that the type of feed given affects the methane production by the cows. The better quality of feed given, the lower CH_4 emitted by the cows. The description of air quality in three villages based on CH_4 exposure is presented in Figure 2.

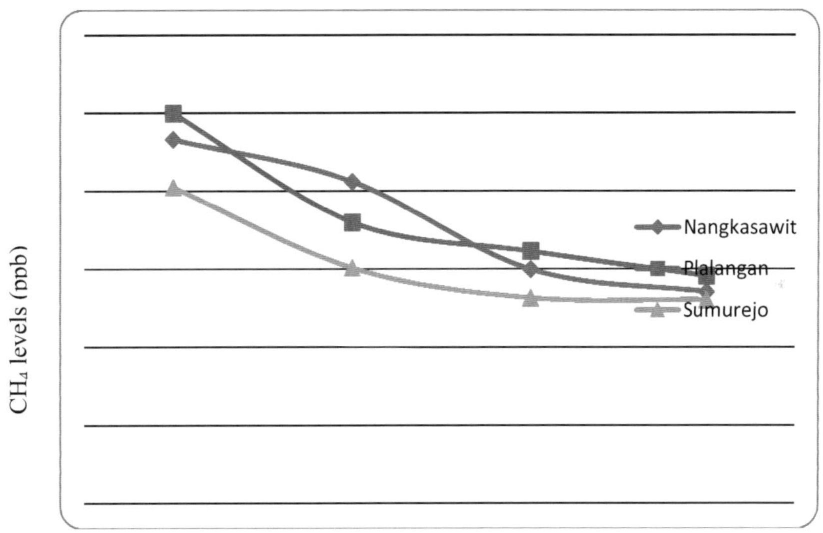

Sampling points (m)

Figure 2. The Levels of CH_4 in Three Villages at Different Sampling Points

In addition to feeding, husbandry and management practice also affects the emitted CH_4. Indeed, husbandry and management practice (waste management) in Sumurejo village was better compared to that in Nangkasawit and Plalangan villages. This was in accordance with Sarwanto and Tuswati (2011) suggesting that the differences in environmental conditions and management practice of farmers in waste management will affect the level of CH4, NH_3 and H_2S.

Based on our observations, it appeared that the highest NH_3 level was found in Plalangan village, followed by Nangkasawit and Sumurejo villages. Similar results were seen for CO_2 and CH_4 emissions. With regard to the distance from the livestock activities, there was a tendency that further distance point of sampling from the location of farm, the levels of contaminants decreased, which means the air quality was better. This was in accordance with the Decree of the Director General of Livestock Nr. 776 / kpts / DJP / Deptan / 1982, stating that the location of dairy farms must not be located in the city center and the settlement with a distance of at least 250 m from the settlement.

4. Conclusion

Air pollution in three observational villages was still within the permitted quality standard threshold. Of all parameters observed, the lowest pollution was found in Sumurejo village (due to good husbandry and management practice), followed by Nangkasawit and Plalangan villages. Another conclusion was that the air quality was better (lower contamination) with the farther distance of locations from the livestock activities.

References

[1] Budinuryanto, D.C., 2010. Restrukturisasi sistem produksi usaha peternakan sapi perah rakyatdalam pembangunan berkelanjutan (Kasus di daerah hulu sungai Citarum). Seminar Nasional Pembangunan Peternakan Berkelanjutan Universitas Padjadjaran, Bandung.

[2] Widyastuti, F.R., Purwanto dan Hadiyanto. 2013. Upaya Pengelolaan Lingkungan Usaha Peternakan Sapi di Kawasan Usahatani Terpadu Bangka Botanical Garden Pangkalpinang. Dalam :S.P. Hadi, Purwanto, H. R. Sunoko, H. Purnaweni (Ed.). Prosiding Seminar Nasional Pengelolaan Sumberdaya Alam dan Lingkungan "Optimasi Pengelolaan Sumberdaya Alam dan Lingkungan dalam Mewujudkan Pembangunan Berkelanjutan", Semarang, 27 Agustus 2013. Hal 80 -84.

[3] Sarwanto, D. dan S.E. Tuswati. 2011. Pengaruh pengelolaan limbah sapi perah terhadap tingkat kebauan. Media Peternakan **(14) 2** : 1

[4] Ruslinda, Y. 2005. Metode Sampling dan Analisis Udara. Jurusan Teknik Lingkungan Fakultas Teknik Universitas Andalas, Medan.

[5] Kementrian Lingkungan Hidup RI. 1996. Surat Keputusan Menteri Lingkungan Hidup No 50/MenLH/11/1996 tentang Kebauan. Kementrian Lingkungan Hidup RI, Jakarta.

[6] Latief, R., E. Sutrisno dan M. Hadiwidodo. Pengaruh jumlah kotoran sapi terhadap konsentrasi gas amonia (NH_3) di dalam rumah : Studi Kasus Desa Dalangan Kelurahan Sumogawe, Kecamatan Getasan, Kabupaten Semarang. Jurnal Teknik Lingkungan **(3) 1** : 1.

[7] Setiawan, H. 1996. Amonia Sumber Pencemar yang Meresahkan. Dalam Informasi Dunia Kesehatan Hewan. Edisi 037 Agustus 1996. Asosiasi Obat Hewan, Indonesia.

[8] Haryanto, B. dan A. Thalib. 2009. Emisi metan dari fermentasi enterik : kontribusi secara Nasional dan faktor-faktor yang mempengaruhinya pada ternak. Wartazoa **19** (4) : 157.

[9] Herawati, T. 2012. Refleksi sosial dari mitigasi emisi gas rumah kaca pada sektor peternakan di Indonesia. Wartazoa **22** (1) : 3

IOP Conf. Series: Earth and Environmental Science **119** (2018) 012065 doi:10.1088/1755-1315/119/1/012065

Risk Preference of Farmer Beef Cattle Smallholder in West Java

H. Arief[1*] and A. Fitriani[1]

[1]Faculty of Animal Husbandry, Universitas Padjadjaran,
Jl Raya Bandung - Sumedang Km. 21 Jatinangor 45363, Indonesia

[*]Coresponding E-mail: hasni.arief@unpad.ac.id

Abstract

Beef cattle farm is an economic activity that its correlation relatively high with ecology condition, ever more its activity be dominated by smallholder with farm scale of 1—3 head beef cattle, so the capital of feeder cattle are component production costs relatively large. By doing so, lose one head of cattle means losing some of the capital that had been invested in its farm. This condition has implications for the behavior of farmers in managing its farm. The farmers in the decision-making act reluctant to risk so there is a bit of a program or package of new technologies to improve the performance of farm rejected. Therefore, this study analyzed the preferences of farmers against risks, and find out the socio-economic conditions that be the deciding factor on farmer preference for risk.The method used was a survey with multistage random sampling. The numbers of samples in this study were 150 people from three different areas, namely: District of Bandung, Subang, and Pangandaran (South Ciamis). Data analysis model used in relation to this research problem was the model's utility function and Component Factor Analysis (CFA). The results showed that the overall breeders had a reluctant attitude to risk (risk averter). This is indicated by the value of x_3, namely -63,692.693 < 0; and socio-economic factors that determine the preference of farmers against risk are age, education, and experience for farm management; while the business scale factor, number of dependents, and ownership determine the size of the gross margin.

1. Introduction

The agricultural sector, including livestock, is an economic activity which the production process is always confronted with risk and uncertainty. This is reasonable because the output of this business sector is a living creature that is relatively high dependence on nature. Under these conditions, it can be presumed level of risk faced by farmers is also relatively high.

The risks faced by farmers certainly have implications on earned income, which in turn affects the sustainability of the business itself. Beef cattle business in Indonesia, especially in West Java, is dominated by people who owned relatively small number of livestock (1-3 head cattle), where the largest production cost component is beef feeder. Thus, losing a cattle means losing most of the capital that has been invested in their business. This condition affects farmers in decision-making.

Decision-making risk associated with certain actions selected and recognized by decision makers. It was alleged that in general farmers are reluctant to risk. Farmer's behaviour in decision making according to [1] is caused by the presence of a central economic dilemma faced by most households' farmers. Focused on the above conditions, the general purpose of this study was to get a description of the risks involved in beef cattle business in the context of the current economic environment changes based on the information obtained in the current study. Thus it is possible to design risk management strategies for future studies. To achieve the goal, this study has several specific objectives, which were: 1) to analyse farmer's preferences against risks on smallholder beef cattle business in West Java; and 2) to analyse the determinants on the farmer's preference of farmers against the risk in smallholder beef cattle business in West Java.

2. Materials and Methods

2.1. Research Design

This study had been conducted in beef cattle production center in West Java province. These areas include the District Pangandaran (South Ciamis), Bandung and Subang. Dimensions in the design of this study are as follows:

2.2 Methods

The research method used was survey method (sampling). Survey is a method of research that collects information from a sample obtained from a population; aim to generalize the population from which the sample was taken.

2.3. Sampling Technique

In line with the survey methodology, then sampling was conducted by multistage random sampling. The steps of the sampling technique in this study were as follows:

a. Step one: West Java is divided into three clusters, that are: northern territory which is represented by Subangregency; the central territory is represented by the Bandung District; and the southern territory is represented by the Pangandaran District (South Ciamis);
b. Step two: Out of three clusters, select an areas which become the production centers of beef cattle breeding (groups of farmers); and
c. Step three: All members of a group of farmers are included in the sample based on the assumption that characteristics of beef cattle breeders are relatively homogeneous, although the scale of livestock ownership varies. However, for smallholder beef cattle business is rooted to livestock breeding which are generally dominated by small-scale (1-3 head cattle). According to this, stratification of livestock ownership is not taken into account.

The numbers of samples as the respondent in this research were consisting of 150 respondents. Below, a more detailed data of groups of farmers are presented:

Table 1. The Number of Sample (Respondents)

Province West Java	District	Subdistrict	Group of Farmers	Number of Member
Northern Territory	Subang	Pagaden Barat and Cikaum	Mandiri Jaya and Mitra Karya	40
Central Territory	Bandung	Ciakncung	Mekar Laksana 1 and 2	55
Southern Territory	Ciamis	Parigi, Cijulang and Cimerak	Mitra Usaha Tani, Lumba-lumba Karya Jaya, Witu nunggal and Rahayu Jaya Mandiri	55
Total Sample (Corespondent)				150

Coefficient Criteria of utility functions:
When $\tau_3 > 0$, meaning decision makers dare to risk (risk lover)
When $\tau_3 < 0$, means that decision makers are reluctant to risk (risk averter)
When $\tau_3 = 0$, meaning neutral decision maker to risk (risk neutral)

Related to the above, the test on farmer's behaviour towards risk, use below hypotheses:
Ho: $x_3 = 0$
Ha: $x_3 < 0$

Hypothesis testing is conducted by using the t-test using this formulation:

$$t_{hitung} = \frac{\tau 3}{Se(\tau 3)}$$

Rule hypothesis testing:

If $t_{count} > t_{table}$, then Ho is rejected, means that farmers is reluctant towards risk
If $t_{count} < t_{table}$, then Ho accepted, means farmers is not reluctant to risk

From the equation above, the independent variables are symbolized by the business scale x_3 measured in the tail and the dependent variable is the gross receipts, which is accumulated from the livestock sale and livestock value changes measured in Rupiahs.

To analyze the determinants of farmers' preference against the risk, Component Factor Analysis (CFA) is used. The formulation is as follows:

$$X_i = A_{i1}F_i + A_{i2}F_2 + A_{i3}F_3 + \dots\dots + A_{in}F_m + V_iU_i$$

Where:

Xi = Standards variable / indicator-i
Aij = Standardization Multiple regression coefficient in the variable i in a common factor j
F = Common factors
Vi = Standardization regression coefficient to the variable i in a unique factor
Ui = unique factor for the variables/indicators i
m = number of common factors

Linear combinations of the observed variables can be formulated as follows:

$$Fi = WiiXi + Wi2X2 + Wi3X3 + \dots\dots + WinXk$$

Where:

Fi = Estimated factor of i
Wi = The weight or score of factor coefficient
k = Number of variables/indicators

Stages in conducting factor analysis are as follows:

a. Create a correlation matrix for all variables / indicators
b. Factors extraction, which summarizes the core factors in selecting variables / indicators that have eigenvalue > 1. The extraction method used is Principal Component
c. Rotation to the final with the varimax method
d. Sort to put the factors in order according to the value of Percentage of Variance ranging from the highest to lowest.
e. Use of analytical results for interpretation and conclusion

3. Results and Discussion

3.1. Overview Location Research

West Java is a province that has a strategic position for agribusiness development. This is reasonable because geographically, the province has close proximity to Jakarta and Bodetabek which is a potential market for agricultural products, including livestock, produced by the regions in West Java. This condition affects farm typology of this area, which leads to growing production patterns (on farm) and commercial business orientation [2]. Related to this research, the chosen location is in the northern part of the West Java region represented by Subang district; the central part represented by Bandung District; and the southern part represented by Ciamis District. All three regions have different characteristics which have the potential of different resources, which its condition will affect livestock business management systems and risk.

Subang district is one of regencies in West Java province who has the area divided into three parts with each characteristic. The northern part of Subang district characterized with its low-lying area that is filled with technical irrigation paddy fields and ponds and its area bordering Java Sea, so the temperature in this region is relatively hot. The middle part of Subang district consists of rubber plantation area, sugarcane, and fruits in agriculture and factories in the industrial field. In addition, this region is the government, education, and services centre. The south of Subang district is an area of highland covered by plantations of government and smallholders with relatively cool/cold temperatures. Based on the above description, it can be inferred that the most apt area for beef cattle breeding is northern territory, where the temperature conditions are very suitable for beef cattle physiology and the availability of agricultural wastes that can be used as an alternative of feed source. Nonetheless, it appears that in fact, Jalancagak district become a centre of beef cattle breeding region in the south. This district is directly adjacent to the Sagalaherang and Ciater district which is relatively cool temperature and suitable for dairy cows. Apart from the issue of climate, the main reason for the development of beef cattle farms in this district is to bring farm closer to the market area, including the Bandung city and Bandung district (feedloters). In its development, Department of Animal Husbandry of Subang district is developing farm area to other districts adjacent to Jalancagak District. These regions are Pagaden subdistrict and West Cikaum subdistrict. Both regions are symbolize with people populated with livestock livelihoods dependant on the first place, so it is reasonable if these regions were developed for smallholder beef cattle centre by government state, especially breeding. In addition, both regions also filled with paddy and corn fields, thus it can ensure the feed availability that will sustain beef cattle farming.

Bandung District became the centre of small holder beef cattle farm in a mountainous region with a relatively cool climate, precisely located in Cikancung subdistrict. Cikancung subdistrict monographs shows physical condition of this region have a number of days with rainfall of between 4 to 6 days. The average rainfall is 800 mm/year with a daily average temperature of 22°C - 24°C. Soil types consist of latosol and andosol, which are the result of the volcanic mineral deposition of Mandalawangi mount that is fairly fertile for various types of crops. This condition will definitely support the performance of small holder beef cattle farm. Physiologically, beef cattle require a relatively tropical climate conditions. Nonetheless, beef cattle in fact can grow well in this region. The beef cattle farm growth of in the region cannot be separated from the feed availability. It's filled with an agricultural area of vegetables which helps farmers for providing alternative feed if the dry season has arrived. Pangandaran district, is a new district formed out of Ciamis District. Most of the district is a coastal / beach so it is plausible if the regions became a centre for small holder beef cattle breeding. Beef cattle breeding centre is concentrated in Parigi subdistrict, Cijulang subdistrict, and Cimerak subdistrict. Residents in these regions are dominated by fishermen, who at a certain time cannot go to sea so they switch with ranching activities. Theoretically, the condition of coastal areas is physiologically suitable for beef cattle growth. Nonetheless, the risk is relatively similar to that in the highlands.

3.2. Farmer's preferences towards risk

Model used to seek preference of farmer risk is utility function in quadratic one. Underlying conditions for choosing quadratic function are: 1) quadratic function is apt on its topic and process that that on other non linier regression for its handy in transformation to linier one; 2) regression coefficient is also its elasticity; and 3) regression line is projected to be non linier for it relative and varieties behavior [3]. The quadratic regression obtained was as follows:

Table 2. Quadratic Reggression Coefficient

	Unstandardized Coefficients		Standardized Coefficients	t	Sig
	B	Std. Error	Beta		
Scale	4178895.687	1316093.270	.538	3175	.002
Scale^2	-63692.693	41870.228	-258	-1.521	.130
(Constant	1175184.496	461671.264		2.537	.012

Based on the output, the value of Xj, x_2, and x_3 are 11.751.843,496, 4.178.895,687, and -63.692,69 so the quadratic model in this study is as follows:

$$U = 11,751,843.496 + 4,178,895.687\ M - 63692.693\ M^2$$

The quadratic model above shows that the estimated utility function shows negative sign (X_3 coefficient = -63692.693). It gives an overall indication that farmers tend to be reluctant to risk (risk averter). This study revealed similar results to [4] which mentioned that most farmers tend to refuse or are reluctant to face a risky situation.

Basically, Farmer's willingness to take risk depends on the psychic nature and the satisfaction or utility received by farmers from the output [5]. According to [6], the farmer's aversion is negatively related to their welfare measured by income level. Higher income would make someone want to sacrifice income for gambling or speculating in order to get a larger acquisition than that have been given.

3.3. Factors Determining Risk Preferences of Farmers

To evaluate the factors that determine the preference of farmers against risk, the appropriate analysis model to answer that is the analysis of Component Factor Analysis (herein after abbreviated CFA) with six indicators studied: 1) the business scale; 2) age; 3) the number of dependants; 4) education; 5) the cattle business experience; and 6) the status of the business ownership. Furthermore, in order to see the effects of factors that form the gross revenue multiple linear regression analysis was conducted. Some assumptions have to be fulfilled for CFA method.

Table 3. KMO and Bartlett's Tests KMO and Bartlett's Test

Kaiser-Meyer-Olkin Measure of Sampling Adequacy	.547
Bartlett's Test of Sphericity Approx. Chi-Square	70.828
Df	15
Sig.	.000

Table 3 shows that the value of KMO MSA is 0,547. This is bigger than the minimum requirement (0,500) therefore it can be concluded that the number of sample used is sufficient and qualified; whilst Barlett Testshows that the probability is as much as 0,000 which is smaller than the alpha (0,05), it can be considered that the six indicators have correlation one another and these facts takes the results to another method which is factor analysis. Furthermore there is another test, anti-images matrices, that shows as follows:

Table 4. Sampling Adequacy (MSA)

	The Business Scale	Age	The Number of Dependamts	Education	Experience	Status Of The Business
Anti-image Correlation: the business scale	.539[a]	-.069	-.134	-038	-.061	-.149
Age	-.069	.560[a]	-.169	.239	-.372	.050
The Number of Dependamts	-.134	-.169	.541[a]	.100	.133	-.064
Education	-.038	.239	.100	.589[a]	.083	.161
Status od The Business	-.149	.050	.-064	.161	.235	.516[a]

[a]Measures of Sampling Adequacy(MSA)

Table 4 shows that the value of MSA for each indicator (sign[a)]) has fulfilled the minimum requirement for significance (0,500) therefore the six indicators are available for factor analysis. The three tests have given result that availability of the data which describe the variables in the regression model is well done therefore it can be usedto predict other factors that determine farmer preference on cattle business risk in West Java. Hence, the next step is factor analysis to seek for the determinant of the sixth socio-economic variable in the model.

Factor analyses showed that the number of factors formed were determined based on the Eigen values. A factor formed if the Eigen value is greater than 1. The analysis indicated that there were two factors that had Eigen values > 1, i.e.: factor 1 (1,650) and factor 2(1,386). Thus, the sixth indicator can be reduced to just two. The first factor accounted for 27.51% and the second factor contributed 23.11% so that the accumulation of both factors contribute to absorb the information attributed in the six indicators of 50.61%. Grouping the six indicators to 2 factors can be seen from the results of the rotation matrix components as follows:

Table 5. Analysis of Factor

Component	Initial Eigenvalues			Extraction Sums of Squared Loadings		
	Total	% of Variance	Cumulative %	Total	% of Variance	Cumulative %
1	1.650	27.505	27.505	1.650	27.505	27.505
2	1.386	23.106	50.611	1.386	23.106	50.611
3	0.956	15.932	66.543			
4	0.853	14.214	80.757			
5	0.632	10.538	91.295			
6	0.522	8.705	100.000			

Extraction Method: Principal Component Analysis

Table 6. Groups of factorsRotated Component Matrix

	Component	
	1	2
The business scale	.067	.522
Age	.783	.251
The number of depedamts	.079	.649
Education	.-436	-.476
Experience	.800	-.181
Status of the business	-.402	.637

Extraction Method: Principal Component Analysis.
Rotation Method: Varimaxwith Kaiser
Normalization.
a. Rotation converged in 3 iterations.

Based on the output it can be referred that factor 1 consists of: 1) age (0.783); 2) education (-0.436); and business experience (0,800), while the Factor 2 consists of: 1) the business scale (0.522); 2) the number of dependamts (0.649); and 3) the status of business ownership (0.637). Classifying factors into group is done by using the maximum value of each indicator.

The above results indicated that factor determining preferences of farmers against the risk was 1: age, education and business experience. Factor 2: the scale of business, number of dependants, and the status of business ownership was a major contributing factor to the gross revenues from livestock business. Associated with the results of the analysis for factor 1: age, education, and livestock business experience can be justified that age is closely associated with a person's attitude, including farmers, where the older the farmers, the more reluctant they are to risk; while the relatively young farmers more willing to take the risk. This is in line with the opinion of [7] stated that farmers who are older may have better ability in livestock business because they have better experienced and skills, but usually more conservative and easily tired. Young farmers may be poorer in experience and skill, but they usually have a progressive attitude towards new innovations and relatively stronger. In related with the risk taking behavior, the progressive attitude towards new innovation is what tends to establish the young farmer behavior to be more willing to take risks.

The results of the above analysis also show that livestock business experience is the deciding factor of farmer's risk preference. This is similar with the opinion of [8] that among several factors that affect the formation of a person's attitude is a personal experience. Based on experience, sometimes farmer use prediction of forthcoming event as considerations in taking risky decisions. Schultz (1964) in [4] also mentioned that the more experience, the better in solving problem. The formal education have a little effect (-0, 436) as it is not taught a specialized knowledge of farming/running livestock business. In practice, farming and livestock business knowledge and skills are obtained by farmers' own experience and observation, both the experience gained from generation to generation as such his parents and from his observations of the surrounding environment and counseling they have been participated.

4. Conclusion

According to the problem formulation, results, and discussions, it can be concluded as follows:
1. Overall farmers are reluctant to risk (risk averter). This is indicated by the value of x_3 -63,692.693<0;
2. Social and economic factors that determine preferences of farmers against risk are age, education, and cattle rearing experience; whereas the scale factor of business, number of dependents, and cattle ownership status determine the size of the gross revenue/income.

Acknowledgments

The research is finally done with so many contributions from so many people and institution facilities therefore we would like to thanks:
1. Rector of Universitas Padjadjaran that concern the needs of research by allocating the fund annually which will take lecturers to create better future related to application of their science;
2. Directorate of Research and Community Service of Universitas Padjadjaran that has given the opportunity for granted;
3. Dean of Animal Husbandry Faculty of Universitas Padjadjaran that has allow the lecturers to conduct the research;
4. Farmers groups and extension workers in each location of the research that cannot be explicitly named, but particularly Pak Didin (Head of Parigi Village, Ciamis) that has given so much concern for the research data collection;
5. The researches and the students for contributing to the writing.

References

[1] Scott. 1977. *The Moral Economy of The Peasant: Rebellion and Subsistence in Southeast Asia.* London: Yale University Press.

[2] Arief, Hasni., Achmad Firman, Lizzah Khaerani, dan Romi Zamhir Islami. 2011. "Inventarisasi dan PemetaanWilayah Sapi Potong dan Lumbung Pakan di Jawa Barat". Laporan Akhir Penelitian Dinas PeternakanProvinsi Jawa Barat.

[3] Soekartawi, Rusmadi, dan Effi Damaijati. 1993. *Risiko dan Ketidakpastian dalam Agribisnis, Teoridan Aplikasi.*Jakarta: PT. Raja Grafindo Persada.

[4] Sabrani, M. 1988. "Perilaku Petani Peternak Domba dalam Alokasi Sumberdaya. DisertasiS-3. Yogyakarta:UGM.

[5] Simanjuntak, S. 1990. "Analisis Risiko Produksi *(Production Risk)* dan Efisiensi AlokasiSumberdaya dalam UsahaPengembangan Budidaya Tambak di Kotamadya Surabaya, Jawa Timur". Tesis S2. Malang: FakultasPascasarjana KPK UGM-Unibraw.

[6] Hammal, A.M., dan Saefuddin, A.M. 1983. *Tata Niaga HasilPerikanan.* Jakarta: UI-Press.

[7] Widodo, S. 1986. "An Econometric Study of Rice Production Efficiency Among Rice Farmers inIrrigated LowLand Villages in Java, Indonesia". Dessertation S3. Tokyo: Tokyo University of Agriculture.

[8] Rusmadi dan Soekartawi. 1992. "Efisiensi Relatif dan Pengaruh Sikap Petani terhadap Risiko dalam UpayaPengembangan Komoditas Kedelai (Studi Kasus di Desa Ngabar Kabupaten Pasuruan, Jawa Timur)". TesisS-2. Malang: Program Pascasarjana KPK UGM - Unibraw.

IOP Conf. Series: Earth and Environmental Science **119** (2018) 012066 doi:10.1088/1755-1315/119/1/012066

Identification of Social Capital on Beef Cattle Farmers Group

V. S. Lestari[1], S. N. Sirajuddin[1] and A. Abdullah[1]

Department of Socio-Economics, Faculty of Animal Science, Hasanuddin University, Makassar 90245,South Sulawesi, Indonesia

Corresponding e mail: veronicasarilestari@unhas.ac.id

Abstract

Social capital plays an important role in the development of beef cattle farms in South Sulawesi. The aim of this research was to know the social capital of beef cattle farmers in South Sulawesi. Population of this research was 31 beef cattle farmers. Variable of social capital was mutual trust, norms and linkage. The data were collected from observation and depth interview by using questionnaire. There were 10 questions which were adopted from Australian Center for International Agriculture Research. The answer was scored by using Likert scale ranging from 1 refer to strongly disagree; 2 refer to disagree; 3 refer to not sure; 4 refer to agree and 5 refer to strongly agree. The data were analyzed descriptively by using frequency distribution. The research revealed that the social capital of beef cattle farmers was categorized as "high".

1. Introduction

Beef cattle has an important role as a source of animal protein. The development of beef cattle industry was supported by Indonesian government, private sectors and beef cattle farmers, however meat production in Indonesia was low and cannot fulfill the domestic demand. It is because most of beef cattle farmers were managed traditionally in a small size business. Beef cattle farmers have cattle only as a savings that at any time can be sold for school or for religious party. The beef cattle management is not professional. Beside that, knowledge in terms of technical livestock breeders was low, beef cattle farmers were lack of education and lack of extension.

One important factor in growing the economy of society is social capital. The role of social capital is no less important than other economic infrastructure, so efforts to build social capital need to be prioritized for the success of economic development. The formation of social capital can contribute to economic development because of the network, norms and trust in which the collaboration (coordination and coorporation) social for the common interest of an institution [1].

Reference [2] argued that social capital is a characteristic of social organization such as trust (trust), norm, and networking that facilitate the occurrence of coordination and cooperation for mutual benefit. While [3] stated that social capital is productive that allows achievement of goals. Social capital is attached to the structure of relationships among people. According to [4], social capital in a society can be strengthened but requires the support of certain resources. In order to create good social and institutional relationships so community members must support it. Social capital acts as a glue that binds all the people in the society. In order for social capital to grow well there is a need for shared values, beliefs, and organizing roles expressed in personal relationships.

The major elements of social capital include norms, reciprocity, trusts and networking [5]. Reference [6] added that social capital is formed and developed through the existence and life experiences of human beings. Each person in a different social environment such as a different family or location will have different views. Further, each community has its own concepts and values that

IOP Conf. Series: Earth and Environmental Science **119** (2018) 012066 doi:10.1088/1755-1315/119/1/012066

are formed through the accumulation of beliefs, moral standards and other individual contacts inside that community.

Social capital is anything related to cooperation in society or nation. To achieve a better purpose of life, sustained by the values and norms that become the main elements such as trust, community participation, reciprocity, collective rule in a society or nation and the like [7] .According to [8], the existence of social capital has an equally important role with the existence of economic capital in the empowerment process, but is still often ignored. Social capital consisting of trust, norms and networks. Mutual trust among members will increase the cohesiveness of group members which will further ensure group sustainability. Reference [9] argued that one of the elements of social capital needed in the group is the nature and attitude to trust and trustworthy in the relationship so that among the group members have a high level of trust. Trust is a very important capital to build a network.

According to [10], norms that must be obeyed with clear and effective sanctions. However, there is no mechanism to impose sanctions because there is no mutually agreed norm regarding the sanctions.Social capital as the ability to command scarce resources by virtue of membership in network and broader social structures [11]. Furthermore, [12] stated that, "Social network are among the most important types of structures in which economic transactions are embedded. Networks are important in economic life because they are sources for acquisition of scarce means, such as capital and information, and because they simultaneously impose effective constraints on the unrestricted pursuit of personal gain".

Beef cattle breeding in Luwu regency is not much different from elsewhere. The breeders are grouping themselves in beef cattle farming groups. The number of beef cattle increased more than 2 times from 6,659 heads in 2014 to 17,886 head by 2015 [13]. Harapan Village, Walenrang sub District, Luwu Regency, is one of the areas whose people come from Java island since colonialsm periode. Therefore, the relation between each other is very strong. The main job are farmers. Beside that they maintain beef cattle. The smallest scale size is 2 heads and the biggest scale size is 11 heads. The population of beef cattle in Walenrang subDistrict, Luwu Regency can be seen in Table 1.

Table 1. Beef Cattle Population in Harapan Village, Walenrang sub District, Luwu Regency

No	Village	Beef cattle population (head)		Total
		Male	Female	(head)
1	Tombang	75	85	160
2	Baramamase	43	57	100
3	Kalibamamase	5	8	13
4	Lalong	121	218	339
5	Walenrang	87	101	188
6	Batusitanduk	32	43	75
7	Bulo	5	11	16
8	Saragi	9	15	24
9	Harapan	103	211	314
	Total	480	749	1,229

Source: BP3K Walenrang (2016)

Based on the data in Table 1, it can be seen that Harapan Village is one of the villages that have the second largest population of beef cattle in sub District Walenrang, Luwu regency. There was 314 head. The purpose of this research was to identify social capital on beef cattle farmers group in Luwu Regency, South Sulawesi Province.

2. Materials and Methods

This research was conducted on beef cattle farmers group in Luwu Regency, South Sulawesi. As a case study total sampel was 31 beef cattle farmers which was located in Harapan Village, Walenrang sub district, Luwu regency. Group leader was asked a series of questions concerning group interaction and group member interaction with the broader community (Table 2). They were also asked attitudinal questions concerning the level of homogeneity within the group with respect to wealth and power. Group leaders was asked to state whether or not he agreed with a range of statements. These were compiled into measures of agreement that ranged from 'disagree strongly' (scored as 1) to 'agree strongly' (scored as 5). The data were analyzed descriptively using frequency distribution.

3. Results and Discussion

Beef cattle farmers mostly were dominated by men (83.87%), because to manage beef cattle, people need a lot of work and power. On average, the age of beef cattle farmers was 43.49 years. This indicated that they were in productive age. Based on their education, beef cattle farmers spent 7.839 year in schooling. This mean that their education was low because they studied only until Junior High School. According to the family size, beef cattle farmers came from small families. Number of beef cattle on average was 4 head, this mean that beef cattle farmers have small scale business. With regard to their experience, on average beef cattle farmers spent 10 years in their farms. They got the experience to manage their cattle from their parent.

Table 2. Beef Cattle Farmer Atitudes to Community and Connectedness

No	Atitude and Connectedness	Average
1	Trust people outside the group	3.3
2	Trust other members of the group	4.5
3	Rely on other in community to provide help when needed	3.7
4	Coorporate with what the group decides	4.0
5	Will make disappointment known if othe group member fails to coorporate	3.5
6	Respect each other's opinion even when different from their own	3.5
7	Avoid saying anythingthat goes against custom and tradition	3.5
8	Respect one another as equals	3.3
9	Have approximately the same level of wealth	3.5
10	Have approximately the same level of power	3.0

1 = strongly disagree, 5 = strongly agree

Table 2. showed that the highest score level was 4.5 refer to trust other members of the group. This indicated that beef cattle farmers more trust to other member of the group rather than people outside the group. Furthermore, the lowest score was 3.0 refer to every member of the group have the same level of power.

IOP Conf. Series: Earth and Environmental Science **119** (2018) 012066 doi:10.1088/1755-1315/119/1/012066

Table 3. Category of Social Capital on Beef Cattle Farms

No	Variable	Score	Category
1	Trust	3.6	High
2	Norm	3.6	High
3	Linkage	3.5	High
	Total Social Capital		High

Based on Table 3, total social capital for beef cattle farmers was categorized as "high". This research agree with that of [14] who found that social capital of beef cattle farmers in Jambi province were categorized as high with the average score was 3.8.

Although social capital was high, it is also important to undertstand whether or not this social capital can be turned into action (i.e. whether or not there is a high level of "agency" within the group). This group agency variable was determined from an understanding of how group would respond to issues or difficulties facing the group and individuals.

Table 4. Beef Cattle Farmers response to beef cattle difficulties

Response	Percentage (%)
Group members would help each other	58.8
Group would ask for help from government	41.2

Based on Table 3, if there was an issue facing the group (for example beef cattle difficulties), 58.8% of beef cattle farmers would like to help each other rather than asking for help from government. Helping each other was the characteristics of beef cattle farmers. This result of the research was supported by [14] who found that beef cattle farmers in Bali and Lombok do the same respond.

Table 5. Expected farmer group response to a disaster affecting one member

Response	Percentage (%)
Group member would come together to help	88.2
Group leader would organise members to help	11.8

Table 4 showed that if an individual group member was to suffer, in most cases the group would come together to assist (88.2%). This agree with that of [14] who found that beef cattle farmers in Bali and in Lombok do the same respond.

4. Conclusion

Social capital on beef cattle farmers in South Sulawesi was categorized as high. The implication of this research was the social capital of beef cattle farmers should be maintained in order to increase the development of beef cattle farms.

Acknowledgments

Authors would like to appreciate to Hasanuddin University who funding this research.

References

[1] Inayah 2012 Peranan Modal Sosial dalam Pembangunan. Paper Jurusan Administrasi Niaga Politeknik Negeri Semarang, Semarang.

[2] Putnam R D 1993 Making Demoracy Work : Civic Traditions in Modern Italy Priceton University Press. USA.

[3]Coleman J S 2008 Dasar-Dasar Teori Sosial Nusa Media. Bandung.

[4] Syahyuti 2008 The role of social capital in agricultural trade. Agro Economic Forum. Vol. 26, No. 1. Pusat Sosial Ekonomi dan Kebijakan Pertanian. Badan Penelitian dan Pengembangan Pertanian. Kementerian Pertanian RI.

[5] Subejo 2004 Peranan Social Capital dalam Pembangunan Ekonomi: Suatu Pengantar untuk Studi Social Capital di Pedesaan Indonesia. Majalah Agro Ekonomi, 11(1): 79.

[6] Anh N T N 2014. Social capital, leadership reputation and labour force motivation in enterprice. The Journal of Comparative Economic Studies. 9:89-97.

[7] Hasbullah J 2006 Social Capital (Menuju Keunggulan Budaya Manusia Indonesia). Jakarta MR-United Press.

[8] Hadi S 2014 Profil modal sosial dan tingkat partisipasi peternak pada pengembangan sapi potong di kabupaten Tebo, propinsi Jambi. Kanal. Vol 2(2): 107-206.

[9] Pratishita R N M. Munandar., S Homzah 2014 Peran modal sosial dalam menunjang dinamika kelompok peternak sapi perah (Studi kasus di kelompok 3 TPK Pulosari Pengalengan). J. Ilmu Ternak. **1**(10): 52.

[10] Coleman J S 1988 'Social capital in the creation of human capital' American Journal of Sociology 94: S95-S120.

[11] Portes, A. (1995a) Children of Immigrants: Segmented Assimilation and Its Determinant. In A. Portes (Ed.), The Economic Sociology of Immigration. New York: Russel Sage Foundation

[12] Portes, A. (1995b). Economic Sociology and the Sociology of Immigration : A Conceptual Overview. In A. Portes (Ed.), The Economic Sociology of Immigration (pp. 1-41). New York: Russel Sage Foundation.

[13] Central Bureau Statistics of South Sulawesi. 2015. Sulawesi in Figure. Makassar.

[14] Patrick I W, G R Marshall., I G A A Ambarawati and M Abdurachman 2010. Social Capital and Cattle Marketing Chains in Bali and Lombok, Indonesia. ACIAR Technical Reports No. 74 Australian Centre for International Agricultural Research Canberra: 78 pp.

IOP Conf. Series: Earth and Environmental Science **119** (2018) 012067 doi:10.1088/1755-1315/119/1/012067

Factors Causing Farmers Not to Ferment Rice Straw as Cattle Feed

S. N. Sirajuddin[1*], I. M. Saleh[1], S. Syawal[2], Syamsinar[3]

[1]Department of Socio Economic, Faculty of Animal Science Hasanuddin,University,Macassar,South Sulawesi,Indonesia
[2]Department of Livestock Production,Faculty of Animal Science, Hasanuddin,University,Macassar,South Sulawesi,Indonesia
[3]Department of Agribusiness, Faculty of Agriculture, Islamic University of Macassar,Macassar,South Sulawesi,Indonesia

*Corresponding e mail: sitti.nurani@unhas.ac.id

Abstract

This study aimed was to identify the factor of breeders have not done fermentation of rice straw as cattle feed. This research was conducted on August-September 2017 in Patampanua village, Marioriawa sub-district, Soppeng district. This research is descriptive quantitative with Delbeq method. Data collection is qualitative and quantitative. Data sources are primary and secondary data. Data analysis used is frequency distribution. The results showed that farmers have not done the fermentation of rice straw as animal feed that is the motivation of farmers, intensity counseling and lack of knowledge of farmers

1. Introduction

Cattle ranching in an area can make optimal use of local resources and by-products such as rice straw from rice plants and the joint production of rice bran that can be utilized by cattle as cattle feed. Feed is a major component for the success of the cattle business [1]. According to [2] the weakness that lay in the livestock production systems are not exactly feeding management. The nutrient content of forage depends on the composition of materials. Forage comes from native pasture, grasslands, improved pasture, legume-grass pasture and agricultural waste. The nutrient content of legume-grass pasture is 21%, grassland is 10, 20% [3] and agricultural waste content is 3.6%. Naturally, ruminants like cattle, buffalo, goat and sheep need grass and foliage as feed. Forage from native pasture and improved pasture the availability of which as forage is affected by season and while in the dry season is very limited,necessitates that an effort to find alternative feed sources that are inexpensive, economical, easy to obtain and do not compete with human need be made. Utilization of waste agricultural and plantation is choice the right to address the feed problems. The wet season is plentiful and yet the dry season has a limited stock of foliage so during the dry season, breeders mostly use agriculture waste like rice straw.Rice generates a relatively large amount of crop residue known as straw. These residues are the leftover vegetative parts after harvesting the grains. Rice straw is made up of panicle rachis, leaf blades, leaf sheath, and the stem. The average ratio of the rice grain to rice straw is 1:1.25 [4].Only about 20% of these straws are used for industrial (e.g., ethanol, paper, and fertilizer) and domestic (e.g., fodder) purposes [5]. Most of the remaining straws can be removed from the field, left undisturbed to serve as mulch, ploughed into the ground to add nutrients to the soil, or burnt. Burning of rice straw is the fastest mode of straw disposal but causes environmental pollution by increasing the amount of greenhouse gas into the air [6]. Rice straw is used as part of the nutritional requirements of ruminant animals in most rice-producing countries [7]. However, low protein content, possession of phenolic properties, and high level of silica and lignin are the primary limiting factors in rice straw digestibility in ruminant animals [8].

To maintain the availability of feed especially during the dry season it is necessary to find alternative feed to substitute the field grass so that the nutritional intake of feed on the cattle is still guaranteed [9]. One alternative for the provision of cheap and competitive feed is through the utilization of waste both agricultural waste, livestock waste and industrial waste [10]. Rice straw is one of the agricultural wastes that quite a lot and not yet utilized. Production of rice straw can reach 12-15 tons per ha / one time harvest or 4-5 tons of dry material depending on the location and variety used [11]. Utilization of rice straw as fodder is constrained mainly due to the limiting factor with low nutritional value of low protein content, as well as high roughage and low digestibility. The low nutrient content of rice straw and the difficulty of digestibility of straw then in its utilization need to get treatment so that the nutrients increase. One way to increase the content of food substances is by processing rice straw through fermentation.

Soppeng Regency is one of the main rice-producing areas in South Sulawesi Province so that the amount of agricultural waste that is rice straw is very high as well as in the District of Marioriawa. However, people in the village of Patampanua, Marioriawa sub district still do not make the fermentation of rice straw as cattle feed so research needs to be done to identify the factors that cause farmers have not fermented rice straw as cattle feed.

2. Materials and methods

This research was conducted from July to August 2017 in Patampanua Village, Marioriawa Subdistrict, Soppeng Regency. The selection of this location is done purposively with consideration that in Desa Patampanua enough rice fields are planted with rice so that the amount of rice straw is also quite high and is the location of Science for Region (IbW). This type of research is a type of descriptive quantitative research that aims to describe the factors that cause farmers have not fermented rice straw as animal feed. Population in this research is all breeder cattle in the District Marioriawa, Soppeng District. Samples are farmers who have not fermented the rice straw as livestock feed, which is a member of the Tonronge group and members of the Lapagiong group. The type of data is qualitative and quantitative. Sources of data used in this study are primary data and secondary data. Data collection method is by observation and interview by using questionnaire. Data analysis used in this research is descriptive analysis is explorative by using method of Delbeq.

3. Results and Discussion

3.1. Characteristic of Respondent

As the subject in the development of beef cattle breeding, the success of the effort dependscrucially on the ability of farmers in managing it. Some important factors are seen in this study were age, education, experience farming, family size, scale of business.

Table 1 shows that the age of farmers in Patampanua sub-district is mostly at productive level (83.3%). This is in accordance with [12] that the productive breeder's age usually has a dynamic mindset and excellent physical ability in handling the business

Farmers in the district of Patampanua are mostly educated at primary school (53.4%), with low level of education indicates that the level of education influences the acceptance of new innovations, this is in the opinion of [12] that the level of education of farmers is an indicator of population quality and is a variable key in the development of human resources. Adequate breeder education will facilitate the acceptance process of innovation and technology of beef cattle farming.

The majority of respondents' livestock ownership experience is greater than 11 years (53.3%) and this has an effect on the management of beef cattle business, it is in accordance with [12] that the experience of raising beef cattle is a very important role in determining the success of farmers in enhancing the development business of cattle and at the same time increase the income of farmers. Breeding experience is a good teacher, with the experience of raising more cattle farmers will be more careful in trying and can fix deficiencies in the past.
Business scale of beef cattle breeder in District of Patampanua on scale of 3-7%, with scale of ownership of beef cattle big enough will influence management of beef cattle business, this is in accordance with opinion of [12] that amount of beef cattle ownership is indicator of success cattle breeding business.

Table 1. Characteristics of Respondents Based on Frequency Distribution

No	Parameter	Number(people)	Percentage
1	Age		
	< 20	0	0
	21-55	25	83,3
	>55	5	16,7
2	Education		
	Finished primary school	16	53,4
	Finished junior high school	6	20
	Graduated from high school	7	23,3
	College	1	3,3
3	Long breeding (year)		
	2-10	14	46,7
	11-22	10	33,3
	23-33	4	13,3
	34-50	2	6,7
4	Scale enteprises		
	<2	10	33,3
	3-7	14	46,7
	>7	6	20

Source: Primary data, 2017

3.2. Breeders have not Fermented Rice Straw as Animal Feed

To identify the beef breeder has not done the fermentation of rice straw as animal feed that is doing four stages:
The first stage is data retrieval using questionnaire using open and focused questionnaire format in the respondents given the freedom to write down the factors of breeders have not done the rice straw fermentation and the results are:

1. Education Level

2. Motivation
3. Intensity of Extension
4. Economic Factors
5. Experience making rice straw fermentation

In the second stage is to do ranking of respondents' answers and the results obtained are:

1. Intensity of extension
2. Level of education
3. Motivation
4. Experience making fermentation
5. Economic Factors

In the third stage of doing a large 3 ranking of respondents' answers and obtained results are:

1. Motivation
2. Intensity of Extension
3. Education Level

Furthermore, in the fourth stage is done ranking and obtained the same results with the third stage, namely:

1. Motivation
2. Intensity of Extension
3. Education Level

Motivation occupies the first rank, due to the motivation is divided into 2, namely the motivation of the self and the motivation of the environment. If breeders have a strong motivation from within and influenced also from the outside environment then the farmers will make the fermentation of rice straw as animal feed, this is in accordance with the opinion of [13] that motivation is the thing that pushes someone to something the business wants to know and also in accordance with the opinion of [14] that one of the internal factors of motivation is very influential on the adoption of innovation.

Furthermore the second factor is the intensity of counseling where the farmers only very little have followed the counseling about the making of rice straw fermentation as beef cattle feed so that the information obtained is also less, it is in accordance with the opinion [15] that the intensity of extension is very important in the process of technology adoption. The higher the frequency of counseling then the extension of agricultural extension delivered higher. And in the third rank indicates that low level of education also influences knowledge and also can be seen from breeder characteristic with highest education (53,3%) cause farmer not easy to adopt a technology, this is according to [16] education can affect a farmer's ability to adopt a technology, the higher the level of education of farmers, then in understanding a technology will be easier.

4. Conclusion

Cattle breeders in the district of Patamanua, Soppeng District Have not done the fermentation of rice straw as beef cattle feed caused by motivation factor, extension intensity factor and level of education The intensity of counseling and motivation improvement for beef cattle farmers in Patampanua sub-district in making the fermentation of rice straw as animal feed can be improved in the sub-district can be improved so that the productivity of beef cattle can be increased.

Acknowledment

To the Ministry of Research and Technology of Higher Education who has Provided the Grant of Science and Technology for the Region (IbW)

References

[1] Sirajuddin,S.N.A.Asnawi,S.Syawal,M.Jamal. 2016. "Response of cattle breeders to ricestraw silage in Soppemg Regency, South Sulawesi Province." J. America-Eurasian Sustainable Agriculture, **10** (3):33.

[2] Priyanto, D., 2008. Peranan usaha ternak kambing local sebagai penunjang perekonomian petani di pedesaan. Prosiding Seminar Nasional Teknologi Peternakan dan Veteriner. Bogor 17- 18 September,pp 418-426

[3] Rukmana, R., 2005.Rumput Unggul Hijauan Makanan Ternak.Kanisius,Yogyakarta(in Indonesia

[4] Haefele, S.M., Y. Konboon, W. Wongboon et al., 2011. "Effects and fate of biochar from rice residues in rice-based systems," Field Crops Research, 121(3): 430-440. View at Publisher • View at Google Scholar • View at Scopus

[5] Kadam, K.L., L.H. Forrest and W.A. Jacobson, 2000. "Rice straw as a lignocellulosic resource: collection, processing, transportation, and environmental aspects," Biomass and Bioenergy,18(5): 369.

[6] El-Gammal, M.I. and A.A. Shakour, 2001. "Emission of pollutants from harvest and burning of rice straw in Egypt villages (North East of Nile Delta)," J. Union Arab Biol, **15**: 191.

[7] Dong, C.-F., Q.-S.Cai, C.-L. Wang, J. Harada, K. Nemoto and Y.-X. Shen, 2008. "QTL analysis for traitsassociated with feeding value of straw in rice (Oryza sativa L.)," Rice Sci., **15**(3): 195.

[8] Van Soest, P.J., 2006. "Rice straw, the role of silica and treatments to improve quality," Animal Feed Sci. Technol, **130**(3-4): 137.

[9] I.N.Sugama and N.L.G.Budiari.2002. Pemanfaatan jerami padi sebagai pakan alternative untuk sapi bali dara.Majalah Ilmiah Peternakan:.**15**(1):21.

[10] Mastika, I.M.1991.Potensi limbah pertanian dan industry pertanian serta pemanfaatannya untuk makanan ternak. Makalah pengukuhan guru besar ilmu makanan ternak pada fakultas peternakan UNUD-Denpasar

[11] Yunilas.2009. Karya Ilmiah. Bioteknologi Jerami Padi melalui Fermentasi sebagai Bahan PakanTernak ruminansia. Fakultas pertanian,Universitas Sumatera Utara,Medan

[12] Murwanto,AG. 2008. Karakteristik peternak dan tingkat masukan teknologi peternakan sapi potong di lembah Prafi, Kabupaten Manokwari. Jurnal Ilmu Peternakan. 3(1): 8-15

[13] Mardikanto, T.1993. Penyuluhan Pembangunan Pertanian.Universitas Negeri Sebelas Maret.Surakarta Press.Surakarta

[14] Subagiyo,R dan Sekarningsih,R.2005. Kajian faktor-faktor social yang berpengaruh terhadap adopsi inovasi usaha perikanan laut di Desa Pantai Selatan Kabupaten Bantul,DIY.Pengkajian dan Pengembangan Teknologi Pertanian.8(2):1-7

[15] Sumbayak.J.B.2006. Materi,metode dan media penyuluhan.Fakultas Pertanian.Univesitas Sumatera Utara,Medan

[16] Soekartawi.2005. Prinsip Dasar Komunikasi Pertanian.UI Press.Jakarta

Mukhabarah as Sharia Financing Model in Beef Cattle Farm Entrepise

A. Asnawi[1*], A.A. Amrawaty[1] and Nirwana[2]

[1]Department of Socio Economicsin Animal Science Faculty, Hasanuddin University, South Sulawesi, Indonesia
[2]Department of Accounting, Faculty of Economics and Business, Hasanuddin University, South Sulawesi, Indonesia

*Corresponding E-mail: aslinaasnawi@unhas.ac.id

Abstract

Financing constraints on beef cattle farm nowadays have received attention by the government through distributed various assistance programs and program loans through implementing banks. The existing financing schemes are all still conventional yet sharia-based. The purpose of this research is to formulate financing pattern for sharia beef cattle farm. A qualitative and descriptive approach is used to formulate the pattern by considering the profit-sharing practices of the beef cattle farmers. The results of this study have formulated a financing pattern that integrates government, implementing banks, beef cattle farmers group and cooperative as well as breeders as its members. This pattern of financing is very accommodating of local culture that develops in rural communities. It is expected to be an input, especially in formulating a business financing policy Sharia-based beef cattle breeding.

1. Introduction

Livestock business is one of the sub sectors in the agricultural sector that has a very important role, especially in the fulfillment of animal protein requirements such as meat, eggs, milk, cheese, and others. Therefore, livestock sub-sector contributes to improving the quality of human resources better. Beef cattle breeding business today is important and needs special attention because as a contributor to the availability of meat is increasing along with the increase in the population in Indonesia.

Most of the beef cattle farm in Indonesia is still managed traditionally and classified as a micro-size farm. A lot of challenges faced by rural beef cattle farmers such as low capital ownership and the level of accessibility of financing to formal financial institutions are still relatively low. The low capital ownership itself requires farmers to seek external financing. The external funding source is basically divided into two, namely from formal financing institutions such as banking and government, while informal financing comes from neighbors, friends, relatives and Micro Finance Institutions [3].

Financing sourced from informal institutions is relatively more accessible to beef cattle farmers because it does not require long procedures, it does not need collateral enough with trust capital, and the required funds can be obtained anytime. However, this financing from the side of creditors is relatively risky because precisely in the absence of collateral then there is no risk coverage in case of things that are not desirable. While the formal financing model of the formal institution that has been run nowadays is considered mostly by the beef cattle farmers is not effective because the procedure is long so it takes a relatively long time, the requirements of collateral that is considered burdensome and not well socialized about the information of the financing to the beef cattle farmers[4]. The realization of credit sometimes does not match the time when the funds are needed by the beef cattle farmers. In addition, other disadvantages are the repayment of loan installments is still based on the amount of the loan when the revenue from the sale of cattle fluctuate for example during religious holidays and certain times breeders earn considerable revenue but smaller outside the season [5].

Breeder financing access is also determined by the extent to which the application of local cultural values in everyday life such as honesty, smart, decency, firmness, persistence and self-esteem/embarrassment. The application of the local culture is closely linked to the formation of a character which is one of the creditor's assessments to the prospective borrower in order to approve the proposed credit proposal for capital; business capacity, collateral and economic conditions [9].

The government has provided various assistance to micro, small and medium enterprises in order to help finance farmers such as Food Security and Energy and People's Business Credit. To overcome the collateral constraints for farmers, then its loantoday, in addition, there are those requiring collateral also exists given to prospective borrowers without collateral but enough guarantee with his business. Therefore, the farmer must show seriousness in running his beef cattle farm in order to gain trust from the bank as an executing bank.

On the other hand, some areas in Indonesia including South Sulawesi are quite developed for profit sharing where livestock owners hand over their livestock to farmers who have time and energy but do not have livestock. Where there is anagreement of cooperation for the results of different forms depending on the custom in each region [3], [11]. For many years ago, this practice in the agricultural sector was known as the muzara'ah which had been done between the landowner and the tiller of the land with the share of harvest by half, one-third, or by mutual consent. This is an implication of the concept of sharia-based financing. Recognizing this, shariah-based financing can be an alternative financing to ease the rate of return on installments from financing sources where returns are set based on the results obtained not during harvest (sale of livestock).

In Islamic economics, also known as mukhabarah. Between muzara'ah and mukhabarah is almost the same that is a practice for the results where the landowner gives up his land to others to manage. But the difference is in muzhara'ah, capital is issued by the land owner while mukhabarah, the capital comes from the manager. The implications of muzara'ah and mukahabrah are also found in thelivestock business. In practice, mukhabarah is mostly applied by the farmer community where the livestock owner is analogized as the land owner, while the farmer who keeps and provides the cage and the feed and forage analogized as the manager or tiller of the land. Therefore, the author raised the title of Islamic financing model on the beef cattle farm.

2. Materials and Methods

Research located in Bone regency of South Sulawesi. This area is very potentialas a beef cattle development area. The large population, the largest area in South Sulawesi, the number of breeders and a large number of beef cattle populations, and the very large agricultural land strongly support it. Local cultural values are highly respected, including in relation to veterinary business. The descriptive approach used is descriptive-analytic, is research that aims to reveal, describe and describe an objective problem of the object being investigated. This descriptive study aims to create a description, description systematically, factually and accurately on the facts, properties, and relationships between the phenomena investigated.

To collect data also use some informants as people who will provide information about pattern or model of financing that can be formulated related to accessibility of financial. .More informant in this research are some parties like Bank of The People of Indonesia, Department of animal and Health Service, Village Head, public figure, cultural figure, community, and beef cattle farmers. It's Bank was selected from the banking element because the banking institution has a business network to the rural areas, especially remote villages which are the area where most of the farmers live, and also has channeled loan to abeef cattle farm.

Sources of data in this study can be grouped into two namely:

1) Primary data is a source of data obtained directly from the object of study (not through intermediaries). In this study was conducted by using direct interviews to informants to obtain data and formulate patterns and financing models tailored to the conditions of the community, breeders, local culture, characteristics and capabilities of the breeder.
2) Secondary data, is a source of data obtained indirectly but through intermediary media (obtained and recorded by other parties), which is obtained from the Provincial Livestock Service Office of South Sulawesi, Department of animal and Health Service in Bone District, including data on the number of livestock population, and other relevant data for the business profile of beef cattle farms.

Data collection method using in-depth interview, which is a method of collecting data qualitatively by conducting more in-depth interviews about the pattern or model financing that has been running and the pattern or model that cool and in accordance with the needs and capabilities of farmers in accessing financing.

Descriptive analysis is done in the early stages after the data collection process is done that is about the pattern or model of financing in accordance with the characteristics of farmers, culture, and beef cattle farmers needs. The results are depicted in the form of a chart or pattern.

3. Results and Discussion

The alternative livestock financing scheme proposed by researchers is to integrate the involvement of government, banks, farmer groups and breeders and profit sharing practices in the community. According to Law No. 7 of 1992 concerning banking that has been changed into Law no. 10 of 1998 concerning banking in Article 1 Number 12[10], sharia-based financing is the provision of money or equivalent claims based on an agreement or agreement between the Bank and another party requiring the party financed to refund the money or invoice after a certain period of time in return or profit sharing. It refers to Shariah principles as a rule of covenant under Islamic law between other party banks for the deposit of funds and / or financing of business activities, or other activities stated in accordance with sharia, such as financing based on profit sharing principles (mudharabah), financing based on the principle of inclusion Capital (musharakah), the principle of sale and purchase of goods by obtaining profit (murabahah), or financing of capital goods based on the principle of pure lease without choice (ijarah), or with the option of transferring ownership of leased goods from the bank by another party (ijarawaiqtina).

The results of this study modify the pattern of financing proposed by Nugraha [8] and consider the input of some informants as well as practice for the results that have been running in the community. The scheme of financing pattern of beef cattle farmers can be seen in Figure 1. It is divided into two stages:

1) First phase, describes the linkage between livestock farming groups and cooperatives, the government, and the implementing bank. A member summit Beef cattle farmers and cooperatives apply for financing to implementing banks. But first get recommendations from the Regional Government, in this case, the Department of Animal Husbandry and Animal Health in each region. In addition to providing recommendations, the role of the government also provides guidance or assistance related to the preparation of business recording and feasibility studies of Livestock business so feasible and assessed bankable at the time of proposals proposal.

   The bank then conducts surveys to summits/cooperatives or even to their respective members to assess and evaluate the feasibility of prospective borrowers. The Bank will always coordinate with the relevant agencies before deciding to approve or reject the

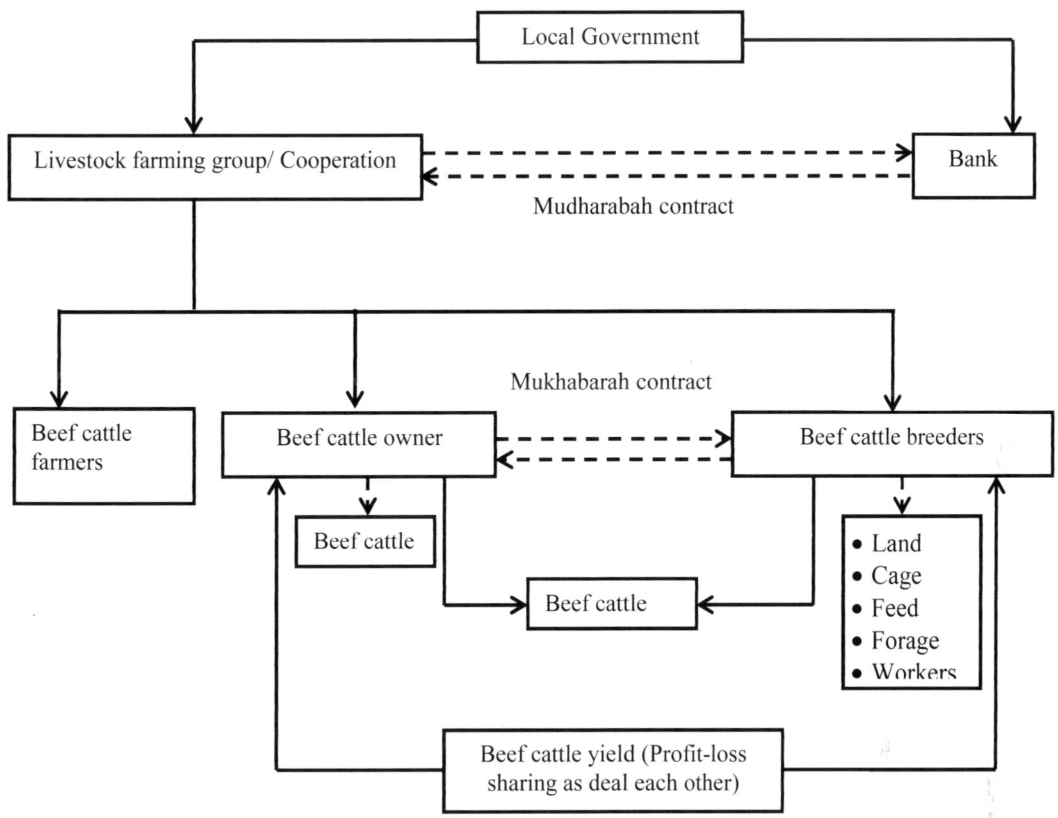

Figure 1. Sharia-based Beef Cattle Farm Financing Scheme

proposed financing. At this stage, the bank will explain to summits and cooperatives about the amount of acceptable fund platform, financing risk, required collateral, term, and method of payment and sharing of revenue sharing and others related to it. If the local government recommends recommending the summit and cooperative, and between the summit/cooperative has agreed the bank then approves and provides financing.

At this stage, the contract between the summit and the cooperative is mudharabah because in this contract the bank is the provider of capital (shahibulmaal) and the Cattle Beef and Cooperative Summit as the capital manager (mudharib) responsible for its management [12] used in beef cattle breeding business. It is asserted by [7] that the profit mudharabah is divided according to the agreement set forth in the contract, while the loss borne by the owner of capital during the loss is not due to the negligence of the manager. If the loss is due to the manager's fraud or negligence, then the manager shall be liable for the loss.

2) The second phase describes the relationship of cooperation contract between the summit and cooperative with the farmer. Breeders in terms of grouped in three namely: individual breeders, breeders as livestock owners and ranchers as cattle rearing. A number of funds available at summits and cooperatives will be distributed to each member according to their needs.

Especially for the practice of sharing the results that have been developed in the community may be said to use the contract mukhabarah. This is because livestock owners only provide and give away their livestock to be reared but the farmers who raise the livestock provide labor, feed, forage, and cage. Revenue shares are set depending on the deal.

Basically, this mukhabarah agreement is much explained for the profit sharing on your agricultural land between the landowner and the manager or the cultivator. Where landowners give up their land to be managed and their capital such as seeds, fertilizers and labor come from managers or cultivators. This is in contrast to the muzara'ah[1] which is a contract of cooperation between the landowner and the tenant in which the rewards of a certain part correspond to the preceding one whether each is one-third, one-third or two-thirds, smaller or greater. Differences with mukhabarah that for muzara'ah all necessary capital in the form of seeds, fertilizers and so forth issued to the land owner.

Looking at the explanation, the practice of sharing in the concept of sharia has not been explained in detail other than on agricultural land. However, the implications of mukhabarah are almost the same as those that developed in the midst of asociety so far.

At this stage, the livestock owner will hand over the cattle to the breeder initiated by the profit-sharing agreement and capitalize on trust [11]. With this concept, it will alleviate the burden of breeders which in conventional financing the amount of credit interest is determined based on the amount of credit proposed by the debtor with the percentage of interest that has been determined with certainty. Unlike the case with the concept of Islamic financing were interest-free, principled profit sharing and calculation for the results are done at the end of the transaction [2]. This relieves beef cattle farmers, especially in paying the financing installment due to deposits through Livestock Farmers Group and Cooperatives depending on when the beef cattle farmers receives the proceeds from the sale of cattle.

4. Conclusion

Sharia-based beef cattle breeding is an alternative financing pattern that combines local government involvement, implementing banks, Livestock Farming Group and Cooperatives and farmers. Between executing banks and livestock farmers' groups and cooperatives using mudharabah contracts. Contracts between livestock owners and cattle ranchers who are entrusted to raising cattle use the implications of a mukhabarah contract because the livestock owners only have and give away their livestock to be maintained while the nursing farmer must provide forage, feed, water, land, and labor.

Acknowledgements

Thanks to the Ministry of Technology and Higher Education which has provided opportunities to researchers through National Strategy research. To the community and Farmers groups in Bone Regency who have provided information related to local culture in the area.

References

[1] Antonio, S. 2001. Bank Syariah Theory and Practice. Jakarta: Gema Insani.

[2] Ashari, and Saptana. 2005. The Prospect of Sharia Financing for Agricultural Sector. Agro Economic Research Forum.(Online), Vol. 23, No.2, (http://ekonomisyariah.info/wp-content/uploads/2013/10/ Financing-Financing-Sharia-for-Sector-Agriculture Prospects.pdf, accessed May 17, 2015).

[3] Asnawi,A., 2013a. Financing, Working Capital Management, Cash Conversion Cycle and Business Performance in Beef Cattle Breeding Business in South Sulawesi.Dissertation.Faculty of Economics and Business, Airlanga University, Surabaya.

[4] Asnawi,A., 2013b. Determinant of Funding Accessibility and its Impacts to The Performance of Beef-Cow Breeding Enterprises in South Sulawesi Province, Indonesia. J. European Business Manag, 5(29).77-83.

[5] Asnawi, A., Sirajuddin, N., Lestari, V.S. 2014. How do Lending Relationship Affect to Credit Accessibility in Cattle Farm in Indonesia? J. European Sustainable Dev. **3** (4): 359.

[6] Fachruddin, F.M. 1993. Riba in Banks, Cooperatives, Companies, and Insurance. Bandung: Al Ma'arif.

[7] Muhammad. 2005. Management of Bank Syariah Financing. Yogyakarta, YKPN.

[8] Nugraha, J.P. 2016. Muzara'ah System as an Alternative of Agricultural Financing in Indonesia.Iqtishodia.Journal of Sharia Economics.Vol. 1 No. September 2nd.

[9] Rivai, V., Veithzal, A.P., Idroes, F.N. 2007. Bank and Financial Institution Management. Jakarta, PT. RajaGrafindo.

[10] Law of the Republic of Indonesia Number 10 of 1998 concerning Amendment to Law Number 7 The year 1992 concerning Banking. 1998. Jakarta, State Ministry of the Republic of Indonesia.

[11]Sirajuddin, S.N. Aminawar, M. Amrawaty, A. Nurlaelah, St. 2015. The Application of Tesang Sharing System at Cattle Farm in Indonesia.Proceedings of the 3rd International Seminar on Animal Industry, Bogor, 17-18 September 2015. Faculty of Animal Science Bogor Agricultural University.

[12] Susyanti, J. 2016. Management of Sharia Financial Institutions. Malang, Publisher Four Two.